Metagenomics

Current Innovations and Future Trends

Edited by

Diana Marco

Microbiology Department
Estación Experimental del Zaidín (CSIC)
Profesor Albareda 1
Granada
Spain

and CONICET (Argentina)

Caister Academic Press

Caister Academic Press
Norfolk, UK

www.caister.com

British Library Cataloguing-in-Publication Data
A catalogue record for this book is available from the British Library

ISBN: 978-1-904455-87-5

Printed and bound in Great Britain

Contents

Contributors

Josefa Antón
División de Microbiología
Departamento de Fisiología, Genética y Microbiología
Universidad de Alicante
Alicante
Spain

anton@ua.es

Rolene Bauer
Institute for Microbial Biotechnology and Metagenomics
Department of Biotechnology
University of Western Cape
Cape Town
South Africa

rbauer@uwc.ac.za

Valeria Bianciotto
Istituto per la Protezione delle Piante del CNR (Consiglio Nazionale delle Richerche)
Torino
Italy

v.bianciotto@ipp.cnr.it

Paola Bonfante
Istituto per la Protezione delle Piante del CNR (Consiglio Nazionale delle Richerche)
Torino
Italy

p.bonfante@ipp.cnr.it

Roberto Borriello
Istituto per la Protezione delle Piante del CNR (Consiglio Nazionale delle Richerche)
Torino
Italy

roberto.borriello@unito.it

Jing Chen
The CAMERA Project
Center for Research on Biological Systems and California Institute of Telecommunication and Information Technology
University of California San Diego
La Jolla, CA
USA

jic002@ucsd.edu

Charles Y. Chiu
UCSF-Abbott Viral Diagnostics and Discovery Center
Departments of Laboratory Medicine and Medicine
University of California
San Francisco, CA
USA

charles.chiu@ucsf.edu

Donald A. Cowan
Institute for Microbial Biotechnology and Metagenomics
Department of Biotechnology
University of Western Cape
Cape Town
South Africa

dcowan@uwc.ac.za

Katerina Demnerova
Department of Biochemistry and Microbiology
Faculty of Food and Biochemical Technology
Institute of Chemical Technology Prague
Prague
Czech Republic

katerina.demnerova@vscht.cz

Chirayu Desai
International Center for Public Health
Public Health Research Institute
New Jersey Medical School
University of Medicine and Dentistry New Jersey
Newark, NJ
USA

chirayurdesai@gmail.com

Yves Dessaux
Institut des Sciences du Végétal
Centre National de la Recherche Scientifique
Gif sur Yvette CEDEX
France

dessaux@isv.cnrs-gif.fr

Robert L. Dorit
Smith College
Northampton, MA
USA

rdorit@smith.edu

Graham J. Etherington
The Sainsbury Laboratory
Norwich Research Park
Norwich
UK

graham.etherington@tsl.ac.uk

Denis Faure
Institut des Sciences du Végétal
Centre National de la Recherche Scientifique
Gif sur Yvette CEDEX
France

denis.faure@isv.cnrs-gif.fr

Rashel V. Grindberg
J. Craig Venter Institute
La Jolla, CA
USA

rgrindberg@jcvi.org

Zhili He
Institute for Environmental Genomics
University of Oklahoma
Norman, OK
USA

zhili.he@ou.edu

Reia Hosokawa-Okamoto
Bioproduction Research Institute
National Institute of Advanced Industrial Science and Technology (AIST)
Higashi
Tsukuba, Ibaraki
Japan;
Department of Medical Genome Sciences
Graduate School of Frontier Sciences
The University of Tokyo
Tsukuba, Ibaraki
Japan

reia-hosokawa@aist.go.jp

Kunal Jain
BRD School of Biosciences
Sardar Patel University
Vallabh Vidyanagar
Gujarat
India

qnal.jain@gmail.com

Roger S. Lasken
J. Craig Venter Institute
La Jolla, CA
USA

rlasken@jcvi.org

Weizhong Li
The CAMERA Project
Center for Research on Biological Systems and California Institute of Telecommunication and Information Technology
University of California San Diego
La Jolla, CA
USA

weizhong@soe.ucsd.edu

Mary-Jane Lombardo
J. Craig Venter Institute
La Jolla, CA
USA

mlombard@jcvi.org

Erica Lumini
Istituto per la Protezione delle Piante del CNR (Consiglio Nazionale delle Richerche)
Torino
Italy

erica.lumini@ipp.cnr.it

Tomas Macek
Department of Biochemistry and Microbiology
Faculty of Food and Biochemical Technology
Institute of Chemical Technology Prague
Prague
Czech Republic;
Institute of Organic Chemistry and Biochemistry
Czech Academy of Sciences
IOCB & ICT Joint Laboratory
Prague
Czech Republic

tomas.macek@vscht.cz

Martina Mackova
Department of Biochemistry and Microbiology
Faculty of Food and Biochemical Technology
Institute of Chemical Technology Prague
Prague
Czech Republic;
Institute of Organic Chemistry and Biochemistry
Czech Academy of Sciences
IOCB & ICT Joint Laboratory
Prague
Czech Republic

martina.mackova@vscht.cz

Dan MacLean
The Sainsbury Laboratory
Norwich Research Park
Norwich
UK

dan.maclean@sainsbury-laboratory.ac.uk

Datta Madamwar
BRD School of Biosciences
Sardar Patel University
Vallabh Vidyanagar
Gujarat
India

datta_madamwar@yahoo.com

Corinne F. Maurice
Harvard FAS Center for Systems Biology
Cambridge, MA
USA

corinne.maurice@etud.univ-montp2.fr

Tracy L. Meiring
Institute for Microbial Biotechnology and Metagenomics
Department of Biotechnology
University of Western Cape
Cape Town
South Africa

tracy.meiring@gmail.com

Kentaro Miyazaki
Bioproduction Research Institute
National Institute of Advanced Industrial Science and Technology (AIST)
Higashi, Tsukuba
Ibaraki
Japan;
Department of Medical Genome Sciences
Graduate School of Frontier Sciences
The University of Tokyo
Tsukuba, Ibaraki
Japan

miyazaki-kentaro@aist.go.jp

Samuel Mondy
Institut des Sciences du Végétal
Centre National de la Recherche Scientifique
Gif sur Yvette CEDEX
France

samuel.mondy@isv.cnrs-gif.fr

Vincent Montoya
British Columbia Centre for Disease Control
Department of Pathology and Laboratory Medicine
University of British Columbia
Vancouver, BC
Canada

vincent.montoya@bccdc.ca

Lucie Musilova
Department of Biochemistry and Microbiology
Faculty of Food and Biochemical Technology
Institute of Chemical Technology Prague
Prague
Czech Republic

lucie.musil@centrum.cz

Mark Novotny
J. Craig Venter Institute
La Jolla, CA
USA

mnovotny@jcvi.org

Colin Ohlhoff
Institute for Microbial Biotechnology and Metagenomics
Department of Biotechnology
University of Western Cape
Cape Town
South Africa

colin.ohlhoff@gmail.com

Alberto Orgiazzi
Istituto per la Protezione delle Piante del CNR (Consiglio Nazionale delle Richerche)
Torino
Italy;
Dipartimento di Biologia Vegetale dell'Università degli Studi di Torino
Torino
Italy

alberto.orgiazzi@unito.it

Margaret A. Riley
Biology Department
University of Massachusetts
Amherst, MA
USA

riley@bio.umass.edu

Fernando Santos
División de Microbiología
Departamento de Fisiología
Genética y Microbiología
Universidad de Alicante
Alicante
Spain

fernando.santos@ua.es

Ilana Scheepers
Institute for Microbial Biotechnology and Metagenomics
Department of Biotechnology
University of Western Cape
Cape Town
South Africa

ilana.ackermann@gmail.com

Varun Shah
BRD School of Biosciences
Sardar Patel University
Vallabh Vidyanagar
Gujarat
India

varun9999@gmail.com

Jian Shen
Shanghai Center for Systems Biomedicine
Shanghai Jiao Tong University
Shanghai
China

shenjian1981@googlemail.com

Shulei Sun
The CAMERA Project
Center for Research on Biological Systems and California Institute of Telecommunication and Information Technology
University of California San Diego
La Jolla, CA
USA

s2sun@ucsd.edu

Patrick Tang
British Columbia Centre for Disease Control
Department of Pathology and Laboratory Medicine
University of British Columbia
Vancouver, BC
Canada

patrick.tang@bccdc.ca

Mélanie Tannières
Institut des Sciences du Végétal
Centre National de la Recherche Scientifique
Gif sur Yvette CEDEX
France

melanie.tannieres@isv.cnrs-gif.fr

Marla I. Tuffin
Institute for Microbial Biotechnology and Metagenomics
Department of Biotechnology
University of Western Cape
Cape Town
South Africa

ituffin@uwc.ac.za

Peter J. Turnbaugh
Harvard FAS Center for Systems Biology
Cambridge, MA
USA

pturnbaugh@mcb.harvard.edu

Ondrej Uhlik
Department of Biochemistry and Microbiology
Faculty of Food and Biochemical Technology
Institute of Chemical Technology Prague
Prague
Czech Republic;
Institute of Organic Chemistry and Biochemistry
Czech Academy of Sciences
IOCB & ICT Joint Laboratory
Prague
Czech Republic

ondrej.uhlik@vscht.cz

Joy D. Van Nostrand
Institute for Environmental Genomics
University of Oklahoma
Norman, OK
USA

joy.vannostrand@ou.edu

John Walshaw
John Innes Centre
Department of Computational and Systems Biology
Norwich Research Park
Norwich
UK

john.walshaw@bbsrc.ac.uk

John C. Wooley
The CAMERA Project
Center for Research on Biological Systems and California Institute of Telecommunication and Information Technology
University of California San Diego
La Jolla, CA
USA

jwooley@ucsd.edu

Joyclyn Yee-Greenbaum
J. Craig Venter Institute
La Jolla, CA
USA

jyee@jcvi.org

Liping Zhao
School of Life Sciences and Biotechnology
Shanghai Jiao Tong University
Shanghai Center for Systems Biomedicine
Shanghai
China

lpzhao@sjtu.edu.cn

Jizhong Zhou
Institute for Environmental Genomics
University of Oklahoma
Norman, OK
USA

jzhou@ou.edu

Current Books of Interest

Extremophiles: Microbiology and Biotechnology	2012
Bacillus: Cellular and Molecular Biology	2012
Microbial Biofilms: Current Research and Applications	2012
Bacterial Glycomics: Current Research, Technology and Applications	2012
Non-coding RNAs and Epigenetic Regulation of Gene Expression	2012
Brucella: Molecular Microbiology and Genomics	2012
Molecular Virology and Control of Flaviviruses	2012
Bacterial Pathogenesis: Molecular and Cellular Mechanisms	2012
Bunyaviridae: Molecular and Cellular Biology	2011
Emerging Trends in Antibacterial Discovery: Answering the Call to Arms	2011
Epigenetics: A Reference Manual	2011
Metagenomics: Current Innovations and Future Trends	2011
Nitrogen Cycling in Bacteria: Molecular Analysis	2011
The Biology of Paramyxoviruses	2011
Helicobacter pylori	2011
Microbial Bioremediation of Non-metals: Current Research	2011
Lactic Acid Bacteria and Bifidobacteria: Current Progress in Advanced Research	2011
Viruses and Interferon: Current Research	2011
Essentials of Veterinary Parasitology	2011
Hepatitis C: Antiviral Drug Discovery and Development	2011
Streptomyces: Molecular Biology and Biotechnology	2011
Alphaherpesviruses: Molecular Virology	2011
Recent Advances in Plant Virology	2011
Vaccine Design: Innovative Approaches and Novel Strategies	2011
Salmonella: From Genome to Function	2011
PCR Troubleshooting and Optimization: The Essential Guide	2011
Insect Virology	2010
Environmental Microbiology: Current Technology and Water Applications	2011
Sensory Mechanisms in Bacteria: Molecular Aspects of Signal Recognition	2010
Bifidobacteria: Genomics and Molecular Aspects	2010
Molecular Phylogeny of Microorganisms	2010
Nanotechnology in Water Treatment Applications	2010
Iron Uptake and Homeostasis in Microorganisms	2010
Caliciviruses: Molecular and Cellular Virology	2010
Epstein-Barr Virus: Latency and Transformation	2010
Anaerobic Parasitic Protozoa: Genomics and Molecular Biology	2010
Lentiviruses and Macrophages: Molecular and Cellular Interactions	2010
Microbial Population Genetics	2010
Borrelia: Molecular and Cellular Biology	2010
Influenza: Molecular Virology	2010
RNA Interference and Viruses: Current Innovations and Future Trends	2010

Preface

Metagenomics is one of the fastest advancing fields in biology. By allowing access to the genomes of entire communities of bacteria, virus and fungi that are otherwise inaccessible, metagenomics is extending our comprehension of the diversity, ecology, evolution and functioning of the microbial world, as well as contributing to the emergence of new applications in many different areas. The continued and dynamic development of faster sequencing techniques, together with the advancement of methods to cope with the exponentially increasing amount of data generated are expanding our capacity to analyse microbial communities from an unlimited variety of habitats and environments. The synergism with the new emerging 'omics' approaches is leading the way to functional metagenomics, and to adopting integrative, wider viewpoints like systems biology.

This book intends to cover in a consistent way the most innovative and recent advances in theoretical, methodological and applied areas of metagenomics. The first chapter provides a comprehensive review of the current methodologies, applications and challenges of metagenomics, acts as an introduction for newcomers to the field and presents a brief view of new methods available to experienced researchers. The next three chapters deal with the involvement of metagenomics in theoretical aspects of the microbial world. Chapter 2 deals with the still controversial problem of whether bacterial genomes can be viewed as discrete, discernible and discontinuous packages that retain their integrity over evolutionary time in spite of the widespread horizontal gene transfer cases documented. Chapter 3 explores the new field of viral metagenomics and the role viruses have in the regulation of prokaryotic communities. Chapter 4 deals with the relationship between metagenomics and new, integrative approaches like systems biology, involving complementary 'omics' approaches.

Although all the chapters in the book provide insights into methodology, Chapters 5, 6, 7 and 8 are directly devoted to recently developed methods. Chapter 5 accounts for the next-generation sequencing approaches that allow researchers to exponentially increase the sequence data available compared to recent years. Chapter 6 deals with the recent use of stable isotope probing in metagenomics, allowing the reconstruction of individual genomes of key uncultured microbes as well as to detect genes of interest. Chapter 7 accounts for the development of a revolutionary method to sequence DNA from a single cell, reviewing current methodologies, the strengths and limitations of this new approach and its application to microbial metagenomics. As the advent of new sequencing methods has enormously increased the amount of data, more sophisticated, but at the same time, user friendly methods of analysing them are needed. Chapter 8 describes the new version of the Community Cyberinfrastructure Resource for Metagenomics Research, CAMERA 2.0, a platform that allows analytical and computational history to be archived, providing accurate reproducibility and validation of sequencing analysis.

As a plethora of new applications of metagenomics have recently arisen, the selection of some applications to be included in this volume has not been easy. However, some of the most innovative are reviewed in the following chapters. Chapter 9 explores how metagenomics-based tools such as pan-viral microarrays and ultra-high-throughput sequencing have significantly improved the ability to detect and characterize divergent as well as novel viruses, showing the potential medical applications. Chapter 10 deals with the recent application of metagenomics to the world of fungi, in particular of Arbuscular Mycorrhizal Fungi, the most widespread symbionts in many ecosystems. Chapter 11 explores the microbial communities that form part of our bodies, describing in particular the links between obesity and the gut microbiome from a systems biology perspective. Chapter 12 focuses on the synergy between metagenomics and other 'omics' approaches to the development of new and better bioremediation strategies and techniques. Chapter 13 deals with the use of of *Escherichia coli* as a host for efficient functional screening of metagenomic libraries, an approach that can lead to the discovery of new enzymes. Chapter 14 discusses the application of metagenomics in the identification of yet-uncultivable phytopathogens and rhizosphere microflora, as well as the identification of novel density-dependent regulatory functions in these microbial communities and discusses the development of biological control strategies of plant pathogens. In Chapter 15 the authors show the advantages of performing metagenomics analysis of below-ground microbial communities using microarrays and the development of functional gene arrays, in particular the recently launched GeoChip 3.0.

The book is aimed, among other users, at researchers performing studies in metagenomics, students interested in starting research in this field, teachers interested in the new developments in the field, and persons involved in biotechnological applications such as bioproducts obtention and bioremediation.

Finally, I would like to thank all the authors for their invaluable contributions. I also would like to thank Caister Academic Press for trusting me again with the role of editor of a book in one of the most innovative and dynamic scientific fields today.

Diana Marco
Granada, Spain
2011

Metagenomics and Beyond: Current Approaches and Integration with Complementary Technologies

1

Tracy L. Meiring, Rolene Bauer, Ilana Scheepers, Colin Ohlhoff, Marla I. Tuffin and Donald A. Cowan

Abstract

Metagenomics is the cultivation independent analysis of the collective genomes of microbes within a given environment, using sequence- and function-based approaches. Early metagenomic studies, aimed at cataloguing the phylogenetic diversity of different habitats, revealed the vast size and richness of the microbial and viral realms. Access to previously unexplored volumes of sequence space from uncultured organisms has since opened up many new avenues of research, with the number of projects and potential applications ever expanding. The rapid growth of the research area is furthermore a direct reflection of the advances in the throughput, as well as the reduction in cost, of sequencing and screening technologies. In this chapter we present a comprehensive introduction to the current methodologies, applications and challenges of metagenomics. We review the major improvements in the technologies that form the metagenomic toolkit, as well as the limitations they currently place on research. We assess the achievements made thus far in the exploration of novel sequence space. Finally, we consider future research trends in light of recent novel metagenomic approaches and the incorporation of complementary technologies.

Introduction

It is well established that microorganisms shaped the earth's biosphere and continue to be major players in key geochemical and biological processes essential for life (Rodriguez-Valera, 2004; Xu, 2006). Considering the dominance of microorganisms for most of the planet's history, it is not surprising that the evolutionary diversity and numbers of microbes far exceed that of higher organisms. This enormous heterogeneity represents a vast and mostly untapped source of novel genetic and metabolic information. Capturing, understanding and exploiting this information is a technically and intellectually challenging endeavour.

Classical microbiology relied largely on the culturing and analysis of pure isolates from environmental samples in the laboratory. In the mid to late 1990s; however, it emerged that the currently cultivatable microorganisms represent only a small fraction (estimated to be less than 1%) of the total microbes within a given habitat (Amann *et al.*, 1995; Hugenholtz *et al.*, 1998; Whitman *et al.*, 1998). Traditional clonal culture is therefore biased and cannot access the vast majority of organisms within a community. Metagenomics is a means of capturing and analysing the genomes of the entire microbial community present within a given habitat (i.e. the metagenome), based on the cultivation-independent extraction of total environmental DNA (Handelsman, 2004). By using metagenomics coupled to other emerging technologies, such as high-throughput sequencing and screening, we now have the potential to probe previously unexplored sequence realms.

Current metagenomic approaches and applications

The defining strength of metagenomics is the ability to gain access to genetic information independently of cultivation-based biases. This review assesses various metagenomic approaches for their ability to access genetic diversity and novelty. As metagenomics is very much a methods-limited discipline, the major achievements and shortcomings of current metagenomic tools are evaluated and recent novel studies which improve upon the basic approach are highlighted.

The basic metagenomic approach

All metagenomic studies start with the isolation of genomic material from the cellular organisms and/or viruses present in a particular environmental sample. The concept of directly isolating environmental DNA, and thereby bypassing the need for clonal culture, was first put forward by Pace and co-workers (1985) as the next logical extension of molecular biomarker analysis. The majority of metagenomics projects involve the construction and analysis of clone libraries. This followed the seminal research by DeLong's group, in which they successfully constructed a metagenomic clone library from prokaryotes in seawater (Stein *et al.*, 1996). The construction of libraries involves the cloning of the DNA into an appropriate large or small insert vector and transforming the cloned DNA into a surrogate host, typically *Escherichia coli*. Large insert libraries can be generated in bacterial artificial chromosomes (BAC) (Rondon *et al.*, 2000), fosmids (Kim *et al.*, 2006) or cosmids (Courtois *et al.*, 2003). These vector systems can accommodate DNA fragments varying from 20 kb to 200 kb in size. Small insert libraries are typically constructed using plasmid vectors, such as pUC19 (Ranjan, 2005), pBluescript SK$^+$ (Stratagene, San Diego, CA, USA) or pSMART (Lucigen, Middleton, WI, USA), and on average contain ten-fold more clones than their large insert counterparts. Resulting clone libraries may be explored using sequence- and/or function-based approaches (Fig. 1.1). More recently, with the advent of next generation sequencing (NGS), many projects bypass cloning and sequence the extracted environmental DNA directly (Edwards *et al.*, 2006).

Sequence-driven metagenomic approaches

Metagenome sequencing

Recent advances in DNA sequencing technologies have made sequencing of the total environmental DNA complement in a community economically feasible, and offer rapid access to the metagenome. Genomes can now be sequenced at a rate that is 50,000 times faster than in the year 2000 at a greatly reduced cost (Venter, 2010). This approach requires no prior sequence knowledge for data generation and the discovery of novel gene sequence is certain. However, our ability to

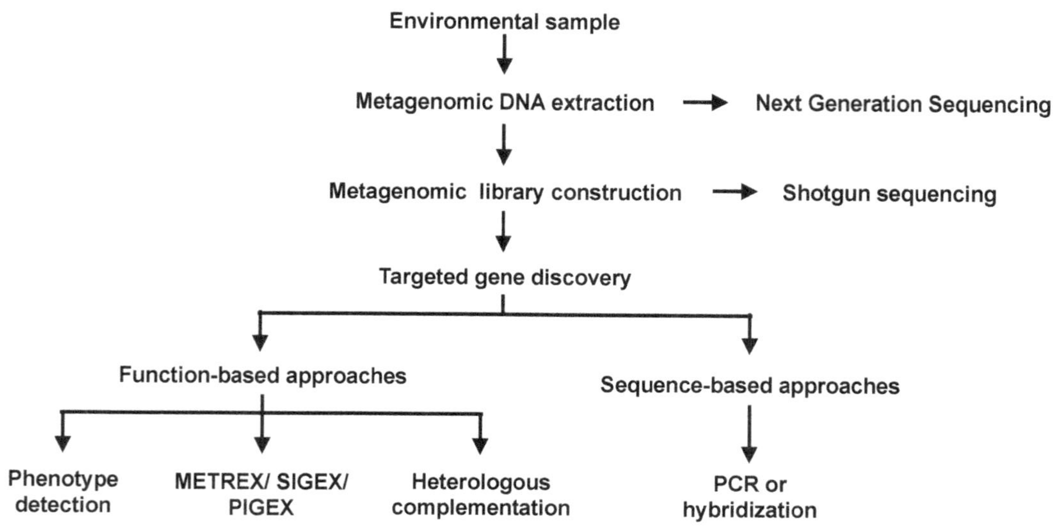

Figure 1.1 The basic metagenomic approach (adapted from Simon and Daniel, 2009).

correctly assign functions to newly discovered genes has not kept pace with advances in sequencing technologies.

As of mid-2010, 242 microbial metagenomes, in various stages of completion, sampled from diverse natural, artificial and host-associated environments such as the ocean, freshwater lakes, saline water lakes, hot springs, hydrothermal vents, soil, mouse gut, bovine rumen, acid mine drainage and the human body, have been reported (http://www.genomesonline.org/). The metagenome sequence-based approach has also recently been expanded to include eukaryotes, such as insects (Pond *et al.*, 2009) and pico-prymnesiophytes (Cuvelier *et al.*, 2010). The complexity and scale of many of these projects has required the establishment of international collaborative efforts aimed at sequencing complete metagenomes and cataloguing reference genomes. There are currently three major microbial metagenome sequencing efforts in progress; the Human Microbiome Project (HMP) (http://nihroadmap.nih.gov/hmp), the marine microbiome or Global Ocean Sampling (GOS) expedition (http://www.jcvi.org/cms/research/projects/gos/overview/) and the soil metagenome or terragenome (http://www.terragenome.org/). The Genomics Standards Consortium (http://gensc.org/) is an international initiative working to standardize genome and metagenome description and the exchange of (meta)genomic data and associated environmental metadata. The CAMERA Project (Community Cyberinfrastructure for Advanced Microbial Ecology Research and Analysis; http://camera.calit2.net/) was initiated in 2006 to serve as comprehensive catalogue of metagenomic sequence data and annotations, as well as providing bioinformatics tools for metagenome analysis. Such initiatives will become increasingly important as the use of NGS becomes routine. For a comprehensive account on the CAMERA Project, see Chapter 8.

Metagenomic sequencing projects seek to answer fundamental questions such as: how many different species inhabit a particular environment, what is the genomic potential of that community (i.e. which genes, functions or pathways are present), which species are responsible for which activities, and how does the community change over time or under different environmental pressures? Metagenome sequencing have revolutionized our understanding of the microbial world. The vast amounts of data generated has provided insight into the diversity and genomic potential of the yet uncultured bacterial, archaeal and viral majority, led to the discovery of literally millions of genes, and deduced metabolic pathways and interactions among bacteria. The largest metagenomic sequencing project to date, the GOS expedition, predicted 6.12 million proteins representing all previous known families and adding 1700 more families with upwards of 20 members each. New protein families continue to be discovered at a near linear rate as new sequence information is generated (Yooseph *et al.*, 2007). Metagenomic sequencing projects have in most environments, therefore, as yet only provided a fragmentary glimpse into total genetic diversity. The exception being in highly specialized environments characterized by low species richness, such as acidic mine-drainage at Iron Mountain (Tyson *et al.*, 2004; Ram *et al.*, 2005) or deep subsurface drainage waters in an African gold mine (Chivian *et al.*, 2008).

The majority of studies have used the 'shotgun sequencing' approach. This entails Sanger dideoxy sequencing of small-insert metagenomic clone libraries, producing sequence reads of 600–900 bp. More recently, NGS technologies such as Roche 454 (Margulies *et al.*, 2005) are gaining in popularity for direct cost effective sequencing of metagenomic DNA (Edwards *et al.*, 2006). While bypassing cloning biases, and offering lower costs, increased speed, higher accuracy and volumes of data, the incorporation of NGS technologies into metagenomic studies has limitations. Smaller reads are generated, currently up to approximately 400–500 bp with Roche 454, 50 bp with SOLiD, 75–100 bp with Solexa, and 25–35 bp with Heliscope (McPherson, 2009). Generated from complex metagenomes, short reads are usually *de novo* assembled to up to 5000 bp, with the result that much of the data remains fragmented (Wooley *et al.*, 2010). These problems are particularly evident in data sets from species-rich environments, where the reconstruction of genomes, or even whole metabolic pathways or transcriptional units, is only possible for the dominant organisms

(Yooseph *et al.*, 2007). As the average gene length in a prokaryotic genome is approximately 900 bp, gene prediction and functional assignment on unassembled short reads is difficult. Nonetheless, NGS technologies continue to advance rapidly and Roche 454 read lengths are soon expected to reach those achieved by the Sanger method (Mardis, 2008). For a detailed account on NGS methodology, see Chapter 5.

The enormous volume of data generated by environmental sequencing, which is several orders of magnitude larger than that acquired in single organism genomics, represent a significant computational challenge and has required the development of new algorithms specifically aimed at metagenomic data analysis (Wooley *et al.*, 2010). The fragmentary nature of the data has also led to a shift from thinking of communities in terms of genomes or species to a description of the gene complement present, i.e. the gene-centric approach (Hugenholtz and Tyson, 2008).

The methodology for metagenomic sequence analysis has made rapid advances and now includes a number of different approaches for taxonomic profiling (or binning) of complex samples (for recent excellent reviews, see McHardy and Rigoutsos, 2007; Raes *et al.*, 2007; Kunin *et al.*, 2008; Pop and Salzberg, 2008; Huson *et al.*, 2009; Wooley *et al.*, 2010). Homology-based methods compare sequence reads against phylogenetic markers (Ludwig and Klenk, 2001; Rusch *et al.*, 2007; Wu and Eisen, 2008) or reference genomic data (Angly *et al.*, 2006; DeLong *et al.*, 2006; Huson *et al.*, 2007), and rely on homology information obtained by database searches with tools such as BLAST (Basic Alignment Search Tool; Altschul *et al.*, 1990). Composition-based methods compare sequence features, such as codon usage (Noguchi *et al.*, 2006), GC content (Foerstner *et al.*, 2005) and oligonucleotide composition (McHardy *et al.*, 2007; McHardy and Rigoutsos, 2007).

Functional annotation of the metagenome is challenging, since a large fraction of genes will have no known homologues or are present as partial sequences. Annotation usually involves combining evidence from multiple sources. The identification of new genes or gene families requires *ab initio* gene prediction tools, such as Genemark.hmm (Azad and Borodovsky, 2004). Putative protein function is typically assigned based on blastp searches against the NCBI non-redundant protein database. Sequence signatures are identified by comparison to HMM (hidden Markov models) protein family databases, for example Pfam (Finn *et al.*, 2008) and TIGR-fam (Haft *et al.*, 2003) using searches such as FastHMM (http://www.microbesonline.org/fasthmm/), or HMMer (http://hmmer.janelia.org/). InterPro (http://www.ebi.ac.uk/interpro/) is an integrated protein database that is also widely used for the classification and annotation of proteins. As gene prediction tools require full-length ORFs, many metagenomic annotation pipelines bypass gene calling. Six-frame translations, translated product cut-off sizes, and incremental clustering by sequence similarity are used instead. Annotation of the predicted peptides are based on conserved motifs, HMM profiles and other sequence signatures that may indicate function (Wooley *et al.*, 2010). These types of strategies have been incorporated into the development of automated annotation pipelines such as MG-RAST (Meyer *et al.*, 2008; metagenomics.nmpdr.org/) and RAMMCAP (Li, 2009).

Functional annotation relies for the most part on sequence features and homology to characterized proteins. A recent review of protein diversity in shotgun metagenome projects showed that between 30% and 60% of proteins could not be assigned functions (Vieites, 2009). While this suggests a high degree of novelty, it highlights the fact that the majority of genes in public databases are not functionally annotated. There is a growing concern over the ever increasing gap between hypothetical and characterized proteins, one of the major limitations to future advances (Hugenholtz and Tyson, 2008). While progress in computational methods that allow for the prediction of protein structure and catalytic activities will certainly aid *in silico* functional assignment, even the most thorough annotations create only hypotheses. The characterization of encoded proteins will still require time consuming experimental wet lab testing using classical physiological and biochemical approaches (Chen and Murrell, 2010). As will be discussed later in this chapter, the comparison of information generated by metagenomic sequencing with RNA and protein expression data, generated from metatranscriptomics and metaproteomic studies, should also prove invaluable in this regard.

Gene targeting in metagenomes

Metagenomic studies often follow a gene-targeted approach as opposed to random sequencing of the entire metagenome. Clone libraries are screened by hybridization or polymerase chain reaction (PCR) based on DNA consensus with conserved regions of known genes and gene products (Daniel, 2005). Genes such as 16S rRNA, RecA, EF-Tu, EF-G, HSP70 and RNA polymerase B (RpoB) are often targeted as phylogenetic markers. While this approach is frequently applied in diversity studies with great success, its applicability to the discovery of novel genes or enzymes is limited (Tuffin *et al.*, 2009). Prior sequence information pertaining to the target gene is pivotal and the approach is therefore, only suitable to the discovery of members belonging to known gene families. This method overcomes limitations associated with heterologous functional expression of genes (Jiang *et al.*, 2006). In combination with various genome walking techniques, sequence information flanking the fragment of interest may be acquired and full-length genes or functional clusters obtained (reviewed by Kotik, 2009).

Recent novel metagenomic gene discovery methods include pre-amplification inverse PCR (PAI-PCR) (Yamada *et al.*, 2008) and *in vitro* metagenomic DNA shuffling (Boubakri *et al.*, 2006). PAI-PCR involves the random amplification of the template DNA using phi29 DNA polymerase followed by gene-specific inverse PCR. This technique has the potential to isolate novel genes from small amounts of environmental DNA or from rare community members. PAI-PCR was successfully used to isolate five putative xylanases from horse and termite guts with only 35–43% identity to closest homologues (Yamada *et al.*, 2008). Metagenomic DNA shuffling simulates horizontal gene transfer and recombination processes using metagenomic DNA as the parent sequence. By applying stringent selection, enzymes that are particularly suited to industrial application may be generated. This approach has been used to generate novel lindane-degrading genes by shuffling metagenomic DNA extracted from soil (Boubakri *et al.*, 2006).

While the sequence-driven approach has recovered some novel functional enzymes, such as lipases (Bell *et al.*, 2002), newly discovered sequences have typically above 50%, and usually around 80–98%, identity to known proteins (Tuffin *et al.*, 2009). Function-based screening still holds more potential for the discovery of novel genes encoding novel biocatalysts (Tuffin *et al.*, 2009). The development of new *in silico* tools for protein function prediction based not only on primary sequence homology, but the ability to predict protein structures and associated catalytic activities, will no doubt greatly improve sequence-based screening in the future.

Metagenomic microarrays

Microarrays have been applied to profile metagenomic libraries and allow for the rapid characterization of thousands of clones in one experiment (Sebat *et al.*, 2003). The technology was developed from the working principle of Southern hybridization, but differs in that it allows for the simultaneous examination of a multitude of genes on a single array (Coppée, 2008; Park *et al.*, 2008). DNA samples are extracted from metagenomic library clones and fluorescently tagged before being applied to the microarray platform. Following the hybridization event, excess DNA strands are washed away, with only the hybridized DNA remaining attached to the array. The fluorescent illumination pattern of the array spots can be used to detect clones of interest. Technological progression has allowed for high-density arrays to be constructed using various physical delivery techniques, including sophisticated robotic systems that offer advanced liquid handling capabilities. Robotic systems allow for the accurate dispensing and spotting of picolitre volumes onto the designated microarray test site and significantly increase its throughput potential.

Microarray screening procedures have been suitably adapted to screen for specific genes within a given environment. Loy and co-workers (2004) designed a 16S rRNA gene-targeted oligonucleotide microarray constituting 79 probes for the congruent detection of all bacteria of the order 'Rhodocyclales'. They based the probe architecture on phylogenetic analysis of all the then-available 16S rRNA sequences from both cultured and as yet uncultured members. The resulting RHC-Phylochip was applied towards culture-independent 'Rhodocyclales' diversity analysis and proved sensitive enough to detect populations with relative abundance of less than 1% of all bacteria present within the sample.

Few studies to date have employed microarray-based techniques to identify industrially valuable biocatalysts. The most recent progress in metagenome-linked applications have been directed towards the development of a powerful GeoChip (He *et al.*, 2007) or sensitive 'genome-proxy' arrays (Rich *et al.*, 2008). GeoChip was designed as an *in situ* tool to aid bioremediation studies by tracing the dynamics of various metal-reducing bacteria and associated communities. It has subsequently also been applied to analyse microorganisms from a variety of habitats including bioreactors, soils and marine sediments. This advocates its use for high-throughput analysis of various biogeochemical, ecological and environmental processes linked with the associated microbial communities (He *et al.*, 2007). The genome-proxy microarray system targets microbial gene fragments sourced directly from the environment. This shows tremendous potential for further application particularly as it could distinguish between organisms having as much as 75% genome identity. Although these methods may not be suited to functional analysis, they do offer an alternative means to track and analyse microbial communities (Pathak *et al.*, 2009).

The comprehensive screening of large metagenomic libraries is dependent on the availability of sizable microarray platforms, able to accommodate the vast numbers of DNA targets (library clones). The implementation cost and speed of microarray data processing is directly proportional to the number of spots, which in turn is dependent on the available array surface area. This presents a significant limitation for the commercial application of hand-held microarray-based biosensors (Dai *et al.*, 2009). The foremost constraint of hybridization-based technology lies in the detection and quantification of vast amounts of fluorescing spots on a single chip. Non-specific hybridization of numerous DNA samples to more than one spot may generate ambiguous results (Yang *et al.*, 2009). The efficiency of such a screening approach, where thousands of clones are processed only for a few functionally active clones detected, is also questioned (van Hellemond *et al.*, 2007). Furthermore, the rate at which sequencing technology has evolved, has also created a level of redundancy surrounding the use of microarrays as an analytical tool to evaluate gene expression levels. Simply by counting the number of sequence reads corresponding to a particular RNA/DNA molecule, the resulting level of expression can be more accurately determined than through conventional analysis of a fluorescent spot (Yang *et al.*, 2009). For a more detailed account on applications of microarrays in metagenomics, see Chapter 15.

Function-driven metagenomic approaches

Functional screening is an alternative approach to sequence-based discovery that is not dependant on prior sequence information. Metagenomic clone libraries are screened directly for a desired phenotypic trait that can be detectably expressed in the host, typically *Escherichia coli* (Handelsman, 2004). The central advantage to this approach is the potential to identify novel full-length functional genes or gene clusters from uncultivated microbes with no significant homology to sequences within the GenBank database. Functional assignment is furthermore unequivocal; this is in contrast to sequence-based approaches where predicted functional annotations are often based on only weak similarities to characterized gene products. The main application of function-driven metagenomic approach has been in the mining of metagenomes for novel biocatalysts for biotechnological purposes (Cowan *et al.*, 2005; Ferrer *et al.*, 2005).

Function-driven screens of metagenomic libraries currently use three different strategies for identifying clones of interest or 'hits': (a) direct phenotypic detection (Handelsman, 2004; Steele *et al.*, 2009), (b) heterologous complementation of host strains or mutants (Simon *et al.*, 2009) and (c) detection following induction of biosensors (METREX; Williamson *et al.*, 2005) or reporter gene expression (e.g. substrate- and product-induced gene expression screening; SIGEX and PIGEX) (Uchiyama *et al.*, 2005; Uchiyama and Miyazaki, 2009, 2010). Readers are referred to Chapter 13 for a more comprehensive review of functional metagenomics for enzyme discovery.

Hit yield and rates have been typically low. Lorenz and Eck (2005) reported that fewer than 2 out of 10,000 hits have led to the discovery of industrially relevant biocatalysts. This limitation may be overcome by (a) sample enrichment prior to library construction (Cowan *et al.*, 2005), (b) development of vector–host systems (Cowan *et al.*, 2005; Ferrer *et al.*, 2005), (c) new sensitive assays for high throughput analysis, (d) automated robotic high-throughput screening and (e) informatics support for data management (Simon and Daniel, 2009; Tuffin *et al.*, 2009). In particular, new enrichment strategies for organisms or genes with the desired catalytic activities, prior to DNA extraction, have been implemented with success (Cowan *et al.*, 2005). New PCR-independent amplification techniques that use multiple displacement amplification (MDA) with phi29 DNA polymerase for whole genome amplification (WGA) (Abulencia *et al.*, 2006; Binga *et al.*, 2008) have allowed access to rare microbes, present in low numbers in bacterial communities (Sogin *et al.*, 2006). Enrichment of target genes by stable isotope probing (SIP) technologies is also gaining in popularity. Microbial communities are provided with an isotope-labelled substrate, and the 'heavier' labelled DNA or RNA separated by density gradient centrifugal separation (Dumond *et al.*, 2006).

Activity based screens are routinely performed in *E. coli*, a well-characterized expression host, and numerous sophisticated genetic tools for cloning and expression are commercially available. The success of this approach, however, depends on several factors (Brondyk, 2009). Host-specific constraints include transcription factors, tRNAs for rare codons, co-factors, posttranslational modification systems, protein folding, proteolytic processing, chaperones and components for multisubunit assembly. Functional expression in heterologous hosts is also typically only attainable for simple monomeric or homo-oligomeric soluble enzymes that do not require trans-acting factors normally provided by the producer strain (Brondyk, 2009). Other factors that may affect protein expression and activity include temperature and pH. For example, it is estimated that a large proportion (more than 40%) of the proteins encoded within the genomes of thermophilic bacteria and Archaea cannot be functionally expressed in *E. coli* (Jenney *et al.*, 2008). Several researchers have thus investigated alternative host-vector systems for the screening of metagenomic libraries. Various Gram-positive bacilli strains have proved particularly useful for the expression of enzymes such as amylases and proteases (Demain and Vaishnav, 2009). *Ralstonia metallidurans*, a β-Proteobacteria, was recently successfully used as a vehicle for the screening of environmental DNA (Craig *et al.*, 2009). The development of shuttle vectors with wider host ranges, such as an *E. coli–Actinomycetes* shuttle (Courtois *et al.*, 2003) and vector systems capable of replicating in *E. coli*, *Streptomyces lividans* and *Pseudomonas putida* (Martinez *et al.*, 2004), should increase the probability of detecting genes encoding novel enzymes. A *Thermus thermophilus–E. coli* fosmid system for functional screening of DNA from thermophiles has recently been described (Angelov *et al.*, 2009). Such systems offer increased potential for the discovery of biotechnologically relevant thermotolerant enzymes capable of withstanding the harsh conditions of industrial processing. Lämmle *et al.* (2007) reported on a plasmid vector that contains a dual-orientation lac-promoter, located on either side of the multiple cloning site. The advantage of a bidirectional system is that gene expression is independent of possible insert-borne promoters as well as the cloning orientation.

Merging sequence and function-based screening efforts might be an effective approach to comprehensively mine metagenomes for novel biocatalysts. This was highlighted by a metagenome study of the microbiome of the tammar wallaby foregut (Morrison *et al.*, 2009). Putative β-1,4-exoglucanases were identified using two different approaches. Functional screening of a fosmid library yielded an enzyme belonging to the glycosyl hydrolase family 6. A sequence-directed screen identified more than 800 putative glycosyl hydrolase sequences. Interestingly, none of these genes coded for the enzyme discovered using the functional screen (Morrison *et al.*, 2009).

Functional screening currently represents the only means for discovering novel classes of genes with either known or new functions (Rees *et al.*, 2003; Heath *et al.*, 2009), and has facilitated the discovery of a number of new enzymes, antibiotics and other products for biotechnological and therapeutic applications (recently reviewed in Simon and Daniel, 2009; Tuffin *et al.*, 2009). However, it is clear that current screening assays result in the discovery of novel sequences rather than novel activities. There is an urgent need for the development of novel assays coupled with increased synthetic chemistry capacity and throughput (Tuffin *et al.*, 2009). An exciting prospect is the potential incorporation of developing technologies such as micro- and nano-fluidics into the functional screening workflow. The use of droplet microfluidic technology in single cell-based functional assays is predicted to greatly improve our ability to perform ultra-high-throughput screens (Brouzes *et al.*, 2009; Schaerli and Hollfelder, 2009; Schoenfeld *et al.*, 2009).

Future trends in metagenomics

The ability to generate better and more comprehensive datasets have enabled researchers to shift the emphasis of their studies from a 'reductionist' approach to a more wide-ranging exploration of complex systems, or 'systems' approach. This shift, especially true for genomics, is a direct result of enormous technical advances in sequencing technologies (Bertin *et al.*, 2008; Raes and Bork, 2008). The 'systems approach' investigates molecules, organisms and their interactions, as a whole. While metagenomics is an excellent technique for discovery and giving direction to experiments, advances in functional tools, such as transcriptomics and proteomics, are called for to link genomic potential with activity. The technologies that complement the metagenomic approach are shown in Fig. 1.2. Integration of the different disciplines is now the biggest challenge.

Complementary 'meta-omics' approaches

Metagenomic sequencing has been invaluable to a greater understanding of the diversity and dynamics of microorganisms within environments. However, quantitative measurements of gene expression, protein levels, and metabolites are required to complete the picture of an environmental system. Linking functionality with a genome or species or even measurements of activity in single cells is furthermore desirable.

Metatranscriptomics

The techniques employed to obtain a sequence-based profile of the set of expressed genes within a microbial community under certain environmental conditions are referred to as metatranscriptomics (Warnecke and Hess, 2009). Such studies give us the opportunity to study the biology of uncultivated but ecologically important organisms. They also provide the opportunity to identify possible biological activities and to study gene regulation in microbial communities. Metatranscriptomics is especially suited for comparative studies without prior knowledge of community composition (Moran, 2009). This can shed light on the combinations of genes required to survive in a specific niche.

Techniques for extraction of bacterial, archaeal and eukaryotic RNA directly from environmental samples, and subsequent cDNA synthesis are well established (Bailly *et al.*, 2007; Poretsky *et al.*, 2005; Gilbert *et al.*, 2008; Shi *et al.*, 2009). In the past, metatranscriptomic studies were mostly based on DNA microarrays (Poretsky *et al.*, 2005) or cDNA clone libraries (Parro *et al.*, 2007). Although both approaches contribute greatly to metatranscriptome analysis, they have limitations. DNA microarrays provide information based only on the sequences it was designed for, and are ineffective in the discovery of novel genes. Detection sensitivity is low and quantification is not very reliable (Warnecke and Hess, 2009). Cloned cDNA libraries circumvent the mentioned problems but may be biased due to abundant transcripts, choice of host organism, and toxic products, to name but a few.

Next-generation sequencing (NGS) revolutionized the analysis of transcriptomes, as

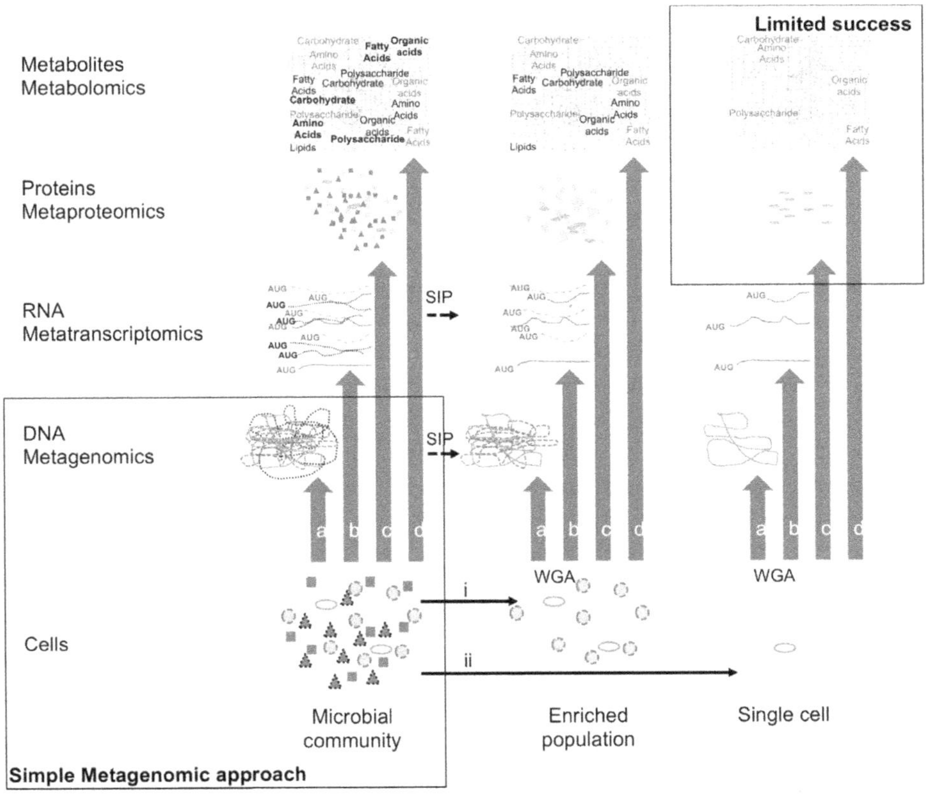

Figure 1.2 Technologies that complement the metagenomic approach. Metagenomic analysis of microbial communities (arrow a) is enhanced by the recognition and examination of transcripts (metatranscriptomics, arrow b), proteins (metaproteomics, arrow c) and metabolites (metabolomics, arrow d). Moreover, the ability to enrich for a group of cells, and even select single cells, is desirable to decrease population complexity and metabolic noise. The technologies to enrich for, and select groups or single cells, include for example (i) fluorescence activated cell sorting, filtration, and (ii) microfluidics and stable isotope probing (SIP). Whole genome amplification (WGA) is one approach used in metagenomic and metatranscriptomic analysis of such samples. The technological tools for metaproteomic and metabolomic studies at single-cell level are still inadequate, but rapidly developing (adapted from Warnecke and Hugenholtz, 2007).

maximum resolution is achieved without transcript size influencing efficiency. Most sequenced transcripts have no significant similarity to protein sequences previously published in public databases (Bailly *et al.*, 2007; Poretsky *et al.*, 2005), demonstrating the enormous potential metatranscriptomics holds for uncovering novel proteins (Warnecke and Hess, 2009). In a metatranscriptomic study on ocean microbes, a large fraction of unidentified transcript sequences were shown to be novel small untranslated RNAs (sRNAs), which have regulatory roles in important biological processes in prokaryotes and eukaryotes (Shi *et al.*, 2009). Microarrays alone, due to limited resolution, could not detect the short sequences of small regulatory RNAs. Not surprisingly, groups studying sRNAs were first to realize the potential and apply NGS technologies.

RNA-Seq is a promising application of NGS that involves direct high-throughput sequencing of cDNA without cloning and any prior knowledge of the genomes they were derived from (Wang *et al.*, 2009). Because the focus is only on expressed sequences, less sequencing capacity is required than analysis of metagenomes. The level of expression can be quantified from the number of corresponding reads. RNA-Seq revealed a much more extensive and complex metatranscriptome than expected. In addition to RNA expression profiling, RNA-Seq uncovered novel classes of RNA, detected point mutations, and identified novel post translational modifications e.g. transcript fusions or alternative splicing (Blow, 2009). Metatranscriptomic sequences can serve as a scaffold for metagenome sequencing and can direct the researcher to the part of the (meta) genome of interest (Moran, 2009; Warnecke and Hess, 2009). Over the past few years metatranscriptomes from marine (Gilbert *et al.*, 2008; Friaz-Lopez *et al.*, 2008; Poretsky *et al.*, 2009; Shi *et al.*, 2009) and soil (Urich *et al.*, 2008) communities were sequenced following this approach. When applied to complex multi-organism samples, collected in the wild or artificially created in the laboratory, metatranscriptomics has revealed microbial dexterity in response to changes in environmental conditions and interactions within ecosystems (Marguerat and Bähler, 2010).

Ribosomal RNA (rRNA) species dominate a transcriptome while only a fraction of transcripts are mRNAs. Amidst constant improvements to select for mRNAs, or get rid of the overabundant rRNAs, metatranscriptomics is faced with other challenges. The high turn-over rate of RNA and instability of the molecule necessitate the optimization of extraction protocols to capture a specific moment in time. Also, the presence of a transcript does not mean the protein is produced or functional. Proteins can be subject to rapid turnover, or inactivity as a result of post-translational modification. Probably the most challenging factor is the small percentage of important genes that are correctly annotated in sequence datasets. Only the most abundant genes from a small number of microbial communities are included in public databases and their function cannot be assigned with confidence (Moran, 2009).

Metaproteomics

Proteomics is defined as the characterization of the complete protein profile (proteome) of an expressed genome (Keller and Hettich, 2009), and gives information about the function of a protein in relation to metabolic pathways. Proteins are therefore better biological markers of metabolic activity than functional genes or RNA transcripts. In addition to detecting the presence of a protein, proteomics gives information about protein abundance, post-translational modifications, stability, turnover rates and protein–protein interactions (Bastida *et al.*; 2009; Keller and Hettich, 2009). Although the terms 'metaproteomics', describing the profiling and studying of environmental community proteins, and 'metaproteome' were suggested as early as 2004 (Wilmes and Bond, 2004; Rodrigues-Valera, 2004), the field is still very young.

There are a few things to consider when analysing proteins from environmental samples. Distribution of proteins vary and extraction protocols have to take into account the location of proteins, be it intracellular or extracellular, or bound to organic particles or minerals. The low copy number of some proteins remains a challenge and, unlike DNA or RNA, amplification is not possible (Bastida *et al.*, 2009; Keller and Hettich, 2009; Cubillos-Ruiz, 2009). Protein isolates are generally separated and ionized prior to identification and characterization with mass spectrometry (MS) and database searches (Bastida *et al.*, 2009). Traditionally, intact proteins are separated applying two-dimensional gel electrophoresis (2D-GE). Proteins are separated in the first dimension based on their isoelectric point (pI) using isoelectric focusing (IEF), and in the second dimension according to molecular weight with SDS-PAGE. Individual protein spots are digested in the gel with specific proteases and ionized with the matrix-assisted laser desorption/ionization (MALDI) system before MS analysis. Alternatively, proteins, or digests thereof, are separated using liquid chromatography (LC). In the 'top-down' approach, intact proteins are separated, ionized, detected, fragmented and MS analysed, providing information on both the intact protein and its fragments. In the 'bottom-up' approach, proteins are digested with specific proteases, and peptides separated before ionization and MS analysis (Bastida *et al.*, 2009; Keller and Hettich, 2009; Cubillos-Ruiz, *et al.*, 2009). Although this approach is well suited for protein identification, protein sequence coverage is incomplete, leading to loss of information on protein modifications and degradation.

Metaproteomics is a powerful tool for data comparison and the discovery of novel gene products. Several studies have been published and include samples collected from the human gut (Klaassens *et al.*, 2007; Verberkmoes, *et al.*, 2009) and salivary communities (Rudney *et al.*,

2010), ocean (Morris *et al.*, 2010; Schweder *et al.*, 2008) and soil (Bastida *et al.*, 2009) communities, acid mine drainage biofilms (Ram *et al.*, 2005; Wilmes *et al.*, 2009), and contaminated soil and water (Bendorff *et al.*, 2007; Wilmes *et al.*, 2008). Some studies have also investigated proteomic responses in reaction to environmental changes (Lacerda *et al.*, 2006; Williams *et al.*, 2010). Such studies illustrate the powerful ability of environmental proteomics to describe the functional diversity of communities. Proteins were identified as biological markers for specific habitats, population and environmental stresses. Data generated increased our knowledge of microbial cooperation, competition and succession within complex communities. All the studies revealed a large portion of proteins with unknown function, highlighting the need for better databases. It is widely recognized that metagenome sequencing capacities will continue to increase. Faster and more sensitive proteomics instrumentation is crucial for improved data analysis, especially when applied to more complex environments. While MS-based approaches are well developed for generating qualitative data, improvements are required for quantitative measurements and *de novo* peptide sequencing strategies. Special attention needs to be given to bioinformatics development and support to analyse and manage extremely large data sets (Bastida *et al.*, 2009; Keller and Hettich, 2009; Cubillos-Ruiz, 2009).

Metabolomics

Metabolomics is defined as the characterization of the metabolites generated by a single or group of organisms under a certain set of environmental and physiological conditions. The aim of metabolomics is to have a holistic understanding of the workings of metabolic networks (Turnbaugh and Gordon, 2008). The Human Metabolome project was one of the first and largest databases constructed by combining identification and quantification of human metabolites from various samples with data mining from papers and books. Metabolic databases of environmental communities such as plankton and virus–plankton interactions (Paul *et al.*, 2009) and the combined symbiotic human and microbial cell activities (Li *et al.*, 2008) are slowly emerging. Different research groups approach metabolomics from different perspectives. Some focus on one aspect at a time, for example perturbations in lipid biochemistry (German *et al.*, 2007). Others associate certain signals with a specific metabolite, i.e. they try and find patterns, a strategy to discover biomarkers that could be used to diagnose and treat diseases (Jansson *et al.*, 2009). Metabolites may be analysed at a specific point in time (snapshot method), disregarding the dynamic responses of a biological system in response to external influences. On the other hand, by following metabolite traffic over time (flux method), pathways have been identified as well as quantified (Teusink *et al.*, 1998).

As with proteomics, metabolomic analysis is mostly achieved by coupling MS to separation techniques such as gas chromatography (GC), capillary electrophoresis (CE) or liquid chromatography (LC). MS coupled approaches are sensitive and robust. Recent technologies that offer ultra-high-resolution power include hydrophilic interaction liquid chromatography (HILIC)-ESMS (Tolstikov and Fiehn, 2002) and Fourier-transform ion cyclotron resonance (FT-ICR)-MS (Han *et al.*, 2008; Ohta *et al.*, 2010). More recently laser ablation electrospray ionization mass spectrometry (LAESI-MS) has been developed for *in situ*, single-cell metabolic analysis (Shrestha and Vertes, 2009). Nuclear magnetic resonances (NMR)-coupled approaches, on the other hand, are appreciated for its non invasive measuring abilities, reproducibility and selectivity (Kell, 2004; Cubillos-Ruiz *et al.*, 2009; Kim *et al.*, 2010). Unfortunately, most of the metabolites recognized by these technologies cannot be identified because they do not appear in any libraries. This impedes the understanding and reconstruction of metabolic networks (Kell, 2004).

With the emergence of genome-scale sequencing, metabolic pathway reconstruction evolved into genome scale metabolic models. As the metabolome is in constant flux, easily influenced by external stimuli and extremely susceptible to fluctuations, that metabolomics is expected to generate even more data than other systems biology disciplines. This calls for high-throughput measurements using instruments with improved sensitivity. It will also require bioinformatics development and support to analyse and integrate the data sets into metabolic models. Structural identification of the uncharacterized metabolites remains an important and ongoing challenge (Kell, 2004; Cubillos-Ruiz *et al.*, 2009).

For a detailed account of complementary 'omics' applications, see Chapter 12.

SIP-metagenomics

DNA stable isotope probing (SIP) has recently emerged as novel tool for enriching DNA of organisms for specific genes. The technique is based on the cultivation of the bacterial community in the presence of substrate labelled with heavy isotopes (^{13}C and ^{15}N). The substrate is incorporated through anabolic processes, and labelled cell constituents separated from lighter unlabelled constituents. For nucleic acids, this separation is achieved through isopycnic or density gradient centrifugation. Molecular analysis of the labelled DNA may allow for the identification of active microbes (Radajewski *et al.*, 2000; Schloss and Handelsman, 2003). Dumond and co-workers (2006) combined DNA-SIP with metagenomics by constructing a large-insert metagenomic library using ^{13}C-labelled DNA isolated from a $^{13}CH_4$-labelled forest soil. Screening for methylotrophy genes led to the isolation of a clone housing a complete pmoCAB operon, coding for the particulate methane monooxygenase (pMMO) enzyme. This technique offers valuable scope for prospective metagenome mining initiatives.

There are some potential disadvantages to the technique. Labelling does not take place in a natural setting, and the concentration of labelled compound could change gene expression or even community composition. Leaching and cross-feeding of the isotope are further potential pitfalls (Bastida *et al.*, 2009; Chen and Murrell, 2010). Leaching can however be positively employed to elucidate nutrient flow in a community.

Because SIP has the potential to link function to genome(s), it will be ideal in identifying less abundant species that play significant roles in specific community functions. These rare species would be overlooked using conventional approaches (Chen and Murrell, 2010). Combinations with other promising methods, such as microfluidics, fluorescence activated cell sorting (FACS), Raman microscopy, fluorescence *in situ* hybridization (FISH), isotope arrays, and microautoradiography (MAR), may offer exciting prospects in linking single species to functions (Wagner *et al.*, 2006; Neuveld and Murrel, 2007; Chen and Murrel, 2010).

A detailed account on SIP uses in metagenomics is given in Chapter 6.

Single-cell analyses

The goal of most metagenomic studies is to obtain complete genome sequences containing an inventory of metabolic potential. Rapid advances in automation and downscaling of omics technologies will enable researchers to analyse single cells and their subcellular compartments (Chen and Murrel, 2010; Wang and Bodovitz, 2010). Single-cell genomics have been applied for *E. coli* and *Prochlorococcus* by combining new PCR-independent amplification techniques, such as MDA, with whole-genome shotgun sequencing (Zhang *et al.*, 2006). Sequence coverage of 65% with a fourfold depth, on average, was achieved. Rapidly developing next-generation sequencing technologies, especially those involved in single-molecule sequencing, could soon enable single-cell sequencing of more complex genomes (Wang and Bodovitz, 2010). Transcriptomic analysis on a single-cell level has been realized by using quantitative PCR (qPCR) (Taniguchi *et al.*, 2009) and mRNA-Seq (Tang *et al.*, 2009). Although amplification can skew transcript quantification, such methodologies are independent of prior sequence knowledge and enable detection of splicing variants (Wang and Bodovitz, 2010). More than 100 metabolites could be detected in single R2 neurons and its subcellular regions using CE-ESI-MS (Lapainis *et al.*, 2009), while single-cell *in situ* metabolic profiling has been achieved with LAESI-MS (Shrestha and Vertes, 2009). Successful proteomic studies on single-cell level are limited and mainly involves fluorescence-activated cell sorting (Cohen *et al.*, 2008). MS-based studies, using MALDI-MS and LC-MS/MS, show limited success with only a small number of detectable peptides. Unlike DNA and RNA, metabolites and proteins cannot be amplified. New methods of sample manipulation and detection are required to realize proteomic and metabolomic analysis on single-cell level (Wang and Bodovitz, 2010).

It is expected that the ability to analyse single cells will transform the study of whole systems. Because noise created by cellular heterogeneity is decreased, more accurate assumptions can be made of inter- and intracellular pathways, networks and dependencies, as well as the relationship with the extracellular environment.

More details on the single-cell approach in metagenomics are given in Chapter 7.

Biological system engineering

Bioinformatic analysis of integrated high throughput data makes computational building of network models possible. Systems biology aims at studying and understanding complex biological systems as a whole, but does not include creating new functions or products (Tisoncik and Katze, 2010; Serrano, 2007). Knowledge of a system's metabolic possibilities, provided by metagenomic studies and complementary approaches (i.e. metatranscriptomics, metaproteomics and metabolomics) combined with computational modelling, would eventually allow for increasingly complex network reconstructions (Raes and Bork, 2008). Synthetic biology on the other hand draws from this gained knowledge to rationally engineer new biological systems that do not exist in the natural world. This includes re-designing existing biological systems for research and useful purposes. In May 2010, the complete synthesis of the *Mycoplasma mycoides* genome, its successful transplant into a *M. capricolum* cell, followed by normal replication was announced (Gibson, *et al.*, 2010). This event hints at the potential of designing bacteria capable of functions previously thought impossible.

In the next several years metagenomics will certainly become more visible in everyday technology. The most immediate change foreseen is the recruitment of an ever-expanding range of organisms for the production of compounds such as pharmaceuticals and biofuels, and for use in the bioremediation of environmental contaminants. The abundance of alternative enzymes harvested from metagenomic libraries can be complemented by the availability of new suitable hosts capable of for example deriving energy from solar light (such as cyanobacteria), or cheap substrates such as agricultural lignocellulosic waste products. Such a combination will bring metabolic engineering to an entirely new level, allowing the construction of customized organisms for every ecological niche that may convert waste into useful products.

Further insights on the relationship between metagenomics and system biology are given in Chapters 4 and 11.

Conclusion

While there can be no doubt that metagenomics has led to the discovery of a plethora of genetic information and genomic potential, and will continue to do so with ongoing advances in technologies such as NGS and single-cell sequencing, the next challenge lies in gaining a greater understanding and better description of this information. In the apt words of Hugenholtz and Tyson (2008): 'Before the advent of metagenomics, however, we were completely oblivious to what we didn't know (unknown unknowns); now we have the blueprints, although we can't read many of the instructions (known unknowns).' This will depend largely on our ability to link sequence information (genotype) to structural and functional information (phenotype), and will rely not only on the expansion of current bioinformatics tools and computational infrastructures, but also on future advances in the fields of structural and functional biochemistry that expand our ability to predict protein structure and activity from novel sequences. Better annotated metagenomes (and reference genomes) will certainly lead to greater informative power in comparative metagenomics studies and associations with environmental metadata. Our current view on the microbial world will undoubtedly undergo a further paradigm shift in the future as projects begin to relate the genomic potential of the metagenome, and environmental factors, to phenotypic information gathered from other disciplines exploring gene expression and transcription regulation (metatranscriptomics), protein expression and translation regulation (metaproteomics), metabolomics, functional enzyme discovery (fuelled by advances in HT functional screening), and *in vitro* evolution. We are likely to see environmental communities being tackled with a more integrated systems-based approach. Knowledge of a system's metabolic possibilities, provided by integrated high throughput data from the 'omic' approaches, combined with computational modelling will allow for increasingly complex and informative network models (Raes and Bork, 2008).

Web resources

CAMERA: Community Cyberinfrastructure for Advanced Microbial Ecology Research and Analysis: Available http://camera.calit2.net/

FastHMM and FastBLAST: tools for analysing large protein sequence databases: Available http://www.microbesonline.org/fasthmm/

Genomes OnLine Database or GOLD: Available http://www.genomesonline.org/

The Genomics Standards Consortium: http://gensc.org/

Global Ocean Sampling (GOS) Expedition: Available http://www.jcvi.org/cms/research/projects/gos/

The Human Microbiome Project or HMP: Available http://nihroadmap.nih.gov/hmp

HMMER3: a new generation of sequence homology search software: Available http://hmmer.janelia.org/

InterPro protein sequence analysis and classification: Available http://www.ebi.ac.uk/interpro/

MG-RAST (Metagenome Rapid Annotation using Subsystem Technology): an automated service for annotating metagenome sequences: Available metagenomics.nmpdr.org/

Terragenome (soil metagenome project): Available http://www.terragenome.org/

References

Abulencia, C.B., Wyborski, D.L., Garcia, J.A., Podar, M., Chen, W., Chang, S.H., Chang, H.W., Watson, D., Brodie, E.L., Hazen, T.C., and Keller, M. (2006). Environmental whole-genome amplification to access microbial populations in contaminated sediments. Appl. Environ. Microbiol. *72*, 3291–3301.

Altschul, S.F., Gish, W., Miller, W., Myers, E.W., and Lipman, D.J. (1990). Basic local alignment search tool. J. Mol. Biol. *215*, 403–410.

Amann, R.I., Ludwig, W., and Schleifer, K.H. (1995). Phylogenetic identification and *in situ* detection of individual microbial cells without cultivation. Microbiol. Rev. *59*, 143–169.

Angelov, A., Mientus, M., Liebl, S., and Liebl, W. (2009). A two-host fosmid system for functional screening of (meta)genomic libraries from extreme thermophiles. Syst. Appl. Microbiol. *32*, 177–185.

Angly, F.E., Felts, B., Breitbart, M., Salamon, P., and Edwards, R.A. (2006). The marine viromes of four oceanic regions. PLoS Biol. *4*, e368. doi: 10.1371/journal.pbio.0040368.

Azad, R.K., and Borodovsky, M. (2004). Probabilistic methods of identifying genes in prokaryotic genomes: connections to the HMM theory. Brief Bioinform. *5*, 118–130.

Bailly, J., Fraissinet-Tachet, L., Verner, M.-C., Debaud, J.-C., Lemaire, M., Wesolowski- Louvel, M., and Marmeisse, R. (2007). Soil eukaryotic functional diversity, a metatranscriptomic approach. ISME J. *1*, 632–642.

Bastida, F., Moreno, J.L., Nicolas, C., Hernandez, T., and Garcia, C. (2009). Soil metaproteomics: a review of an emerging environmental science. Significance, methodology and perspectives. Eur. J. Soil Sci. *60*, 845–859.

Benndorf, D., Balcke, G.U., Harms, H., and Von Bergen, M. (2007). Functional metaproteome analysis of protein extracts from contaminated soil and groundwater. ISME J. *1*, 224–234.

Bell, P.J.L., Sunna, A., Gibbs, M.D., Curach, N.C., Nevalainen, H., and Bergquist, P.L. (2002). Prospecting for novel lipase genes using PCR. Microbiology *148*, 2283–2291.

Bertin, P.N., Medigue, C., and Norman, P. (2008). Advances in environmental genomics: towards an integrated view of micro-organisms and ecosystems. Microbiology *154*, 347–359.

Binga, E.K., Lasken, R.S., and Neufeld, J.D. (2008). Something from (almost) nothing: the impact of multiple displacement amplification on microbial ecology. ISME J. *2*, 233–241.

Blow, N. (2009). The digital generation. Nature *458*, 239–243.

Boubakri, H., Beuf, M., Simonet, P., and Vogel, T.M. (2006). Development of metagenomic DNA shuffling for the construction of a xenobiotic gene. Gene *375*, 87–94.

Brondyk, W.H. (2009). Selecting an appropriate method for expressing a recombinant protein. In Methods in Enzymology Volume *463*, Burgess, R.R., and Deutscher, M.P., eds. (Academic Press), pp 131–147.

Brouzes E., Medkova, M., Savenelli, N., Marran, D., Twardowski, M., Hutchison, J.B., Rothberg, J.M., Link, D.R., Perrimon, N., and Samuels, M.L. (2009). Droplet microfluidic technology for single-cell high-throughput screening. Proc. Natl. Acad. Sci. U.S.A. *106*, 14195–200.

Chen, Y., and Murrell, J.C. (2010). When metagenomics meets stable-isotope probing: progress and perspectives. Trends Microbiol. *18*, 157–163.

Chivian, D., Brodie, E.L., Alm, E.J., Culley, D.E., Dehal, P.S., DeSantis, T.Z., Gihring, T.M., Lapidus, A., Lin, L.H., Lowry, S.R. *et al.* (2008). Environmental genomics reveals a single-species ecosystem deep within Earth. Science *322*, 275–8.

Cohen, D., Dickerson, J.A., Whitmore, C.D., Turner, E.H., Palcic, M.M., Hindsgaul, O., and Dovichi, N.J., (2008). Chemical cytometry: Fluorescence-based single-cell analysis. Ann. Rev. Anal. Chem. *1*, 165–190.

Coppée, J.-Y. (2008). Do DNA microarrays have their future behind them? Microbes Infect. *10*, 1067–1071.

Courtois, S., Cappellano, C.M., Ball, M., Francou, F.X., Normand, P., Helynck, G., Martinez, A., Kolvek, S.J.,

Hopke, J., Osburne, M.S. *et al.* (2003). Recombinant environmental libraries provide access to microbial diversity for drug discovery from natural products. Appl. Environ. Microbiol. *69*, 49–55.

Cowan, D., Meyer, Q., Stafford, W., Muyanga, S., Cameron, R., and Wittwer, P. (2005). Metagenomic gene discovery: Past, present and future. Trends Biotechnol. *23*, 321–329.

Craig, J.W., Chang, F.-Y., and Brady, S.F. (2009). Natural Products from Environmental DNA Hosted in *Ralstonia metallidurans*. ACS Chem. Biol. *16*, 23–28.

Cubillos-Ruiz, A., Junca, H., Baena, S., Venegas, I., and Zambrano, M.M. (2009). Beyond metagenomics: Integration of complementary approaches for the study of microbial communities. In Metagenomics; Theory, methods and applications, Marco, D., ed. (Norfolk: Caister Academic Press), pp. 1–15.

Cuvelier, M.L., Allen, A.E., Monier, A., McCrow, J.P., Messié, M., Tringe, S.G., Woyke, T., Welsh, R.M., Ishoey, T., Lee, J.H. *et al.*, (2010). Targeted metagenomics and ecology of globally important uncultured eukaryotic phytoplankton. Proc. Natl. Acad. Sci. U.S.A. *107*, 14679–14684.

Dai, W.D., Sheikh, M.A., Milenkovic, O., and Baraniuk, R.G. (2009). Compressive Sensing DNA Microarrays. EURASIP J. Bioinform. Syst. Biol. Article ID 162824, 12 pages. doi:10.1155/2009/162824.

Daniel, R. (2005). The metagenomics of soil. Nat. Rev. Microbiol. 3, 470–478.

DeLong, E.F., Preston, C.M., Mincer, T., Rich, V., Hallam, S.J., Frigaard, N.U., Martinez, A., Sullivan, M.B., Edwards, R., and Brito, B.R. (2006). Community genomics among stratified microbial assemblages in the ocean's interior. Science *311*, 496–503.

Demain, A.L., and Vaishnav, P. (2009). Production of recombinant proteins by microbes and other organisms. Biotechnol. Adv. *27*, 297–306.

Dumond, M.G., Radajewski, S.M., Miguez, C.B., McDonald, I.R., and Murrell, J.C. (2006). Identification of a complete methane monooxygenase operon from soil by combining stable isotope probing and metagenomic analysis. Environ. Microbiol. *8*, 1240–1250.

Edwards, R.A., Rodriguez-Brito, B., Wegley, L., Haynes, M., Breitbart, M., Peterson, D.M., Saar, M.O., Alexander, S., Alexander, E.C. Jr., and Rohwer, F. (2006). Using pyrosequencing to shed light on deep mine microbial ecology. BMC Genomics *7*, 57. doi:10.1186/1471-2164-7-57.

Ferrer, M., Martínez-Abarca, F., and Golyshin, P.N. (2005). Mining genomes and 'metagenomes' for novel catalysts. Curr. Opin. Biotechnol. *16*, 588–593.

Finn, R.D., Tate, J., Mistry, J., Coggill, P.C., Sammut, S.J., Hotz, H.R., Ceric, G., Forslund, K., Eddy, S.R., Sonnhammer, E.L., and Bateman, A. (2008). The Pfam protein families database. Nucleic Acids Res. *36*, D281–288.

Foerstner, K.U., von Mering, C., Hooper, S.D., and Bork, P. (2005). Environments shape the nucleotide composition of genomes. EMBO Rep. *6*, 1208–1213.

Friedrich, M.W. (2006). Stable-isotope probing of DNA: insights into the function of uncultivated microorganisms from isotopically labeled metagenomes. Curr. Opin. Biotechnol. *17*, 59–66.

Frias-Lopez, J., Shi, Y., Tyson, G.W., Coleman, M.L., Schuster, S.C., Chisholm, S.W., and DeLong, E.F. (2008). Microbial community gene expression in ocean surface waters. Proc. Nat. Acad. Sci. U.S.A. *105*, 3805–3810.

German, J.B., Gillies, L.A., Smilovitz, J.T., Zivkovic, A.M., and Watkins, S.M. (2007). Lipidomics and lipid profiling in metabolics. Curr. Opin. Lipidomics *18*, 66–71.

Gibson, D.G., Glass, J.I., Lartigue, C., Noskov, V.N., Chuang, R.-Y., Algire, M.A., Benders, G.A., Montague, M.G., Ma, L., Moodie, M.M. *et al.* (2010). Creation of a bacterial cell controlled by a chemically synthesized genome. Science *329*, 52–56.

Gilbert, J.A., Field, D., Huang, Y., Edwards, R., Li, W., Gilna, P., and Joint, I. (2008). Detection of large numbers of novel sequences in the metatranscriptomes of complex marine microbial communities. PLoS ONE 38, e3042. doi:10.1371/journal.pone.0003042.

Haft, D.H., Selengut, J.D., and White, O. (2003) The TIGRFAMs database of protein families. Nucleic Acids Res. *31*, 371–373.

Han, J., Danell, R.M., Patel, J.R., Gumerov, D.R., Scarlett, C.O., Speir, J.P., Parker, C.E., Rosyn, I., Zeisel, S., and Barchers, C.H. (2008). Towards high-throughput metabolomics using ultrahigh-field fourier transform ion cyclotron resonance mass spectrometry. Metabolomics *4*, 128–140.

Handelsman, J. (2004). Metagenomics: application of genomics to uncultured microorganisms. Microbiol. Mol. Biol. Rev. *68*, 669–685.

He Z., Gentry, T.J., Schadt, C.W., Wu, L., Liebich, J., Chong, S.C., Huang, Z., Wu, W., Gu, B., Jardine, P. *et al.* (2007). Geochip: a comprehensive microarray for investigating biogeochemical, ecological and environmental processes. ISME J. *1*, 67–77.

Heath, C., Hu, X.-P., Cary, C., and Cowan, D. (2009). Isolation and characterization of a novel low-temperature-active alkaliphilic esterase from an Antarctic desert soil metagenome. Appl. Environ. Microbiol. *75*, 4657–4659.

Hugenholtz, P., Goebel, B.M., and Pace, N.R. (1998). Impact of culture-independent studies on the emerging phylogenetic view of bacterial diversity. J Bacteriol. *180*, 4765–74.

Hugenholtz, P., and Tyson, G.W. (2008). Microbiology: metagenomics. Nature *25*, 481–483.

Huson, D.H., Auch, A.F., Qi, J., Schuster, S.C. (2007). MEGAN analysis of metagenomic data. Genome Res *17*, 377–386.

Huson, D.H., Richter, D.C., Mitra, S., Auch, A.F., and Schuster, S.C. (2009). Methods for comparative metagenomics. BMC Bioinform. *10*, S12. doi:10.1186/1471-2105-10-S1-S12.1190719

Jansson, J., Willing, B., Lucio, M., Fekete, A., Dicksved, J., Halfvarson, J., Tysk, C., and Schmitt-Kopplin, P. (2009). Metabolomics reveals metabolic biomarkers of Chrohn's disease. PLoS ONE *4*, e6386. doi:10.1371/journal.pone.0006386.

Jenney, F.E., Jr., and Adams, M.W. (2008). The impact of extremophiles on structural genomics (and vice versa). Extremophiles *12*, 39–50.

Jiang, Z., Wang, H., Ma, Y., and Wei, D. (2006). Characterisation of two novel lipase genes isolated directly from environmental sample. Appl. Microbiol. Biotechnol. *70*, 327–332.

Kell, D.B. (2004). Metabolomics and systems biology: making sense of the soup. Curr. Opin. Microbiol. *7*, 296–307.

Keller, M., and Hettich, R. (2009). Environmental proteomics: a paradigm shift in characterizing microbial activities at the molecular level. Microbiol. Mol. Biol. Rev. *73*, 62–70.

Kim, H.K., Choi, Y.H., and Verpoorte, R. (2010). NMR based metabolomic analysis of plants. Nat. Protoc. *5*, 536–549.

Kim, Y.-J., Choi, G.-S., Kim, S.-B., Yoon, G.-S., Kim. Y.-S., and Ryu, Y.-W. (2006). Screening and characterisation of a novel esterase from a metagenomic library. Protein Expr. Purif. *45*, 315–323.

Klaassens, E.S., de Vos, W.M., and Vaughan, E.E. (2007). A metaproteomics approach to study the functionality of the microbiota in the human infant gastrointestinal tract. Appl. Environ. Microbiol. *73*, 1388–1392.

Kotik, M. (2009). Novel genes retrieved from environmental DNA by polymerase chain reaction: Current genome-walking techniques for future metagenome applications. J. Biotechnol. *144*, 75–82.

Kunin, V., Copeland, A., Lapidus, A., Mavromatis, K., and Hugenholtz, P. (2008). A bioinformatician's guide to metagenomics. Microbiol. Mol. Biol. Rev. *72*, 557–578.

Lacerda, C.M.R., Choe, L.H., and Reardon, K.F. (2007). Metaproteomic analysis of a bacterial community response to Cadmium exposure. J. Proteome Res. *6*, 1145–1152.

Lämmle, K., Zipper, H., Breuer, M., Hauer, B., Buta, C., Brunner, H., and Rupp, S. (2007). Identification of novel enzymes with different hydrolytic activities by metagenome expression cloning. J. Biotechnol. *127*, 575–592.

Lapainis, T., Rubakhin, S.S., and Sweedler, J.V. (2009). Capillary electrophoresis with electrospray ionization mass spectrometric detection for single-cell metabolomics. Anal. Chem. *81*, 5858–5864.

Li, W. (2009). Analysis and comparison of very large metagenomes with fast clustering and functional annotation. BMC Bioinform. *10*, 359.

Li, M., Wang, B., Zhang, M., Rantalainen, M., Wang, S., Zhou, H., Zhang, Y., Shen, J., Pang, X., Zhang, M. *et al.*, (2008). Symbiotic microbes modulate human metabolic phenotypes. Proc. Nat. Acad. Sci. U.S.A. *105*, 2117–2122.

Lorenz P., and Eck, J. (2005). Metagenomics and industrial applications. Nat. Rev. Microbiol. *3*, 510–516.

Loy, A., Schulz, C., Lücker, S., Schöpfer-Wendels, A., Stoecker, K., Baranyi, C., Lehner, A., and Wagner, M. (2004). 16S rRNA gene-based oligonucleotide microarray for environmental monitoring of the Betaproteobacterial order 'Rhodocyclales'. Appl. Environ. Microbiol. *71*, 1373–1386.

Ludwig, W., and Scheifer, K.H. (1999). Phylogeny of Bacteria beyond the 16S rRNA standard. ASM News 65, 752–757.

Mardis, E.R. (2008). Next-generation DNA sequencing methods. Ann. Rev. Genomics Hum. Genet. *9*, 387–402.

Marguerat, S., and Bähler, J. (2010). RNA-seq: from technology to biology. Cell. Mol. Life Sci. *67*, 569–579.

Margulies M., Egholm, M., Altman, W.E., Attiya, S., Bader, J.S., Bemben, L.A., Berka, J., Braverman, M.S., Chen, Y.J., Chen, Z. *et al.* (2005). Genome sequencing in microfabricated high-density picolitre reactors. Nature *437*, 376–380.

Martinez, A., Kolvek, S.J., Yip, C.L.T., Hopke, J., Brown, K.A., MacNeil, I.A., and Osburne, M.S. (2004). Genetically modified bacterial strains and novel bacterial artificial chromosome shuttle vectors for constructing environmental libraries and detecting heterologous natural products in multiple expression hosts. Appl. Environ. Microbiol. *70*, 2452–2463.

McHardy, A.C., and Rigoutsos, I. (2007). What's in the mix: Phylogenetic classification of metagenome sequence samples. Curr. Opin. Microbiol. *10*, 499–503.

McHardy, A.C., Martín, H.G., Tsirigos, A., Hugenholtz, P., and Rigoutsos, I. (2007). Accurate phylogenetic classification of variable-length DNA fragments. Nat. Methods *4*, 63–72.

McPherson, J.D. (2009). Next-generation gap. Nature Methods *6*, S2–5.

Meyer, F., Paarmann, D., D'Souza, M., Olson, R., Glass, E.M., Kubal, M., Paczian, T., Rodriguez, A., Stevens, R., Wilke, A. *et al.*, (2008). The metagenomics rast server – a public resource for the automatic phylogenetic and functional analysis of metagenomes. BMC Bioinform. *9*, 386.

Moran, M.A. (2009). Metatranscriptomics: Eavesdropping on Complex Microbial Communities. Microbe *4*, 329–335.

Morris, R.M., Nunn, B.L., Frazar, C., Goodlett, D.R., Ting, Y.S., and Rocap, G. (2010). Comparative metaproteomics reveals ocean-scale shifts in microbial nutrient utilization and energy transduction. ISME J. *4*, 673–685.

Morrison, M., Pope, P.B., Denman, S.E., and McSweeney, C.S. (2009). Plant biomass degradation by gut microbiomes: more of the same or something new? Curr. Opin. Biotechnol. *20*, 358–363.

Neufield, J.D., and Murrell, J.C. (2007). Witnessing the last supper of uncultivated microbial cells with Raman-FISH. ISME J. *1*, 269–270.

Noguchi, H., Park, J., and Takagi, T. (2006). MetaGene: Prokaryotic gene finding from environmental genome shotgun sequences. Nucl. Acids Res. *34*, 5623– 5630.

Ohta, D., Kanaya, S., and Suzuki, H. (2010). Application of Fourier-transform ion cyclotron resonance mass spectrometry to metabolic profiling and metabolite identification. Curr. Opin. Biotechnol. *21*, 35–44.

Pace, N.R., Stahl, D.A., Lane, D.J., and Olsen, G.J. (1985). Analyzing natural microbial populations by rRNA sequences. ASM News *51*, 4–12.

Park, S.-J., Kang, C.-H., Chae, J.-C., and Rhee, S.-K. (2008). Metagenome microarray for screening of

fosmid clones containing specific genes. FEMS Microbiol. Lett. *284*, 28–34.
Parro, V., Moreno-Paz, M., and González-Toril, E. (2007). Analysis of environmental transcriptomes by DNA microarrays. Environ. Microbiol. 9, 453–464.
Pathak, G.P., Ehrenreich, A., Losi, A., Streit, W.R., and Gärtner, W. (2009). Novel blue light-sensitive proteins from a metagenomic approach. Environ. Microbiol. *11*, 2388–2399.
Paul, C., Barofsky, A. Vidoudez, C., and Pohnert, G. (2009). Diatum exudates influence metabolism and cell growth of co-cultured diatom species. Mar. Ecol. Prog. Ser. *389*, 61–70.
Pond, S.K., Wadhawan, S., Chiaromonte, F., Ananda, G., Chung,W.-Y., Taylor, J., Nekrutenko, A., and The Galaxy Team. (2009). Windshield splatter analysis with the Galaxy metagenomic pipeline. Genome Res. *19*, 2144–2153.
Pop, M., and Salzberg, S.L. (2008). Bioinformatics challenges of new sequencing technology. Trends Genet. *24*, 142–149.
Poretsky, R.S., Bano, N., Buchan, A., Lecleir, G., Kleikemper, J., Pickering, M., Pate, W.M., Moran, M.A., and Hollibaugh, J.T. (2005). Analysis of microbial gene transcripts in environmental samples. Appl. Environ. Microbiol. *71*, 4121–4126.
Poretsky, R.S., Hewson, I., Sun, S., Allen, A., Zehr, J.P., and Moran, M.A. (2009). Comparative day/night metatranscriptomic analysis of microbial communities in the North Pacific subtropical gyre. Environ. Microbiol. *11*, 1358–1375.
Radajewski, S., Ineson, P., Parekh, N.R., and Murrell, J.C. (2000) Stable-isotope probing as a tool in microbial ecology. Nature *403*, 646–649.
Raes, J., Foerstner, K.U., and Bork, P. (2007). Get the most out of your metagenome: Computational analysis of environmental sequence data. Curr. Opin. Microbiol. *10*, 490–498.
Raes, J., Korbel, J.O., Lercher, M.J., von Mering, C., Bork, P. (2007). Prediction of effective genome size in metagenomic samples. Genome Biol. *8*, R10.
Raes, J., and Bork, P. (2008). Molecular eco-systems biology: towards an understanding of community function. Nat. Rev. Microbiol. *6*, 693–699.
Ram, R.J., VerBerkmoes, N.C., Thelen, M.P., Tyson, G.W., Baker, B.J., Blake, R.C., Shah, M., Hettich, R.L., and Banfield, J.F.. (2005). Community proteomics of a natural microbial biofilm. Science *308*, 1915–1920.
Ranjan, R., Grover, A., Kapardar, R.K., and Sharma, R. (2005). Isolation of novel lipolytic genes from uncultured bacteria of pond water. Biochem. Biophys. Res. Comm. *335*, 57–65.
Rees, H.C., Grant, S., Jones, B., Grant, W.D., and Heaphy, S. (2003). Detecting cellulase and esterase enzyme activities encoded by novel genes present in environmental DNA libraries. Extremophiles *7*, 415–421.
Rich, V.I., Konstantinidis, K., and Delong, E.F. (2008). Design and testing of 'genome-proxy' microarrays to profile marine microbial communities. Environ. Microbiol. *10*, 506–521.
Rodriguez-Valera, F. (2004). Environmental genomics, the big picture? FEMS Microbiol. Lett. *231*, 153–158.
Rondon, M.R., August, P.R., Bettermann, A.D., Brady, S.F., Grossman, T.H., Liles, M.R., Loiacono, K.A., Lynch, B.A., MacNeil, I.A., Minor, C. *et al.* (2000). Cloning the soil metagenome: a strategy for accessing the genetic and functional diversity of uncultured microorganisms. Appl. Environ. Microbiol. *66*, 2541–2547.
Rudney, J.D., Xie, H., Rhodus, N.L., Ondrey, F.G., and Griffin, T.J. (2010). A metaproteomic analysis of the human salivary microbiota by three-dimensional peptide fractionation and tandem mass spectrometry. Mol. Oral Microbiol. *25*, 38–49.
Rusch, D.B., Halpern, A.L., Sutton, G., Heidelberg, K.B., Williamson, S., Yooseph, S., Wu, D., Eisen, J.A., Hoffman, J.M., and Remington, K. (2007). The Sorcerer II Global Ocean Sampling Expedition: Northwest Atlantic through Eastern Tropical Pacific. PLoS Biol. *5*, e77. doi: 10.1371/journal.pbio.0050077.
Schaerli Y., and Hollfelder, F. (2009). The potential of microfluidic water-in-oil droplets in experimental biology. Mol. Biosyst. *5*, 1392–404.
Schloss P.D., and Handelsman, J. (2003). Biotechnological prospects from metagenomics. Curr. Opin. Biotechnol. *14*, 303–10.
Schoenfeld, T., Liles, M., Wommack, K.E., Polson, S.W., Godiska, R., and Mead, D. (2009). Functional viral metagenomics and the next generation of molecular tools. Trends Microbiol. *18*, 20–29.
Schweder, T., Markert, S., and Hecker, M. (2008). Proteomics of marine bacteria. Electrophoresis *29*, 2603–2616.
Sebat, J.L., Colwell, F.S., and Crawford, R.L. (2003). Metagenomic profiling: microarray analysis of an environmental genomic library. Appl. Environ. Microbiol. *69*, 4927–4934.
Serrano, L. (2007). Synthetic biology: promises and challenges. Mol. Sys. Biol. 3, 158.
Shi, Y., Tyson, G.W., and DeLong, E.F. (2009). Metatranscriptomics reveals microbial small RNAs in the ocean's water column. Nature *459*, 266–269.
Shrestha, B., and Vertes, A. (2009). In situ metabolic profiling of single cells by laser ablation electrospray ionization mass spectrometry. Anal. Chem. *81*, 8265–8271.
Simon, C., Herath, J., Rockstroh, S., and Daniel, R. (2009). Rapid identification of genes encoding DNA polymerases by function-based screening of metagenomic libraries derived from glacial ice. Appl. Environ. Microbiol. 75, 2964–2968.
Simon, C., and Daniel, R. (2009). Achievements and new knowledge unraveled by metagenomic approaches. Appl. Microbiol. Biotechnol. *85*, 265–276.
Sogin, M.L., Morrison, H.G., Huber, J.A., Mark Welch, D., Huse, S.M., Neal, P.R., Arrieta, J.M., and Herndl, G.J. (2006). Microbial diversity in the deep sea and the underexplored 'rare biosphere'. Proc. Natl. Acad. Sci. U.S.A. *103*, 12115–12120.
Steele, H.L., Jaeger, K.-E., Daniel, R., and Streit, W.R. (2009). Advances in recovery of novel biocatalysts

from metagenomes. J. Mol. Microbiol. Biotechnol. *16*, 25–37.

Stein, J.L., Marsh, T.L., Wu, K.Y., Shizuya, H., and DeLong, E.F. (1996). Characterization of uncultivated prokaryotes: isolation and analysis of a 40-kilobase-pair genome fragment front a planktonic marine archaeon. J. Bacteriol. *178*, 591–599.

Tang, F., Barbacioru, C., Wang, Z., Nordman, E., Lee, C., Xu, N., Wang, X., Bodeau, J., Siddiqui, A., Lao, K., and Surani, A. (2009). mRNA-Seq whole-transcriptome analysis of a single cell. Nat. Methods *6*, 377–382.

Taniguchi, K., Kajiyama, T., and Kambara, H. (2009). Quantitative analysis of gene expression in a single cell by qPCR. Nat. Methods *6*, 503–506.

Teusink, B., Baganz, F., Westerhoff, H.V., and Oliver, S.G. (1998). Metabolic control analysis as a tool in the elucidation of the function of novel genes. Methods Microbiol. *26*, 297–336.

Tisoncik, J.R., and Katze, M.G. (2010). What is systems biology? Future Microbiol. *5*, 139–141.

Tolstikov, V.V., and Fiehn, O., (2002). Analysis of highly polar compounds of plant origin: Combination of hydrophilic interaction chromatography and electrospray ion trap mass spectrometry. Anal. Biochem. *301*, 298–307.

Tuffin, M., Anderson, D., Heath, C., and Cowan, D. (2009). Metagenomic gene discovery: How far have we moved into novel sequence space? Biotechnol. J. *4*, 1–13.

Turnbaugh P.J., and Gordon, J.I. (2008). An invitation to the marriage of metagenomics and metabolomics. Cell *134*, 708–713.

Tyson, G.W., Chapman, J., Hugenholtz, P., Allen, E.E., Ram, R.J., Richardson, P.M., Solovyev, V.V., Rubin, E.M., Rokhsar, D.S., Banfield, J.F. (2004). Community structure and metabolism through reconstruction of microbial genomes from the environment. Nature *428*, 37–43.

Uchiyama, T., Abe, T., Ikemura, T., and Watanabe, K. (2005). Substrate-induced gene expression screening of environmental metagenome libraries for isolation of catabolic genes. Nat. Biotechnol. *23*, 88–93.

Uchiyama, T., and Miyazaki, K. (2009). Functional metagenomics for enzyme discovery: challenges to efficient screening. Curr. Opin. Biotechnol. *20*, 616–622.

Uchiyama, T., and Miyazaki, K. (2010). Product-induced gene expression (PIGEX): a product-responsive reporter assay for enzyme screening of metagenomic libraries. Appl. Environ. Microbiol. *6*, 7029–7035.

Urich, T., Lanzén, A., Qi, J., Huson, D.H., Schleper, C., and Schuster, S.C. (2008). Simultaneous assessment of soil microbial community structure and function through analysis of the meta-transcriptome. PLoS ONE 3, e2527. doi: 10.1371/journal.pone.0002527.

van Hellemond, E.W., Janssen, D.B., and Fraaije, M.W. (2007). Discovery of a novel styrene monooxygenase originating from the metagenome. Appl. Environ. Microbiol. *73*, 5832–5839.

Venter, J.C. (2010). Multiple personal genomes await. Nature *464*, 676–677.

Verberkmoes, N.C., Russell, A.L., Shah, M., Godzik, A., Rosenquist, M., Halfvarson, J., Lefsrud, M.G., Apajalahti, J., Tysk, K., Hettich, R.L., and Jansson, J.K. (2009). Shotgun metaproteomics of the human distal gut microbiota. ISME J. 3, 179–189.

Vieites, J.M., Guazzaroni, M.E., Beloqui, A., Golyshin, P.N., and Ferrer, M. (2009). Metagenomics approaches in systems microbiology. FEMS Microbiol. Rev. *33*, 236–255.

Wagner, M., Nielsen, P.H., Loy, A., Nielsen, J.L., and Daims, H. (2006). Linking microbial community structure with function: fluorescence *in situ* hybridization-microautoradiography and isotope arrays. Curr. Opin. Biotechnol. *17*, 83–91.

Wang, Z., Gerstein, M., and Snyder, M. (2009). RNA-Seq: a revolutionary tool for transcriptomics. Nat. Rev. Genet. *10*, 57–63.

Wang, D and Bodovitz, S. (2010). Single cell analysis; the new frontier in 'omics'. Trends Biotechnol. *28*, 281–290.

Warnecke, F., and Hugenholtz, P. (2007). Building on basic metagenomics with complementary technologies. Genome Biol. *8*, 231. doi:10.1186/gb-2007-8-12-231.1190719.

Warnecke, F., and Hess, M. (2009). A perspective: Metatranscriptomics as a tool for the discovery of novel biocatalysts. J. Biotechnol. *142*, 91–95.

Whitman, W.B., Coleman, C.D., and William, J.W. (1998). Prokaryotes: The unseen majority. Proc. Natl. Acad. Sci. U.S.A. *95*, 6578–6583.

Williams, M.A., Taylor, E.B., and Mula, H.P. (2010). Metaproteomic characterization of a soil microbial community following carbon amendment. Soil. Biol. Biochem. *42*, 1148–1156.

Williamson, L.L., Borlee, B.R., Schloss, P.D., Guan, C., Allen, H.K., and Handelsman, J. (2005). Intracellular screen to identify metagenomic clones that induce or inhibit a quorum-sensing biosensor. Appl. Environ. Microbiol. *71*, 6335–6344.

Wilmes, P., and Bond, P.L. (2004). The application of two-dimensional polyacrylamide gel electrophoresis and downstream analyses to a mixed community of prokaryotic microorganisms. Environ. Microbiol. *6*, 911–920.

Wilmes P., Wexler, M., and Bond, P.L. (2008). Metaproteomics provides functional insight into activated sludge wastewater treatment. PLoS ONE 3, e1778. doi:10.1371/journal.pone.0001778.

Wilmes, P., Remis, J.P., Hwang, M., Auer, M., Thelen, M.P., and Banfield, J.F. (2009). Natural acidophilic biofilm communities reflect distinct organismal and functional organization. ISME J. 3, 266–270.

Wooley, J.C., Godzik, A., Friedberg, I. (2010). A Primer on Metagenomics. PLoS Comput. Biol. *6*, e1000667. doi:10.1371/journal.pcbi.1000667.

Wu, M., and Eisen, J.A. (2008). A simple, fast, and accurate method of phylogenomic inference. Genome Biol. *9*, R151. doi: 10.1186/gb-2008-9-10-r151.

Xu, J. (2006). Microbial ecology in the age of genomics and metagenomics: Concepts, tools, and recent advances. Mol. Ecol. *15*, 1713–1731.

Yamada, K., Terahara, T., Kurata, S., Yokomaku, T., Tsuneda, S., and Harayama, S. (2008). Retrieval of entire genes from environmental DNA by inverse PCR with pre-amplification of target genes using primers containing locked nucleic acids. Environ. Microbiol. *10*, 978–987.

Yang, M.Q., Athey, B.D., Arabnia, H.R., Sung, A.H., Liu, Q., Yang, J.Y., Mao, J., and Deng, Y. (2009). High-throughput next-generation sequencing technologies foster new cutting-edge computing techniques in bioinformatics. BMC Genomics *10*, I1. doi:10.1186/1471–2164–10-S1-I1.

Yooseph, S., Sutton, G., Rusch, D.B., Halpern, A.L., Williamson, S.J., Remington, K., Eisen, J.A., Heidelberg, K.B., Manning, G., Li, W. *et al.*, (2007). The Sorcerer II Global Ocean Sampling expedition: expanding the universe of protein families. PLoS Biol. *5*, e16. doi:10.1371/journal.pbio.0050016.

Zhang K., Martiny, A.C., Reppas, N.B., Barry, K.W., Malek, J., Chisholm, S.W., and Church, G.M. (2006). Sequencing genomes from single cells by polymerase cloning. Nat. Biotechnol. *24*, 680–686.

Bacterial Genealogy: Not Dead

2

Robert L. Dorit and Margaret A. Riley

Abstract

The importance of horizontal gene transfer (HGT) in bacterial evolution has led many bacteriologists to question the very existence of bacterial species. If gene transfer is as rampant as comparative genomic and metagenomic studies suggest, how could bacterial species survive such genomic fluidity? Indeed, some go so far as to propose the metagenome as the appropriate unit of evolutionary distinction. The coherence and continuity of the metagenome remains, for us, an open question. Whatever the ultimate fate of metagenome-based phylogenetic reconstructions, we contend that genetic information in the bacterial world still comes in discrete, discernible and discontinuous packages that retain their integrity over evolutionary time. Despite the many fascinating instances of horizontal gene transfer that have been documented, the genetic coherence of bacterial species has not devolved into a continuous smear of promiscuously shared genetic information. The distinctions that we see and name in the microbial world are not arbitrary slicings of a continuous distribution, but natural breakpoints and boundaries that reflect the operation of time, history and selection on bacterial populations.

Introduction

The diversity of bacteria is simply staggering (Torsvik *et al.*, 1990; Torsvik and Ovreas, 2002). The censuses that have been carried out suggest that soil harbours the highest level of diversity, with potentially over one million bacterial species per gram of soil (Gans *et al.*, 2005). Most of this variety is entirely uncharacterized at the molecular level: the NCBI sequence database contains information about a mere 15,111 bacterial or archaeal species, suggesting that 98–99% of microbial taxa are simply undescribed (Sayers *et al.*, 2009).

Until relatively recently, the sparse coverage of bacterial diversity was blamed squarely on the near impossibility of culturing most bacteria. Of late, however, this particular roadblock has been if not overcome, at least bypassed. The advent of genomic and metagenomic methods now enables us to construct a snapshot of bacterial communities. In the process, the bias introduced by 'culturability' in our representation of bacterial diversity has finally been overcome (Wu *et al.*, 2009). While these technical and analytic breakthroughs hold enormous promise for microbiology, the data they supply have raised as many questions as they have answered. Intoxicating as the prospect of tens of thousands of complete genome sequences is, the burden of interpreting these data still falls on us (Raes *et al.*, 2007).

The widespread appeal of comparative genomics and metagenomics has triggered an unexpected crisis in microbial biology about the reality, utility and nature of bacterial species. While the history of biology is marked by cyclical paroxysms concerning the biological meaning of species, the re-emergence of this topic at the beginning of the 21st century comes as something of a surprise. But the data emerging from genomic studies – specifically concerning the magnitude of lateral gene transfer, the enormous variability in genome size and the fluidity of gene content – have led to questions about the appropriate unit of evolution and of classification in bacterial biology (Doolittle and Zhaxybayeva, 2010).

While some discussion was always bubbling on the back burner, most microbiologists accepted, at least until recently, that species level descriptions were both useful and necessary (for philosophical, practical, and legal reasons). To be sure, the applicability at the bacterial level of the 'universal species concept', with its emphasis on reproductive isolation was not always clear (Staley, 2009). Nonetheless, related definitions that emphasized the evolutionary, genetic or ecological coherence over time of bacterial species still seemed useful (Cohan, 2002; Riley and Lizotte-Waniewski, 2009). These definitions in no way dismissed the emerging portrait of an active genetic conversation across species boundaries. Nonetheless, while the concept of what constitutes a bacterial species (or evolutionary unit) is itself an evolving concept, the argument for abandoning it altogether seems premature.

Early attempts to use molecular data to delineate bacterial species involved the use of DNA–DNA hybridization, in which bacterial species were defined as those isolates sharing at least 70% hybridization under standard conditions (Stackebrandt *et al.*, 2002). Given the enormous range of genetic variation detected in different clearly recognized species, it became obvious that a variability cut-off, such as is imposed with hybridization methods, was not appropriate. Levels of variability will vary over the lifetime of a species and will reflect aspects of its life history, particularly the processes by which it adapts to its habitat (Riley and Lizotte-Waniewski, 2009).

More recent investigations into microbial species distinctions have sought to incorporate more robust estimates of molecular diversity into the process of species identification. The assumption is that this molecular diversity will fall into discrete clusters that correspond closely with phenotype-based clusters. In essence, the question has been raised whether DNA sequence variability can be employed to inform the division of a genus into species, to distinguish among similar species, or to address whether bacterial species exist at all (Godoy *et al.*, 2003; Priest *et al.*, 2004; Baldwin *et al.*, 2005; Hanage *et al.*, 2005; Thompson *et al.*, 2005b).

While taxonomic definitions and evolutionary reality are necessarily tightly linked (at least one hopes so), it might be useful for this discussion to separate the two issues. The taxonomic discussion – what names we use to describe microbial diversity – is important, and in microbiology, has always been fraught (Van Niel, 1946; Gest, 1999). Traditions in different subfields of microbiology have evolved in virtual splendid isolation, giving rise to a series of ad hoc taxonomic ranks: serovars, strains, types, etc. each defined by one or more 'unique' attributes. Not coincidentally, the attributes deemed essential to the identity of the taxonomic unit reflect the interests and technical competence of the subfield advancing the definition. Thus, if your forte is pulse field gel electrophoresis, it is PFGE profile that defines the relevant unit; if you can screen antibiotic resistance profiles with the best of them, then differences in resistance profiles define the meaningful units of analysis. While this backdrop may seem anarchic, it is both empirically grounded, and in many cases useful for the purposes it is meant to serve.

Recent improvements in sequencing technology now provide us with a new set of markers that we can use to define empirically-based units. In an effort to characterize the apparent diversity of players in microbial communities, rapid, reproducible methods that characterize a short stretch of diagnostic DNA ('barcoding') hold obvious appeal (Andersson *et al.*, 2008). While it is easy to caricature such efforts as reductive or altogether irrelevant, proponents are well aware of the fact that 10 kb of sequence cannot stand in for a lifetime of study of a particular group (Will *et al.*, 2005). Their plea to define taxonomic units by their barcodes is driven by pragmatism: a snapshot of diversity, based on arbitrary, but not capricious, data that can be verified and reproduced, is better than nothing at all. But there is more at stake in taxonomy than just the name. The use of genetic information- short barcoding stretches or whole genomes – as a way to define meaningful evolutionary units is not assumption-free. By making some of these assumptions explicit, we may be able to anticipate some of the challenges that lie ahead for the field.

At the limit, of course, no amount of data-molecular, morphological, biochemical or behavioural – can be said to be a complete description of the species or population of interest. It seems useful, therefore, to ask two separate questions about modern genomic and metagenomic approaches to taxonomy: are they useful, and are they valid? As a number of studies have illustrated, rapid methods of characterizing overall diversity have clearly proven useful. Such studies, for instance, have revealed far greater diversity in most microbial ecosystems than previously thought, have elucidated microbial guilds and associations that were previously uncharacterized, and have examined the consequences of a variety of perturbations on the diversity of microbial communities (Rodriguez-Valera, 2004; Andersson *et al.*, 2008; Singh *et al.*, 2009).

A more nuanced picture emerges when we examine the validity of such tagging or barcoding approaches (Siddall *et al.*, 2009). The answers here will depend in part on the objectives for which such approaches are deployed. Given the potential loose coupling between gene genealogies and species genealogies, for instance, short stretches of sequence cannot easily stand in for the evolutionary units themselves. A cautious interpretation of phylogenetic hypotheses generated using sampled sequences would seem warranted, at least until multiple genes (or sampled sequences) begin to converge on a single reconstruction. In this regard, the debate about the uses and abuses of genomic approaches to taxonomy and classification mirror earlier controversies about phenetic methods in systematics. Now, as then, the appeal of a rigorous, empirical and quantitative approach was pitted against the nuanced knowledge accumulated by specialists over a lifetime of work. And now, as then, the dichotomy is a false one: detailed molecular data is an enormously valuable source of taxonomic and phylogenetic information, but not until and unless we understand the forces that shape its evolution. Stated differently, the blind use of sequence information is about as useful as the blind rejection of such data.

The one place where the flood of new genomic data has truly stirred the waters is the discussion about the reality of bacterial species (Doolittle and Zhaxybayeva, 2010). At one extreme, certain authors have argued that the unexpectedly high level of horizontal gene transfer, coupled with the ease with which bacterial genomes shed or add large segments of functional information, makes any notion of species simply a platonic vestige from a less informed age. At the other extreme, the discovery of the active traffic in genetic information occurring across vast phylogenetic distances is dismissed as mere noise, irrelevant to our conception of evolving bacterial entities. This may be a case where neither extreme captures the interesting aspects of the emerging picture. Instead, we might be best served by asking three interrelated questions: (1) What is the extent and pattern of non-genealogical sharing of genetic information? (2) Are particular classes of genetic information more or less likely to be involved in such exchanges, reflecting biases exerted by the transport vector, by the recipient genome, or both? and (3) Is the extent of such non-genealogical sharing sufficient to overwhelm the signature of genealogical inheritance?

Horizontal transfer and non-genealogical traffic

The ability of bacteria to acquire genetic information in unconventional ways was well established long before the advent of genomics. The traffic in plasmids, the shuttling of genetic material by phage and other vectors, and the direct take-up of nucleic acids from the environment have all been extensively described and characterized (Zaneveld *et al.*, 2008). In addition, the selective shaping of mobile cassettes – clusters of genes and accessory regulatory elements that correspond to one or more related functions – is now well established. These mobile operons, some of them capable of catalysing their own movement (excision and insertion), are associated with a number of important phenotypes, most notably in the form of pathogenicity islands and multiresistance cassettes (Kobayashi, 2001). The functional coherence of these mobile genetic elements is noteworthy, and certainly suggests strong evolutionary forces at work, both in the donors and recipients of such elements (Cohan, 1994; Hacker and Carniel, 2001; Dobrindt *et al.*, 2004). Furthermore, in cases where these elements can influence the rate and location of excision and insertion, selection for increased mobility is certainly evident.

The expectation that phylogenetic barriers would limit the exchange of genetic material to closely related strains may have proven optimistic. Instead, a number of whole genome and metagenomic sequencing projects reveal genetic exchange among very distantly related organisms: propinquity, rather than phylogenetic adjacency, may prove to be the more relevant parameter (Davison, 1999; Beiko *et al.*, 2005). The surprise that the genomic revolution has revealed is thus not the existence of such genetic traffic, but its unexpected range and frequency. Virtually every genome characterized to date contains the signature of multiple non-genealogical events. Importantly, the observable (or retrievable) events of HGT are an underestimate of the true frequency of such events. The signals that allow for their identification – altered codon usages, deviant GC content, etc. – fade over time as these events acquire the characteristics of their host genome. At that point, only phylogenetic approaches that reveal incongruence are likely to detect these events. The implications of HGT on our understanding of bacterial function and phenotype are hard to overstate. Similarly, the extent and range of HGT events have very significant clinical and public health consequences. But are these events sufficiently ubiquitous that they require us to abandon the fundamental unit of biological organization, the species?

While it is easy to cast this as a semantic or philosophical question, we prefer to cast it as a quantitative, empirical challenge. Genealogical inheritance is still taking place, leaving behind a genomic signal that we can, in principle, retrieve. Horizontal gene transfer, too, leaves behind a retrievable signal, in its wanton disregard of the rules of phylogenetic etiquette. From a strict phylogenetic perspective, HGT is noise, resulting in incongruent trees and weakly supported consensus trees. We are living in an age where more and more sequence data is coming at us. Are the phylogenetic reconstructions made possible by these rich datasets descending into chaos, suggesting that the noise is more than sufficient to swamp the signal?

Genealogies alive and well

In our view, the signal of genealogy is still loud and clear. In the overwhelming majority of cases where multiple genes have been used for phylogenetic reconstruction, the resulting trees are resolved and largely congruent (Wertz *et al.*, 2003; Lerat *et al.*, 2005; Riley and Lizotte-Waniewski, 2009). Furthermore – with some interesting exceptions- these reconstructions match the previously established patterns of classification. Despite the evidence that bacterial genomes are both more plastic and more dynamic than we had previously expected, the stately unfolding of the tree of life still appears to be the prevailing mode of bacterial evolution.

The availability of whole genome sequences provided our first glimpse into the dynamic nature of a species genome. Glasner and Perna (2004) and Mau *et al.* (2006) compared six complete genomes of *Escherichia coli* and revealed a highly conserved genome backbone with greater than 98% sequence similarity among the isolates. Further, Mau *et al.* (2006) detected a high level of homologous recombination among these shared genes, confirming earlier studies by Roger Milkman (Milkman *et al.*, 1999; Milkman *et al.*, 2003), that the level of within species recombination for what was then considered a 'clonal' species was significant. Even more surprising was the observation that the conserved genome backbone of *E. coli* was interrupted by hundreds of strain specific 'sequence islands'. Edwards *et al.* (2002) provided a similar comparison of *Salmonella enterica* strains. As in *E. coli*, a backbone of highly conserved genes was identified, each with an average of greater than 99% sequence similarity and a similar pattern of strain specific unique sequence islands.

This pattern of shared and unique sequences appears to be common among many bacterial species (Waterfield *et al.*, 2003; Coleman *et al.*, 2006; Juhas *et al.*, 2007). A comparison of eight genomes of group B streptococcus revealed a core of 1806 genes present in every genome and 907 genes absent in one or more. In contrast, eight genomes of *Bacillus anthracis* revealed very few strain-specific genes. However, the general pattern that emerges is that members of a bacterial species share some large fraction of their genomes, but often carry unique, strain-specific sequences. The relative fraction of the genome that is shared (or unique) varies greatly from one bacterial species to the next.

The determination of multiple, complete genome sequences of numerous additional bacteria has enabled a new view of the genomic plasticity of several 'well-defined' bacterial species. Such an analysis, referred to as genome-tree building, is based on genome-level phylogenetic analysis. This approach involves scaling-up the traditional tree-building approach and analysing the phylogenetic trees for multiple gene families (in principle, all families represented in many genomes), in an attempt to derive a consensus, 'organismal' phylogeny (Brown and Volker, 2004). It is often not practical to determine the complete genome sequence of multiple isolates within a species. In lieu of complete genomes, a sample of housekeeping genes (referred to as multi-locus sequence typing)(MLST) has become the norm for characterizing genetic diversity within a bacterial species (Maiden *et al.*, 1998; Feil *et al.*, 2003). This method permits the analysis of large numbers of related bacterial isolates, which is essential to the determination of species designations (Feil *et al.*, 2003; Hanage *et al.*, 2005). Such studies have confirmed that species designations based upon phenotypic criteria have a corresponding underlying MLST-based genotypic clustering (Woodward *et al.*, 2000; Godoy *et al.*, 2003; Thompson *et al.*, 2005a,b; Whitaker *et al.*, 2005).

One of the first multi-gene based investigations specifically designed to address the bacterial species concept was conducted by Wertz *et al.* (2003), who sequenced seven housekeeping genes from a sample of environmental bacteria representing seven species of Enterobacteriaceae (Gordon and FitzGibbon, 1999). Molecular phylogenies for each of the genes were inferred and the branching patterns of the resulting trees compared. In each case, isolates from a species formed a monophyletic group, which corresponded precisely with the clusters identified by phenotypic data, and upon which species distinctions were initially delineated (Holt *et al.*, 1994; Rossello-Mora and Amann, 2001). An enteric species phylogeny was inferred from the composite data by concatenating the housekeeping gene sequences (Wertz *et al.*, 2003), which contained enough phylogenetic signal to resolve all of the interspecies nodes and thus provided a robust estimate of the enteric phylogeny corresponding with the existing phenotype-based phylogeny. The authors concluded that, at least for this sample of enteric lineages, bacterial species clearly do exist and the same species emerge from phenotypic and genotypic data. More recently, a highly robust phylogenetic tree was constructed for 13 gamma-proteobacteria using a concatenated alignment of several hundred conserved orthologous proteins (Lerat *et al.*, 2003). Only two of the proteins had incongruent tree topologies in this analysis.

A similar type of investigation was undertaken with *Neisseria*, in which a sample of housekeeping genes was obtained from nearly one thousand strains and 11 named species (Hanage *et al.*, 2005). Phylogenetic trees were inferred to investigate whether genotypic clusters can be resolved among these highly recombinogenic bacteria and, if so, the extent to which they correspond to named species. The analysis suggests that the use of concatenated sequences largely buffered the distorting effect of recombination events and resulted in the resolution of clusters corresponding to the three most abundant species in the sample, *N. meningitidis*, *N. lactamica* and *N. gonorrhoeae*. Comas *et al.* (2006) go so far as to propose which genes should be incorporated into these genome-wide surveys for phylogenetic signals. They define 'essential' genes, as opposed to 'universal' genes, as those providing the greatest phylogenetic signal to noise ratio.

These earlier, human-pathogen focused studies have since been expanded to cover a much broader sampling of bacterial diversity. One such investigation involved an *Agrobacterium* species complex (Popoff *et al.*, 1984; Mougel *et al.*, 2002; Portier *et al.*, 2006). Genome sequence comparisons among members of the genus highlighted a broad range of intra-species divergence within very closely related but distinct species of *Agrobacterium*. A subsequent study of the same species compared the sensitivity of recombination to DNA sequence divergence across the species complex. Their data supported earlier claims by Majewski (2001) that 'bacterial species experience a degree of sexual isolation from genetically divergent organisms since recombination occurs more frequently within species than between species'. A genome-based study of the Actinobacteria, a dominant bacterial phylum in the soil, revealed a surprising level of gene synteny and genome conservation. Prior 16S rRNA gene sequence comparisons had recognized 39 families and 130 genera, all of which share a unique molecular synapomorphy: an insertion in the 23S rRNA gene (Roller *et al.*, 1992). How this ancient phylum has retained such a high degree of genome conservation remains a mystery. Similar outcomes have been observed for *Haemophilus, Pseudomonas* and *Streptococcus*, where as little as 9% divergence at recombination marker genes results in a drop of three orders of magnitude in recombination efficiency. Konstantintidis and Tiedje (2004) compared the gene content of 70 closely related and fully sequenced bacterial genomes to identify whether species boundaries exist, and to determine the role of the organism's ecology on its shared gene content. They found the levels of sequence similarity on the order of 94% corresponded to the traditional 70% DNA–DNA reassociation standard of the current species definition. Notably, a large fraction – up to 65% – of the differences in gene content within species is associated with bacteriophage and transposable elements, revealing an important role for these elements in bacterial speciation. Their findings are consistent with a species definition that would include a more homogeneous set of strains and would incorporate the ecology of the strains in addition to their evolutionary distance.

Prior to the molecular era, bacterial phylogenetic reconstructions relied primarily on phenotypic characters, including biochemical characteristics, which raised concerns about homology. The advent of genomic data has certainly helped address that concern, although early studies often relied (by necessity) on one or a small number of genes (or gene fragments). As more extensive whole genome data come on line, we can expect phylogenetic reconstructions to stabilize, although methods for extracting phylogenetic signal from the data continue to be hotly debated. Some existing phylogenetic reconstructions will stand, others will fall. In addition, many unresolved nodes are likely to yield to the quantity and quality of new genomic data.

Traditional species designations have traditionally been based upon extensive phenotypic characterization of a large number of isolates. Now, such designations also require the use of 16s rRNA sequence comparisons to identify the closest relatives of a proposed species (Rossello-Mora and Amann, 2001). The most frequently employed bacterial species concept – the so-called phylo-phenetic bacterial species concept – posits that a bacterial species is 'a monophyletic and genomically coherent cluster of individual organisms that show a high degree of overall similarity with respect to many independent characteristics, and is diagnosable by a discriminative phenotypic property' (Rossello-Mora and Amann, 2001). Numerous studies have revealed clusters of bacterial isolates that share complex phenotypes and these clusters are often designated as species (Shute *et al.*, 1984; Sneath and Stevens, 1985; Mauchline and Keevil, 1991; Barrett and Sneath, 1994; Kirschner *et al.*, 2001). In fact, Cohan uses the mere existence of these clusters as prima facie evidence of the existence of bacterial species (Cohan, 2002).

Core genome hypothesis

How then to reconcile this quantized distribution of genetic information with the well-characterized homogenizing effects of gene flow across species boundaries? The answer may lie in a more nuanced view that suggests that the kind of genetic information most likely to flow outside of established genealogy is not a random sample of all genetic information. Phrased differently, the bacterial genome may consist of regions that retain their integrity over evolutionary time and regions that can tolerate (or may in fact depend) on recurrent transfer across phylogenetic lines. This concept is captured by the distinction between the core genome and the accessory genome.

Lan and Reeves (1996) were the first to recognize the potential link between the observation of shared versus unique sequences in bacterial genomes and its implication for discriminating bacterial species. They proposed the core genome hypothesis (CGH), which distinguishes between that fraction of the genome (the core) shared by all members of a species and that fraction found in only a subset of the population (the auxiliary). Core genes encode essential metabolic housekeeping and informational processing functions (Feil, 2004). They are ubiquitous in a species and define the species-specific characteristics. Auxiliary genes may or may not be present in a strain and are generally genes that encode supplementary biochemical pathways, are associated with phage or other mobile elements, or encode products that serve to interact with the external environment. Thus, auxiliary genes serve in the adaptation of strains to local competitive or environmental pressures (Cohan, 2002). They are likely to encode antibiotic resistance, novel metabolic functions, toxin production, and the like (Dobrindt and Reidl, 2000; Karlin, 2001; White *et al.*, 2001).

Conclusions

The CGH has dramatically influenced how bacteriologists think about the nature of bacterial species. Prior to the CGH, the strongest argument against the recognition of bacterial 'species' was the simple observation of horizontal gene transfer (HGT) between bacterial lineages. The fact that bacterial species gene pools may not be tightly closed was reason enough for many bacteriologists to conclude that bacterial species cohesion could not survive such exchanges. This contradicts the clearly demonstrated fact that bacteria exist in phenotypic clusters, which many bacteriologists recognize as species. Even more compelling, it is becoming clear that these well-defined phenotypic clusters correspond to underlying genotype clusters (Woodward *et al.*, 2000; Godoy *et al.*, 2003; Thompson *et al.*, 2005a; Whitaker *et al.*, 2005).

Although the subset of genes involved in HGT events are not a random sampling of all genes in the bacterial genome, the underlying causes of this phenomenon are not obvious. The pattern could in effect be arising at the source of these mobile events, at the destination, or as a combination of both. The source bias can depend on the modularity of certain genes or gene complexes: their ability to direct and regulate their expression in a context independent manner. In addition, constraints imposed by the particular vehicle mediating the horizontal gene transfer could further bias the character of the genetic information likely to be traded. At the destination, the successful instances of HGT are shaped by the location of integration, by the regulatory elements that can affect the regulation of newly integrated genetic information, by the discrepancy in macro-level properties (GC bias, codon utilization, etc.) between incoming and recipient sequences, and, most importantly, by the effects on organismal fitness conferred by the newly arrived genetic material. Superimposed on these deterministic forces, stochastic effects play a critical role in determining the fate of HGT events.

According to the CGH, a bacterial species is comprised of groups of strains that frequently exchange, or could exchange, core genes, but which are relatively restricted from such exchange with other groups. This view is a natural extension of the Biological Species Concept, first articulated by Mayr (1942). The BSC should not be taken to imply any particular process of speciation, merely that the observation of more gene 'sharing' (via recombination and/or lateral gene transfer) is observed within versus between putative bacterial species. The CGH provides a perfect backdrop from which to articulate this modification.

The CGH predicts that a subset of genes, the core, is present in all, or nearly all, individuals within the species. These are the genes that provide the defining characteristics of the species and are assumed to experience primarily purifying selection, to remove deleterious mutations and maintain existing functions. As a species evolves, its core genome will evolve as a complex of co-evolved functions. When transferred between species, such genes will most likely experience a selective disadvantage, as this will disrupt co-evolved functions. Such transfer will rarely survive. Thus, core genes will diverge as the species diverge.

In contrast, auxiliary genes will be found in only a subset of individuals within a species. The CGH predicts that these genes experience intermittent positive selection, when their function enhances survival in a varied and ever-changing environment. When such genes are exchanged between species, their functions will often provide a selective advantage to the recipient. Frequent successful transfer between species will serve to limit the divergence of auxiliary genes, relative to the core.

The most specific prediction that emerges from the CGH concerns the rate at which core and auxiliary genes accumulate variability. Core genes will, on average, display a neutral rate of evolution, while auxiliary genes will experience a variety of selective pressures, including diversifying selection (acting to increase levels of variation), directional selection (acting to decrease levels of variation), balancing selection (acting to maintain particular alleles in the species), and purifying selection (acting to weed out deleterious mutations). Thus, the average rate of evolution for auxiliary genes could be just about anything, and the variance around this rate should be extreme. These expectations, based upon the neutral theory (Kimura, 1968), are quite useful for testing predictions from the CGH (Fay *et al.*, 2002). However, such tests require population-based samples of multiple genomes per species and, unfortunately, most existing species-based genome samples are chosen to represent the diversity of clinical isolates of human pathogens and thus will often underestimate standing levels of genome diversity. The appropriate data is in the pipeline and should enable us to bring population genomics to bear on this complex and fascinating matter.

Future trends

The demonstration of extensive horizontal transfer does raise the intriguing possibility that a microbial community can now be considered a discrete unit of selection (Doolittle and Zhaxybayeva, 2010). The argument rests (among other criteria) on the demonstration that a microbial assemblage retains sufficient coherence over time to allow 'selection' or sorting amongst such units, as well as a birth and death process that controls the longevity and fecundity of such an assemblage. While a full discussion of the extent to which microbial communities meet such criteria are outside the scope of this article, the coherence (or unity) of microbial communities has been greatly strengthened by the data flowing from metagenomic studies. What is likely true of all ecological communities, but may be particularly visible in microbial ecosystems, is the ecological cohesion they display (DeLong and Karl, 2005; DeLong, 2009). In virtually all microbial communities examined to date, the biochemical interdependence of members of the community is evident. Carbon sources, limiting nutrients and rare metabolites are taken in, processed and exploited by tightly interdependent guilds of bacteria that ensure the efficient breakdown of inputs and the relative homeostasis of the community. Perhaps less appreciated is the extent to which particular tasks within the metabolic network can and are fulfilled by a number of ecologically interchangeable agents (Friedmann, 2004; Falkowski *et al.*, 2008). This interchangeability in itself supports the notion that something other than the individual agents within the community is the target of selection.

This notion of ecological coherence can now be supplemented with the concept of genetic cohesion. Proponents of this view suggest that importance of lateral gene transfer has been vastly underestimated in more conventional definitions of bacterial species. Instead, it is the metagenome – the collection of all genes and regulatory elements derived from all participants in a microbial community – that evolves under the watchful eye of natural selection (Acinas *et al.*, 2004; Doolittle and Zhaxybayeva, 2010). In this formulation, the genome that we find in an individual organism formerly known as '*E. coli*' is no more than an evolutionarily transient collection of functioning genetic instructions. Importantly, the lifespan of this specific genome is far shorter than that of the metagenome from whence it is sampled. Genomes are thus no more than marriages of convenience, and the consistent association between genomes and their bearers a coincidence of little evolutionary impact. In contrast, the collection of genes and genetic elements harboured by all members is considered a cohesive and coherent evolutionary unit, making possible a community level phenotype that can be shaped by selection.

The coherence and continuity of the metagenome remains, for us, an open question. Nonetheless, even if it should prove to exhibit the necessary features of an emergent unit of selection, this would in no way require us to abandon the notion of bacterial species. A considerable literature addresses the possibility of selection occurring, either simultaneously or at different timescales, at multiple levels (Green and Bohannan, 2006; Ward *et al.*, 2008).

Whatever the ultimate fate of metagenome-based phylogenetic reconstructions, we contend that genetic information in the bacterial world still comes in discrete, discernible and discontinuous packages that retain their integrity over evolutionary time. Despite the many fascinating instances of horizontal gene transfer that have been documented, the genetic congruence of bacterial species has not devolved into a continuous smear of promiscuously shared genetic information. The distinctions that we see and name in the microbial world are not arbitrary slicings of a continuous distribution, but natural breakpoints and boundaries that reflect the operation of time, history and selection on bacterial populations.

References

Acinas, S.G., Klepac-Ceraj, V., Hunt, D.E., Pharino, C., Ceraj, I., Distel, D.L., and Polz, M.F. (2004). Fine-scale phylogenetic architecture of a complex bacterial community. Nature *430*, 551–554.

Andersson, A.F., Lindberg, M., Jakobsson, H., Backhed, F., Nyren, P., and Engstrand, L. (2008). Comparative analysis of human gut microbiota by barcoded pyrosequencing. PLoS One 3, e2836.

Baldwin, A., Mahenthiralingam, E., Thickett, K.M., Honeybourne, D., Maiden, M.C., Govan, J.R., Speert, D.P., Lipuma, J.J., Vandamme, P., and Dowson, C.G. (2005). Multilocus sequence typing scheme that provides both species and strain differentiation for the *Burkholderia cepacia* complex. J. Clin. Microbiol. *43*, 4665–4673.

Barrett, S.J., and Sneath, P.H. (1994). A numerical phenotypic taxonomic study of the genus *Neisseria*. Microbiology *140*, 2867–2891.

Beiko, R.G., Harlow, T.J., and Ragan, M.A. (2005). Highways of gene sharing in prokaryotes. Proc. Natl. Acad. Sci. U.S.A. *102*, 14332–14337.

Brown, J.R., and Volker, C. (2004). Phylogeny of gamma-proteobacteria: resolution of one branch of the universal tree? Bioessays *26*, 463–468.

Cohan, F.M. (1994). Genetic exchange and evolutionary divergence in prokaryotes Trends Ecol. Evol. *9*, 175–180.

Cohan, F.M. (2002). Sexual isolation and speciation in bacteria. Genetica *116*, 359–370.

Coleman, M.L., Sullivan, M.B., Martiny, A.C., Steglich, C., Barry, K., Delong, E.F., and Chisholm, S.W. (2006). Genomic islands and the ecology and evolution of *Prochlorococcus*. Science *311*, 1768–1770.

Comas, I., Moya, A., Azad, R.K., Lawrence, J.G., and Gonzalez-Candelas, F. (2006). The evolutionary origin of *Xanthomonadales* genomes and the nature of the horizontal gene transfer process. Mol. Biol. Evol. *23*, 2049–2057.

Davison, J. (1999). Genetic exchange between bacteria in the environment. Plasmid *42*, 73–91.

DeLong, E.F. (2009). The microbial ocean from genomes to biomes. Nature *459*, 200–206.

DeLong, E.F., and Karl, D.M. (2005). Genomic perspectives in microbial oceanography. Nature *437*, 336–342.

Dobrindt, U., Hochhut, B., Hentschel, U., and Hacker, J. (2004). Genomic islands in pathogenic and environmental microorganisms. Nat. Rev. Microbiol. *2*, 414–424.

Dobrindt, U., and Reidl, J. (2000). Pathogenicity islands and phage conversion: evolutionary aspects of bacterial pathogenesis. Int. J. Med. Microbiol. *290*, 519–527.

Doolittle, W.F., and Zhaxybayeva, O. (2010). Metagenomics and the units of biological organization. BioScience *60*, 102–112.

Edwards, S.V., Fertil, B., Giron, A., and Deschavanne, P.J. (2002). A genomic schism in birds revealed by phylogenetic analysis of DNA strings. Syst. Biol. *51*, 599–613.

Falkowski, P.G., Fenchel, T., and Delong, E.F. (2008). The microbial engines that drive Earth's biogeochemical cycles. Science *320*, 1034–1039.

Fay, J.C., Wyckoff, G.J., and Wu, C.I. (2002). Testing the neutral theory of molecular evolution with genomic data from *Drosophila*. Nature *415*, 1024–1026.

Feil, E.J. (2004). Small change: keeping pace with microevolution. Nat. Rev. Microbiol. *2*, 483–495.

Feil, E.J., Cooper, J.E., Grundmann, H., Robinson, D.A., Enright, M.C., Berendt, T., Peacock, S.J., Smith, J.M., Murphy, M., Spratt, B.G. *et al.* (2003). How clonal is Staphylococcus aureus? J. Bacteriol. *185*, 3307–3316.

Friedmann, H.C. (2004). From 'butyribacterium' to '*E. coli*': an essay on unity in biochemistry. Perspect. Biol. Med. *47*, 47–66.

Gans, J., Wolinsky, M., and Dunbar, J. (2005). Computational improvements reveal great bacterial diversity and high metal toxicity in soil. Science *309*, 1387–1390.

Gest, H. (1999). Bacterial classification and taxonomy: a 'primer' for the new millennium. Microbiol. Today *26*, 70–72.

Glasner, J.D., and Perna, N.T. (2004). Comparative genomics of *E. coli*. Microbiol. Today *31*, 125.

Godoy, D., Randle, G., Simpson, A.J., Aanensen, D.M., Pitt, T.L., Kinoshita, R., and Spratt, B.G. (2003). Multilocus sequence typing and evolutionary relationships among the causative agents of melioidosis and glanders, *Burkholderia pseudomallei* and *Burkholderia mallei*. J. Clin. Microbiol. *41*, 2068–2079.

Gordon, D.M., and FitzGibbon, F. (1999). The distribution of enteric bacteria from Australian mammals: host and geographical effects. Microbiology *145*, 2663–2671.

Green, J., and Bohannan, B.J. (2006). Spatial scaling of microbial biodiversity. Trends Ecol. Evol. *21*, 501–507.

Hacker, J., and Carniel, E. (2001). Ecological fitness, genomic islands and bacterial pathogenicity. A Darwinian view of the evolution of microbes. EMBO Rep. *2*, 376–381.

Hanage, W.P., Kaijalainen, T., Herva, E., Saukkoriipi, A., Syrjanen, R., and Spratt, B.G. (2005). Using multilocus sequence data to define the pneumococcus. J. Bacteriol. *187*, 6223–6230.

Holt, J.G., Krieg, N.R., Sneath, P.H.A., Staley, J.T., and Williams, S.T. (eds.) (1994), Bergey's Manual of Determinative Bacteriology (Baltimore, Lippincott Williams & Wilkins).

Juhas, M., Crook, D.W., Dimopoulou, I.D., Lunter, G., Harding, R.M., Ferguson, D.J., and Hood, D.W. (2007). Novel type IV secretion system involved in propagation of genomic islands. J. Bacteriol. *189*, 761–771.

Karlin, S. (2001). Detecting anomalous gene clusters and pathogenicity islands in diverse bacterial genomes. Trends Microbiol. *9*, 335–343.

Kimura, M. (1968). Genetic variability maintained in a finite population due to mutational production of neutral and nearly neutral isoalleles. Genet. Res. *11*, 247–269.

Kirschner, C., Maquelin, K., Pina, P., Ngo Thi, N.A., Choo-Smith, L.P., Sockalingum, G.D., Sandt, C., Ami, D., Orsini, F., Doglia, S.M. *et al.* (2001). Classification and identification of enterococci: a comparative phenotypic, genotypic, and vibrational spectroscopic study. J. Clin. Microbiol. *39*, 1763–1770.

Kobayashi, I. (2001). Behavior of restriction-modification systems as selfish mobile elements and their impact on genome evolution. Nucleic Acids Res. *29*, 3742–3756.

Konstantinidis, K.T., and Tiedje, J.M. (2004). Trends between gene content and genome size in prokaryotic species with larger genomes. Proc. Natl. Acad. Sci. U.S.A. *101*, 3160–3165.

Lan, R., and Reeves, P.R. (1996). Gene transfer is a major factor in bacterial evolution. Mol. Biol. Evol. *13*, 47–55.

Lerat, E., Daubin, V., and Moran, N.A. (2003). From gene trees to organismal phylogeny in prokaryotes: the case of the gamma-Proteobacteria. PLoS Biol. *1*, E19.

Lerat, E., Daubin, V., Ochman, H., and Moran, N.A. (2005). Evolutionary origins of genomic repertoires in bacteria. PLoS Biol. *3*, e130.

Maiden, M.C., Bygraves, J.A., Feil, E., Morelli, G., Russell, J.E., Urwin, R., Zhang, Q., Zhou, J., Zurth, K., Caugant, D.A. *et al.* (1998). Multilocus sequence typing: a portable approach to the identification of clones within populations of pathogenic microorganisms. Proc. Natl. Acad. Sci. U.S.A. *95*, 3140–3145.

Majewski, J. (2001). Sexual isolation in bacteria. FEMS Microbiol Lett *199*, 161–169.

Mau, B., Glasner, J.D., Darling, A.E., and Perna, N.T. (2006). Genome-wide detection and analysis of homologous recombination among sequenced strains of *Escherichia coli*. Genome Biol. *7*, R44.

Mauchline, W.S., and Keevil, C.W. (1991). Development of the BIOLOG substrate utilization system for identification of *Legionella* spp. Appl. Environ. Microbiol. *57*, 3345–3349.

Mayr, E. (1942). Systematics and the Origin of Species (New York, Columbia University Press).

Milkman, R., Jaeger, E., and McBride, R.D. (2003). Molecular evolution of the *Escherichia coli* chromosome. VI. Two regions of high effective recombination. Genetics *163*, 475–483.

Milkman, R., Raleigh, E.A., McKane, M., Cryderman, D., Bilodeau, P., and McWeeny, K. (1999). Molecular evolution of the *Escherichia coli* chromosome. V. Recombination patterns among strains of diverse origin. Genetics *153*, 539–554.

Mougel, C., Thioulouse, J., Perriere, G., and Nesme, X. (2002). A mathematical method for determining genome divergence and species delineation using AFLP. Int. J. Syst. Evol. Microbiol. *52*, 573–586.

Popoff, M.Y., Kersters, K., Kiredjian, M., Miras, I., and Coynault, C. (1984). Taxonomic position of *Agrobacterium* strains of hospital origin. Ann. Microbiol. (Paris) 135A, 427–442.

Portier, P., Fischer-Le Saux, M., Mougel, C., Lerondelle, C., Chapulliot, D., Thioulouse, J., and Nesme, X. (2006). Identification of genomic species in *Agrobacterium* biovar 1 by AFLP genomic markers. Appl. Environ. Microbiol. *72*, 7123–7131.

Priest, F.G., Barker, M., Baillie, L.W., Holmes, E.C., and Maiden, M.C. (2004). Population structure and evolution of the *Bacillus cereus* group. J. Bacteriol. *186*, 7959–7970.

Raes, J., Foerstner, K.U., and Bork, P. (2007). Get the most out of your metagenome: computational analysis of environmental sequence data. Curr. Opin. Microbiol. *10*, 490–498.

Riley, M.A., and Lizotte-Waniewski, M. (2009). Population genomics and the bacterial species concept. In Horizontal Gene Transfer: Genomes in Flux, Gogarten, P., Olendzenski, L., and Gogarten, M., eds (New Jersey, USA: Humana Press), pp. 367–377.

Rodriguez-Valera, F. (2004). Environmental genomics, the big picture? FEMS Microbiol. Lett. *231*, 153–158.

Roller, C., Ludwig, W., and Schleifer, K.H. (1992). Gram-positive bacteria with a high DNA G+C content are characterized by a common insertion within their 23S rRNA genes. J. Gen. Microbiol. *138*, 1167–1175.

Rossello-Mora, R., and Amann, R. (2001). The species concept for prokaryotes. FEMS Microbiol. Rev. *25*, 39–67.

Sayers, E.W., Barrett, T., Benson, D.A., Bolton, E., Bryant, S.H., Canese, K., Chetvernin, V., Church, D.M., Dicuccio, M., Federhen, S. *et al.* (2009). Database resources of the National Center for Biotechnology Information. Nucleic Acids Res. *37*, D5–15.

Shute, L.A., Gutteridge, C.S., Norris, J.R., and Berkeley, R.C. (1984). Curie-point pyrolysis mass spectrometry applied to characterization and identification of selected Bacillus species. J. Gen. Microbiol. *130*, 343–355.

Siddall, M.E., Fontanella, F.M., Watson, S.C., Kvist, S., and Erseus, C. (2009). Barcoding bamboozled by bacteria: convergence to metazoan mitochondrial primer targets by marine microbes. Syst. Biol. *58*, 445–451.

Singh, A.H., Doerks, T., Letunic, I., Raes, J., and Bork, P. (2009). Discovering functional novelty in metagenomes: examples from light-mediated processes. J. Bacteriol. *191*, 32–41.

Sneath, P.H., and Stevens, M. (1985). A numerical taxonomic study of Actinobacillus, Pasteurella and Yersinia. J. Gen. Microbiol. *131*, 2711–2738.

Stackebrandt, E., Frederiksen, W., Garrity, G.M., Grimont, P.A., Kampfer, P., Maiden, M.C., Nesme, X., Rossello-Mora, R., Swings, J., Truper, H.G. *et al.* (2002). Report of the ad hoc committee for the re-evaluation of the species definition in bacteriology. Int. J. Syst. Evol. Microbiol. *52*, 1043–1047.

Staley, J.T. (2009). Universal species concept: pipe dream or a step toward unifying biology? J. Ind. Microbiol. Biotechnol. *36*, 1331–1336.

Thompson, F.L., Gevers, D., Thompson, C.C., Dawyndt, P., Naser, S., Hoste, B., Munn, C.B., and Swings, J. (2005a). Phylogeny and molecular identification of vibrios on the basis of multilocus sequence analysis. Appl. Environ. Microbiol. *71*, 5107–5115.

Thompson, J.R., Pacocha, S., Pharino, C., Klepac-Ceraj, V., Hunt, D.E., Benoit, J., Sarma-Rupavtarm, R., Distel, D.L., and Polz, M.F. (2005b). Genotypic diversity within a natural coastal bacterioplankton population. Science *307*, 1311–1313.

Torsvik, V., Goksoyr, J., and Daae, F.L. (1990). High diversity in DNA of soil bacteria. Appl. Environ. Microbiol. *56*, 782–787.

Torsvik, V., and Ovreas, L. (2002). Microbial diversity and function in soil: from genes to ecosystems. Curr. Opin. Microbiol. *5*, 240–245.

Van Niel, C.B. (1946). The classification and natural relationships of bacteria. Cold Spring Harbor Symp Quant Biol *11*, 285–301.

Ward, D.M., Cohan, F.M., Bhaya, D., Heidelberg, J.F., Kuhl, M., and Grossman, A. (2008). Genomics, environmental genomics and the issue of microbial species. Heredity *100*, 207–219.

Waterfield, N.R., Daborn, P.J., Dowling, A.J., Yang, G., Hares, M., and ffrench-Constant, R.H. (2003). The insecticidal toxin makes caterpillars floppy 2 (Mcf2) shows similarity to HrmA, an avirulence protein from a plant pathogen. FEMS Microbiol. Lett. *229*, 265–270.

Wertz, J.E., Goldstone, C., Gordon, D.M., and Riley, M.A. (2003). A molecular phylogeny of enteric bacteria and implications for a bacterial species concept. J. Evol. Biol. *16*, 1236–1248.

Whitaker, R.J., Grogan, D.W., and Taylor, J.W. (2005). Recombination shapes the natural population structure of the hyperthermophilic archaeon *Sulfolobus islandicus*. Mol. Biol. Evol. *22*, 2354–2361.

White, P.A., McIver, C.J., and Rawlinson, W.D. (2001). Integrons and gene cassettes in the enterobacteriaceae. Antimicrob. Agents Chemother. *45*, 2658–2661.

Will, K.W., Mishler, B.D., and Wheeler, Q.D. (2005). The perils of DNA barcoding and the need for integrative taxonomy. Syst. Biol. *54*, 844–851.

Woodward, M.J., Sojka, M., Sprigings, K.A., and Humphrey, T.J. (2000). The role of SEF14 and SEF17 fimbriae in the adherence of *Salmonella enterica* serotype Enteritidis to inanimate surfaces. J. Med. Microbiol. *49*, 481–487.

Wu, D., Hugenholtz, P., Mavromatis, K., Pukall, R., Dalin, E., Ivanova, N.N., Kunin, V., Goodwin, L., Wu, M., Tindall, B.J. *et al.* (2009). A phylogeny-driven genomic encyclopaedia of Bacteria and Archaea. Nature *462*, 1056–1060.

Zaneveld, J.R., Nemergut, D.R., and Knight, R. (2008). Are all horizontal gene transfers created equal? Prospects for mechanism-based studies of HGT patterns. Microbiology *154*, 1–15.

Viral Metagenomics and the Regulation of Prokaryotic Communities

3

Fernando Santos and Josefa Antón

Abstract

Since the publication of the first genomic analysis of a marine uncultured viral community in 2002, the analysis of the viral metagenomes from a wide variety of natural environments has provided a wealth of information on the diversity, abundance and metabolic capabilities of viruses. Here, we summarize how such studies have shed light into the control than viruses exert on prokaryotic communities. We have focused on three main aspects of this control: (i) what metagenomes tell about the dynamics of virus–host interaction and how the well established kill-the-winner theory fits into new metagenomic data; (ii) which picture the analysis of viral metagenomes provides on the extend of lysogeny in the microbial communities; and (iii) which are the new findings on the role of viruses as gene transfer agents. Finally, we give some hints on the analysis of expression of virome genes ('metavirotranscriptomes') and which are, in our opinion, the likely future trends in the 'omics' analysis of viruses in nature.

Introduction

During decades, the study of viruses infecting prokaryotes has relied on the isolation of their hosts. Given the well known limitations of the standard culture techniques (Rappé and Giovannoni, 2003) this approach can only provide information on a minor fraction of the naturally occurring viruses. In addition, there is no universal marker present in all viral genomes that would allow for the monitoring of viral diversity in a way analogous to the rRNA approach (Amann *et al.*, 1995). So far, the most effective way to study uncultured viral diversity is the characterization of viral metagenomes, that have been some times called metaviromes (Paul and Sullivan, 2005).

Viral metagenomics is a relatively young discipline that began in 2002 with the publication of the paper by Breitbart *et al.* 'Genomic analysis of uncultured viral communities', a work developed in the group of Forest Rohwer at San Diego State University. The analysis of viral metagenomes allows not only the cataloguing of viral diversity but also the characterization of the relative abundance of different viral lineages within the viral assemblage. However, this kind of analysis has also some limitations (see below), being the very high number of unknowns one of the most pervasive.

As stated by Kristensen *et al.* (2009), there are two ways to access the viral metagenome: by directly sequencing nucleic acids extracted from viral concentrates of natural samples (targeted approach) or by detecting virus-specific sequences in microbial metagenomes (non-targeted), although this later approach requires a larger sequencing effort (Reyes *et al.*, 2010) and is less suited for the discovery of completely unknown viruses.

Although metaviromic studies have also brought light into the world of eukaryal viruses [such as ssRNA marine viruses (Culley *et al.*, 2006), ssRNA and dsRNA viruses in freshwater (Djikeng *et al.*, 2009), the recently found large nucleocytoplasmic DNA viruses (Kristensen *et al.*, 2009) and the herpes-like viruses infecting corals (Thurber *et al.*, 2008), for instance] we will focus here on the metagenomic studies on viruses infecting prokaryotes and how these studies have shed light into the viral mechanisms of prokaryotic control. Viruses infecting bacteria are called bacteriophages. Some authors also call phages those viruses infecting *Archaea* (Paul *et al.*, 2002; Duhaime, 2011; Sorokin *et al.*, 2010; Thurber, 2009), although they do not display always 'normal' phage-like behaviour and structure. For this reason, other authors (see Forterre, 2010, and references therein) propose the classification of viruses into three categories: archaeoviruses, bacterioviruses and eukaryoviruses. Since this issue is submitted to controversy, we will avoid this terminology, although the reader should keep in mind that, in metagenomics, there is no a straightforward way to distinguish between viruses infecting *Archaea* and *Bacteria*. So, when we refer to phages, we could be meaning virus infecting either of these prokaryotic groups. Indeed, analysis of different spindle-shaped archaeal viruses do not reveal any sequence similarity (Sime-Ngando *et al.*, 2010) which could have been expected due to the high similarity in shape.

When the microbial world was first studied using culture-independent techniques (Rapé and Giovannoni, 2003) one of the main conclusions was that the culturable assemblage represented only a minor fraction of the total community and that the most frequently recovered microorganisms were neither the most abundant nor the most relevant in a given environment. A similar scenario has been unveiled by the analysis of viral metagenomes since most of the metaviromic sequences retrieved so far do not resemble previously isolated viruses. However, the focus here will not be the wide and unknown diversity unveiled by viral metagenomics but the aspects of such studies that have brought some light into the ways in which virus control their hosts in 'natural' communities.

Phages in nature

In most environments that have been studied so far, there are from 3 to 10 viruses for every microbial cell (Paul *et al.*, 2002) although in some cases this number is considerably lower like in Yellowstone hot springs (Schoenfeld *et al.*, 2008) or can reach even more than several tens, like in some hypersaline environments (Santos *et al.*, 2011) or in some polar waters (Säwström *et al.*, 2007), making viruses the most abundant 'biological entities' on the planet. Notice that many authors avoid the question of whether virus are a life-form or not, since there is still a dispute on this issue (Brussard *et al.*, 2008; Forterre, 2010; Moreira and López-García, 2009).

With a few exceptions, such as the viral community from an Antarctic lake that is dominated by eukaryotic viruses (López-Bueno *et al.*, 2009), phages dominate viral metagenomes. The meta-viromes of eight biomes, including marine and freshwater samples, high salt pond from solar salterns, animal (coral, aquaculture fish, terrestrial animal and mosquito) associated, were dominated by phage (Dinsdale *et al.*, 2008). Phages are also dominant in marine sediments where the virus-like particles (VLP) concentration ranges from 10^9 to 10^{13} VLP/kg of sediment, most of the phages, that are 'produced' *in situ* rather than arriving through sedimentation (Breitbart *et al.*, 2004). The marine viromes of four oceanic regions (Angly *et al.*, 2006) were dominated by prokaryotic viruses, including an abundant group of ssDNA phage, that had been previously overlooked. Analysis of the Global Ocean Sampling (GOS) expedition has also highlighted the abundance of phages, especially cyanophages, in marine surface waters (Williamson *et al.*, 2008). Phage dominated DNA viral metagenomic libraries in reclaimed and potable waters, while RNA metaviromes from the same samples were dominated by eukaryotic viruses (Rosario *et al.*, 2009). Unique communities of viruses were found in modern stromatolites and trombolites, with the microbialites from Highborne Cay (Bahamas) dominated by native ssDNA microphages that had not been previously detected and were persistent across time (Desnues *et al.*, 2008). Single-stranded DNA phages have also been found to be very abundant in Sargasso Sea samples (Angly *et al.*, 2006). The majority of the DNA viruses in infant (Breitbart *et al.*, 2008) and adult (Breitbart *et al.*, 2003; Reyes *et al.*, 2010) faecal samples are phages, although the number of RNA phages was very low compared to that found among plant viruses (Zhang *et al.*, 2006). The virome of the human oropharynx is also dominated by phages, some of them containing streptococcal virulence factors (Willner *et al.*, 2010).

Most of the viral metagenomes mentioned above are most likely dominated by bacterial viruses (since the microbial communities are dominated by *Bacteria*). However, there are also some examples of viral metagenomes from some *Archaea* dominated environments that should therefore be highly enriched in archaeal viruses. Such is the case of the viral metagenome of Yellowstone hotsprings (Schoenfeld *et al.*, 2008). The low virus–microbe ratios found in this system may be related to the observation that cultured viruses from thermophilic *Crenarchaeota* do not proliferate via lytic infections and/or to the harsh extracellular conditions (Prangishvili *et al.*, 2006). Unexpectedly, the viral metagenome of the Yellowstone hotsprings (Schonenfeld *et al.*, 2008) contained genes related to the lytic cycle, in contrast to the cultured thermophilic crenarchaeal viruses, that are all non-lytic. This finding may be indicating that also for archaeal viruses, cultured dependent and independent techniques can be unveiling different portions of the diversity, as has happened with prokaryotic communities (Rappé and Giovannoni, 2003). Other viral metagenomes that should be dominated by archaeal viruses are those from hypersaline environments such as crystallizer ponds from solar salterns (Santos *et al.*, 2007; 2010) or the Lake Rebka in Senegal (Sime-Ngnado *et al.*, 2010). Given the absence of predators feeding on prokaryotes, the control of populations by viruses may be more relevant in these and other extreme environments than in temperate systems (Guixa-Boixareu *et al.*, 1996; López-Bueno *et al.*, 2009).

Viral infection and lysis of host cells is a very efficient transformation of microbial biomass into dissolved organic matter (DOM), thus disrupting the biological pump (Fuhrman, 1999) and influencing food web interactions. When viruses cause the lysis of their host, both host and progeny viruses became dissolved organic matter that cannot go into higher trophic levels nor can sink to the bottom. Besides their effects on biogeochemical cycles, that definitely will have an indirect effect over microbial community structure, there are at least three ways in which phages can act on prokaryotic communities: (i) controlling prokaryotic numbers, (ii) controlling their diversity, (iii) modifying prokaryote genomes through lateral gene transfer (LTG) in processes such as lysogenic conversion or as vectors of LTG, thus driving the diversification and evolution of microbial lineages. Indeed, in the very heart of the microbial genomes there is a strong viral imprint, not only in the presence of viral genes but also in the evolution of CRISPR systems (a prokaryotic defence system against cell 'invaders', such as viruses and plasmids). This is also true for phage genomes that have been shaped through the interaction with their hosts, as discussed below.

The kill-the-winner hypothesis and the Red Queen dynamics

One of the most important issues related to viral function is how viruses control the number and diversity of microbial hosts. The dynamic of this prey–predator interaction has been described trough the so-called kill-the-winner hypothesis, proposed by Thingstad and Lignell in 1997. According to this model, the increase in numbers of a specific microbial species (strain, populations... see below) leads to an increase in the numbers of phages infecting this species (or strain, or population...) which is followed by a decrease in abundances of both host and virus (Fig. 3.1A).

A first prediction of this model (summarized in Fuhrman, 2009) is that there is a succession of dominant hosts over time. However, data showed that dominant OTUs (operational taxonomic units, normally detected by ribosomal operon targeting tools) tend to remain stable for long periods of time. One of the most likely explanation for the observed OTU stability is supported by both genomic and metagenomic data (Peña *et al.*, 2010 and references therein) and also by experiments with 'pure' cultures (Middelboe *et al.*, 2009), that indicate that OTUs normally encompass a suite of different 'strains' that are infected by different suites of phages (Fig. 3.1B). Another prediction of the kill-the-winner is that the most abundant phage genotype will not be the same at different time points.

To integrate the kill-the-winner model with the new wealth of genomic and metagenomic data,

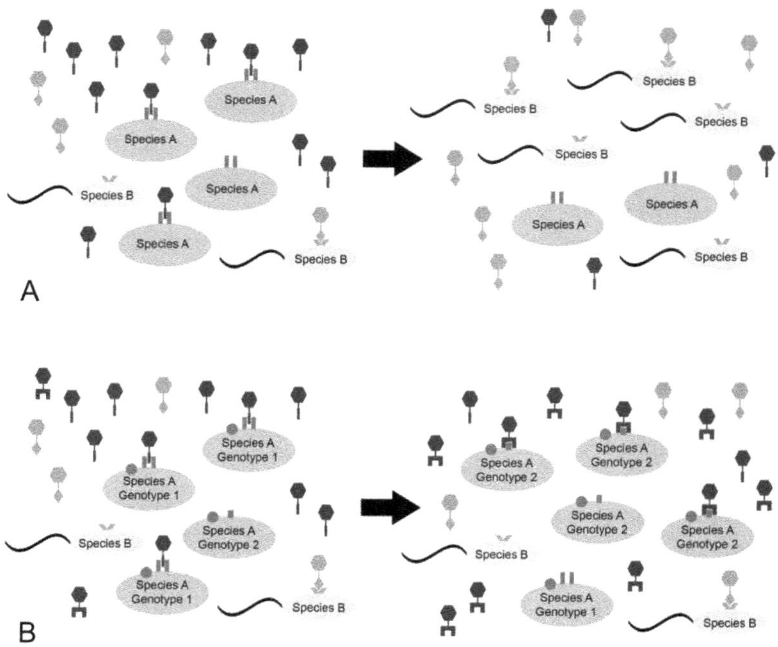

Figure 3.1 Kill-the-winner dynamics. (A) A classical view in which the winner is a prokaryotic 'species'. (B) Here the winner is a strain (with a specific genotype) within the species.

Rodríguez-Valera *et al.* (2009) have proposed the constant-diversity dynamics (CDD) model according to which the diversity of prokaryotic populations is preserved by phage predation. Indeed, comparative genomics indicate that very closely related strains (definitely, members of the same species) harbour different suites of potential phage targets in the areas of the genomes that are specific for each strain (i.e. not shared among different strains). This is a pervasive phenomenon in the open ocean, and it has been interpreted as a way of avoiding phage infection (Coleman *et al.*, 2006; Wilhelm *et al.*, 2007). This diversity of receptors could constitute the first line of defence against phage infection. Accordingly, metagenomic studies in salterns (Rodríguez-Brito *et al.*, 2010) and in sludge (Kunin *et al.*, 2008) showed that many of the changes observed with time can be attributed to *loci* related with phage attachment. In addition, many genomes indicate the presence of a second intracellular line of phage defence (Rodríguez-Valera *et al.*, 2009): their metagenomic islands showed numerous examples of restriction-modification systems and the CRISPRS (see below) involved in phage interference.

The novelty of the CDD model compared to the kill-the-winner is that it proposes that interaction with viruses is responsible for maintaining diversity among closely related lineages, avoiding the dominance of a given lineage within a population. Thus phage infection would make possible the simultaneous presence of cell types that otherwise would have been eliminated by other ecological forces. According to the authors of this model, the availability of different resources alone cannot grant a high bacterial diversity since only the presence of phages can give different cell lineages the opportunity to reach a similar average density. Obviously, we would need to know whether indeed different cell lineages (way below the detection of ribosomal targeting techniques) reach a similar average density and how they change with time. In agreement with this model, data indicate that there are no 'dominant' viral genomes (Table 3.1) in most of the natural environments analysed so far. However, this picture is incomplete without information on how the dominant virotypes evolve, which has been studied only in a few examples discussed in this chapter.

Table 3.1 Characteristics of some studied metaviromes

Environment	% Unknown[a]	Diversity (Shannon/ viral genotypes)	Most abundant genome	Reference
Marine waters (Scripps Pier and Mission Bay, San Diego, USA)	65	–/374–7144	2–3%	Breitbart *et al.* (2002) (first paper on viral metagenomics), Angly *et al.* (2005)
Infant gut	66	1.69/8	43.6%	Breibart *et al.* (2008)
Adult human faeces	Most	6.4/1200	4%	Breitbart *et al.* (2003), Angly *et al.* (2005)
Near-shore viral sediment	75	7.4–9.2/7340	0.0136%	Breitbart *et al.* (2004), Angly *et al.* (2005)
Antarctic lake	87.6	–/5130–9730	–	López-Bueno *et al.* (2009)
184 marine assemblages	89.5–95.7	6.05–10.8/–	–	Angly *et al.* (2006)
Yellowstone hot springs	41.2–62.8	5.29–6.88/–	–	Schoenfeld *et al.* (2008)
Crystallizer pond	75.3	2.1/32	45.6%	Santos *et al.* (2010); this work
Human oropharynx	51–64[b]	–	–	Willner *et al.* (2010)
Cheaspeake Bay water	61	7.4–8.31/1650–4110[c]	0.065–0.153%	Bench *et al.* (2007)
Modern stromatolites and trombolites	97.7–99.3	–	–	Desnues *et al.* (2008)
Induced metavirome	93.4	9.13/15400	4.43%	McDaniel *et al.* (2008)
Faecal microbiota	81	3.32/35	–	Reyes *et al.* (2010)
Potable and reclaimed DNA metaviromes	44–70	–	–	Rosario *et al.* (2009)

[a]The number of unknown sequences in a viral metagenome also depends on the method used for sequencing; normally metagenomes analysed by pyrosequencing have a higher number of unknowns. No similarity to nt or nr database.

[b]Depending on the method used for sample preparation.

[c]Depending on the contig size used for calculating contig spectrum.

One of the few works addressing the temporal changes of host and virus genomes in natural systems by culture independent methods is the study by Rodriguez-Brito *et al.* (2010). The authors used a metagenomic approach to characterize the viral and microbial community dynamics in four human-controlled aquatic environments (a fish farm and three ponds of 8%, 14% and 30% salinity in a solar saltern). This is the most in depth study describing virus-microbe dynamics over time using metagenome characterization that has been published so far. The authors found that each microbial community had metabolic potential profiles that were very stable with time. Most abundant microbial 'species' persisted although displayed some 'shuffling' in their relative abundance. However, microbial strains encompassed in a given species would come and go, showing a fine-scale variation in the microdiversity of the samples. On the other side, dominant viral taxa also persisted although viral genotypes were rapidly changing. According to the authors, this phenomenon could be included in the so-called Red Queen dynamics, in which both hosts and virus have to 'run fast' in order to keep in the same place. In addition, this cycling of hosts (preys) and viruses (predators) would follow the 'fine tuned' kill-the-winner model, in which the winner would not be a given species, but a host-sensitive microbial strain, among many others in the same species (Fig. 3.1). These observations also fit into the CDD model explained above.

A very recent work (Reyes *et al.*, 2010) on the viruses in the human faecal microbiota has provided an example in which the predatory kill-the-winner dynamics is remarkably absent.

These authors found that the viromes present in the faecal microbiota collected from four pairs of adult female monozygotic twins and their mothers during 1-year period (sampling at $t = 0$, $t = 2$ months, and $t = 1$ year) had a very low intrapersonal diversity, with more than 95% of viriotypes remaining constant in the viromes, that were highly enriched in temperate phages. Since the analysed metaviromes had been constructed from purified VLPs, the presence of such a high amount of prophage-related sequences in the free viral fraction could be due to an 'emergency' mechanism of 'survival', that would make viruses leave their host due to the energy limitations in the distal intestine, where energy from dietary nutrients is no longer available (Reyes *et al.*, 2010).

Lytic versus lysogenic

Viruses infecting prokaryotes normally replicate through two main kinds of cycles: lytic and lysogenic. In the lytic cycle, the viruses infect its host and, as a consequence of viral particle production, the host cell lyses and dies. During the lysogenic cycle, the virus (called temperate) integrates its genome into the host genome becoming a prophage that remains inactive until the induction of the lytic cycle. In addition, there are other situations, such as pseudolysogeny in which (i) the viruses are not segregated equally to all progeny cells, (ii) neither integrates nor replicates or lyses the host and (iii) the chronic infection in which the virus is released without lysing the cell (Wommack and Colwell, 2000; Sime-Ngando and Colombet, 2009). Prophages can influence their host by, for instance, expressing their genes provoking a change of phenotype in a phenomenon known as lysogenic conversion, which is relatively frequent in human pathogens such as the causative bacterial agents of cholera, botulism, and diphtheria among others (Davis and Waldor, 2002).

In oligotrophic environments lysogeny could be more advantageous than the lytic cycle for virus survival although the relationship between lysogeny and the trophic status has not been properly elucidated yet (Weinbauer, 2004). In some cases, it has been observed that bacterioplankton with low host growth favour lysogeny. For instance Long *et al.* (2008) when analysing lysogeny in the Gulf of Mexico and Mississippi river plume, found a strong correlation between productivity and lysogeny both for *Synechococcus* and for heterotrophic bacteria. This could also explain the increase in the lysogenic fraction in deeper waters compared to the photic zone (see below). Since viruses cannot persist in the water column for long periods of time, especially due to UV-induced decay, lysogenic bacteria can be regarded as a reservoir of bacteriophages under adverse conditions.

One of the first works providing metagenomic information on the extent of lysogeny in natural communities is that of DeLong *et al.* (2006) 'Community genomics among stratified microbial assemblages in the ocean's interior'. In our opinion, this is one outstanding example of how to draw biological meaning form metagenomic data. This study showed a vertical distribution of putatively lytic and temperate viruses, with large numbers of lytic viral sequences in the photic zone and a prevalence of prophage and transposase sequences in deeper samples. Transposases have recently been found to be the most abundant and most ubiquitous genes in nature (Aziz *et al.*, 2010).

Besides marine waters, putative lysogenic (or temperate) viruses have been found in other environments. Metagenomic studies (Breitbart *et al.*, 2004) of marine sediments indicated that their viral diversity was extremely high, and was dominated by novel viruses, most of them phages with the potential for temperate lifestyles. The same phenomenon was observed in adult human faeces (Breitbart *et al.*, 2003): most the phages that could be annotated (i.e. identified by sequence homology with known viruses) were siphophages and prohages, both with the ability to be temperate. These findings made the authors suggest that temperate lifestyles may be important in these environments. Breitbart *et al.* (2004), suggested that the production (i.e. the release from the cell) of temperate phages would be favoured in environments with higher microbial densities, since lysogeny is thought to be more important when host densities or resources are low; they found temperate phage signatures in the free phage fraction in sediments; these lysogens could have been induced by high host and nutrient availability or by unknown environmental stress signals. In contrast, the analysis of the metavirome form a nutrient rich environment (a crystallizer pond from a solar saltern) did not indicate a relevant presence of lysogeny at the time of sampling (Santos *et al.*, 2010).

However, the extent of lysogeny has been little investigated using metagenomic approaches, with the exception of the pioneer work by McDaniel *et al.* (2008) who succeeded at connecting metaviromic sequence data with an ecologically relevant function in a coastal marine environment. These authors sequenced the metavirome in water samples from Tampa Bay after treating the cellular fraction with mytomycine C that, in some examples of cultured virus–host systems, is known to induce the lytic cycle in virus present in lysogens. In other words, viral genomes that would be either integrated or stabilized into bacterial genomes would be forced by this treatment to move into a lytic cycle, provoking the release of new viral particles into the environment. In addition, the expression of some selected genes coding for integrases was also analysed by real-time PCR. Although, as stated by the authors, this work did not provide a control of a non-induced metavirome, very interesting conclusions could be drawn from the data. For instance, although the number of unknowns was as high as in most viral metagenomes (Table 3.1), the number of sequences that could be clearly identified as viral was considerable higher than in other marine water samples, indicating that the mitomycin C treatment was indeed causing a shift in the viral assemblage, provoking and enrichment in 'identifiable' sequences, most likely due to an induction of prophages. Among the induced viruses identifiable by BLASTx analysis, cyanophages were the most abundant. However, when the metagenomic sequences were assembled and analysed again, 9 of the 15 top hits corresponded to experimentally verified prophage or temperate phages, further indicating that the treatment with mitomycin C was indeed provoking the release of prophages. Among the induced viral fraction there was also a lytic cyanophage that, therefore, should also be very abundant in the sample, as shown in many marine environments (see below). The induced metagenome was then searched for the presence of genes known to be involved in lysogeny in well known host–viruses systems, such as cI-type repressors and phage integrases. Unexpectedly, cI-type repressors, that normally are expressed constitutively, could not be detected in the induced metavirome, which could again be an example of how environmental 'communities' differ from laboratory 'models'. Integrases could be readily detected and four of them were selected for quantifying lysogens in a water sample taken from the same environment, indicating than one of them (a *Vibrio*-like integrase) was present in 0.005% of the ambient population, two could not be detected and the remaining (a *Oceanicola*-like integrase) was present in 0.00008% population, just at the detection limit. Although these figures would indicate a rather modest presence of lysogens, one must keep in mind that they are obtained with probes for only four integrases and, most likely, there are many more integrase-like genes in the environment. In addition, the presence of these genes in a different sample taken from Tampa Bay indicated that they were persistent in the environment and could thus be used as probes for monitoring expression through the annual cycle. Furthermore, expression of two of these integrases (i.e. *Clostridium*-like and *Vibrio*-like) was detected in different time points along an annual cycle, corresponding with what the authors called the lysogenic season. The presence of these two integrases was further investigated in water samples from microbial and viral marine metagenomic libraries to get an estimate of their prevalence in prophages and in the free virus fraction, respectively. The frequency of integrases in the metagenomes was calculated by comparison with the proportion of the single-copy gene recA in the samples available in the CAMERA database. For these samples, the authors estimated that around 3.14% and 0.58% of the microbial population carried the *Clostridum*-like and the *Vibrio*-like integrases respectively, without any clear trend with respect to the type of environment where these genes were found. The *Vibrio*-like integrase could also be detected in four additional marine viral metagenomes (from the following oceanic regions: British Columbia, Gulf of Mexico, Sargasso Sea and Arctic; Angly *et al.*, 2006) with a highest prevalence in the Arctic samples.

Interestingly, the Arctic free virus metagenome had a high number of identifiable prophage genes compared to the rest of environments. As expected, since integrases main function is to integrate viral DNA into microbial genomes, these genes were more frequently observed in microbial samples than in free virus samples.

Virus as a gene transfer agents

One of the most efficient and persistent ways that viruses have to impact prokaryotic communities is through lateral gene transfer (LTG). It has been estimated (Paul *et al.*, 2002) that 10^{28} bp of DNA are transduced by phages per year in the ocean.

In spite of the difficulty of deciphering the functioning of natural viral communities and their role in LTG, some recent works on cyanophages (i.e. cyanobacteria-infecting viruses) have been very successful in this regard. Without doubt, the study of marine cyanophages is the most advanced area of research in environmental phages. Oxygenic photosynthesis system II genes as well as stress response genes have been found in the genomes of phages infecting *Prochlorococcus* and *Synechococcus* strains (revised in Clokie and Mann, 2006), the main components of the prokaryotic phytoplankton, that contribute up to 50% of photosynthetic primary productivity in oligotrophic marine waters. The presence and the expression of these genes upon host infection has been explained as a mechanisms to support photosynthetic activity in infected (and thus damaged) cells and ensure the success of viral replication. In addition, analysis of psbA (a gene coding for the photosystem II protein D1) points out to a reshuffling of psbA genes between *Synechococcus* and *Prochlorococcus*, thorough cyanophage mediated LTG (Sharon *et al.*, 2007). This underscores the importance of viruses in the evolution of microbial processes, in this case photosynthesis, as a source of innovation. The non-targeted metagenomic study carried out by DeLong *et al.* in 2006 (see above) also detected cyanophage-related sequences that seem to be viral versions of genes coding for proteins involved into different host-specific functions.

In 2007, Sharon *et al.* carried out a survey of psbA genes in surface water from the GOS transect (Rusch *et al.*, 2007), that covers diverse oceanic regions from the North Atlantic to the Equatorial Pacific, finding that around of 60% of psbA genes in these samples were of phage origin that were undergoing selection from D1 proteins distinct from those present in their hosts genomes. Furthermore, these authors found that psbA viral genes were being expressed in the marine environment. Cyanophages were also found to be important components of the viral assemblage in Chesapeake Bay (Bench *et al.*, 2007), a highly productive estuarine ecosystem, where the photosynthesis core gene psbA was also found to be universally present in the cyanophage genomes.

As stated above, viral psbA genes differ from those of their hosts, a fact that could be relevant in the global photosynthetic activity in the biosphere: since viral D1 proteins are less susceptible to photodamage and turnover than their host counterparts, the support of photosynthesis during infection might constitute a gain of several hours of light energy that otherwise would be lost during infection.

Later, Sharon *et al.* (2009), in a brilliant example of how analysis of pre-existing data can bring amazing discoveries, found that photosystem I genes were also present in the genomes of viruses infecting *Prochlorococcus* and *Synechococcus*. The use of a new search scheme to analyse the previous metagenomic data from the GOS expedition and from the metaviromes described in Dinsdale *et al.* (2009) revealed that the genomes of cyanophages contained a cluster of genes of cyanobacterial origin coding for proteins with the potential to form an intact monomeric plant-like PSI complex (as opposed to trimeric cyanobacterial-like complexes). In addition to this cluster of photosynthetic genes, these authors found a viral gene fusion that constitutes the first phage 'engineered' gene involving a structural membrane protein. In view of these findings, Sharon *et al.* (2009) anticipate that the marine viral assemblage could be 'an almost unlimited source of naturally bioengineered gene cassettes'.

Analysis of the abundance of PSI and PSII in marine viral metagenomes (Sharon *et al.*, 2009) indicated the possibility that some cyanophages carry both types of genes. This was later confirmed by the work of Alperovitch-Lavy *et al.* (2010) that finally 'reconstructed the puzzle' using a unique crossing of the metagenomic datasets. This analysis showed that there are cyanophages that contain PSII and PSI genes, as in cyanobacteria and plants, as well as NADH dehydrogenase genes. This fact strongly suggests that, in order to maximize virus survival during infection, viruses might disturb host photosynthetic electron flow, switching the host from producing both NADH and ATP to only ATP, to use this extra energy for virus replication. The unveiling of the metabolic capabilities encoded in the genomes of cyanophages provides a paradigmatic example of how the analyses of viral metagenomes can bring information of how viruses can control prokaryotic communities.

Thus, the genomes of cyanophages and their hosts have evolved together through their interaction and gene interchange. This co-evolution, that has also been experimentally addressed (Lindell *et al.*, 2007), has probably influenced the overall distribution of cyanobacterial ecotypes in the global ocean. In addition, the high abundance of cyanophages in marine samples suggests an important role of these viruses in the high levels of cyanobacterial microdiversity in this environment (Williamson *et al.*, 2008).

In addition to genes related to photosynthesis, many other metabolic genes of host origin have been detected in viral genomes using different metagenomic approaches. Williamson *et al.* (2008), using a non-targeted approach, identified and analysed viral sequences present in the metagenomes from microbial fractions along the GOS transect. They found from hundreds to thousands of viral genes coding for metabolic and cellular functions, that had undergone a distinct evolution compared to that of their host counterparts (coding for, for instance, S-adenosylmethionine decarboxylase, transaldolases, phosphate metabolism enzymes, vitamin B_{12} biosynthesis, small heat shock proteins, antibiotic resistance and nitrogen fixation, in addition the other genes mentioned above). The amount of some of these genes in the natural samples was measured, yielding up to 10^7 copies per litre for some genes in some of the points analysed. Considering that the amount of VLPs in these samples was around 10^7–10^8 VLP/l, the presence of host genes in viral genomes seems to be a very frequent phenomenon. In addition, the amount of these genes in different samples had very wide variations, indicating that different environmental pressures are driving the incorporation of these genes by viruses. These analyses suggest that the role of viruses in ecologically relevant metabolic processes can be more important than previously thought. These processes would not be restricted to conversion of light energy but also would include, for instance, phosphate acquisition and metabolism (Williamson *et al.*, 2008).

As in the examples given above, the analysis of 42 viral metagenomes from seven different biomes (Dinsdale *et al.*, 2008; see above) indicated a high functional diversity, with very little restriction to the kinds of genes present in the viral communities, including, for instance, the unexpected high presence of genes related to motility and chemotaxis. In other words, viruses seem to pick up genes from their hosts in any of the analysed biomes. This fact, together with the high abundance of viruses in nature, underscores the magnitude of the potential of viruses as gene transfer agents and the subsequent influence at a global scale.

Metabolic genes have also been found in the viral metavirome form an Antarctic lake (López-Bueno *et al.*, 2009). These authors found an increase of genes involved in carbohydrate and amino acid metabolism respiration and stress response in the Arctic summer virome. This, again, could be interpreted as a way of increasing the survival rate of infected hosts during the changing environmental conditions that microbes encounter after the melting of the ice cover.

Finally, the analysis of the virus assemblage in human faecal microbiota has also unveiled the presence in viral genomes of genes coding for proteins involved in processes associated with anaerobic gut microbiota (Reyes *et al.*, 2010). As in the examples discussed above, some of these proteins are also undergoing evolution different from that of their microbial homologues. Given that the faecal virome is dominated by temperate phages, this is also an example of how they can change the metabolic capacity (and virulence) of their hosts.

Genes move from hosts to viruses where they experience an evolution independent from that of their host counterparts and, in their 'new version', they can go back to their host thus increasing their genomic diversity. Since viruses can move and propagate between different biomes (Sano *et al.*, 2004), there is also a realistic possibility that they can transfer DNA between microbes from very different ecosystems. Studies (Thurber *et al.*, 2009 and references therein) showed that some phages are cosmopolitan and can move between major biomes which would allow the transference of viral encoded genes between them.

According to Kristensen *et al.* (2009) the analysis of metaviromic sequences indicates that there is a realistic possibility than part of the virus-like particles present in natural samples (that are concentrated for metavirome characterization) are not bona fide viruses but gene transfer agents or similar entities that derive from defective prophages. Gene transfer agents (GTAs) were first discovered in the alpha proteobacterium *Rhodobacter capsulatus* and subsequent analyses have revealed their presence in other hosts, as well as GTA features in many bacterial genomes (Paul, 2008). They are small virus-like particles released by prokaryotes that contain random genome pieces of their hosts (Zhao *et al.*, 2009). The 'genomes' of these entities would thus not harbour 'real' virus gene but rather packages of host genes. If metaviromes indeed contained a considerable proportion of GTAs, the VLP mediated gene transfer would be much more relevant than previously imagined.

In addition to their involvement in LTG, viruses leave their imprint in prokaryotic genomes also in the CRISPR systems, that are proposed to provide resistance to extrachromosomal elements. This constitutes a new example of the Red Queen Dynamics between prokaryotes and their viruses (Reyes *et al.*, 2010). As mentioned above, CRISPR/Cas are prokaryotic defence systems against viruses and plasmids. They consist of one or more clusters of regularly interspaced short, often palindromic, repeats (hence the name CRISPR), and a variable set of associated (Cas) genes. When exposed to foreign elements, cells harbouring a functional CRISPR/Cas system may become specifically 'immunized' by the acquisition of new spacers (i.e. sequences intervening the repetitions) that derive from invader DNA sequences (Karginov and Hannon, 2010; Mojica and Díez-Villaseñor, 2010). Heidelberg *et al.* (2009) have proposed the name 'viriotopes' for the spacer, due to their analogy with epitopes in proteins, to emphasize their role in viral immunity. These authors analysed the genomes of two thermophilic *Synechococcus* isolates and prokaryotic and viral metagenomes obtained from microbial mats collected from hotspring in Yellowstone National Park (USA). Their work provides insight of how prokaryotes acquire phage immunity through CRISPR organization and how, in turn, viruses can co-evolve with their hosts and evade host immunity by changing the sequences targeted by the CRISPR spacers without affecting infectivity. Changes within CRISPRs due to the interaction between host and viruses had been previously suggested based on metagenomic analysis of virus and prokaryotes in two natural acidophilic biofilms from Richmond Mine (CA, USA) (Anderson and Banfield, 2008). Analysis of the GOS metagenome samples (Sorokin *et al.*, 2010) indicates that CRISPR cassettes retain the memory of the local virus population at a particular ocean location, which is also an indirect evidence of the difference among phage assemblages between different locations. Something similar had been observed in two geographically remote sludge bioreactors highly enriched in *Candidatus* Accumulibacter phosphatis for which imprints of past phage infection (i.e. CRISPR spacer acquisition) could be observed mostly between phage and host from the same bioreactor (Kunin *et al.*, 2008).

Metavirotranscriptomics: measuring the *in situ* expression of viral genes

The analysis of viral metagenomes has certainly provided a wealth of information on virus diversity in natural systems as well as on relevant aspects related with the structure of virus assemblages and the effects they exert on prokaryotic communities, as discussed above. However, the analysis of viral metagenomes does not normally provide direct information on the activity of viruses and whether the metabolic capabilities encoded in their genomes are effectively expressed under given conditions. Viral metagenomes do provide information on the abundance of given viral genomes (i.e. the longest contigs correspond to the most abundant genomes), but, as pointed out by Fuhrman and Schwalbach (2003) 'abundance alone does not tell how active the viruses may be in infecting hosts'. Thus, in order to study the viral activity in natural systems, other analyses have to be carried out. Obviously, we refer here to the analysis based on the expression of the metagenome (-omic approaches) since there are well established methods that provide information on the activity of virus assemblages (such as TEM counting of infected cells, of measuring viral production or decay rates, among others, which are out of the scope of this chapter).

To understand the activity of a viral community a step further after metagenomic characterization can be the analysis of the expression of the viral genomes (the viral metatranscriptome). Some examples of analysing expression of certain viral genes in natural communities have already been discussed above when describing work on cyanophages and lysogeny. However, we refer here to the analysis of the complete viral metatranscriptome. This can be accomplished by directly sequencing all the mRNAs present in a given sample (these mRNAs include those derived from the transcription of cellular genes as well as from viral genes undergoing expression inside cells) or by 'separating' the viral mRNAs from the bulk RNA in the analysed sample. These strategies would be analogous to the non-targeted and targeted approaches described above for analysing viral metagenomes. An example of sequencing the whole metatranscriptome is the recent work by Gilbert *et al.* (2010). These authors used a 'multi-omic' approach to characterize seasonal and dial changes in the microbial community in the western English Channel. Although this study was focused on the microbial fraction, it also provides information on the virus assemblages. For instance, these authors found that while the amount of prophage genes increased at night in the metagenome, there was no such an increase in the metatranscriptome, most likely indicating that hosts divide at night without prophage induction.

In our opinion, the non-targeted approach has two main drawbacks: (i) there is no straightforward, unambiguous way to extract viral mRNA from bulk mRNAs and (ii) it requires the sequencing and analysis of all mRNAs which can be very expensive and time consuming. One way of overcoming this problem is to construct microarrays with immobilized metavirome fragments and used them as probes for hybridization against the bulk RNA. This approach has been used to analyse viral expression in sludge bioreactors (Kunin *et al.*, 2008) and crystallizer ponds from solar salterns (Santos *et al.*, in press). In the first case, microarrays were constructed from predicted genes from both bacterial and phage metagenomes while the second study used a microarray containing the whole metavirome, including non coding regions. In both case, virus genes were highly expressed. In addition, the analysis from the hypersaline environment showed high expression for many virus genomes that, according to metagenomic data, were probably minor components of the virus assemblage.

Future trends

Although still in its infancy, the analysis of viral metagenomes has provided a completely new view on the diversity and function of viruses in the environment, which probably will go on growing in the near future. These studies, however, have some caveats that will have to be overcome in the future in order to get the maximum information from the gathered sequences. In our opinion, the main drawback of viral metagenomes is the high number of unknowns that are found by using homology-based tools (some examples are provided in Table 3.1) as well as the problems encountered when assembling viral genomes from metagenomic data. These issues have been discussed elsewhere (Edwards and Rohwer, 2005; McDaniel *et al.*, 2008; Rodríguez-Brito *et al.*, 2010) and are out of the scope of this chapter. However, we anticipate that, to overcome these problems, the sequence of genomes from new isolated viruses could help to retrieve more information from metagenomic data. In addition, the use of analysis tools that are not based on homology but in signature characteristics of viral genomes, could also be a great advance in the field. To this regard, it is worth mentioning the work by Willner *et al.* (2009), that characterize the metagenomic signatures of 86 microbial and viral metagenomes and found that dinucleotide analyses defined a metagenomic signature which could explain as much difference among biomes as functional genomics (based on metagenome annotation) could provide. This approach was lately used to analyse the metavirome of a crystallizer pond in an attempt to sort viral sequences and hypothesize on the nature of their corresponding hosts (Santos *et al.*, 2010). The identification of the putative hosts for the viruses unveiled trough metagenomics is also one of the future challenges in the field. In fact, the use of genome sequence signatures has recently been proposed 'to compare phage to their host genome signature in order to obtain host–phage relationships and information on their lifestyle' (Deschavanne *et al.*, 2010). Again, for testing these non-homology-based tools, more sequences of viral genomes isolated from different environments and host are needed.

A second aspect that, in our opinion, future metaviromic studies will have to clarify is the relationship between abundance and activity for different types of viruses: viruses infecting hosts with different life strategies, viruses infecting *Archaea* or *Bacteria*, virus undergoing different replication strategies. This will require the characterization of not only viral metagenomes but also metatranscriptomes and metaproteomes. In turn, the identification of new viral proteins from environmental viral metaproteomes will certainly help to identify new proteins in the metagenomes and will probably minimize the amount of unknowns that, in fact, are related to structural proteins.

So far we are able to count viruses in a natural sample and to sequence their metagenome. In order to link numbers and identification of viruses, a technique like FISH (fluorescence *in situ* hybridization) would be a big step towards the complete opening and sorting of the virus black-box. In an ideal scenario, probes against virus sequences could be labelled and hybridized against the fixed VLPs so the virions carrying the sequences could be counted. However, we do not even know now whether this is technically feasible.

Whatever new technical advances the future will bring, there is no doubt that the field of viralomics will keep on mightily growing and that these trends we anticipate (Fig. 3.2) and other we

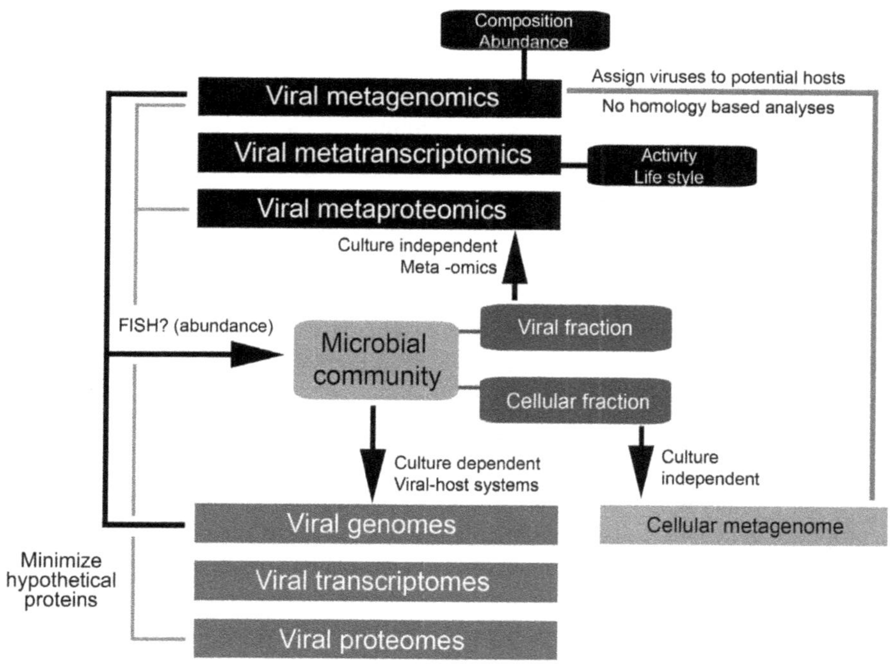

Figure 3.2 Summary of the aspects that future viral metagenomics will be able to address (for a detailed explanation see text).

cannot even imagine will fully develop within the next years.

Web resources

http://biome.sdsu.edu/phaccs/
http://camera.calit2.net/
http://www.broadinstitute.org/annotation/viral/Phage
http://www.ebi.ac.uk/genomes/archaealvirus.html
http://www.ebi.ac.uk/genomes/phage.html
http://www.ictvdb.org/
http://scums.sdsu.edu/phage/

References

Alperovitch-Lavy, A., Sharon, I., Rohwer, F., Aro, E.-M., Glaser, F., Milo, R., Nelson, N., and Béjà, O. (2010) Reconstructing a puzzle: existence of cyanophages containing both photosystem-I and photosystem-II gene suites inferred from oceanic metagenomic datasets. Environ. Microbiol. (Epub ahead of print).

Amann, R.I., Ludwig, W., and Schleifer, K.H. (1995) Phylogenetic identification and *in situ* detection of individual microbial cells without cultivation. Microbiol Rev. *59*, 143–69.

Andersson, A.F., Banfield, J.F. (2008). Virus population dynamics and acquired virus resistance in natural microbial communities. Science *320*, 1047–1050.

Angly, F., Rodríguez-Brito, B., Bangor, D., McNairnie, P., Breitbart, M., Salamon, P., Felts, B., Nulton, J., Mahaffy, J., and Rohwer, F. (2005). PHACCS, an online tool for estimating the structure and diversity of uncultured viral communities using metagenomic information. BMC Bioinformatics *6*, 41.

Angly, F.E., Felts, B., Breitbart, M., Salamon, P., Edwards, R.A., Carlson, C., Chan, A.M., Haynes, M., Kelley, S., Liu, H. *et al.* (2006). The marine viromes of four oceanic regions. PLoS Biology *4*, e368.

Aziz, R.K., Breitbart, M., and Edwards, R.A. (2010). Transposases are the most abundant, most ubiquitous genes in nature. Nucleic Acids. Res. *38*, 4207–4217.

Bench, S.R., Hanson, T.E., Williamson, K.E., Ghosh, D., Radosovich, M., Wang, K., and Wommack, K.E. (2007). Metagenomic characterization of Chesapeake Bay virioplankton. Appl. Environ. Microbiol. *73*, 7629–7641.

Breitbart, M., Hewson, I., Felts, B., Mahaffy, J.M., Nulton, J., Salamon, P., and Rohwer, F. (2003). Metagenomic analyses of an uncultured viral community from human feces. J. Bacteriol. *185*, 6220–6223.

Breitbart, M., Salamon, P., Yresen, B., Mahaffy, J.M., Segall, A.M., Mead, D., Azam, F., and Rohwer, F. (2002). Genomic analysis of uncultured marine viral communities. Proc. Natl. Acad. Sci. U.S.A. *99*, 14250–14255.

Breitbart, M., Felts, B., Kelley, S., Mahaffy, J.M., Nulton, J., Salamon, P., and Rohwer, F. (2004). Diversity and population structure of a near-shore marine-sediment viral community. Proc. Roy. Soc. B: Biol. Scie. *271*, 565–574.

Breitbart, M., Haynes, M., Kelley, S., Angly, F., Edwards, R.A., Felts, B., Mahaffy, J.M., Mueller, J., Nulton, J., Rayhawk, S. *et al.* (2008). Viral diversity and dynamics in an infant gut. Res. Microbiol. *159*, 367–373.

Brussard, C.P.D., Wilhelm, S.W., Thingstad, F., Weinbauer, M.G., Bratbark, G., Heldal, M., Kimmance, S.A., Middelboe, M., Nagasaki, K., Paul, J.H., Schroeder, D.C., Suttle, C.A., Vaqué, D., and Wommack, K.E. (2008). Global-scale processes with a nanoscale drive: the role of marine viruses. ISME J. *2*, 575–578.

Clokie, M.R., and Mann, N.H. (2006). Marine cyanophages and light. Environ. Microbiol. *8*, 2074–2082.

Coleman, M.L., Sullivan, M.B., Martiny, A.C., Steglich, C., Barry, K., DeLong, E.F., and Chisholm, S.W. (2006). Genomic islands and the ecology and evolution of *Prochlorococcus*. Science *311*, 1768–1770.

Culley, A.I., Lang, A.S., and Suttle, C.A. (2006). Metagenomic analysis of coastal RNA virus communities. Science *312*, 1795–8.

Davis, B.M., and Waldor, M.K. (2002) Mobile genetic elements and bacterial pathogenesis. In: Mobile DNA II. Craig, N.L. ed. (Washington, DC: American Society for Microbiology Press), pp. 1040–1059.

DeLong E.F., Preston, C.M., Mincer, T., Rich, V., Hallam, S.J., Frigaard, N.U., Martínez, A., Sullivan, M.B., Edwards, R., Brito, B.R. *et al.* (2006) Community genomics among stratified microbial assemblages in the ocean's interior. Science *311*, 496–503.

Deschavanne, P., DuBow, M.S., andRegeard, C. (2010). The use of genomic signature distance between bacteriophages and their hosts displays evolutionary relationships and phage growth cycle determination. Virology J. *7*, 163.

Desnues, C., Rodriguez-Brito, B., Rayhawk, S., Kelley, S., Tran, T., Haynes, M., Liu, H., Furlan, M., Wegley, L., Chau, B. *et al.* (2008). Biodiversity and biogeography of phages in modern stromatolites and thrombolites. Nature *452*, 340–343.

Dinsdale, E.A., Edwards, R.A., Hall, D., Angly, F., Breitbart, M., Brulc, J.M., Furlan, M., Desnues, C., Haynes, M., Li, L. *et al.* (2008). Functional metagenomic profiling of nine biomes. Nature *452*, 629–632.

Djikeng, A., Kuzmickas, R., Anderson, N.G., and Spiro, D.J. (2009). Metagenomic analysis of RNA viruses in a fresh water lake. PLoS One, e7264.

Duhaime, M.B., Wichels, A., Waldmann,J., Teeling, H., and Glöckner, F.O. (2011). Ecogenomics and genome landscapes of marine Pseudoalteromonas phage H105/1. ISME J. *5*, 107–121.

Edwards, R.A., and Rohwer, F. (2005). Viral metagenomics. Nat. Rev. Microbiol. *3*, 504–510.

Forterre, P. (2010). Defining life: The virus viewpoint. Orig. Life Evol. Biosph. *40*, 151–160.

Fuhrman, J.A. (1999). Marine viruses and their biogeochemical and ecological effects. Nature *399*, 541–548.

Fuhrman, J.A. (2009). Microbial community structure and its functional implications. Nature *459*, 193–199.

Fuhrman, J.A., and Schwalbach, M. (2003). Viral influence on aquatic bacterial communities. Biol. Bull. *204*, 192–195.
Gilbert, J.A., Field, D., Swift, P., Thomas, S., Cummings, D., Temperton, B., Weynberg, K., Huse, S., Hughes, M., Joint, I. *et al.* (2010). The taxonomic and functional diversity of microbes at a temperate coastal site: a 'multi-omic' study of seasonal and diel temporal variation. PLoS One *5*, e15545.
Guixa-Boixareu, N., Calderón-Paz, J.I., Heldal, M., Bratbak, G., and Pedrós-Alió, C. (1996). Viral lysis and bacterivory as prokaryotic loss factors along a salinity gradient. Aquat. Microb. Ecol. *11*, 215–227.
Heidelberg, J.F., Nelson, W.C., Schoenfeld, T., and Bhaya, D. (2009). Germ warfare in a microbial mat community: CRISPRs provide insights into the co-evolution of hosts and viral genomes. PLoS One*4*, e4169.
Karginov, F.V., and Hannon, G.J. (2010). The CRISPR system: small RNA-guided defense in bacteria and archaea. Mol Cell. *37*, 7–19.
Kristensen, D.M., Mushegian, A.R., Dolja, V.V., and Koonin, E.V. (2009). New dimension of the virus world discovered through metagenomics. Trends Microbiol. *18*, 11–19.
Kunin, V., He, S., Warnecke, F., Peterson, S.B., García Martín, H., Haynes, M., Ivanova, N., Blackall, L.L., Breitbart, L., Rohwer, F. *et al.* (2008). A bacterial metapopulation adapts locally to phage predation despite global dispersal. Genome Res. *18*, 293–297.
Lindell, D., Jaffe, J.D., Coleman, M.L., Futschik, M.E., Axmann, I.M., Rector, T., Kettler, G., Sullivan, M.B., Steen, R., Hess, W.R. *et al.* (2007). Genome-wide expression dynamics of a marine virus and host reveal features of co-evolution. Nature *499*, 83–86.
Long, A., McDaniel, L.D., Mobberley, J., and Paul, J. (2008). Comparison of lysogeny (prophage induction) in heterotrophic bacterial and Synecchococcus populations in the Gulf of Mexico and Mississippi river plume. ISME J. *2*, 132–144.
López-Bueno, A., Tamames, J., Velázquez, D., Moya, A., Quesada, A., and Alcamí, A. (2009). High diversity of the viral community from an Antarctic lake. Science *326*, 858–861.
McDaniel, L., Breitbart, M., Mobberley, J., Long, A., Haynes, M., Rohwer, F., and Paul, J.H. (2008). Metagenomic analysis of lysogeny in Tampa Bay: implications for prophage gene expression. PLoS One, e3263.
Middelboe, M., Holmfeldt, K., Riemann, L., Nybroe, O., and Haaber, J. (2009). Bacteriophages drive strain diversification in a marine Flavobacterium: implications for phage resistance and physiological properties. Environ. Microbiol. *11*, 1971–1982.
Mojica, F.J., and Díez-Villaseñor, C. (2010). The on-off switch of CRISPR immunity against phages in *Escherichia coli*. Mol. Microbiol. *77*, 1341–5.
Moreira, D., and López-García, P. (2009). Ten reasons to exclude viruses from the tree of life. Nat. Rev. Microbiol. *7*, 306–11.
Paul, J.H. (2008) Prophages in marine bacteria: dangerous molecular time bombs or the key to survival in the seas? ISME J. *2*, 579–589.
Paul, J.H., and Sullivan, M.B. (2005). Marine phage genomics: what have we learned? Curr. Op. Biotechnol. *16*, 299–307.
Paul, J.H., Sullivan, M.B., Segal, A.M., and Rohwer, F. (2002). Marine phage genomics. Comp. Biochem. Phys. *133*, 463–476.
Peña, A., Teeling, H., Huerta-Cepas, J., Santos, F., Yarza, P., Brito-Echevarría, J., Lucio, M., Schmitt-Kopplin, P., Meseguer, I., Schenowitz, C. *et al.* (2010). Fine-scale evolution: genomic, phenotypic and ecological differentiation in two coexisting *Salinibacter ruber* strains. ISME J. *4*, 882–895.
Prangishvili, D., Forterre, P., and Garrett, R.A. (2006). Viruses of the Archaea: a unifying view. Nat. Rev. *4*, 837–848.
Rappé, M.S., and Giovannoni, S.J. (2003). The uncultured microbial majority. Annu. Rev. Microbiol. *57*, 369–94.
Reyes, A., Haynes, M., Hanson, N., Angly, F.E., Heath, A.C., Rohwer, F., and Gordon, J.I. (2010) Viruses in the faecal microbiota of monozygotic twins and their mothers. Nature. *466*, 334–8.
Rodríguez-Brito, B., Li, L., Wegley, L., Furlan, M., Angly,. F., Breitbart, M., Buchanan, J., Desnues, C., Dinsdale, E., Edwards, R. *et al.* (2010). Viral and microbial community dynamics in four aquatic environments. ISME J. *4*, 739–751.
Rodríguez-Valera, F., Martín-Cuadrado, A.B., Rodríguez-Brito, B.-, Pasić, L., Thingstad, T.F., Rohwer, F., and Mira, A. (2009). Explaining microbial population genomics through phage predation. Nat. Rev. Microbiol. *7*, 828–36.
Rosario, K., Nilsson, C., Lim, Y.W., Ruan, Y., and Breitbart, M. (2009). Metagenomic analysis of viruses in reclaimed water. Environ. Microbiol. *11*, 2806–2820.
Rusch, D.B., Halpern, A.L., Sutton, G., Heidelberg, K.B., Williamson, S., Yooseph, S., Wu, D., Eisen, J.A., Hoffman, J.M., Remington, K. *et al.* (2007). The Sorcerer II Global Ocean Sampling expedition: I, The northwest Atlantic through the eastern tropical Pacific. PLos Biol. *5*, e77.
Sano, E., Carlson, S., Wegley, L., and Rohwer, F. (2004). Movement of viruses between biomes. Appl. Environ. Microbiol. *70*, 5842–5846.
Santos, F., Meyerdierks, A., Peña, A., Rosselló-Mora, R., Amann, R., and Anton, A. (2007). Metagenomic approach to the study of halophages: the environmental halophage 1. Environ. Microbiol. *9*, 1711–1723.
Santos, F., Yarza, P., Briones, C., Parro, V., and Antón, J. (2010). The metavirome of a hypersaline environment. Environ. Microbiol. *12*, 2965–2976.
Santos, F., Moreno-Paz, M., Meseguer, I., López, C., Rosselló-Mora, R., Parro, V., and Antón, J. (2011). Metatranscriptomic analysis of extremely halophilic viral communities. ISME J. (In press).
Säwström, C., Lisle, J., Anesio, A.M., Priscu, J.C., and Laybourn-Parry, J. (2007). Bacteriophage in polar inland waters. Extremophiles *12*, 167–175.

Schoenfeld, T., Patterson, M., Richardson, P.M., Wommack, K.E., Young, M., and Mead, D. (2008). Assembly of viral metagenomes from Yellowstone hot springs. Appl. Environ. Microbiol. *74*, 4164–4174.

Sharon, I., Tzahor, S., Williamson, S., Shmoish, M., Man-Aharonovich, D., Rusch, D.B., Yooseph, S., Zeidner, G., Golden, S.S., Mackey, S.R. *et al.* (2007). Viral photosynthetic reaction center genes and transcripts in the marine environment. ISME J. *1*, 492–501.

Sharon, I., Alperovitch, A., Rohwer, F., Haynes, M., Glaser, F., Atamna-Ismaeel, N., Pinter, R.Y., Partensky, F., Koonin, E.V., Wolf, Y.I. *et al.* (2009). Photosystem I gene cassettes are present in marine virus genomes. Nature *461*, *258*,262.

Schoenfeld, T., Patterson, M., Richardson, P.M., Wommack, E.R., Young, M., Mead, D. (2008). Assembly of viral metagenomes from Yellowstone hot springs. Appl. Environ. Microbiol. *74*, 4164–4174.

Sime-Ngando, T., and Colombet, J. (2009). Virus et prophages dans les écosystèmes aquatiques. Rev. Can. Microbiol. *55*, 95–109.

Sime-Ngando, T., Lucas, S., Robin, A., Tucker, K.P., Colombet, J., Bettarel, Y., Desmond, E., Gribaldo, S., Forterre, P., Breitbart, M., *et al.* (2010). Diversity of virus–host systems in hypersaline Lake Retba, Senegal. Environ. Microbiol. DOI 01.1111/j.1462-2920.2010.02323.x.

Sorokin, V.A., Gelfand, M.S., and Artamonova, I.I. (2010). Evolutionary dynamics of clustered irregularly interspaced short palindromic repeat systems in the ocean metagenome. Appl. Env. Microbiol. *76*, 2136–2144.

Sorokin, V.A., Gelfand, M.S., and Artamonova, I.I. (2010). Evolutionary dynamics of clustered irregularly (sic) interspaced short palindromic repeat systems in the ocean metagenome. Appl. Environ. Microbiol. *76*, 2136–2144.

Thingstad, T.F., and Lignell, R. (1997). Theoretical models for the control of bacterial growth rate, abundance, diversity and carbon demand. Aquat. Microb. Ecol. J, *13*, 19–27.

Thurber, R.V. (2009). Current insights into phage biodiversity and biogeography.Curr. Op. Microbiol. *12*, 582–587.

Thurber, R.L.V., Barott, K.L., Hall, D., Liu, H., Rodriguez-Mueller, B., Desnues, C., Edwards, R.A., Haynes, M., Angly, F.E., Wegley, L. *et al.* (2008). Metagenomic analysis indicates that stressors induce production of herpes-like viruses in the coral Porites compressa. Proc. Natl. Acad. Sci. U.S.A.*105*, 18413–18418.

Weinbauer, M. (2004). Ecology of prokaryotic viruses. FEMS Microbiol. Rev. *28*, 127–181.

Wilhelm, L.J., Tripp, H.J., Givan, S.C., Smith, D.P., and Giovannoni, S.J. (2007). Natural variation in SAR11 marine bacterioplankton genomes inferred from metagenomic data. Biol. Direct *2*, 27.

Williamson, S.J., Rusch, D.B., Yooseph, S., Halpern, A.L., Heidelberg, K.B., Glass, J.I., Andrews-Pfannkoch, C., Fadrosh, D., Miller, C.S., Sutton, G. *et al.* (2008). The Sorcerer II Global Ocean Sampling expedition: Metagenomic characterization of viruses within aquatic microbial samples. PLoS One *1*, e1456.

Willner, D., Thurber, R.V., and Rohwer, F. (2009). Metagenomic signatures of 86 microbial and viral metagenomes. Environ. Microbiol. *11*, 1752–1766.

Willner, D., Furlan, M., Schmieder, R., Grasis, J.A., Pride, D.T., Relman, D.A., Angly, F.E., McDole, T., Mariella, R.P. Jr., Rohwer, F., *et al.* (2011). Metagenomic detection of phage-encoded platelet-binding factors in the human oral cavity. Proc. Natl. Acad. Sci. U.S.A. *108*, 4547–4553.

Wommack, K.E., and Colwell, R.R. (2000). Virioplankton: viruses in aquatic ecosystems. Microbiol. Mol. Biol. Rev. *641*, 69–114.

Zhang, T., Breitbart, M., Lee, W.H., Run, J.-Q., Wei, C.L., Soh, S.W.L., Hibberd, M.L., Liu, E.T., Rohwer, F., and Ruan, Y. (2006). RNA viral community in human feces: Prevalence of plant pathogenic viruses. PLoS Biol. *4*, e3.

Zhao, Y., Wang, K., Budinoff, C., Buchan, A., Lang, A., Jiao, N., and Chen, F. (2009). Gene transfer agent (GTA) genes reveal diverse and dynamic *Roseobacter* and *Rhodobacter* populations in the Chesapeake Bay. ISME J. 3, 364–373.

Functional Metagenomics and Systems Biology: Understanding the Human Organismal Complexity in Disease and Health

4

Liping Zhao and Jian Shen

Abstract

A devastating epidemic of chronic diseases is threatening the public health worldwide. Preventative healthcare systems require novel types of health assessment technologies which focus on the early warning biomarkers before the clinical onset of chronic diseases. In light of the systems theory, emergent functions of the human body should be measured for health evaluation. Humans are superorganisms harbouring two integrated genomes, the human genome and the microbiome which is the collective genomes of all symbiotic microorganisms, particularly those inhabiting the gut. The gut microbiota and the host interact intimately. The structure and functions of the gut microbiota, together with the host metabolism as reflected for example in urine metabolite profiles, are emergent functions of the human body. Metagenomics and metabonomics can be used to monitor the dynamics of gut microbiota and host metabolism. Large-scale cohort studies in which urine and faecal samples are analysed by the whole body systems approaches may lead to the discovery of patterns of gut microbiota and host metabotypes which can in turn be used as a biomarker for diagnosis or target for developing new therapeutics for chronic diseases. The application of these systems approaches in traditional Chinese medicine and nutritional studies may lead to a significant paradigm shift in modern medicine and nutritional sciences.

Introduction

A devastating epidemic of chronic diseases, particularly obesity and related metabolic disorders, has become the most important threat to the public healthcare system in the developed world, for example in USA and Europe, and rapidly developing countries such as China and India (Consultation, 2000). These diseases, e.g. diabetes, are diet and lifestyle related and characterized by a long-term, deranged metabolism which often leads to various complications such as cardiovascular diseases or kidney failure, etc. By 2015, the cost from heart disease, stroke and diabetes treatments alone will cost China 558 billion US dollars (WHO, 2005), a real threat to national stability and security.

Although prevention has been deemed as the only viable solution to this public healthcare disaster (Campbell and Campbell, 2005; Willett, 2008), the actual implementation of any effective preventative measures prove to be rather difficult due to the typically poor compliance of the general public to following the advice of healthcare providers. One important reason for this meagre compliance lies in the poor performance of current health assessment technology to detect very early signs and conditions that reliably predict the health trajectory of a person towards one of the disease attractors. Early warning is obviously vital for the success of preventative healthcare, which calls for completely novel types of health assessment technologies.

Health assessment is a fundamental task of health management science and the foundation of clinical medicine as well (German *et al.*, 2005). To face the challenges of chronic diseases, biomarkers of the pre-disease transition stage must be used instead of biomarkers related to the disease stage, if the health assessment technology's focus is to usher implementation of preventative healthcare strategies. New health assessment technology must facilitate the discovery of such pre-disease biomarkers.

Targeting emergent functions at the organismal level for managing human health

Human knowledge that accumulated from two seemingly unrelated fields as traditional Chinese medicine (TCM) and engineering has laid a philosophical and methodological foundation for developing this new health assessment technology. Insights from TCM into alternative medicine and structural health monitoring into engineering can be very helpful for developing a new type of health assessment technology under the framework of systems biology (Chang and Yang, 2004). TCM doctors always assess the health of their patients and intervene at the whole-body level. These physicians rely on palpation of the patient's pulse and observation of the patient's tongue for assessing human pathological conditions and determine prescriptions based on these measurements. Their methodologies are comparable to the rapidly growing field in engineering science, called 'structural health monitoring', in which pattern changes in vibration, noise, and exhaust gas can be used for on-line, non-invasive, quantitative, and predictive assessment of engine health (Chang and Yang, 2004).

Artificial systems such as airplanes, cars, and robots, and natural systems like the human body share many fundamental features (Csete and Doyle, 2002). They must be robust enough to maintain their functional and structural integrity despite environmental changes. They follow modular design rules so that the effects of a local problem can be curbed within one module to avoid whole system failure. Several strategies have been evolutionarily developed in humans for ensuring the same key functions. This is called functional redundancy.

Global regulation is needed to integrate and coordinate functions of individual modules and parts. The most important feature of a complex system is its emergent properties, i.e. 'the whole is bigger than the sum of the parts', because we cannot deduce the functions of the whole system by simply adding the functions of all the parts together. The formation of one emergent property is the result of the integrated functions of many if not all parts of a complex system. Understanding the structure and function of each part of a system will not lead automatically to an understanding of the system's behaviour as a whole. 'To understand the whole, one must study the whole' (Kell, 2004).

Both TCM doctors and engineers rely on the emergent functions of their complex systems for health assessment, no matter if they are dealing with natural complex systems such as human bodies or artificial complex systems such as robots, engines, or skyscrapers. The founder of the general systems theory, Ludwig von Bertalanffy, proposed the concept of 'emergent functions', and considered that the ultimate goal of general systems theory was to develop methods to describe emergent functions and to understand why functions of the whole are bigger than the summation of those of all the parts (Hammond, 2003). At variance from functional genomics, systems biology should focus on developing methods for measuring emergent functions of biological systems, and for understanding why functions of the whole body (or cell) are more than simply the summation of those of all component organs (or organelles).

Establishment of a classification and identification framework to organize research subjects is the fundamental task of science disciplines. Systems science has complex systems as its research subjects; thus, systems science needs to provide both a conceptual framework and a technological platform for classification of complex systems. Complex systems should be classified based on their emergent functions as a prerequisite for understanding these systems as a whole. Classification and identification of natural complex systems can be based on the measurements of their emergent functions before each component of the system has been understood. The foundation of biomedical science and the healthcare industry is to stratify individuals quantitatively and predictively according to their health status. In other words, classification is the very fundamental task of health assessment technology. For natural complex systems such as the human body, classification in an unsupervised learning mode with no prior knowledge required is the only viable method of classification. Hence, the measurement and determination of the health status of the human body should be based on emergent functions of this complex biological system.

What are the emergent functions of a living cell? A living cell can actively excrete metabolites and proteins into the outside space. When you detect changes of one particular metabolite in the extracellular space, you can be quite confident that all of the genes that are needed for synthesis, transportation, and excretion of this metabolite must be present and expressed under specific environmental conditions. Thus, one parameter can actually reflect the function of many genes and their interactions with the environment. Therefore, the complete repertoire of metabolites in the extracellular space can serve as the emergent function that reflects the functional status of most, if not all, genes in the cell under specific environmental conditions (Nicholson and Wilson, 2003).

What are the emergent functions at the whole-body level? Pulse types, colour, and texture of the patient's tongue are emergent functions of the human body employed by TCM doctors for health assessment. However, these types of emergent functions of the human body are difficult if not impossible to be explained in molecular terms such as genomics, proteomics, or metabolomics, etc. In this era of genomics, biological research that cannot find its root in genes will be difficult to be integrated with mainstream medicine.

What kinds of emergent functions at the whole-body level can be measured with molecular methods in a non-invasive, global, and dynamic way comparable to TCM approaches? These emergent functions need to be digital, quantitative, and highly reproducible to support preventative, predictive, and personalized healthcare. Metabolites, proteins and DNAs and RNAs in extracellular spaces, such as urine, faecal matter, and blood, are emergent functions at the whole-body level which can be assessed with molecular profiling approaches. Many of these emergent functions are related with the composition and activity of gut microbiota which have been shown recent years to play pivotal roles in human health; particularly in the onset and progression of chronic diseases such as obesity and related metabolic diseases.

A vastly diverse and highly populated microbial ecosystem exists in human body to make it a 'superorganism' (Lederberg, 2000). Actually, human cells are overwhelmed by 10 times more symbiotic microbial cells in our body. A recent metagenomic sequencing effort of 124 European faecal DNA samples revealed 3 million new genes in gut microbiota, 100 times of the total genetic repertoire of human genome (Qin *et al.*, 2010). Obviously, the genetic information relevant to human health should include both the human genome and the microbiome, the collective genomes of all symbiotic microbes in human body, particularly those in human gut. Human health phenotype at the whole body level is the result of interactions between the two integrated genomes and our living environment. Diverse arrays of metabolites and signalling molecules can be produced by gut microbiota. These molecules can get into our bloodstream via a normal route enterohepatic circulation or through partially impaired gut barrier. Human cells in different parts of the body are being constantly dosed by these products from gut microbiota as so succinctly put by McFall-Ngai (2008): 'As nutrients are brought to cells and tissues by the circulatory system, so are the metabolic products from our microbial symbionts'.

Human beings are like walking bioreactors. Our intestinal system is essentially a continuous culture system, a chemostat. Foods, drinks and drugs are feeding medium to the system to maintain the populations of gut microbiota. Change of diet would lead to changes of composition of bacteria living in our gut as different members of gut microbiota have different nutritional preferences. Change of human physiology would also lead to change in gut microbiota as this would change its immediate living environment. Variations of gut microbiota would change metabolites produced by it and eventually change human health (Nicholson *et al.*, 2005).

Recent work shows that gut microbiota may exhibit significant early changes in its structure when the host experiences important but clinically not yet detectable health changes such as development of precancerous lesions in the gut. Aberrant crypt foci (ACF) are precancerous lesions induced by carcinogens in the mucosa of human and animals gut. These lesions are very early stage for cancer development. Within humans, from ACF to real cancer may take 10–15 years. Obviously, individuals who are carrying ACFs in their gut are at high risk for developing colon cancer yet there is no clinically available method to pinpoint these individuals from the general populations. Biomarkers associated with this early stage of cancer development may have a significant potential in preventative management of colon cancer. We collected faecal samples intermittently after dosing rats with the carcinogen 1,2-dimethylhydrazine (DMH). DGGE and 454 barcoded pyrosequencing of faecal 16S rRNA genes showed that gut microbiota between treated and control animals segregated into two distinct spaces in PCA or UniFrac analysis when ACFs were fully developed in the gut of carcinogen dosed animals. PLS-DA modelling identified several key species whose population changes were associated with ACF formation and may be used as biomarkers for identifying animals carrying precancerous lesions in a non-invasive way (Wei *et al.*, 2010). This work indicates that structural shifts of gut microbiota may be a potential source for early disease biomarkers for prevention as these bacteria may be highly sensitive to the early changes of host health upon environmental perturbations such as exposure to carcinogens.

In some other cases, variations of gut microbiota may play an aetiological role in disease development. This makes gut microbiota both a biomarker for health changes and a target for nutritional/medicinal interventions of the disease. This can be best illustrated by the role of gut microbiota in the development of obesity and related metabolic diseases.

Changing patterns of food consumption has been closely linked with the dramatic increase in the incidence of obesity, diabetes, and cardiovascular diseases in China as well as in other countries (Campbell and Campbell, 2005). Recent developments have shown that disruption of gut microbiota by a high fat-diet may play a pivotal role in the onset and progression of obesity and insulin resistance, early stage of metabolic diseases.

Gnotobiotic mice models have been highly instrumental in the elucidation of the mechanisms for gut microbiota getting involved in obesity development. It was shown that germ-free mice are resistant to high fat diet-induced obesity (Backhed *et al.*, 2007). Transfer of gut microbiota from genetically obese mice or from high-fat diet-induced obese mice to germ-free wild-type lean recipients lead to significant accumulation of fat in the latter, indicating that gut microbiota are not only necessary but also sufficient to induce obesity in the host animals (Turnbaugh *et al.*, 2006). Gut microbiota participate in obesity development in two ways: (1) ferment plant polysaccharides into short chain fatty acids, thus helping the host extract more calories from otherwise indigestible food components; (2) distort host energy metabolism by directly regulating relevant genes. Compelling data show that gut microbiota can suppress the expression of the fasting-induced adipogenic factor (*fiaf*) gene, whose expression is required for mobilizing genes involved in fatty acid oxidation, the process for burning stored fat. Gut microbiota can also stimulate genes involved in lipid synthesis in the liver such as *acc1* and *ras.* This effect on both side of the energy equation by gut microbiota transforms the host animals into a highly efficient fat-making and storage machine (Backhed *et al.*, 2004, 2007).

Another distinct but complementary mechanism for gut microbiota getting involved in development of obesity and insulin resistance is by way of provoking a low-grade, systemic and chronic inflammatory condition which has been known for many years as an underlying pathological condition for ageing and age-related diseases (Chung *et al.*, 2009).

Cani *et al.* (2007a) showed that LPS, the endotoxin produced by gram-negative opportunistic pathogens in the gut, was increased 2–3 times in the blood of high-fat diet fed animals, which showed low-grade systemic chronic inflammation comparable to what has been found in human subjects. Subcutaneous injection of comparable amount of purified LPS into the bloodstream of mice fed on normal chow diet made the otherwise lean and healthy animals start to develop inflammation and eventually became obese and insulin resistant. CD14 knockout mice did not respond to LPS injection. This work demonstrates that LPS produced by gut microbiota may play an essential mediator role in the development of inflammation-induced obesity and insulin resistance.

It was hypothesized that high-fat diet may promote the population levels of opportunistic pathogens and lead to increased levels of free antigens such as LPS in the gut. Owing to lack of sufficient dietary fibre in the high-fat diet, many beneficial bacteria such as bifidobacteria start to be washed out from the gut bioreactor. It has been shown that bifidobacteria are among those which can protect gut barrier integrity possibly through a GLP-1 involved mechanism (Cani *et al.*, 2007b). Addition of oligofructose into high-fat diet prevented mice from developing obesity and insulin resistance possibly by maintaining a normal level of bifidobacteria and thus a gut barrier less permeable to LPS (Cani *et al.*, 2007c). The application of 454 barcoded pyrosequencing and PLS-DA modelling enabled us to identify a sulfate-reducing bacterium as the key pathogenic species related with high-fat diet-induced obesity and insulin resistance in a mouse model (Zhang *et al.*, 2010). Our work also showed that 56% of structural variations of gut microbiota can be attributed to diet types while only 12% to host genetic mutation. Animals with no genetic defect can develop severe obesity and insulin resistance by taking excess amount of high-fat diet. In other words, animals do not need to harbour a genetic defect to develop metabolic diseases. The disruption of gut microbiota by a high fat diet would be sufficient to distort the host energy metabolism and provoke inflammation, causing severe adiposity and tissue damage.

These experiments demonstrate that a high-fat diet disruption of gut microbiota plays an essential mediator role in the inflammatory condition responsible for onset and progression of obesity, as well as of related metabolic diseases such as type II diabetes and cardiovascular diseases (Hotamisligil *et al.*, 1996; Wellen and Hotamisligil, 2005; Yudkin *et al.*, 2004).

This central mediating role of the gut microbiota in the onset and progression of chronic diseases, particularly metabolic, has placed it at the centre stage for being an ideal biomarker in monitoring human health at both individual and population levels for preventative healthcare (Nicholson, 2006).

Metabonomics to monitor the urine metabolome

The concept of metabonomics was first defined by Jeremy Nicholson as the 'the quantitative measurement of the dynamic multiparametric metabolic response of living systems to pathophysiological stimuli or genetic modification' (Nicholson *et al.*, 1999; Nicholson and Wilson, 2003). In metabonomics studies, metabolites in body fluids (e.g. blood, urine and saliva) under different genetic/physiological/environmental conditions are globally profiled with chemical screening techniques. Multivariate statistical analysis, such as principal components analysis (PCA), partial least squares discriminant analysis (PLS-DA) and orthogonal partial least squares discriminant analysis (OPLS-DA), of the profiles can help characterize the dynamics in the metabolite composition, identify biomarkers for different physiological or pathological status of the human body, and ultimately model the relationship of the genotype, phenotype and environments.

To maintain the internal homeostasis, the human body eliminates unwanted metabolites, which are produced in the host endogenous metabolism, gut microbiota activities and the co-metabolism of both parts, in the form of urine (Nicholson and Wilson, 2003). The biochemical composition of urine is sensitive to the internal and external perturbations to the body, and thus provides a good window for the host health assessment.

Gas chromatography–mass spectrometry (GC-MS), liquid chromatography–MS (LC-MS) and nuclear magnetic resonance (NMR) are mainly used in metabonomics studies. GC-MS technologies are highly sensitive in metabolite detection, and allow the identification and accurate quantification of metabolites (Fernie *et al.*, 2004). Fancy *et al.* (2006) processed the urine samples from 11 health human individuals by digesting with urease, precipitating proteins, and derivatizing with pyridine and N,O-bis(trimethylsilyl)trifluoroacetamide (BSTFA), to profile the urinary metabolites with GC-MS. They showed that 200–300 urine metabolites were well resolved and quantified in a 90-minute run, and the reproducibility of the signal retention time could be optimized by appropriate baseline correction and time-warping methods, so the intra-sample variability is smaller compared to the variation between samples from different human individuals. GC-MS can be used in metabonomics studies where differences between different sample groups are explored by multivariate statistical analysis (Fancy *et al.*, 2006). Pasikanti *et al.* (2008) performed GC-MS analysis on trimethylsilyl (TMS)-derivatized human urine extracts, and identified 150 putative metabolites of different chemical classes. 95% of the detected metabolites showed good reproducibility, and the comparison of metabolite concentration quantified intra-day and inter-day showed the relative standard deviation was below 15%. The PCA and OPLS-DA analysis on the urinary metabolite profiles of 31 males and 22 female healthy volunteers revealed gender differences. Concentrations of succinic acid, 3-hydroxyhippuric acid, fumaric acid and fucose were higher in female urine, whereas concentrations of 2,4-dihydroxybutanoic acid and 2,3-butanediol were higher in male urine (Pasikanti *et al.*, 2008).

Compared to GC-MS, LC-MS-based techniques (capillary, high-performance and ultra-performance) can profile a wider array of molecules without derivatization, however, LC-MS-based metabolite profiling is being developed and improved to guarantee the data quality for metabonomics studies. Gika *et al.* (2007) evaluated the within-day reproducibility of LC-MS with a standard quality control (QC) sample of pooled human urine. The authors suggested that the first injections of samples, which are required before the main run to achieve the stability of the system, should be discarded due to the poor reproducibility in retention time. They also found that samples could be stable for 20h at 4°C, and the storage of samples at −20°C or −80°C for 1 month did not make difference in the LC-MS profiles (Gika *et al.*, 2007). The LC-MS related methodology for metabonomics research has been well reviewed (Theodoridis *et al.*, 2008).

To explore the biomarkers for the diabetes risk, Chen *et al.* profiled the metabolite of the spot urine samples collected from 23 prediabetic, insulin-resistant (IR) people and 28 healthy, insulin-sensitive (IS) individuals with UltraPerformance LC-MS (UPLC-MS). The PLS-DA analysis showed that the profiles of two people groups formed two distinct clusters and the discriminating variable was the ion with m/z 194.34, of which peak area was significantly lower in IR individuals. Using LC-MS^n experiments, micro preparation, Fourier transform ion cyclotron resonance mass spectrometry (FTICR-MS), GC retention index, database search, and generation of an isotope-labelled standard, the authors unambiguously identified this signal to be 3-hydroxyhippuric acid as the biomarker candidate for the diabetes risk. Because 3-hydroxyhippuric acid is formed by the co-metabolism of the gut microbiome and the liver, these results indicate that the gut microbiome plays an important role in insulin resistance (Chen *et al.*, 2008).

However, the LC/GC-MS techniques have relatively large variation in the chemical shift of signals produced in different batches with the identical instruments or by different labs with different instruments, and the variations in data from different batches/instruments may cover the inherent differences between sample groups, so the instability in data becomes a major liability for the LC/GC-MS-based techniques being applied to large scale samples collected from human populations.

Despite that NMR spectroscopy has lower sensitivity than GC-MS or LC-MS, it is non-invasive, high-throughput, with stable chemical shifts, containing information on both metabolite composition and molecular structure, and requiring no complex sample preparation such as deproteination or derivatization. Thus, NMR spectroscopy-based techniques have been used widely in metabonomics studies (Rezzi *et al.*, 2007).

Dumas *et al.* (2006) evaluated the analytical reproducibility and inter-instrument variability of proton NMR (^{1}H NMR) spectroscopy-based metabonomics for large-scale epidemiological studies by profiling the metabolites in 24-h urine specimens collected from three human populations of Aito Town, Japan ($n=259$), Chicago, USA ($n=315$) and Guangxi, China ($n=278$). To test the repeatability of the NMR over the 7-month data acquisition period, they analysed the profiles of the 48 aliquots of each of three quality control (QC) urine samples, and the PLS regression analysis showed there was no significant correlation between the variances in the QC spectra with time, indicating the analytical approach is stable. By blindly identifying the spectra of 134 pairs of split duplicate aliquots of urine samples with a hierarchical clustering tree analysis, it was found that the analytical reproducibility of the NMR screening was >98%, NMR spectroscopic data from different labs and instruments could be integrated due to the high reproducibility, and most classification errors were caused by inconsistencies in the urine sample handling procedures. The multivariate comparison of the urinary metabolites spectra of the three populations showed that the discrimination was related to genetic, dietary, and gut microbial factors. This study demonstrated that NMR-based metabonomics is highly suitable for large-scale population phenotyping in epidemiological studies.

Holmes *et al.* (2008) applied ^{1}H NMR-based metabonomics techniques in epidemiological research to explore the variation of the urine metabolic phenotype within and across human populations, and to find out the urine metabolites closely associated with the blood pressure (BP), which is a major risk factor for coronary heart disease (CHD) and stroke. They obtained the ^{1}H NMR spectra from two 24-hour urine samples for each of 4,630 human individuals, who aged 40–59 and belonged to 17 populations in China (Guangxi, Beijing and Shanxi), Japan (Aito town, Toyama, Wakayama and Sapporo), UK (Belfast and West Bromwich) and USA (Honolulu, Chicago, Baltimore, Jackson, Minneapolis, Pittsburgh, non-Hispanic and Hispanic in Corpus Christi). These populations have different lifestyles, especially diets, and different CHD/stroke rates or BP levels. Both the hierarchical clustering analysis (HCA) and PCA analysis of the spectra showed that the urine metabolite phenotypes of East Asian and westerners are significantly differentiated, Chinese and Japanese phenotypes are clustered separately, phenotypes of Chinese can be further divided into the north and the south clusters, but those of UK and USA populations overlap. Gender metabolic difference was detected but it was smaller than geographic differences. To be noticed, although Japanese living in Japan and Japanese Americans have similar genetic background, their urine metabolite patterns are discriminated and clustered with the local populations respectively. Besides, both in the eastern and western populations, the spectra can be differentiated between the people consuming different ratio of vegetable and animal protein (high vegetable/low animal protein vs. low vegetable/high animal protein), who have differing BP levels accordingly. The discriminatory metabolites for populations of different geographic locations and dietary habits were identified and quantified. The researchers correlated the mean 24-hour urinary excretion amounts of these metabolites with BP for individuals, and found formate is inversely correlated with the BP level, and alanine (directly) and hippurate (inversely) that reflect diet and gut microbial activities are associated with BP as well. This study proved that NMR spectroscopy-based metabonomics techniques provided powerful tools to identify biomarkers for diseases threatening the public health in epidemiological investigations in which large amounts of samples have to be accurately measured (Holmes *et al.*, 2008).

Recently, Law *et al.* (2008) used GS-MS, LC-MS and ^{1}H NMR-based techniques coupled with PCA and OPLS-DA to study the effects of green tea consumption on the human urine metabolome. Although the GS-MS, LC-MS and ^{1}H-NMR profiles all showed that urine metabolome before and after the green tea ingestion was different, the discriminating metabolites detected by the three techniques were different, with LC-MS and ^{1}H-NMR results having some overlapping, but GC-MS providing a different set of metabolite signatures. The combination of these powerful analytical instruments produces complementary data which can give a more comprehensive picture of the human metabonome (Law *et al.*, 2008).

Metagenomics to monitor the gut microbiome

The evidence showing that the balance of the gut microbiome is an important determinant for the host health is accumulating rapidly. The imbalance of the gut microbiome is involved in the aetiology of diseases not only in the bowel (e.g. inflammatory bowel diseases (Macfarlane *et al.*, 2009) but also other body parts or the whole system (e.g. autism and obesity; Cani *et al.*, 2008; Finegold *et al.*, 2002; Turnbaugh *et al.*, 2006). To reveal the impacts of the gut microbiota on human health, it is necessary to understand the associations between variations of genetic information content in the gut microbiome and host health status.

Metagenomics explores all genes from all members of the sampled microbial communities by directly accessing the total genetic material from environmental samples (Eisen, 2007), which can offer a holistic overview of the genetic and metabolic potentials of the sampled microbial communities without the isolation and cultivation of the constituent microbes (Handelsman, 2004).

Human gut metagenomic investigations treat the gut microbial community as a whole, that is, as a 'macro-organism', and sequence the genomic DNA of the human gut microbiome, seeking to understand how the microbiome varies with population, genotypes, disease, age, nutrition, medication and environment, and ultimately to find ways to modify it to improve human health.

To investigate the role of the gut microbiome in the onset and chronicity of Crohn's disease (CD), Manichanh *et al.* (2006) characterized the diversity of the gut microbiome of healthy donors and CD patients using a comprehensive metagenomic approach. They isolated genomic DNA directly from faecal samples of six healthy donors and six CD patients, and used the pooled DNA to construct one fosmid library for each group of subjects. 1190 clones that contained 16S rRNA gene from two libraries were sequenced, and the phylogenetic analysis showed that the diversity of phylum *Firmicutes* was reduced in the faecal microbiota of Crohn's disease patients when compared to healthy individuals.

Gill *et al.* (2006) carried out the shotgun metagenome sequencing of the distal gut microbiome of two American adults by Sanger sequencer. They analysed approximately 78 Mb of unique DNA sequence, and predicted 50,164 open reading frames. The metabolic function analyses of identified genes showed that the human microbiome has significantly enriched metabolism of glycans, amino acids, and xenobiotics; methanogenesis; and 2-methyl-d-erythritol 4-phosphate pathway-mediated biosynthesis of vitamins and isoprenoids.

Kurokawa *et al.* (2007) sequenced the faecal metagenome of 13 healthy Japanese individuals, including unweaned infant, weaned children and adults. They produced 727 Mb sequences from the shotgun clone libraries by Sanger sequencing, and identified 20,063 – 67,740 potential protein-coding genes. The gut microbiome of unweaned infants were simple but highly divergent among individuals in taxonomy and gene composition, while that of adults and weaned children were more complex in diversity but showed a high functional uniformity regardless of age or sex. The comparative metagenomics analysis revealed 237 gene families commonly enriched in adults/children (e.g. genes encoding enzymes for DNA repair) and 136 families in infants (e.g. gene involved in transport system), with a small overlap (e.g. gene involved in carbohydrate metabolism), suggesting that these gene sets encode the core functions of adult and infant-type gut microbiome. The authors also identified 647 new gene families to be present exclusively in human gut microbiome by analysing the orphan genes. In addition, a conjugative transposon family was found to explosively amplify in human gut microbiomes, indicating that the horizontal gene transfer may be active between intestinal microbes.

The application of conventional clone library construction and Sanger sequencing in metagenomics investigations is of relatively high cost and low throughput, but the second generation sequencing techniques (e.g. 454 pyrosequencing and Illumina-Solexa sequencing) overcomes the disadvantages and greatly advances the human gut metagenomic research.

Turnbaugh *et al.* (2009) characterized the faecal metagenome of adult female monozygotic and dizygotic twin pairs concordant for leanness or obesity, and their mothers. Analysis of 18 individuals yielded 2.14 Gb from their microbiomes by 454 pyrosequencing. A 'core microbiome' which is an array of shared microbial genes among all sampled individuals was identified and included genes involved in transcription, translation, and carbohydrate and amino acid metabolism. The obesity-associated gut microbiome was enriched for phosphotransferase systems for microbial processing of carbohydrates. 75% of the obesity-associated microbial genes were from Actinobacteria (compared with 0% of lean-enriched genes; the other 25% are from Firmicutes) whereas 42% of the lean-enriched genes were from Bacteroidetes (compared with 0% of the obesity-enriched genes).

Most recently, Qin *et al.* (2010) used high throughput Illumina-Solexa short-read-based deep sequencing to characterize the human faecal metagenome of 124 European adults. They obtained 576.7 Gb of the faecal genomic DNA, which is almost 200 times more than in all previous human gut metagenomic studies. After assemble the reads into contigs, 3.3 million non-redundant microbial genes were predicted, which is 150 times larger than the human gene complement. Each individual harboured 536,112 ± 12,167 prevalent genes, 294,110 genes were found in at least 50% of individuals and thus termed as 'common' genes, and 38% of an individual's gut metagenome is shared by others. The authors found the 'minimal gut genome' (functions necessary for a bacterium to thrive in the gut) including housekeeping genes (e.g. central carbon metabolism, amino acid synthesis, DNA polymerase, and secretory apparatus) and genes involved in adhesion to the host proteins and in harvesting sugars from host cell glycolipids. The 'minimal gut metagenome' (functions necessary for the homeostasis of the whole ecosystem) consists of genes encoding (pro)phage-related proteins and genes involved in degradation of complex sugars derived from host diets/intestinal lining and fermentation sugars. Besides, the analysis on the data of inflammatory bowel disease patients from the 124 European adults also confirmed the significant difference in the species composition of the gut microbiota between the patients and healthy individuals, which was previously revealed by 16S-rRNA-gene-based surveys. This study established an extensive catalogue of microbial genes in the human metagenome and also showed the power of massive short tag sequencing in revealing genomic variations in gut microbiota relevant to human disease and health (Qin *et al.*, 2010).

On the other hand, the National Institutes of Health (NIH) is sequencing the genomes of diverse bacteria of pure cultures from human microbiome, which is one important part of the Human Microbiome Project (HMP) (Nelson and White, 2010). The HMP aims to complete the genome of at least 900 bacteria which can provide a catalogue of reference genomes and aids researchers to assign random metagenomic sequences to a known bacterial species. They recently published the assembled and annotated genomes of 178 microbes (Bacteria and Archaea) sequenced with both 454 pyrosequencing and Illumina-Solexa sequencing. 547,968 polypeptides were predicted, and 30,867 of them were defined as unidentified ('novel') polypeptides that had both unmasked sequence length greater than 100 amino acids and no BLASTP match to any non-reference entry in the non-redundant subset. This set of reference microbial genomes allows the assignment of approximately 40% of random sequences from the gut microbiome to microbial organisms. The pan-genome analysis also showed that the current sequencing is far from saturating microbial species genetic data sets. This work set a high quality standard for generating reference genomes for the human microbiome investigations (Nelson *et al.*, 2010).

Actually, with the high throughput next-generation DNA sequencing techniques (e.g. Illumina-Solexa short-read-based deep sequencing) and the expanding reference microbial genome database, we can monitor the dynamics of the gut microbiome in human populations at different healthy/nutritional conditions, and disclose the complex symbiotic relationship between the gut microbiome and the host (Zhao, 2010).

Functional metagenomics

From the large numbers of data produced in metabonomic and metagenomic studies, we aim to recognize the microbial species/functions that are the most important in shaping the host metabotypes and affecting the host health, and to understand the molecular basis for the host–microbiome interactions.

We developed a 'transgenomic' approach to link gut microbiome and human metabolic phenotype (metabotype) variations. The rationale behind this methodology is that if a particular bacterial taxon plays an important role to the host, its interpersonal and intrapersonal variations would be correlated with variations of concentrations of urine metabolites which are affected directly or indirectly by the bacterium. For a seven-member Chinese family cohort, we obtained the fingerprints of the faecal bacterial composition with 16S-rRNA-gene-based PCR-denaturing gradient gel electrophoresis (DGGE), and profiled the urinary metabolites with NMR spectroscopy. To correlate the microbial species (represented by the DGGE bands) to urine metabolites, we constructed Orthogonal projections to latent structures (OPLS) models where ^1H NMR data were used to generate the matrix and individual DGGE band intensity was used as the orthogonal component (data were mean-centred and scaled to unit variance before modelling). The predictive performance of the OPLS models was evaluated by fivefold cross-validation, where the Q^2 value (goodness of prediction) was used to select DGGE bands of interest ($Q^2 > 0.4$). Ten bacterial species were identified to have significant metabolic interactions with the host. For example, *Faecalibacterium prausnitzii* population variation is associated with eight urinary metabolites of diverse structure, indicating that this species is a highly functionally active member of the microbiome, and influences numerous host metabolic pathways. Based on these results, we defined the concept of functional metagenomics as 'the characterization of key functional members of the microbiome that most influence host metabolism and hence health' (Li *et al.*, 2008). Although there are several limitations in this study, e.g. the human cohort in this study is quite small, and the DGGE technique is semiquantitative, has low resolution, and cannot detect bacteria accounting less than 1% of the community, it is still a successful proof of principle study that this transgenomic approach can be applied to the metagenomics and metabonomics data collected from large human cohort and intervention studies. Thus, the general basis of the host–microbiome symbiosis can be elucidated using a systems methodology.

Conclusions

The global function and health of human body is the results of the interactions of the human genome, the microbiome and the external environments. The systems understanding of the biology of the human superorganism is more an ecological than a molecular biology problem (Nicholson and Wilson, 2003). Thanks to the significant technological development in DNA sequencing, analytical chemistry, bioinformatics and statistics, we are now able to quantitatively and globally profile the metabolites in the urine and the microbial composition in the faeces using metabonomics and metagenomics approaches. Multivariate statistical analyses provide a powerful toolbox to mine the relevant biological information from the data matrix, so the health state of the whole human body under different environments can be systematically and accurately characterized at the molecular level. Systems understanding of the biology of the human superorganism with this functional metagenomics methodology may thus lead to a significant paradigm shift in modern medicine.

Future trends

The whole-body systems biology methods (e.g. the metabonomic analysis of the urine metabolites and the metagenomic sequencing of the gut microbiome) can be applied in large-scale epidemiological studies relevant to important issues in public health. The urine and faecal samples will be regularly collected over time from the same individual (longitudinal), or collected from populations with different health status (cross-sectional). By comparing the urine metabonome and faecal metagenome between healthy and diseased people, and by monitoring the changing trajectory of individual people from health to disease, we will discover patterns of gut microbiota and host metabotypes which are associated with the onset and progression of chronic diseases. These patterns can be used as a biomarker for diagnosis and prevention if they are the consequences of the diseases. They may be used as target for developing new therapeutics if gut microbiota works as an essential mediator for disease development. In addition, the health of one individual will be evaluated by comparing his data to those of human populations; if there is any healthy problem, the data can provide the clue on the origin of the abnormality, which may guide the appropriate therapy. Based on these epidemiological data, a preventative, predictive and personalized medicine system will be finally established.

The whole-body systems biology methods will also be used in the nutritional studies to understand the effects of nutritional ingredients in human diets on the regulation of physiology of the human superorganism. By monitoring the changes in the global metabolism and the gut microbiome of tested volunteers before and after the nutritional interventions, and correlating the metabolic changes to the microbiome functions with the trans-genomic approaches, we can determine the role of human gut microbiome in nutrition, i.e. how diets affect the microbial composition or functions of the gut microbiome, and how these microbial changes further influence the global metabolism of the host.

Systems biology research has a great potential to better understand traditional Chinese medicine (TCM), because both systems biology and TCM stress a lot on 'holism' rather than 'reductionism'. In TCM, various diseases and conditions are categorized into different syndromes based on a patient's clinical manifestations such as pulse and tongue images, which are also the emergent functions of the whole human body. However, the pulse and tongue images are determined by the TCM doctors based on their sensory observation and personal experience, and thus cannot be accurately quantified, making it difficult to explain the principles of TCM with modern science. In contrast, systems biology can be used to study the effects of TCM on the human/model animal body by quantitatively analysing the blood, urine and faecal samples with metabonomic and metagenomic techniques. Thus, systems biology will be capable to illustrate the working mechanisms of TCM in the whole body at the molecular level, and hence may bridge the old TCM with the modern medical science.

Web resources

NCBI BLAST of microbial genomes (http://www.ncbi.nlm.nih.gov/sutils/genom_table.cgi)

1000 prokaryotic genomes are now completed and freely available in the Genome database of NCBI. The BLAST search enables a researcher to compare a query sequence (e.g. the amino acid sequences of different proteins or the nucleotides of DNA sequences) with the reference sequences in the database, and identify reference sequences that resemble the query sequence above a certain threshold.

RDP (http://rdp.cme.msu.edu/)

The Ribosomal Database Project (RDP) provides ribosome related data and services to the scientific community, including online data analysis and aligned and annotated Bacterial and Archaeal small-subunit 16S rRNA sequences. Until Aug 30, 2010, there are deposited 1,418,497 16S rRNA gene sequences in RDP database.

Greengenes (http://greengenes.lbl.gov/cgi-bin/nph-index.cgi)

The greengenes web application provides access to the current and comprehensive 16S rRNA gene sequence alignment for browsing, blasting, probing, and downloading. The data and tools presented by greengenes can assist the researcher in choosing phylogenetically specific probes, interpreting microarray results, and aligning/annotating novel sequences. The 16S rRNA gene database and workbench is compatible with ARB, so ARB users can use greengenes to keep their own local database.

Sapelo Island Microbial Observatory (http://simo.marsci.uga.edu/public_db/analysis_tools.asp)

Sapelo Island Microbial Observatory (SIMO) database is designed for archival, synthesis, and distribution of data from SIMO research to the microbial ecology community, consisting of all information about samples, clones, 16S rRNA gene sequences, and other SIMO research products. Information derived from analysis of 16S rRNA gene sequences using standard bioinformatics tools, such as the Ribosomal Database Project's Sequence Match program and sequence alignment tools in the FASTA package, is also stored in the database to provide taxonomic context.

UniFrac (http://bmf.colorado.edu/unifrac/)

The website provides the UniFrac method for computing differences between microbial communities based on phylogenetic information of rRNA (gene) sequences. UniFrac measures the phylogenetic distance between microbial communities, which are represented by sets of taxa, in a phylogenetic tree as the fraction of the branch length of the tree, and compare many communities simultaneously using clustering and ordination techniques.

Kyoto Encyclopaedia of Genes and Genomes (http://www.genome.jp/kegg/)

Kyoto Encyclopaedia of Genes and Genomes (KEGG) is an integrated database resource consisting of 16 main databases, which are classified into genomic information, chemical information, and systems information. Genomic and chemical information represents the molecular building blocks of life in the genomic and chemical spaces, respectively, and systems information represents functional aspects of the biological systems that are built from the building blocks, such as the cell and the organism. KEGG has been used as a reference knowledge base for biological interpretation of the metagenomics sequencing data and for metabolite identification of the signals in metabonomics profiles.

Evolutionary genealogy of genes: Non-supervised Orthologous Groups (http://eggnog.embl.de/)

Evolutionary genealogy of genes: Non-supervised Orthologous Groups (eggNOG) is an integration of the COG and KOG databases and consists of orthologous groups (OGs) of genes. The OGs are annotated with functional description lines (derived by identifying a common denominator for the genes based on their various annotations) with functional categories (i.e. derived from the original COG/KOG categories). The eggNOG's database currently counts 224,847 OGs in 630 species of fungi, metazoa, insects, vertebrates, mammals, bacteria and archaea, covering 2,242,035 proteins (built from 2,590,259 proteins) of which 1,966,709 are annotated.

Human Metabolome Database Version 2.5 (http://www.hmdb.ca/)

The Human Metabolome Database (HMDB) is a freely available electronic database containing detailed information about small molecule metabolites found in the human body. It is intended to be used for applications in metabolomics, clinical chemistry, biomarker discovery and general education. The database contains or links the chemical, clinical and molecular biology/biochemistry data. The database (version 2.5) contains over 7900 metabolite entries.

PubChem (http://pubchem.ncbi.nlm.nih.gov/)

PubChem provides information on the biological activities of small molecules. It contains substance information, compound structures, and BioActivity data in three primary databases, Pcsubstance (more than 69 million records), Pccompound (more than 27 million unique structures), and PCBioAssay (more than 434,000 BioAssays), respectively. PubChem can be used for the identification of metabolites in metabonomics studies.

METLIN (http://metlin.scripps.edu/)

METLIN is a mass spectral metabolite database for metabolomics containing metabolites and their mass, chemical formula, and structure. Over 15,000 structures are included in the database. Each metabolite is conveniently linked to outside resources such as KEGG for further reference and inquiry.

References

Backhed, F., Ding, H., Wang, T., Hooper, L.V., Koh, G.Y., Nagy, A., Semenkovich, C.F., and Gordon, J.I. (2004). The gut microbiota as an environmental factor that regulates fat storage. Proc. Natl. Acad. Sci. U.S.A. *101*, 15718–15723.

Backhed, F., Manchester, J.K., Semenkovich, C.F., and Gordon, J.I. (2007). Mechanisms underlying the resistance to diet-induced obesity in germ-free mice. Proc. Natl. Acad. Sci. U.S.A. *104*, 979–984.

Campbell, C.T., and Campbell, T.M. (2005). The China Study: The Most Comprehensive Study of Nutrition Ever Conducted and the Startling Implications for Diet, Weight Loss and Long-Term Health, 1. (Dallas, TX: Benbella Books).

Cani, P.D., Amar, J., Iglesias, M.A., Poggi, M., Knauf, C., Bastelica, D., Neyrinck, A.M., Fava, F., Tuohy, K.M., Chabo, C. *et al.* (2007a). Metabolic endotoxemia initiates obesity and insulin resistance. Diabetes *56*, 1761–1772.

Cani, P.D., Hoste, S., Guiot, Y., and Delzenne, N.M. (2007b). Dietary non-digestible carbohydrates promote L-cell differentiation in the proximal colon of rats. Br. J. Nutr. *98*, 32–37.

Cani, P.D., Neyrinck, A.M., Fava, F., Knauf, C., Burcelin, R.G., Tuohy, K.M., Gibson, G.R., and Delzenne, N.M. (2007c). Selective increases of bifidobacteria in gut microflora improve high-fat-diet-induced diabetes in mice through a mechanism associated with endotoxaemia. Diabetologia *50*, 2374–2383.

Cani, P.D., Bibiloni, R., Knauf, C., Waget, A., Neyrinck, A.M., Delzenne, N.M., and Burcelin, R. (2008). Changes in gut microbiota control metabolic endotoxemia-induced inflammation in high-fat diet-induced obesity and diabetes in mice. Diabetes *57*, 1470–1481.

Chang, C., and Yang, H. (2004). A correlation between pulse diagnosis of human body and health monitoring of structures. Earthquake Engineering and Engineering Vibration 3, 117–125.

Chen, J., Zhao, X., Fritsche, J., Yin, P., Schmitt-Kopplin, P., Wang, W., Lu, X., Haring, H.U., Schleicher, E.D., Lehmann, R. *et al.* (2008). Practical approach for the identification and isomer elucidation of biomarkers detected in a metabonomic study for the discovery of individuals at risk for diabetes by integrating the chromatographic and mass spectrometric information. Anal. Chem. *80*, 1280–1289.

Chung, H.Y., Cesari, M., Anton, S., Marzetti, E., Giovannini, S., Seo, A.Y., Carter, C., Yu, B.P., and Leeuwenburgh, C. (2009). Molecular inflammation: underpinnings of aging and age-related diseases. Ageing Res. Rev. *8*, 18–30.

Consultation, W.H.O. (2000). Obesity: preventing and managing the global epidemic. World Health Organization Technical Report Series.

Csete, M.E., and Doyle, J.C. (2002). Reverse engineering of biological complexity. Science *295*, 1664–1669.

Dumas, M.E., Maibaum, E.C., Teague, C., Ueshima, H., Zhou, B., Lindon, J.C., Nicholson, J.K., Stamler, J., Elliott, P., Chan, Q. *et al.* (2006). Assessment of analytical reproducibility of 1H NMR spectroscopy based metabonomics for large-scale epidemiological research: the INTERMAP Study. Anal. Chem. *78*, 2199–2208.

Eisen, J.A. (2007). Environmental shotgun sequencing: its potential and challenges for studying the hidden world of microbes. PLoS Biol. *5*, e82.

Fancy, S.A., Beckonert, O., Darbon, G., Yabsley, W., Walley, R., Baker, D., Perkins, G.L., Pullen, F.S., and Rumpel, K. (2006). Gas chromatography/flame ionisation detection mass spectrometry for the detection of endogenous urine metabolites for metabonomic studies and its use as a complementary tool to nuclear magnetic resonance spectroscopy. Rapid Commun. Mass Spectrom. *20*, 2271–2280.

Fernie, A.R., Trethewey, R.N., Krotzky, A.J., and Willmitzer, L. (2004). Metabolite profiling: from diagnostics to systems biology. Nat. Rev. Mol. Cell. Biol. *5*, 763–769.

Finegold, S.M., Molitoris, D., Song, Y., Liu, C., Vaisanen, M.L., Bolte, E., McTeague, M., Sandler, R., Wexler, H., Marlowe, E.M. *et al.* (2002). Gastrointestinal microflora studies in late-onset autism. Clin. Infect. Dis. *35*, S6-S16.

German, J.B., Hammock, B.D., and Watkins, S.M. (2005). Metabolomics: building on a century of biochemistry to guide human health. Metabolomics *1*, 3–9.

Gika, H.G., Theodoridis, G.A., Wingate, J.E., and Wilson, I.D. (2007). Within-day reproducibility of an HPLC-MS-based method for metabonomic analysis: application to human urine. J. Proteome Res. *6*, 3291–3303.

Gill, S.R., Pop, M., Deboy, R.T., Eckburg, P.B., Turnbaugh, P.J., Samuel, B.S., Gordon, J.I., Relman, D.A., Fraser-Liggett, C.M., and Nelson, K.E. (2006). Metagenomic analysis of the human distal gut microbiome. Science *312*, 1355–1359.

Hammond, D. (2003). The Science of Synthesis: Exploring the Social Implications of General Systems Theory (Boulder, CO, University Press of Colorado).

Handelsman, J. (2004). Metagenomics: application of genomics to uncultured microorganisms. Microbiol. Mol. Biol. Rev. *68*, 669–685.

Holmes, E., Loo, R.L., Stamler, J., Bictash, M., Yap, I.K., Chan, Q., Ebbels, T., De Iorio, M., Brown, I.J., Veselkov, K.A. *et al.* (2008). Human metabolic phenotype diversity and its association with diet and blood pressure. Nature *453*, 396–400.

Hotamisligil, G.S., Peraldi, P., Budavari, A., Ellis, R., White, M.F., and Spiegelman, B.M. (1996). IRS-1-mediated inhibition of insulin receptor tyrosine kinase activity in TNF-alpha- and obesity-induced insulin resistance. Science *271*, 665–668.

Kell, D.B. (2004). Metabolomics and systems biology: making sense of the soup. Curr. Opin. Microbiol. *7*, 296–307.

Kurokawa, K., Itoh, T., Kuwahara, T., Oshima, K., Toh, H., Toyoda, A., Takami, H., Morita, H., Sharma, V.K., Srivastava, T.P. *et al.* (2007). Comparative metagenomics revealed commonly enriched gene sets in human gut microbiomes. DNA Res. *14*, 169–181.

Law, W.S., Huang, P.Y., Ong, E.S., Ong, C.N., Li, S.F., Pasikanti, K.K., and Chan, E.C. (2008). Metabonomics investigation of human urine after ingestion of green tea with gas chromatography/mass spectrometry, liquid chromatography/mass spectrometry and (1)H NMR spectroscopy. Rapid Commun. Mass Spectrom. *22*, 2436–2446.

Lederberg, J. (2000). Infectious history. Science *288*, 287–293.

Li, M., Wang, B., Zhang, M., Rantalainen, M., Wang, S., Zhou, H., Zhang, Y., Shen, J., Pang, X., Zhang, M. *et al.* (2008). Symbiotic gut microbes modulate human metabolic phenotypes. Proc. Natl. Acad. Sci. U.S.A. *105*, 2117–2122.

Macfarlane, G.T., Blackett, K.L., Nakayama, T., Steed, H., and Macfarlane, S. (2009). The gut microbiota in inflammatory bowel disease. Curr. Pharm. Des. *15*, 1528–1536.

Manichanh, C., Rigottier-Gois, L., Bonnaud, E., Gloux, K., Pelletier, E., Frangeul, L., Nalin, R., Jarrin, C., Chardon, P., Marteau, P. *et al.* (2006). Reduced diversity of faecal microbiota in Crohn's disease revealed by a metagenomic approach. Gut *55*, 205–211.

McFall-Ngai, M. (2008). Are biologists in 'future shock'? Symbiosis integrates biology across domains. Nat. Rev. Microbiol. *6*, 789–792.

Nelson, K.E., and White, B.A. (2010). Metagenomics and its applications to the study of the human microbiome. In Metagenomics: Theory, Methods and Applications, Marco, D.E., ed. (Norfolk: Caister Academic Press), pp. 171–182.

Nelson, K.E., Weinstock, G.M., Highlander, S.K., Worley, K.C., Creasy, H.H., Wortman, J.R., Rusch, D.B., Mitreva, M., Sodergren, E., Chinwalla, A.T. *et al.* (2010). A catalog of reference genomes from the human microbiome. Science *328*, 994–999.

Nicholson, J.K. (2006). Global systems biology, personalized medicine and molecular epidemiology. Mol. Syst. Biol. 2, 52.

Nicholson, J.K., and Wilson, I.D. (2003). Opinion: understanding 'global' systems biology: metabonomics and the continuum of metabolism. Nat. Rev. Drug Discov. *2*, 668–676.

Nicholson, J.K., Lindon, J.C., and Holmes, E. (1999). 'Metabonomics': understanding the metabolic responses of living systems to pathophysiological stimuli via multivariate statistical analysis of biological NMR spectroscopic data. Xenobiotica *29*, 1181–1189.

Nicholson, J.K., Holmes, E., and Wilson, I.D. (2005). Gut microorganisms, mammalian metabolism and personalized health care. Nat. Rev. Microbiol. *3*, 431–438.

Pasikanti, K.K., Ho, P.C., and Chan, E.C. (2008). Development and validation of a gas chromatography/mass spectrometry metabonomic platform for the global profiling of urinary metabolites. Rapid Commun. Mass Spectrom. 22, 2984–2992.

Qin, J., Li, R., Raes, J., Arumugam, M., Burgdorf, K.S., Manichanh, C., Nielsen, T., Pons, N., Levenez, F., Yamada, T. *et al.* (2010). A human gut microbial gene catalogue established by metagenomic sequencing. Nature *464*, 59–65.

Rezzi, S., Ramadan, Z., Fay, L.B., and Kochhar, S. (2007). Nutritional metabonomics: applications and perspectives. J. Proteome Res .*6*, 513–525.

Theodoridis, G., Gika, H.G., and Wilson, I.D. (2008). LC-MS-based methodology for global metabolite profiling in metabonomics/metabolomics. TrAC Trends Anal.l Chemi. *27*, 251–260.

Turnbaugh, P.J., Hamady, M., Yatsunenko, T., Cantarel, B.L., Duncan, A., Ley, R.E., Sogin, M.L., Jones, W.J., Roe, B.A., Affourtit, J.P. *et al.* (2009). A core gut microbiome in obese and lean twins. Nature *457*, 480–484.

Turnbaugh, P.J., Ley, R.E., Mahowald, M.A., Magrini, V., Mardis, E.R., and Gordon, J.I. (2006). An obesity-associated gut microbiome with increased capacity for energy harvest. Nature *444*, 1027–1031.

Wei, H., Dong, L., Wang, T., Zhang, M., Hua, W., Zhang, C., Pang, X., Chen, M., Su, M., Qiu, Y. *et al.* (2010). Structural shifts of gut microbiota as surrogate endpoints for monitoring host health changes induced by carcinogen exposure. FEMS Microbiol. Ecol. 73, 577–586.

Wellen, K.E., and Hotamisligil, G.S. (2005). Inflammation, stress, and diabetes. J. Clin. Invest. *115*, 1111–1119.

WHO (2005). Preventing chronic diseases: a vital investment (Geneva, World Health Organization).

Willett, W.C. (2008). Overview and perspective in human nutrition. Asia Pac. J. Clin. Nutr. 17 Suppl. *1*, 1–4.

Yudkin, J.S., Juhan-Vague, I., Hawe, E., Humphries, S.E., di Minno, G., Margaglione, M., Tremoli, E., Kooistra, T., Morange, P.E., Lundman, P. *et al.* (2004). Low-grade inflammation may play a role in the etiology of the metabolic syndrome in patients with coronary heart disease: the HIFMECH study. Metabolism *53*, 852–857.

Zhang, C., Zhang, M., Wang, S., Han, R., Cao, Y., Hua, W., Mao, Y., Zhang, X., Pang, X., Wei, C. *et al.* (2010). Interactions between gut microbiota, host genetics and diet relevant to development of metabolic syndromes in mice. ISME J. *4*, 232–241.

Zhao, L. (2010). Genomics: The tale of our other genome. Nature *465*, 879–880.

Next-Generation Sequencing Approaches to Metagenomics

5

John Walshaw, Graham J. Etherington and Dan MacLean

Abstract

Next-generation sequencing approaches enable us to gather many more times sequence data than was possible a few years ago. Next-generation sequencers from the main vendors, Illumina, 454 and ABI SOLiD are distinct and varied technologies with unique approaches to sequencing that produce sequence reads with different strengths and weaknesses. We describe these technologies in detail and also discuss the applicability of the co-evolving new approaches for manipulation of reads and assembly and alignment of the high volume of reads that have also been developed to tackle the particular challenges of the shorter read format. By using next-generation sequence data in metagenomics experiments a wide range of new analyses are possible and we discuss the scope, goals, utility and practicality of these in modern laboratories including analysing environmental communities with partial assembly, clustering of taxonomically related reads to identify community structure, phylogenetic classification of unclustered reads and functional analyses that rely on identification of specific pathways or gene family members in samples. Metagenomic study has an increasingly powerful partner in the next-generation sequence technology and this partnership is likely to get more productive as software and hardware mature.

Introduction

Metagenomics has been energized by the rise of next-generation DNA sequencing technologies. The new platforms have not only enabled a profusion of new environmental studies, but have also changed the complexion of routine metagenomics analysis. The new sequencing technologies have of course greatly benefitted all sequencing enterprises, including the traditional assembly of genomes of single cultured organisms. In this chapter we will discuss the DNA sequencing machines and protocols that have emerged as the front-runners in what has been a frantic race for sequence yield and machine speed and the associated technologies and methods required to get the most from them. We will describe some of the drawbacks of the new technologies, notably the shorter read length, and how these challenges are approached. The new platforms have their own informatics issues associated with base calling, quality assessment and *de novo* assembly.

Environmental sequence sampling, in particular – where breadth and depth is usually paramount – stands to gain greatly from a step-change in the volume of sequence reads produced. Compared to the studies published in very recent years, traditional metagenomics could be considered to only scratch the surface. Previously, sequencing limitations meant that the usual focus was necessarily on a small complement of sequences from a library. Typically, these would result from screening for genes such as 16S rRNA which distinguish phylotypes (Rondon *et al.*, 2000), or from a functional screen such as an enzyme activity of interest (Henne *et al.*, 2000). While such a directed approach is still very powerful, the number of reads provided by next-generation technology now affords genome-wide sampling of many thousands of species, from a minimal number of sequencing runs. Although notable highly resourced projects using Sanger sequencing had earlier demonstrated these capabilities (Venter *et al.*, 2004), the later generation of technologies has made such studies widely affordable and relatively routine.

We will illustrate the impact of the new sequencing platforms by drawing upon a variety of metagenomics projects, which have applied the major technologies to a wide range of sampled environments. Metagenomics encompasses a collection of disciplines and different studies to date have had rather different aims. In some cases a detailed assessment of different phylotypes and biodiversity has been the goal, in others a functional profile of a community has been the priority, sometimes with regard to a narrow range of gene families. Some have focused on novel gene sequences, while others have even succeeded in assembling the almost complete genomes of a small number of new organisms.

The exciting possibilities for metagenomic data analysis has stimulated the production of dedicated software and databases, as well as the consolidated use of established tools and resources. Many of these resources have been designed without expectation of command-line computing or programming skills and are easy to use and accessible to all scientists. In this chapter we will discuss the new and powerful computational tools that have been developed specifically to work with next generations sequencing technologies in metagenomics, and the informatics choices to be made by the researcher when using them.

Next-generation sequencing

Next-generation techniques have delivered on their promise to sequence DNA many orders of magnitude faster than the highest speed capable only a few years ago. Such unprecedented sequencing capability brings with it a tidal wave of new technology and methods (Fig. 5.1). The major next-generation technologies for DNA sequencing are currently Roche's 454 brand, Illumina's Solexa brand and ABI's SOLiD machines which are all well established and have together been used in hundreds of published studies. The much newer SMRT sequencer from Pacific Biosciences is an important emerging technology that brings some exciting innovations to the field.

The 454, Solexa and SOLiD systems have all been around long enough to have gone through various machine updates and protocol improvements variously resulting in longer reads and greater yields and thus there is a range of instruments in the companies catalogues, we will consider only the latest machines from each company, the 454 GS FLX Titanium, the Illumina Solexa GA II and the ABI SOLiD 4hq.

The most mature technology available is Roche's 454 system which first became available in 2005 and even then was capable of producing more sequence data than 50 of the traditional 'Sanger' method ABI 3730 XL machines at around one sixth of the operating cost (Schuster, 2008). 454 is based on pyrosequencing, a sequencing-by-synthesis approach that relies on detection of pyrophosphate released as nucleotides are incorporated into a growing DNA strand.

The first step is to produce a population of DNA molecules to sequence, for 454 this often means fragmentation of long DNA strands into 300 – 800 base pair segments. Fragmentation can be avoided for samples with short DNA fragments, such as small RNAs or PCR amplicons. Once template is prepared oligonucleotide adaptors are ligated onto both the 5′ and 3′ ends of the template DNA strands and the resulting ligation product is bound to specially designed DNA-capture beads that carry oligonucleotides with sequence complementary to the adaptors, a single bead carries a single DNA fragment. The beads are next immobilized into oil–water droplets along with reagents for amplification to create an isolated micro-reactor in which template amplification takes place. By the end of the emulsion PCR amplification, millions of copies of the DNA fragment are present on each bead. The beads are then cleared of amplification reagent and isolated in picolitre volume wells on a specially designed plate for pyrosequencing proper to begin.

A complementary strand is synthesized by the sequential addition of each type of dNTP and DNA polymerase in the presence of luciferase and sulfurylase. Incorporation of a dNTP into the growing DNA strand releases a pyrophosphate

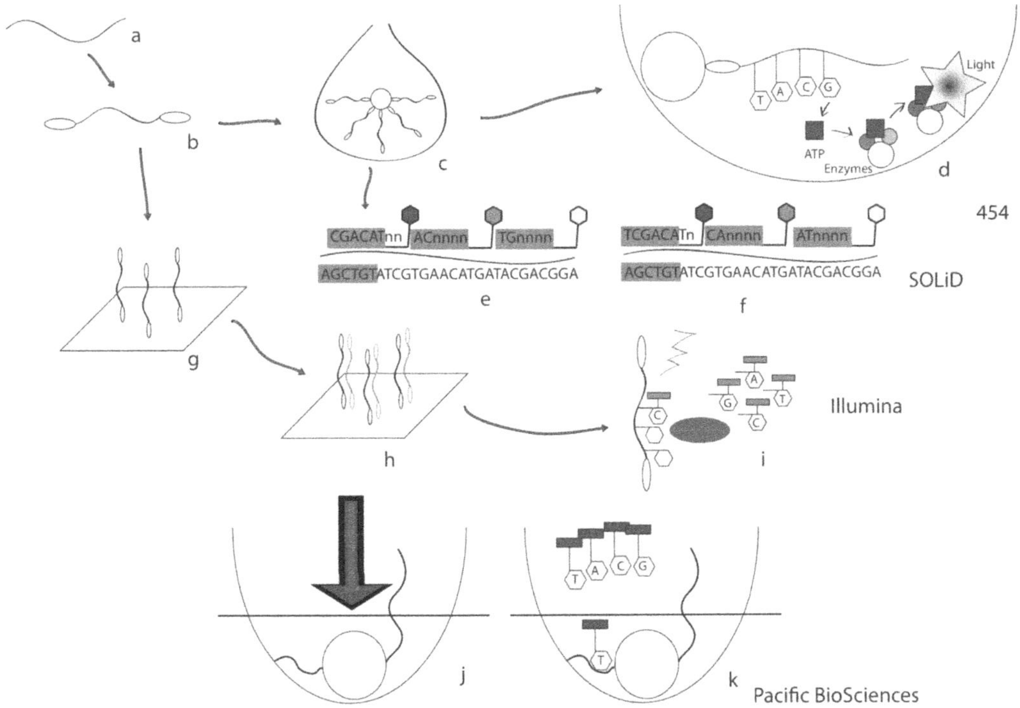

Figure 5.1 Sequencing processes for 454, ABI SOLiD, Illumina and Pacific BioSciences sequencing platforms. DNA (a) is prepared and adaptors are ligated (b). In 454 and SOLiD, sequencing amplification occurs in an oil droplet (c). With 454 sequencing, beads carrying amplified DNA are isolated in picolitre wells where the pyrosequencing reaction is carried out (d). In SOLiD sequencing amplified DNA is immobilized onto a substrate and fluorophore-bearing oligonucleotides, including one complementary to the sequence of the DNA adapter but overhanging the adapted strand, are ligated onto it (e), allowing one colour recording of the DNA bases paired to the end of the adaptors with the fluorophore. Melting away of these adaptors and annealing of new oligonucleotides is carried out offsetting the reading by one base and allowing for a second reading of each base (f). In Illumina sequencing the adapted DNA strands are bound to a substrate (g) and undergo a fold-back PCR amplification *in situ* (h), and then sequencing-by-synthesis using mixed populations of dye-terminator nucleotides (i). With Pacific BioSciences method, DNA is added to a zeptolitre well with an immobilized DNA polymerase molecule in a zone the light detection device can focus upon (j). Nucleotides bearing a fluorophore are incorporated individually and the light can only be detected as they pass through the focusing zone (k).

molecule, a substrate for sulfurylase which converts the pyrophosphate into ATP in the presence of adenosine 5′-phosphosulfate. The ATP then provides energy for the luciferase-based conversion of luciferin to oxy-luciferin, a process that releases light in proportion to ATP availability. The light is recorded using the machines CCD and the resulting image saved. The images constitute the raw data and are analysed computationally. Of all the current crop of next-generation sequencing machines the 454 generates the longest reads, averaging over 400 bp (web resource: About the latest 454 technology).

Like 454, initial amplification of template in ABI's SOLiD system occurs in a water-oil emulsion prior to sequencing. In SOLiD the beads are deposited onto glass slides, which act as support for the sequencing reaction. Sequencing proceeds with addition of the P1 primer, a primer that binds close to the ligation site with the sample DNA. Sequential addition of degenerate 8mer oligonucleotides with known 3′ dinucleotide and a corresponding fluorophore follows. Initially these probes can bind anywhere along the site, but a ligase stabilizes the interactions that occur adjacent to the primer. The fluorescence signal is read and then the probe is cleaved after position five. After five repeats of this process the growing complementary strand is melted away and a new primer is added which offsets by one nucleotide the position at which a new set oligonucleotides will be incorporated into the strand relative to the previous cycle. In this way the entire sequence is read twice. Unique to the SOLiD method, each fluorophore is not associated to a single nucleotide but a group of dinucleotides, because the four fluorophores are represented by 16 different dinucleotides in a degenerate code. ABI call this method their 'colourspace' analysis. The ABI SOLiD generates reads of up to 50 nt long at the moment.

Sequencing with Illumina's Genome Analyzer begins with immobilisation of fragmented DNA sequence onto the Illumina flowcell on which amplification and sequencing-by-synthesis occurs. The surface of the flowcell is covered in a dense lawn of oligonucleotides that have sequence complementarity to adapters ligated to the DNA. The amplification takes place in-situ by a mechanism known as bridge or fold back PCR, the DNA fragments, bound at their tail-end to the flowcell are free to fold over and bind to other flowcell oligonucleotides that have sequence complementarity to the adaptors at the free end of the DNA. Subsequent PCR allows a second strand synthesis that results in a DNA strand rooted to a spot in the vicinity of the first strand. In this manner 'clusters' of clonal DNA are produced for sequencing. The diameter of the clusters are dependent on the fragment size used, typically between 100 and 800 bp. Sequencing is performed by incorporating fluorescently labelled, reversible terminator nucleotides into the strand with a DNA polymerase. Imaging is carried out at the end of the synthesis cycle, after which the terminator is removed and a new cycle begins. Reads are routinely of 76 nt though support for 100 nt reads has been available for some time now.

The Helicos Heliscope method is somewhat different in that it does not require an amplification step, and can therefore lay claim to being a single molecule sequencing technology, although otherwise it has many similarities with the Illumina process. In the Heliscope DNA fragments of 100–200 nt are attached to the surface of a microfluidics flowcell and foregoing the amplification step nucleotides bearing a bright fluorophore that increases detectability are incorporated. The extra-bright fluorophores are introduced one species at a time and incorporated by DNA polymerase into the strand. Images are recorded and analysed at this step to identify the nucleotide that was incorporated into the second strand before the cycle is repeated with the next species of nucleotide. Reads of up to 55 nt can be generated, though 35 nt is the average.

A coming but not yet available single-molecule technology that in many ways promises to be the most revolutionary for some applications is Pacific Biosciences'. They use single molecules of DNA polymerase immobilized in zeptolitre (10^{-21} litre) wells to incorporate labelled nucleotides into a complementary strand from a single-stranded template. The wells, only nanometres in diameter and depth, allow for high concentrations of nucleotides that bear bright fluorophores. Imaging is carried out continuously by maintaining focus on the detection volume that contains the DNA polymerase allowing each newly incorporated base to be read in a short time. The use of phospholinked rather than base-labelled nucleotides means that the preceding fluorophore is removed after incorporation of the next base and diffuses out of the detection volume. Pacific Biosciences claim that this system can generate reads that are thousands of nucleotides long, at the expense of throughput.

Comparison of current next-generation technologies

All of the currently available technologies have limits on the lengths and the quality of the reads that can be produced. Common limiting factors are those that influence the integrity of the chemistry such as the degrading effects of lasers on DNA and enzymes and the effects of cyclical washes, which slowly reduce the amount of DNA sequenced. With the amplified methods in which readings are really being taken from a population of molecules, some of the molecules in the cluster do not get the next nucleotide incorporated. This results in the reading it gives lagging behind. These errors build up through the extension cycles and accumulate to a point where read quality is compromised and reading must terminate. Amplification steps can also introduce errors into the population of DNA that is being sequenced. All the technology providers mentioned above are constantly improving protocols and reagent mixtures to increase read lengths. The 454 method while very reliable overall has an extra limitation, most notably the lack of reversible terminator nucleotides means that more than one nucleotide can be incorporated into the complementary strand in one cycle and can cause problems in resolving homopolymeric stretches of sequence, as runs of n nucleotides are occasionally read as $n - 1$ nucleotides. The Pacific Biosciences technology is expected to suffer from insertion and deletion errors caused by the action of the DNA polymerase itself; continuous reading means that if the polymerase selects a nucleotide, but does not incorporate it, an extra base can be read. Incorporation of a nucleotide can also occur too quickly for the machine to read, resulting in deletion errors.

Image analysis and base calling pipelines

The raw data from sequencing are generally a set of images that record the intensity and wavelength of light at each collection point during the process, these must be analysed to infer the nucleotide, a procedure known as base calling. Base calling is done in different ways on the different platforms. In the 454 pyrosequencing reaction, the collection of images is timed so as to correspond to the addition of each nucleotide. The single wavelength light intensity is measured, since the intensity is proportional to the number of nucleotides, then the number of a single type of nucleotide that is incorporated can be calculated from the intensity. The accuracy of the calculation relies on the precise measurement of the light intensity but the system is prone to error at high values (that is when many of the same nucleotide are incorporated at once) so 454 data sets suffer in the sequencing of homopolymeric sections. Rather than present the result as a sequence, 454 data are usually presented as flowgrams, graphs that indicate the intensity of light measured at each introduction of a new nucleotide. By separating the base call decision for each nucleotide the flowgram allows errors on each to be isolated and called separately.

In the Illumina pipeline, an image is recorded at each cycle for each of the four different fluorophores, these must be registered (slightly scaled and shifted so that they overlie each other properly, differences in filtering and focusing put them slightly off) before base calling proper. The wavelengths of light emitted by the fluorophores upon excitation are not completely distinct, there is correlation between the A and C spectra and the G and T spectra. The spectral overlap means the signal must be corrected, which is done by the application of values from a cross-talk matrix that is estimated from the first and second imaging cycle, assuming that the four nucleotides are present in equal abundance across the clusters. A further complication is caused by the signal dependency of the current cycle on the previous cycle due to phasing, the dropping out of sequence of the growing nucleotide chain due to incomplete removal of terminators and fluorophores from the nucleotides or the skipping of nucleotide incorporation.

The most involved base calling method is the SOLiD colourspace analysis, which uses two-colour encoding for each base. By knowing the first base of the sequence (which is dictated by the sequencing primer) it is straightforward to calculate the rest of the sequence from the colour that is recorded. For example, knowing that the last adaptor nucleotide is T and the colour is red means that the first base to be sequenced must be A. Knowing that the first base is A and the colour is green means that the next base must be C and so on.

The vendor-supplied basecallers are not the only options, the limitations and assumptions inherent in all basecalling methods have led independent research groups to develop their own techniques that try to circumvent one or more of these issues in order to increase the number of bases that can be accurately called and thereby squeeze extra throughput from the machines. As may be expected, by solving one problem other problems are introduced. At present only alternative basecalling methods for the Illumina machine exist. Two of these, Alta-Cyclic (Erlich *et al.*, 2008) and IBIS (Kircher *et al.*, 2009), use a machine learning method and integrate with Bustard (the Illumina basecalling software) to increase yield and accuracy. Rolexa (a set of libraries for Illumina basecalling in the R statistical programming language)(Rougemont *et al.*, 2008), uses Gaussian models and BayesCall (Kao *et al.*, 2009) is based on a Bayesian framework that should be applicable to sequencing platforms other than Illumina. Tempting as extra sequence yield is, when choosing whether to use an alternative basecaller in a production pipeline labs should be wary of the potential difficulties that can be caused simply by changing the software. The software pipelines provided by the vendors are tightly integrated and replacing elements can be a difficult task in itself. Furthermore, the raw data from sequencing are usually image data, which can take up masses of valuable disk space. Often the images are condensed into a text file containing analysed image data on capture and then disposed of. If the external basecaller requires image data to operate, then the images must be stored somewhere, increasing the storage requirement for a run. If the basecaller relies on the intermediate files then the proprietary format of these leaves them prone to being altered by the vendor, leaving a user with a useless pipeline until the group that produced the alternative basecaller can catch up with the change. As a result the most common pipelines in use are the vendor supplied and developed ones. The yields from these are sufficiently large and accurate enough for most uses.

Quantifying sequencing error and pre-processing issues

At each stage of the base calling process, regardless of method, it is never possible to determine absolutely which base is found at that position in the process, rather the most likely base out of all the options is called. This uncertainty means that a measure of the quality of the call has been developed, the quality score. Each base in a sequence read has its own quality score. The concept is not new: long reads from older capillary sequencers have long had a quality scoring system. The *de facto* standard with these machines is that used by the phred base-caller and quality scores are generally in this format. The actual score assigned to a base by phred is a function of the probability of an error in the base call, the actual number of the quality score is generated by a log transformation of the calculated probability (Ewing and Green, 1998). Quality scores generally fall in the range between 0 (very poor confidence in the base called) and 40 (very high confidence in the base called), though there are some practical complications to this scheme (Cock *et al.*, 2010).

The sequence reads and quality scores are usually output from base callers into text files, in one or more formats. The simplest is the popular fasta format for sequence and the qual format for quality scores. Fasta format contains a single header line followed by one or more lines containing sequence, qual format files are comprised of a header line that matches the header of the corresponding fasta file followed by lines of quality scores. A more prevalent but obfuscated format is fastq. Both sequence and quality scores are in the same file, in consecutive entries. The quality scores are converted from a space-separated list of numbers to an unseparated list of characters that encode the number directly. The fastq format encodes the base qualities in single characters based on the ASCII code, a computing convention that assigns a unique number to each of the characters used in computing systems, for example the ! character has the decimal number 33 in the ASCII code (web resource: The ASCII Table). The fastq format is complicated by a number of factors, first the codes must be offset since the first 32 characters of the ASCII set represent non-printable characters like system alerts and the offset is slightly different depending on the origin of the file. Also the range of numbers that the quality score can fall into differs depending on the provenance of the file, even down to the version of the basecalling pipeline that was used. The difficulties in this popular format and how to deal with them are laid out better elsewhere, but it is important that any user of these technologies is aware of the potential for confusion (Cock *et al.*, 2010).

Some sources of DNA can prove difficult to sequence, environmental samples can have chemical contaminants such as protein or carbohydrate that are hard to remove from the DNA and can affect 'readability' of the DNA in the sequencer. Also quality of sequencing runs can vary and sometimes a run returns sequence of lower fidelity than previous or subsequent runs. For these reasons an often necessary first step once sequence has been collected is the pre-processing of reads. Pre-processing steps vary according to provenance and final use of the sequence. For cloned sequences or sequences obtained with the use of amplification or ligation adaptors, vector databases such as NCBI's UniVec (web resource: UniVec) or EMBL's emvec (web resource: EmVec) libraries are useful general resources for identifying artefactual sequences in reads. Homopolymer tracts are also easily identified and have limited use in downstream applications so should also be removed at the discretion of the user, as should reads with ambiguous bases (e.g. sequences carrying the letter N). For sequence with quality scores, some form of quality filtering is often useful, usually this is implemented by trimming reads past a point at which quality falls below some pre-determined threshold or discarding the read completely. Most of these steps can be accomplished easily by a competent bioinformatician and there are many tools for such tasks, including BLAST (Altschul *et al.*, 1997), RepeatMasker (Tarailo-Graovac *et al.*, 2009), DUST (Becher *et al.*, 2009), the EMBOSS tools package (web resource: EMBOSS), various tools in the Open Bioinformatics Foundation toolkits (such as BioPerl and BioPython's fastq parsing features)(web resource: Open Bio Foundation) and the very useful FastX toolkit (web resource: FastX Toolkit). Most of these tools require some command-line experience but there are more accessible web-fronted tools such as SeqTrim and the general and highly useful public Galaxy instance (web resource: Galaxy).

Next steps after sequence acquisition

Once sequence has been acquired the question of what to do next can, for metagenomic applications, be a vexed one. The entire problem is caused by the limited length of the sequence reads that have been generated which do not have sufficient length to be usefully informative on their own and must be used with either a reference sequence (often a fully sequenced genome) or must be assembled into longer sequence *de novo*.

Although there are now over 5010 sequenced (or 'in-progress') prokaryote genomes, and 1594 eukaryote (web resource: Genomes On-Line Database), we are still far from having even a representative sample of either, the sequence that is available in databases is likely to represent only a fraction of the sequence present in environmental populations and many sequences will be completely novel. As a result relying on existing databases to identify sequences is in many cases going to risk losing many of the data that have been acquired. Finding new sequences from short reads relies on the extension of the small fragments into longer contiguous sequence by *de novo* assembly.

Assembly *de novo* of short-read sequences

Sequence assembly *de novo* seeks simply to form longer sequences from short ones, accurately and efficiently without the assistance of any reference sequence. The long sequences that are produced, contigs, should span an entire chromosome but this is not often achievable because most eukaryotic and prokaryotic genomes contain regions that cannot be sequenced or assembled, leaving most genomes assembled into a few (or many) large fragments. *De novo* assembly is not a problem that has emerged with the next-generation sequencing, there are numerous well-established algorithms and tools for assembly of capillary sequence reads. These methods do not work with next-generation reads for several reasons: the distributions of sequencing errors are different making the prior error models inappropriate, they rely on overlaps between sequence reads that do not exist in short sequence reads and are unable to deal with the number of sequence reads.

Alongside the next generation sequencing machines, software that can perform assembly with shorter reads have been developed. These assemblers are based on just three central classes of algorithm and data structure, although the implementations of each can be quite different. The usefulness of each depends on the sequence source and read length to a great extent and the expected results can vary considerably between them.

A brute force approach and the most simplistic is the greedy algorithm, as implemented by SSAKE (Warren *et al.*, 2007), SHARCGS (Dohm *et al.*, 2007) and VCAKE (Jeck *et al.*, 2007). At a basic level these take reads and find overlaps with others, merge them to create contigs and continue to compare reads against the growing contigs until no more can be added. Typically the greedy assemblers suffer from the problem that extension ends when a read error is encountered or some contigs cannot grow because the reads are taken up earlier by other contigs. These methods are usually very sensitive to errors in reads and rely on very high coverage to make extensive contigs.

The most common approach to assembly of the Sanger-style long reads is overlap consensus (OLC) assembly, used by the Celera (Rusch *et al.*, 2007), Arachne (Chaisson and Pevzner, 2008), CAP (Huang and Yang., 2005) and phrap (web resource: Phrap) software. The process of OLC begins by calculating the k-mer content of all sequence reads, k-mers being the unique set of subsequences of length *k* of a read. A seed and extend approach, which identifies reads sharing k-mers and then extends out to find longer regions of similarity is taken to calculate overlaps between reads. From this an overlap graph is formed (a mathematical abstraction very similar to a network in which links are created between reads that overlap) and manipulated to create a basic assembly. Subsequent conventional multiple sequence alignment is used to refine the assembly and create a consensus from the original reads.

The Newbler assembler by 454 was originally an OLC assembler (Margulies *et al.*, 2005), however it is proprietary and closed source and recent updates suggest that the internal process may have changed substantially from that published. Nonetheless, many studies have found the 454 assemblies to be very useful, including human metabiome analyses (Human Metabiome Jumpstart Reference Sequence Consortium 2010) and gut microbiome in twins (Turnbaugh *et al.*, 2009).

The Celera assembler (originally created for Sanger-style reads by the Celera corporation for its WGS human genome project) is another widely used tool and has been revised and modified specifically for 454 data. The new pipeline called CABOG has found utility in a large number of individual genomics studies, including the human genome project (Venter *et al.*, 2001) and metagenomics studies, notably the J. Craig Venter Sorcerer II ocean sampling expedition (Rusch *et al.*, 2007).

The shorter Illumina and ABI SOLiD reads are more difficult to assemble with the OLC approach and are typically assembled using de Bruijn graph assemblers, such as Velvet (Zerbino and Birney, 2008), ABySS (Simpson *et al.*, 2009), Euler (Chaisson and Pevzner, 2008), SOAPdenovo (Li *et al.*, 2010) and ALLPATHS (Butler *et al.*, 2008). In the majority of these assemblers the de Bruijn graph is an internal data structure that represents a network in which the linked entities are (usually) the k-mers and links are made between k-mers that overlap by $k - 1$ with a single nucleotide overhang. In perfect data (i.e. a situation with no read errors and with *k* long enough to include the longest repeat in a single k-mer) then it is theoretically possible to reconstruct the genome by following from k-mer to k-mer along the edges, passing through each k-mer once. Since next-generation sequencing data do contain errors and genomic repeats can be very long (many times longer than read length, let alone the length of *k* used) then the read errors can cause dead ends in the path that terminate extension of the contig, repeats in the genome can cause cycles in the graph. The different ways in which the assembly programs get around these problems is what separates them. A discussion of the details is beyond the scope of this chapter but most allow the application of coverage cut-off to eliminate erroneous reads, real k-mers should be represented multiple times within the sequence data. Making use of paired-end reads to impose a limit on the path length through the graph is also frequently applied. Techniques like bubble collapsing which remove some circuits within the graph and other simplifications that broadly reduce the number of k-mers and paths through the graph are also common. There is no 'one-stop' assembly solution for next-generation sequence and each assembler should be considered on its own merits and the needs of the project at hand. For a much more in-depth technical discussion of the workings of the assemblers see Miller *et al.*, 2010 (Miller *et al.*, 2010) and the primary papers describing the assembler.

Limitations of assembly

Notwithstanding the various levels of sophistication and clever design in the assembly programs all of these assemblers stumble and fail to reconstruct complete genomes from the input sequence for many reasons. Often insufficient coverage is the main factor limiting assembly length, uneven sampling from the genome can result in some regions of the genome not being represented in the sequence set and thus contigs cannot extend fully over them. In many genomes some regions cannot be sequenced in the first place because the DNA does not shear or clone or amplify, dependent on the protocol used so never appear in the sequence set to be assembled. Some regions cannot be assembled, even if they are represented by sequence reads, in particular, regions that contain repeats longer than the length of the read or the k-mer size used in assembly are difficult to disambiguate because they create cycles in graphs that are difficult to resolve. One technique to limit this problem is to use paired-end or mate-pair reads, reads with a partner read that is known to be placed on the genome within a defined distance from its partner. Such information provides an extra constraint for the placement of reads and contigs relative to each other and aids extension.

An additional complication is that the performance of each of the assemblers is sensitive to input parameter values, it is not usually possible to determine *a priori* the best parameter values (indeed, the 'best' parameter values may depend entirely on the final downstream application). Routinely in our laboratory we carry out assemblies at a wide-range of parameter values and compare the resulting contig sets with a view to processing the set of sequences downstream. This can take the form of clustering and removing identical redundant sequences to make a maximal set or using a set of control sequences such as commonly occurring orthologue sets, to aid assessment of the reconstruction of these core genes as an indicator of assembly success.

Benchmarking of assemblers has been carried out in primary publications and independent studies. Typically these are done on small prokaryotic genomes or idealized data, but are rarely compared on identical data. Farrer *et al.* (2009) carried out simulations of paired 30-nucleotide reads at 40 times deep genome coverage of a previously sequenced 6 Mb bacterial genome using a wide range of parameter values and both paired reads and unpaired assemblies. They assessed the resulting assemblies for length and accuracy of each contig. Using the assembler Velvet, they were able to assemble 96% of the genome at an error rate of 0.33% per nucleotide. N50 (a measure indicating the distribution of contig length) was over 8 kb. Using paired reads, N50 contig length was increased to 165 kb, and more than 93% of the annotated protein-coding genes recovered from these assemblies were intact, full length and had no errors.

The next-generation assemblers are still very new and are not yet developed to a stage that they are able to provide 'out-of-the-box' answers so assessment of assembly quality is vital. As for all scientific experiments comparison with positive controls is essential, in terms of genome assembly this may mean comparison with the most closely related genomes sequenced or assessment of the number of expected conserved protein coding genes within the assembly. Without reference sequence, it is possible to detect some types of misassembly, tools like amosvalidate (Phillippy *et al.*, 2008) have been developed to highlight errors in draught genome assemblies and should be used before making conclusions about genome structure. An assembly viewing tool, like Eagleview (Huang and Marth, 2008) can be useful for manual assessment of assemblies. When assembling environmental sequence reads, distinguishing the non-identical but highly similar gene sequences which occur in closely related organisms can create further difficulties. The number of metagenomics reports using assemblies is to date quite small, but growing and the continued development of better assemblers will help to bring this approach into greater use.

Next generation sequence alignment

Assembly of sequences need not be the first option, there is already a wealth of publicly available genome sequence (over 6000 in the GOLD database) (web resource: Genomes On-Line Database) and these can be used as 'reference genomes' with which a set of reads can be compared in the manner of a template. As with assemblers a new generation of software has emerged, some of it being produced by the machine vendors themselves, others from within the research community itself. The sequencer vendor software includes Illumina's GenomeStudio, 454's GS Reference Mapper and SOLiD's BioScope. The Illumina solution, GenomeStudio is part of the analysis pipeline included with the machine and will provide all steps from basecalling through alignment. The actual alignment is handled by the Eland software and variation detection by a component called CASAVA (for 'Consensus Assessment of Sequence and Variation'). The 454 software which comes as stand alone packages rather than as components of a pipeline carries out the same tasks as the Illumina equivalents. The 454 software is available on request and subject to a license.

The vendor supplied software does a great job of many core tasks, but much more complex analyses including parallel analyses, can be achieved programmatically using command-line tools and scripts that wrap or 'glue' the independent processes together. Community developed 'free' or open-source solutions for alignment of reads to a reference are powerful, popular solutions include BWA (Burrows-Wheeler aligner) (Li and Durbin 2009) from the Wellcome Trust Sanger Institute (and its slightly older cousin MAQ), Bowtie (Langmead *et al.*, 2009), NovoAlign (web resource: Novocraft), GMAP (Wu and Watanabe., 2005), RMAP (Smith *et al.*, 2008), Patman (Prufer *et al.*, 2008), SSAHA (Ning *et al.*, 2001) and more traditional solutions like BLAST (Altschul *et al.*, 1997) or BLAT (Kent, 2002). Each of these programs has a slightly different approach to the alignment problem though many of the next-generation specific ones take their speed from the creation of clever indexed databases that allow random access of the sequence within. Output from these aligners can often be huge and recently the community has started to standardize on common formats for alignment storage. The SAM (sequence alignment map) (Li and Durbin., 2009) format is a generic format for alignments of read to a reference, again desktop GUI-based tools like the Integrative Genomics Viewer IGV (web resource: Integrative Genomics Viewer) are able to take advantage of this format but more can be achieved by bioinformatic automation or semi-automation using scripts and command-line tools. The samtools library is a programmer's library of functions that interact with the SAM format and its compressed binary alternative BAM. Samtools is served by a series of wrappers in varied high-level scripting languages. For Perl lovers there is Bio-Samtools, and Picard is available if your favourite flavour of programming is Java, although this seems less fully featured than the Perl variant and is more suited to converting datatypes. Pysam for Python and samtools-ruby for Ruby both allow the calling of samtools functions from within scripts written in these languages (all are obtainable from the samtools home page) (web resource: SAMtools).

Needless to say the informatics threshold to setting up working pipelines and IT infrastructure is quite high, particularly for groups in situations without significant support in this vein already. Thankfully there are alternatives to downloading and installing lots of different applications onto local machines and building them into your own pipeline. Web-based solutions fronting major compute hardware and cutting-edge tools are now available from publicly funded sources. Galaxy (web resource: Galaxy) is a web-based integrated bioinformatics pipeline and graphical user interface with built in support for many of the most widely used next-generation sequence tools. Galaxy has a point and click interface with which users can create their own re-usable workflows for analysis and re-use. With Galaxy it is very straightforward to create in a few clicks a generic metagenomics workflow which could take next generation sequencing reads, assess their quality, select high quality segments, conduct database searches and process taxonomic data. The Galaxy tool was used by Pond *et al.* (2009) to examine the metagenomics of windshield splatters from two separate journeys in the north-east USA. The pipeline involved 16 steps similar to those above; import of data – selection of high-quality reads – megablast against various databases – merge results – filter out suboptimal hits – fetch taxonomic information – find the lowest diagnostic rank and finally build phylogenetic tree. All steps of which could be run consecutively from start to finish without human intervention beyond workflow initialisation. Excitingly, this is the first publication that features a live online supplement providing access to not only the data but also the complete analysis pipeline used by the authors.

The Galaxy workflow can be downloaded and used to do the same analysis on different datasets.

Metagenomics in the next-generation sequencing era

The goals of metagenomics and next-generation sequencing

Despite new assembly capabilities and the growing number of sequenced genomes available as reference, the new metagenomics is, in general, not yet a route to sequencing the complete genome of unculturable organisms. Sparse genome coverage can be expected of the least abundant species in a community. The presence of a spectrum of very closely related microbes results in many near-identical reads whose assembly into distinct parent genomes is difficult or impossible using traditional methods. However, in extremely skewed populations dominated by a very small number of abundant species, almost complete genome sequences can be obtained. For example, community sequencing of a microbial biofilm from a mine drainage conduit resulted in two near-complete genomes of acidophilic bacteria (Tyson *et al.*, 2004). This was possible also because the differences between the small number of related species was limited to local polymorphism and the results of homologous recombination. In fact, these two genomes as well as three others partially assembled were obtained from only 76 Mb resulting from the end-sequencing of 50,000 clones (using capillary technology).

Many environmental communities however consist of a range of taxa that is both broad and deep. Several kingdoms and numerous phyla may be present and simultaneously there could be richly represented genera of closely related species and strains whose phylogeny has witnessed all manner of genomic rearrangements, insertions and deletions, and which may exhibit numerous polymorphisms even within one strain. Consequently, the assembly of complete genome sequences is not usually the aim of metagenomic studies. Instead, the goal will typically be either a detailed community analysis in terms of the abundance of different organisms, or of the genes present and inferred pathways.

It should also be noted that traditional metagenomics using only phylotypic marker sequences remains very much a current technique, either on its own (He *et al.*, 2010) or alongside total community DNA sampling (Warnecke *et al.*, 2007).

Communities analysed by the next-generation metagenomics approach

Communities have been sampled from a very broad range of environments. Those sequenced by next-generation technologies include bulk soil (Leininger *et al.*, 2006), mycorrhizae (Lumini *et al.*, 2009), coral reefs (Wegley *et al.*, 2007), the human gut and stools (Qin *et al.*, 2010), permafrost-preserved mammoth bone (Huson *et al.*, 2007), and even the super-expressway aerial community – sampled by scraping debris from a moving vehicle's front bumper (Pond *et al.*, 2009). Notable metagenomics studies using first-generation methods included open oceans (Venter *et al.*, 2004), coastal waters (Culley *et al.*, 2006), coastal stromatolites (Goh *et al.*, 2009), insect guts (Warnecke *et al.*, 2007), and acid mine biofilms (Tyson *et al.*, 2004). Rock-dwelling bacterial communities have also been sampled both before and after exposure to the vacuum, temperature and solar radiation of low earth orbit (Olsson-Francis *et al.*, 2010).

Inevitably, some studies have focused on particular taxonomic groups of interest. In some cases, the screening takes place at the sample purification stage; for example, filtration and centrifugation has been used to isolate virus particles from ocean samples prior to pyrosequencing (Angly *et al.*, 2006). In other numerous studies limited to ribosomal RNA gene analysis, taxon-specific primers are used to amplify regions of 16S rRNA genes (Rondon *et al.*, 2000). Clearly, in the case of whole-community sampling, a taxonomic focus can also be applied at the post-sampling, post-sequencing, informatics stage. Equally, this applies to screening by gene type; community DNA sequences from ocean surface water have been used in conjunction with large numbers of sequences obtained by more traditional methods, to perform a comprehensive analysis of prokaryote protein kinases (Kannan *et al.*, 2007).

Viral, prokaryote and eukaryote communities

Metagenomics to date has been almost completely a microbial discipline, and mostly prokaryotic. Eukaryote genomes pose sampling problems, since they are large but have a low density of gene-coding sequence. Communities of choice for successful shotgun metagenomic studies, even those using next-generation sequencing, therefore tend to be those dominated by bacteria and archaea, or viruses. The traditional metagenomics approach of targeting taxonomically informative sequence is also available for eukaryote genomes: for example, amplification of a region of the 18S rRNA gene followed by pyrosequencing has been used to sample the arbuscular mycorrhizae fungal community of soil managed under different agricultural regimes (Lumini *et al.*, 2009). A 'windshield splatter' analysis broke ground by specifically addressing the question of whether randomly sampled DNA could be used to identify eukaryote genomes; although the great majority of the unambiguously resolved reads were from prokaryotes, a variety of viridiplantae and metazoan organisms were identified along with microbial eukaryotes (Pond *et al.*, 2009). The identity of unambiguous hits is in all cases determined by the availability of genomes or transcriptomes in the databases, which we will discuss later.

Metagenomics as a complementary methodology

Frequently, metagenomic sampling is coupled with other protocols, some of which raise their own experimental and informatics challenges as well as benefits. Sampling the metatranscriptome does not merely indicate which genes are actually expressed; but transcripts from species present in low abundance may reap more signal (i.e. have greater molecular copy number) than genomic DNA, and so lead to the detection of taxa and genes that would otherwise be missed. A downside of metatranscriptomics is that isolated RNA tends to be swamped by ribosomal RNA (although that can aid greatly in identifying phylotypes, like rRNA gene sequences). Tandem sampling of the metagenome and metatranscriptome of the same community has been demonstrated as a powerful means of detecting the differential composition of several communities, using the metagenome as a reference representing the total expression potential (Frias-Lopez *et al.*, 2008;Gilbert *et al.*, 2008).

Of course, further experimental approaches are used as the study may demand; for example, a metagenomic approach might be coupled with the sequencing of cultured isolates where possible, such as in an analysis of endolithic cyanobacteria in coastal rock samples (Olsson-Francis *et al.*, 2010). The isolation and sequencing of substantial quantities of DNA from environmental communities is itself an important advance in genome sampling, since 90–99% of microbes are estimated to be unculturable (Kellenberger, 2001).

Analysis of metagenomic reads: a variety of approaches

Metagenomics in contrast to genome sequencing

Different metagenomics studies have witnessed markedly different approaches to processing sequence reads of environmental DNA, as best practices are established. To those immersed in traditional genome-sequencing projects using the DNA of a clone or even individual, it may seem counterintuitive that in numerous metagenomics studies, no attempt has been made to assemble the reads. However, in the traditional exercise of sequencing a single haploid genome containing relatively few gene duplications, the identification of reads corresponding precisely to overlapping regions of the same chromosome is – in principle – straightforward. Mismatches occur due only to sequencing errors, highly conserved paralogous genes and repetitive sequence; consequently, the pairwise sequence identity of overlaps exhibits a sharp peak at around 99–100%. Even the publicly funded human genome sequence, famously constructed from the DNA of around 100 individuals (International Human Genome Sequencing Consortium, 2001), was possible by an intensive effort which saw the patching together of a mosaic of genomic data.

The origin of the reads obtained will depend on the purification and amplification protocols used prior to sequencing; but even with a conservative, focused approach, such as amplification of rRNA genes of a limited taxonomic group, and using a fraction of one plate, genomes of a vast number of individuals will be sampled. In the unlimited whole-metagenome approach, the situation is greatly complicated by closely related but different members of gene families as well as related organisms, besides the usual problems of intraspecific polymorphism. Individual microbial species will generally be manifest as many different strains, genera as closely related species, families as numerous genera, and so on. Something of a continuum of sequence similarity is therefore expected, and indeed a distribution of pairwise identity exhibits a much fatter tail than that obtained from single-genome reads. Moreover, widely different coverage levels for different genomes can be expected. Therefore genuine *de novo* assembly of reads, to attempt reconstruction of individual genomes or even fragments thereof, is not possible in general; notable exceptions involving relatively uniform populations exist (Tyson *et al.*, 2004). Even the greater disambiguation offered by longer reads is of limited help for contig construction, as seen in studies that used a large volume of capillary sequencing reads (Venter *et al.*, 2004). The existence of closely related reference genomes aids the assembly process. Near-complete genomes have been assembled in this way, such as several marine Microviridae phages from pyrosequencing of community DNA (Angly *et al.*, 2006); but will still generally result in mixed sequence mapped to the template.

Assembly versus clustering of sequence reads

Meaningful contigs and scaffolds can be constructed, but generally will represent a richer blend of genomic information than their counterparts from classic genome sequencing. Their content will also depend upon the relative abundances of the species and strains sequenced, which in most cases will be far from uniform. This poses problems for previously well-tuned assembly algorithms, which expect the varying levels of coverage of different regions of a single genome to match a predictable distribution. Assembly programs sometimes assume that very high coverage represents repetitive sequence, whereas in the metagenome, the highest coverage can represent the most abundant organisms. To recalibrate the automated analysis of the Celera assembler (Myers *et al.*, 2000), the problem was tackled by manual intervention in the initial assembly of contigs in order to re-estimate the expected levels of coverage (Venter *et al.*, 2004). Manual curation has also been required for reconnection of mixed-species contigs, which exhibit too much variation for traditional assembly tools to consider them as single assemblies.

Often it is neither necessary nor usually possible to reconstruct genomic sequence from metagenomic data, it may be sufficient to use sequence alignment tools, rather than sequence read assembly programs. This may be especially so if the aim of an investigation is to identify gene sequences (including novel ones) rather than produce a detailed taxonomic analysis of a community. Reads may be aligned to produce clusters defined by levels of sequence identity, irrespective of the organism of origin. The CD-HIT package (Li and Godzik, 2006) was originally created for very fast clustering of similar protein sequences, with subsequent programs added for similar alignment of nucleic acid sequences. A tandem study of about 1.5 million metagenomic and metatranscriptomic reads from a coastal ocean mesocosm (Gilbert *et al.*, 2008) used the CD-HIT-EST program to produce clusters of both read types collectively, from which open reading frames were then predicted. Inevitably, many clusters had more than one ORF predicted, often in different reading frames, but also some in-frame, producing a high frequency of putative ORF per read. Some of these could be rejected as spurious since they overlapped with ORFs in other frames which were supported by matches with database sequences; however the supported ORFs, numbering half a million, were still in the minority. Around 2 million additional, unsupported ORFs were predicted. Translated putative protein sequences were then clustered with CD-HIT to produce a non-redundant set of putative proteins, using a high identity threshold; subsequent lower-identity clustering was used to perform a protein family analysis.

Analysis of unclustered read sequences

A contrary approach to handling metagenomic (and metatranscriptomic) sequence reads is to use the individual reads themselves, unclustered, for comparison with known sequences in databases. A simple assumption is that the number of sample reads representing a particular organism or gene is approximately proportional to the abundance of that organism or gene in the community. In this line of enquiry, it is therefore essential to retain redundancy in the set of sequence reads; collapsing many copies of near-identical sequences into a single cluster would lead to spuriously low estimates of the abundance of that moiety. Quantitative results for the incidence of particular gene/protein families in the metagenome can be obtained by straightforward sequence similarity searches of protein sequence databases. For example, another tandem metagenomic/metatranscriptomic study of ocean water (Frias-Lopez *et al.*, 2008) used BLASTX (Altschul *et al.*, 1997) to query approximately 400,000 pyrosequencing reads (mean length 110 bp) against the peptide sequences in the NCBI non-redundant database as well as those predicted from the Global Ocean Sampling data (Yooseph *et al.*, 2007). The metagenomic gene frequencies were further used to normalize the metatranscriptomic data, to elucidate relative expression of genes in known metabolic pathways.

The identification of members of known gene families amongst community DNA is relatively straightforward in principle. Query sequences either closely match database sequences or they do not; those that make weaker matches can be treated as more distant homologues, observing the usual caveats necessitated by such relationships. Many reads will make no match at all, while appropriate search parameters will ensure that misidentification of genes is rare. Where a read very strongly matches two or more database sequences, the gene/protein identification is unaffected; all the best hits will essentially represent the same gene or protein.

In contrast, a taxonomy-accented analysis can be endangered both by highly conserved gene sequences and the biased content of the reference databases, especially if detailed taxonomic assignments are the aim. Many of the metagenomic reads will be of genome-specific sequences which are unknown, and so absent from all databases, even if closely related sequences are present. When almost equally good matches are made with several hits, it is questionable to assume that the best hit must represent the species from which the read originated, or even the most closely related species known in the database. Where highly similar database hits occur, which is the best may even depend on the precise search and scoring regime employed. Background levels of sequencing errors also lower the precision of database matches. Of course some genes are extremely highly conserved across related taxa, and so 100% identical matches with database sequences from several different species may occur. The short lengths of some reads also increase the chances of ambiguous matches.

Such dead heats in sequence matching represent one facet of the problem. Conversely, the sequence database composition may be impoverished in the immediate phylogenetic vicinity of the organism providing a particular metagenomic read. The strongest matching sequence in a database, even if it scores far better than the next hit and has a rather high score in absolute terms, may not be from the same species as the query – or even from the same family or still higher grouping. As noted in a recent study which specifically aimed to identify eukaryote genes as well as prokaryote from splatter community samples (Pond *et al.*, 2009), the breadth of the eukaryotic lineage is rather sparsely impacted by genome-sequencing projects; but the sampling of those taxa which have been focused upon, has thus far been very deep. In the June 2010 release of the EMBL-Bank nucleotide sequence database 26 [essentially the same as GenBank (Benson *et al.*, 2010) and DDBJ (Kaminuma *et al.*, 2010)], 155 million sequence records are placed in divisions representing taxonomic groups. Of these, 98% are in fact eukaryotic sequences (including genomic, expressed tags, etc.). However, almost a quarter of the eukaryote sequences are from two mammalian species, whereas just fewer than 5% are arthropod sequences for instance (the arthropod group being expected to turn up frequently in the splatter samples). Of the arthropod entries 80%

are from insect species, yet around half of these originate from only two families, which importantly include several completely sequenced genomes. Using such a reference database to analyse community DNA reads, there is therefore a danger that many insect – and even some non-insect arthropod – reads would make their best matches with mosquito or fruit fly sequences.

The lowest common ancestor approach

The problems described above apply equally to the case of analysing pre-clustered read sequences, but the dangers are especially acute when processing unclustered reads – where the desirable assumption is that the number of matches to sequences from a taxon is proportional to the incidence of that taxon in the sampled community. Therefore, a well-tuned treatment of matches between metagenomic query sequences and known sequences in databases is required.

One such approach is the Lowest Common Ancestor (LCA) method, which has been implemented in the MEGAN software (Huson *et al.*, 2007), designed to provide a taxonomic profile of a metagenomic sample (Fig. 5.2).

It requires a predefined taxonomic classification tree, in this case the NCBI taxonomy (Sayers *et al.*, 2010). A synonym for LCA might be 'most specific taxon': the method attempts to identify the lowest, that is the narrowest, least inclusive taxonomic grouping to which each sequence read can be exclusively assigned. In principle, the procedure is applicable to the results of querying any taxonomically annotated database by any search method; MEGAN is designed to process the results of BLAST (Altschul *et al.*, 1997) searches.

At the outset, the characteristics of a good

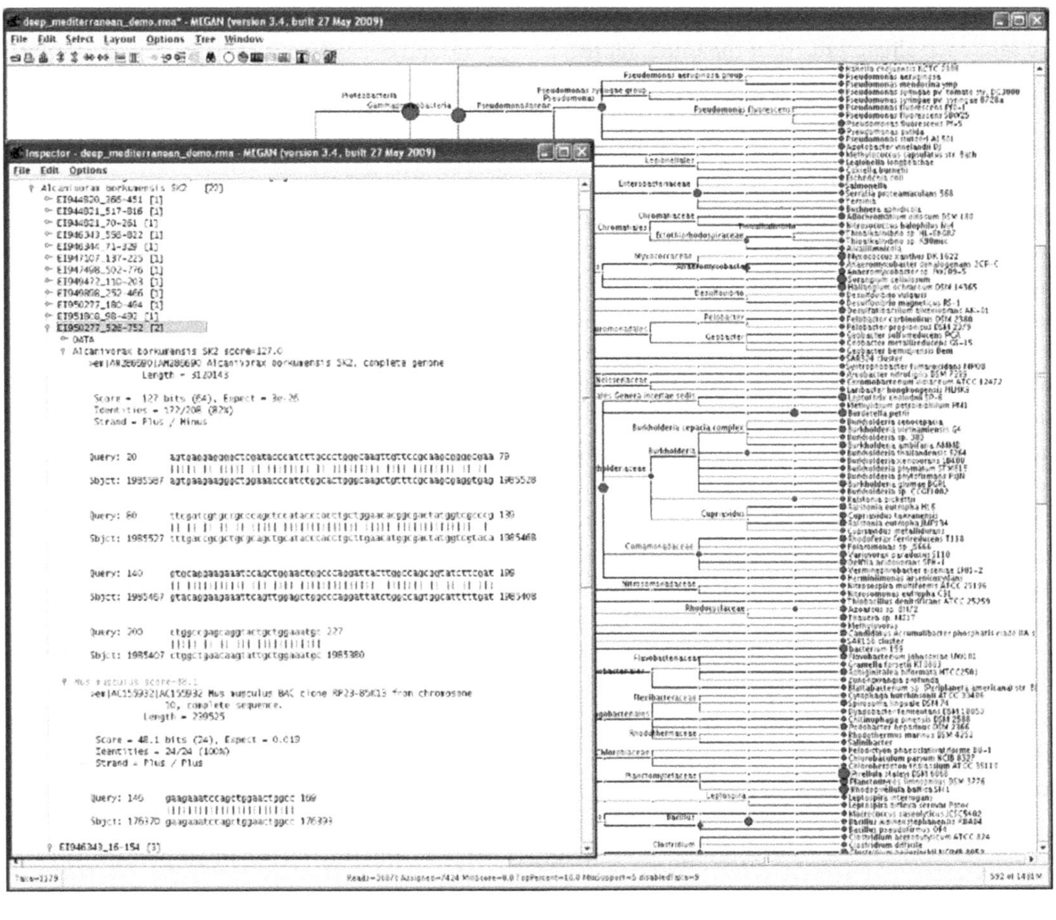

Figure 5.2 MEGAN analysis screenshot.

database match must be defined. This could be any combination of characteristics of a query-hit match. MEGAN uses the BLAST bit score, rather than the E (Expect) value, of the best high-scoring segment pair [for short query sequences, very low expect (N) values, where $N > 1$ can sometimes occur even when the sequence identity is relatively modest]. The bit score must therefore be higher than some defined threshold. Additionally, MEGAN specifies that a good hit must have a bit score within X% of the best score for that query. This criterion addresses the problem of matches with very highly conserved genes from many different species (as is often the case with 16S rRNA genes for example). That is, a database match with an extremely high bit score should not be considered as a taxonomic assignment to the query in the event that numerous other matches from different species have even higher scores. This does mean that care must be taken when setting the parameters of the BLAST search itself. MEGAN uses the bit scores from the sequence alignments (not the initial hit lists), so the limit on the number of alignments to display per hit sequence can determine whether a hit is considered at all, irrespective of bit score. These good hit parameters for MEGAN can be changed by the user (without the software needing to re-parse the BLAST output file); typical values might be a lowest-bit score of 35.0 and X=10. A single query can make a good match with no database sequences, or one database sequence, or more than one. Where good database matches occur, these may originate from one or more species or strains, as determined by the annotation in the FASTA header of the formatted BLAST database. MEGAN matches annotation keywords against a predefined taxonomy and can therefore determine the LCA of all the good hits of a particular query. The LCA may therefore be a strain, subspecies, species, genus, family, or indeed any taxonomic level (in extreme cases 'bacteria', 'eukaryote' or even 'cellular organism'). Every LCA is then scored by the number of queries to which it applies. However, a necessary additional parameter is the minimum number of good hits to a taxonomic group (the 'minimum support'). In order to eliminate the effect of occasional fluke matches which meet the good-hit criteria, a taxonomic group will only be considered as being present in the sample if at least this number of query reads match it. Again this is user-definable, and the default minimum support is 5 in MEGAN.

The result is a tally for every branch and leaf on the taxonomic tree. The leaves of the taxonomic tree are the most specific levels; an ideal analysis, resulting in a detailed resolution will therefore be one possessing a large number of leaves supported by many queries each. The MEGAN software is user-friendly and provides a choice of schematic graphical representation of the results. It also has a powerful tool for calculating and displaying the differences between two different samples' LCA results. MEGAN and its LCA approach are equally applicable to whole-metagenomic read sets and those obtained by amplifying and sequencing limited genes of interest, such as those coding ribosomal RNA subunits. It can also be used to process the results of searches of nucleotide or protein databases, so long as the taxonomic annotation is present; and is equally useful for metatranscriptome analysis.

Phylogenetic classification of unclustered reads

The LCA methodology described above maps reads to a database sequence or sequences, which have themselves already been anchored to a predefined taxonomic tree (by dint of their taxonomic annotation in the database). A variation on this approach is to dynamically construct a phylogenetic tree consisting of one or more reads and the known database sequences that they match. The CARMA algorithm (Krause *et al.*, 2008) uses the established Pfam methodology (Finn *et al.*, 2010) to match translated reads to protein sequence families characterized as profile hidden Markov models (pHMMs). To reduce the processing time, reads which have no chance of matching a Pfam pHMM are first screened out by a liberal BLASTX search of all the reads versus the full set of protein sequences in Pfam. Only those reads that have any hits are used for the next step. The taxonomic origin of each protein contributing to the Pfam pHMM is already known. These sequences and the matching translated read or reads from the query set (there may be several which match the same Pfam family) are used as input to a fast neighbour-joining method for inferring phylogeny (the NEIGHBOR program of the PHYLIP package; Felsenstein *et al.*, 2005). The taxonomy of the queries can then be inferred, to a greater or lesser degree of specificity, depending on the position of the translated reads on the tree relative to the known proteins. Therefore, CARMA returns data on both the taxonomic and functional family composition of the environmental sample. Additionally, a user-friendly web-based implementation has since been made available (Gerlach *et al.*, 2009), which produces graphical results depicting the abundance of several taxonomic ranks and of the protein families.

Functional analysis of metagenome sequence reads

The CARMA software described above provides a quantitative profile of the environmental DNA in terms of the functions of the genes sampled. The most recent versions of the MEGAN software provide similar data – in terms of the distribution of reads amongst known clusters of orthologous groups (COGs) (Tatusov *et al.*, 2003). As with the taxonomic analysis, the MEGAN results are determined by the annotation of the hits in the BLAST output files, and thus in the original database searched. Therefore, COG annotations need to be present in order to use this facility. MEGAN now also provides similar functionality regarding incidence of associated Gene Ontology (GO) (Ashburner *et al.*, 2000) terms in the read set.

Another very useful tool, available via a web interface, is the KEGG Automatic Annotation Server (KAAS) (Moriya *et al.*, 2007). This takes collections of sequence data as input and compares it against the sequences of genes in metabolic pathways, providing a graphical display of the pathway steps present in the query set. Although of great use for the annotation of genomic data, one mode of searching KAAS is designed for processing EST data, which is the most appropriate for the fragmentary nature of metagenomic sequence reads. However, as with the EST data, it is preferable to pre-cluster the reads (at a fairly conservative level of sequence similarity).

Assessing biodiversity using metagenomic data

Whether using assembled contigs or the proportional, unclustered-read approach, the read depth that can be obtained from the experimental sampling limits determination of the abundance of different organisms. Clearly, many species or genes actually present in the community will be omitted completely by any given sampling exercise if they are very scarce in the population. To determine the breadth of biodiversity, a means of estimating the true number of different species is therefore required.

One is the assessment of the discovery rate of the experiment, by repeating the analysis on randomly selected subsamples of incremental proportions, varying between 0% and 100% of the total set of sequences. This produces a rarefaction curve that indicates whether further sampling would result in new species being observed, and if so at what rate compared to the number of new reads. The MEGAN package has this capability and uses the number of leaves as the metric of discovery, recalculating this by the LCA algorithm for each successive subsample.

Determining biodiversity from environmental samples of phylotypic markers, such as 16S rRNA genes from metagenomes, is indeed a well-established technique in microbial ecology and alternative software tools are available for this purpose. Such methods do not necessarily require direct comparison of query data sets with a database by means of a similarity search.

The Ribosomal Database Project (RDP; Cole *et al.*, 2007) maintains a very large curated set of 16S rRNA sequences. These have been used to train a software tool (Wang *et al.*, 2007) which assigns query sequences to prokaryote taxa, this uses a naive Bayesian classifier, which is very fast but does not classify as many query sequences as would a similarity search against the RDP sequence database.

Predefined taxonomic trees such as the NCBI system (Sayers *et al.*, 2010) (used by MEGAN) are invaluable, but necessarily rely upon characterized species, genera etc. (usually of course making some use of DNA sequence information). Whereas, large-scale sampling of environmental communities can be expected to deliver many previously unknown strains or species, meaning that comparison with database sequences has its limitations. All of these considerations mean that from one viewpoint, taxa can define themselves: they are simply those that can be distinguished using the available sequence data. Those individual sequences which can be considered to be the same using an appropriate similarity threshold, therefore belong to one operational taxonomic unit (OTU). The OTU concept is especially pertinent to metagenomic samples; but the breadth and depth of OTUs can be expected to vary massively from one type of environment to another.

Numerous packages then are available for the analysis of biodiversity in terms of OTU richness. Unlike MEGAN, these algorithms are taxonomy independent, and effectively create their own taxonomy as appropriate for the data set. Generally, the methods use phylotypic markers as input, such as 16S rRNA gene sequences. One great challenge posed by such an analysis is that the marker sequences must be aligned. Relatively fast multiple alignment (using MUSCLE; Edgar, 2004) has been used in some analyses of 454 pyrosequencing reads, such as a hypervariable region of 16S rRNA genes of deep ocean and hydrothermal vent communities (Sogin *et al.*, 2006). In this case the alignment was tractable since it was applied to individual samples of a few thousand reads each. Multiple sequence alignment is generally infeasible for the numbers of reads that result from a full next-generation sequencing run. At the very least however, mutually exclusive groups of reads, whose members must fulfil some kind of pairwise similarity criterion, need to be obtained to identify OTUs. The exhaustive comparison of all pairs of sequence reads from the set produces a distance matrix (this is indeed one step in numerous multiple alignment algorithms). Such a matrix can then be used to cluster the reads into OTUs, by various procedures. The DOTUR software (Schloss and Handelsman, 2005) can use nearest neighbour, furthest neighbour or average neighbour (UPGMA) algorithms to do this. Like MEGAN, DOTUR also performs progressive random subsampling; for each of these samples, it calculates metrics variously indicating diversity (Simpson, 1949; Shannon and Weaver, 1949), OTU coverage (Chao *et al.*, 1992, 1993) and estimation of richness (Burnham and Overton, 1979; Chao, 1984; Smith and Vanbelle, 1984).

Obtaining and manipulating such an exhaustive pairwise distance matrix is intractable for a full run of second-generation reads; roughly 1 terabyte of memory (double that currently available to a single process in common cluster compute farms) would be required to hold the matrix (Sun *et al.*, 2009). Therefore, shortcuts must be taken to avoid calculating the distances between irrelevant pairs (i.e. those which are so dissimilar that they could not belong to the same OTU). The ESPRIT program (Sun *et al.*, 2009) does this by using incidences of k-mer sequences in the set of reads, to screen out unwanted pairs. In fact, only about 1–5% of pairs remain for the calculation of pairwise distances. In real terms, this sparse matrix still represents a large number of distances, and the software employs a novel clustering algorithm to process it efficiently with a relatively low memory overhead.

Software and databases for analysis of metagenomic reads: practicalities

Here we outline some possible pipelines that may be applied to different situations. The protocol followed will be determined partly by the nature of the sequences – for example, are they whole-metagenome samples, or amplified 16S rRNA gene sequence, of sequences amplified from a relatively narrow taxonomic range; and what is the line of biological enquiry. However, another important consideration is the power of the computational resources available to the researcher. Given the huge size of next-generation sequencing samples, even estimable computer clusters may now be seriously challenged by the application of some of the protocols described. On the other hand, even with very modest resources, there is usually the opportunity to tailor the analysis to the computer processors and disk storage at hand; some methods will produce somewhat less detailed results than others, but with the benefit of far greater speed.

The following are some suggestions for using a set of metagenomic reads to answer different biological questions. Other approaches may be just as appropriate, while alternative software packages exist to perform similar tasks.

Self-comparison for richness/ diversity calculations

All the analysis methods involve the comparison of the read sets with a database of some kind, even if the database is the set of reads itself (in other words for self-comparison). We have seen that some methods that establish a phylotypic profile, in terms of the numbers of different taxonomic groups, are self-contained and require no external data resource. Although the amount of self-comparison required is usually far from trivial, it is certainly feasible to perform high-quality analysis using everyday resources; for example the MOTHUR pipeline, which incorporates some of the tools described such as DOTUR and CD-HIT, was used to process a quarter of a million sequences of 16S rRNA gene fragments on a laptop computer in a few hours (Schloss *et al.*, 2009).

Detailed taxonomic analysis

In order to establish what the taxonomic groups actually are, the analysis must make recourse to established databases of some kind, although not necessarily directly. To continue on the theme of a sample of 16S rRNA gene sequences, one possibility is to compare all the reads against a curated ribosomal RNA sequence database, such as RDP (Cole *et al.*, 2007) or SILVA (Pruesse *et al.*, 2007). For a detailed analysis, obvious tools of choice are BLASTN or MEGABLAST. In one project for example, we used both programs to query 840,000 reads (mean length 190 nt) obtained by 454 pyrosequencing of a 16S rRNA hypervariable region amplified from an environmental sample, against the Ref edition of the small subunit (SSU, i.e. 16S rRNA) component of SILVA (which consists of approximately 900,000 sequences). Crucially, because most queries have a large number of well-matching hits, this is computationally demanding for BLASTN, which required thousands of CPU hours in total for this task. In general, MEGABLAST takes about a quarter of the time. Being optimized for detecting closely matching sequences quickly at the expense of finding more distant homologues, MEGABLAST is very suitable for taxonomic analysis of a read set.

The database matches so obtained can then be used to analyse the relative abundance of organisms in the community. The BLAST output above (either from BLASTN or MEGABLAST) can be input straight into a tool such as MEGAN. However, the choice of downstream analysis tool means that some thought may be required as to appropriate parameters for the search. When querying 16S rRNA against a 16S rRNA sequence database, a very large number of hits can usually be expected for most reads. Therefore, the default settings for the number of returned BLAST hits and alignments may be very inappropriate. As we have seen in the Lowest Common Ancestor approach that MEGAN uses, if a query is making high-scoring matches with 1000 database sequences, then completely ignoring the lowest 750 might strongly affect the LCA that is identified for that query. This is pertinent given that a significant number of these ignored hits might have a bit score within X% of the best; X% also being a key component of this algorithm.

Therefore, in this process we used essentially unlimited numbers of hits – and alignments, since it is the alignment (HSP) data that MEGAN parses. Although this software is capable of parsing some more compact BLAST output formats, in our case, the default format was used, to include the full annotation string (which is parsed to match to the predefined taxonomy). The key point here is then that some large output files were produced; the set of 0.8 million reads of 160 megabase total sequence, resulted in tens of Gb of BLAST output files which took several hours to load into MEGAN on a PC. 99.9% of the reads had BLAST hits in SILVA SSU-Ref, with almost 90% of these assigned by MEGAN to a particular bacterial taxon. The remainder were almost all assigned 'bacteria' as their lowest common ancestor, with a tiny proportion as 'cellular organism'.

However, a much faster solution for identifying the taxonomy of a set of 16S rRNA reads is available. The RDP classifier (Wang *et al.*, 2007) is available via a web interface with limited input data size, and also for local installation. It is extremely fast; for example, we used a test set of 100,000 reads of hypervariable region 16S rRNA gene sequences, which was processed in 15 minutes. This exercise does illustrate the adage that nothing is gained without cost: the RDP classifier assigns a taxonomy to a much smaller proportion (about 77%; the remainder were almost all denoted 'unclassified bacteria'). The RDP classifier does however avoid the need to store large BLAST output files.

Selecting a database for BLAST searches

If the BLAST route is to be taken for resolving (16S) rRNA gene reads, then a further consideration is the choice of database, which is non-trivial since the size of database will have a significant effect on both the search time and the size of output file, given that most queries will have many matching hits in the database, even if an extremely small E-value cut-off is employed. For bacterial/archaeal small subunit reads, either RDP or SILVA is suitable. Only the small subunit (SSU) component of SILVA need be searched, since SILVA also includes large subunit (LSU) sequences (admittedly these are far fewer in number). There is then a choice as to whether to search the comprehensive version of a database or a less redundant version. RDP is a comprehensive compilation of about 1.5 million archaeal and bacterial 16S SSU sequences; the full ('Parc') release of SILVA SSU is a little larger since it also contains eukaryote SSU (i.e. 18S rRNA) sequences. Both RDP and SILVA-Parc contain a good deal of redundancy, since sequences from closely related organisms may be identical or near identical. This means that searching the comprehensive database may be a waste of resources. A less redundant release of SILVA SSU is available (named 'Ref'), which is about one-third of the size and should result in approximately the same reduction in search time and output volume. The drawback of using the smaller database is that it could result in a less detailed resolution of a read's taxonomy. In general however, an approach like the LCA algorithm should be robust to the absence of sequences intermediate to those in the Ref database, notwithstanding that the total number of hits and those in the top section of the hit list could be affected.

Even with whole-metagenome samples, curated rRNA databases are very useful for processing that complement of the read set which represents ribosomal RNA genes. The curation of the RDP and SILVA databases results in high-quality, detailed annotations, which may be lacking from the original data source. There are many cases of sequence records in EMBL which are annotated at only the genus level, but which have their species, subspecies or strain included in the SILVA annotation.

When faced with a whole-metagenome read set, one approach is to query it against a general nucleotide sequence database such as EMBL, which will, for a proportion of the reads, identify the type of gene represented. The smaller subset identified as ribosomal RNA can then be examined in greater taxonomic detail if required by searching against an rRNA sequence database (however as described below, there are other strategies for quickly screening for rRNA gene sequences). SILVA may be more appropriate here because it contains sequences of both large and small subunit rRNA, as well as eukaryote sequences; but just focusing on the SSU prokaryote sequences may be felt to be sufficient to resolve phylotypic abundance.

When choosing a sequence database to compare whole-metagenome query sets against, there are further considerations. An entire nucleotide sequence resource such as EMBL may be appropriate for some studies. For example, in a soil metagenome, an abundance of prokaryote sequences are to be expected, along with fungal, oomycete, plant and metazoan. If the investigation is of prokaryote soil communities, then a suitable subsection of EMBL could be used. If the focus is on gene function and pathways, then a wide range of database sequences will be desirable. Given that many environments have been subject to metagenomic analysis for the first time relatively recently, it is unsurprising that many of the reads are not readily identifiable. Identification of more remote homologues may therefore be worthwhile in this kind of analysis to assign putative function (probably using a tool like BLASTN instead of MEGABLAST). It will also be recommended to identify putative coding sequences in the metagenome by BLASTX-searching a primary protein sequence database such as UniProt (Wu *et al.*, 2006), or using a specialized tool such as CARMA (Krause *et al.*, 2008).

When searching a primary nucleotide database such as EMBL, the inclusion or exclusion of large sets of 'tag' sequences, such as ESTs, should be weighed up. On the one hand, searching the tag sets can be expected to identify some reads which would otherwise remain unknown. On the downside, using EST sequences can massively increase the size of the database, and therefore search times, without producing a concomitant gain in the resolution of reads; given that many of the tags will be at least rather similar to genomic sequences in the same database. Inclusion of the tag components of EMBL may thus be liable to obfuscate the downstream analysis by MEGAN etc., and so a slimmer reference database is preferred in the general case.

Depending on the environment being sampled, it may be very worthwhile searching databases of other tag sequences – those of previous metagenomic and metatranscriptomic studies. Many publications have accompanying data sets available for download, including sequence reads and sequence clusters where appropriate. Clearly, where a more general database might draw a blank, a source of potential identification of newly sampled reads would be the sequence reads of earlier published studies of a similar environment. Shared repositories in which the research community can exchange data are growing rapidly. Some of these are extensions to established nucleotide sequence databases, such as the ENV (environmental sequence) component of the EMBL database. Others are dedicated to particular types of environment, such as the CAMERA repository, originally, for marine microbial ecology (Seshadri *et al.*, 2007), but since expanded to environmentally sampled microbes in general.

When querying high-volume, unclustered, whole metagenome read sets, the kinds of databases (such as EMBL) that might be required for searching are also very large; this can be computationally very expensive. Depending on the nature of the experiment, it may be worthwhile to use a priori knowledge to take shortcuts and reduce the search space by rapidly identifying the 'complete unknowns'. For example, when sequencing the metagenome of a particular microbial community, a number of essential genes can be expected to be found in significantly larger than trace proportions. These include numerous genes encoding enzymes, structural and other proteins and ribosomal and transfer RNA molecules. Of special use are genes or families of genes that may be expected to be present in multiple copies per genome. A reasonably liberal converse BLAST search using a very small collection of family sequences as queries and the metagenome reads as the database would be expected to identify reads representing each gene at low computational cost. Such an approach is useful for obtaining a subset of reads for further more detailed investigation, such as an enzyme or rRNA gene to be used as a phylogenetic marker; the identified sequences would thus be processed using one of the methods already described. Equally, it is useful for subtracting sequences of no interest from the read set. In a study of gene/protein function and pathways, most of the reads that are fragments of rRNA genes can be quickly eliminated. A judicious choice of query sequences for the converse BLAST can ensure that most of the relevant reads are picked up, but it is not essential to detect all of them; since the goal is to reduce the computational search time overall. This strategy is especially useful in metatranscriptomics, where the distribution of gene coverage is highly skewed (for example bulk RNA would be approximately 90% rRNA). The dividends in metagenomics are smaller, but may be worth pursuing if resources are limited.

With the current volumes of sequences being produced however, the computational cost of querying a whole metagenome read set against a general nucleotide sequence database should not be beyond even a relatively modest setup. In contrast to the BLASTing of rRNA reads against an rRNA database, many whole metagenome reads will have no hits, while many others will have only a very small number. By using k-mer incidence and other shortcuts, the BLAST algorithm discards many database sequences very quickly and does not waste time by aligning query–database pairs which are poor matches. For example, to process some metagenomic data we use a 'tag-less' subset of the EMBL database, currently consisting of 10 million sequences. We have found that whole metagenome read sets of mean length ~ 200 bp can be searched against this at the rate of about 10 reads per CPU second using MEGABLAST.

Future trends

Metagenomics is in a very exciting place right now; next-generation sequencing has brought about great new opportunities for analyses of greater depth and breadth than was conceivable a few years ago. Firstly, the prospects for metagenomics are tied tightly to the prospects of next-generation sequencing technologies and it is clear that the future is very bright indeed. Sequence machine vendors operate in a highly competitive high-stakes market and are driven to produce machines that provide greater numbers of ever-longer reads. The increasing coverage and length that can be achieved bodes well for all kinds of sequence related metagenomics analyses. An interesting development is the appearance of the USEARCH software (Edgar, 2010). This permits extremely fast searches compared with BLAST, of comparable accuracy, but at the cost of finding only the single best hit for each query, and of large memory requirements. This tool is likely to enable some alternative, fast approaches to database searching for metagenomics.

Currently, the breadth and depth of sampling of DNA from an environment is sufficiently high to afford the sound, cutting-edge metagenomic research projects which have proliferated recently. Sampling in metagenomics is often very far from comprehensive. Currently, we do not have a detailed understanding of the reproducibility of a particular metagenomics exercise, and of the full implications of differences and overlaps between the sequence sets obtained from two experiments. The situation will be greatly improved by the ever increasing output capacity of sequencing platforms. Broader and deeper sequencing could at least reduce the random sampling element of metagenomics. We may be in a position to resolve differences between different environments which are rather nuanced and impossible to assess soundly currently and this can be expected to greatly increase the power of environmental sequencing projects.

Exhaustive sequencing of a whole complex metagenome is likely to remain some way off, due to the potentially colossal numbers of different species, strains and sequences present. The range of environments whose sequencing can be 'almost complete' is likely to expand. Coupled with advances in algorithms designed to assemble distinct sequences from a jumble of closely related ones, this should permit the deconvolution of metagenomic DNA into individual complete (or near-complete) genomes, to a greater extent than is now possible.

One of the current barriers to making sense of a metagenomic sequence sample is the incompleteness and bias in the reference databases. The onward march of genome sequencing will improve this situation. One obvious area that needs correcting is the highly skewed complement of eukaryote genome sequences in the databases. A broader range of model organism sequencing will help to make more realistic identifications of a much larger proportion of eukaryote reads. Indeed, plant and metazoan metagenomics may come of age, with the reproducible detection of forensic-style traces of the DNA of larger organisms, as well as elucidation of genes from small non-microbial animals and plants, and from spores, pollen and seed. Environmental territory hitherto relatively untouched by DNA sequencing, may be opened up spectacularly.

Metagenomics itself will likely be instrumental in the strengthening of sequence databases. Given the huge diversity and abundance of microbes, and the difficulty in their culture, we will soon reach a situation where the larger proportion of the organisms known to the databases are derived from environmental samples which may be expected to greatly reduce the proportion of completely unidentifiable sequences.

In the same way that the study of a single genome sequence is greatly illuminated by a comparative genomic approach, it is likely that the real power of environmental sampling will lie in comparative metagenomics, coupled with transcriptomics. We will witness a significant increase in our ability to correlate differences in gene complement and expression, with differences in the characteristics of the ecosystem concerned. This will place the spotlight on the function of novel sequences and organisms which influence and are influenced by particular environmental characteristics.

Traditional metagenomics, focusing on one or a few particular kinds of genes which are already well characterized, will itself continue to advance as a result of the improving technologies. With an increasing knowledge of diversity between individuals, strains and species, we may for example be able to identify phylotypes from marker genes to a considerably greater degree of precision than we can now do. This will be facilitated by improvements in sequencing error rates and their assessment, as well as richer reference databases.

The improving technologies will of course bring future difficulties as well as benefits. One is the increased storage and computational power and bioinformatics expertise required to deal with the sequence, for now this remains a considerable barrier, particularly for smaller groups wishing to make a first entry into the field. However, the bar is being lowered. Core and external sequencing facilities, maintained by large core grants from central funding bodies are popping up to provide sequencing capacity and expertise as a service to the communities. Computational power and expense drops quickly and core facilities often provide this as part of, or in addition to sequencing. The cutting-edge software required is starting to come of age and amazing tools such as Galaxy (web resource: Galaxy) lower the perceived technical skills barrier, while the software itself becomes more standard and easier to use.

Eventually next-generation sequencing will appear to come full circle, average sequence read lengths will have increased to that similar to (or, more likely, greater than) that of capillary sequencers of old, while software and compute power, currently in a lag-phase, will have matured to be something that is in practise very much like a current common desktop computer task (even if in reality this is backed up by behind the scenes cluster or cloud computing hardware). Current metagenomicists would do well to think big now as these technologies have a habit of being upon us sooner than we expect and the current fervour in the industry is showing no signs of abating.

Web resources

About the latest 454 technology – http://454.com/about-454/index.asp
EMBOSS – http://emboss.sourceforge.net
EmVec – http://www.ebi.ac.uk/Tools/blastall/vectors.html
FastX Toolkit – http://hannonlab.cshl.edu/fastx_toolkit
Galaxy – http://main.g2.bx.psu.edu/
Genomes On-Line Database – http://www.genomesonline.org/
Integrative Genomics Viewer – http://www.broadinstitute.org/igv/
Novocraft – http://www.novocraft.com/
Open Bio Foundation – http://www.open-bio.org
Phrap – http://www.phrap.org/
SAMtools – http://samtools.sourceforge.net/
The ASCII Table – http://www.asciitable.com/
UniVec – http://www.ncbi.nlm.nih.gov/VecScreen/UniVec.html

References

Altschul, S.F., Madden, T.L., Schaffer, A.A., Zhang, J., Zhang, Z., Miller, W., and Lipman, D.J. (1997). Gapped BLAST and PSI-BLAST: a new generation of protein database search programs. Nucleic Acids Res. *25*, 3389–3402.

Angly, F., Felts, B., Breitbart, M., Salamon, P., Edwards, R., Carlson, C., Chan, A., Haynes, M., Kelley, S., Liu, H. *et al.* (2006). The marine viromes of four oceanic regions. PLoS. Biol. *4*, e368.

Ashburner, M., Ball, C.A., Blake, J.A., Botstein, D., Butler, H., Cherry, J.M., Davis, A.P., Dolinski, K., Dwight, S.S., Eppig, J.T. *et al.* (2000). Gene ontology: tool for the unification of biology. The Gene Ontology Consortium. Nat. Genet. *25*, 25–29.

Becher, V., Deymonnaz, A., and Heiber, P. (2009). Efficient computation of all perfect repeats in genomic sequences of up to half a gigabyte, with a case study on the human genome. Bioinformatics *25*, 1746–1753.

Benson, D., Karsch-Mizrachi, I., Lipman, D., Ostell, J., and Sayers, E. (2010). GenBank. Nucleic Acids Res. *38*, D46-D51.

Burnham, K., and Overton, W. (1979). Robust estimation of population-size when capture probabilities vary among animals. Ecology *60*, 927–936.

Butler, J., MacCallum, I., Kleber, M., Shlyakhter, I.A., Belmonte, M.K., Lander, E.S., Nusbaum, C., and Jaffe, D.B. (2008). ALLPATHS: *de novo* assembly of whole-genome shotgun microreads. Genome Res. *18*, 810–820.

Chaisson, M.J., and Pevzner, P.A. (2008). Short read fragment assembly of bacterial genomes. Genome Res. *18*, 324–330.

Chao, A. (1984). Non-parametric estimation of the number of classes in a population. Scand. J. Stat. *11*, 265–270.

Chao, A., and Lee, S. (1992). Estimating the number of classes via sample coverage. J. Am. Stat. Assoc. *87*, 210–217.

Chao, A., Ma, M., and Yang, M. (1993). Stopping rules and estimation for recapture debugging with unequal failure rates. Biometrika *80*, 193–201.

Cock, P.J.A., Fields, C.J., Goto, N., Heuer, M.L., and Rice, P.M. (2010). The Sanger FASTQ file format for sequences with quality scores, and the Solexa/Illumina FASTQ variants. Nucleic Acids Res. *38*, 1767–1771.

Cole, J.R., Chai, B., Farris, R.J., Wang, Q., Kulam-Syed-Mohideen, A.S., McGarrell, D.M., Bandela, A.M., Cardenas, E., Garrity, G.M., and Tiedje, J.M. (2007). The ribosomal database project (RDP-II): introducing myRDP space and quality controlled public data. Nucleic Acids Res. *35*, D169–72

Culley, A., Lang, A., and Suttle, C. (2006). Metagenomic analysis of coastal RNA virus communities. Science *312*, 1795–1798.

Dohm, J.C., Lottaz, C., Borodina, T., and Himmelbauer, H. (2007). SHARCGS, a fast and highly accurate short-read assembly algorithm for *de novo* genomic sequencing. Genome Res. *17*, 1697–1706.

Edgar, R.C. (2004). MUSCLE: a multiple sequence alignment method with reduced time and space complexity. BMC Bioinformatics *5*, 113.

Edgar, R.C. (2010). Search and clustering orders of magnitude faster than BLAST. Bioinformatics *26*, 2460–2461.

Erlich, Y., Mitra, P.P., delaBastide, M., McCombie, W.R., and Hannon, G.J. (2008). Alta-Cyclic: a self-optimizing base caller for next-generation sequencing. Nat Methods *5*, 679–682.

Ewing, B., and Green, P. (1998). Base-calling of automated sequencer traces using phred. II. Error probabilities. Genome Res *8*, 186–194.

Farrer, R.A., Kemen, E., Jones, J.D.G., and Studholme, D.J. (2009). *de novo* assembly of the *Pseudomonas* syringae pv. syringae B728a genome using Illumina/Solexa short sequence reads. FEMS Microbiol. Lett. *291*, 103–111.

Felsenstein, J. (2005). PHYLIP (Phylogeny Inference Package). http://evolution.gs.washington.edu/phylip.html.

Finn, R.D., Mistry, J., Tate, J., Coggill, P., Heger, A., Pollington, J.E., Gavin, O.L., Gunasekaran, P., Ceric, G., Forslund, K., Holm, L., Sonnhammer, E.L., Eddy, S.R., and Bateman, A. (2010). The Pfam protein families database. Nucleic Acids Res. *38*, D211–22.

Frias-Lopez, J., Shi, Y., Tyson, G., Coleman, M., Schuster, S., Chisholm, S., and Delong, E. (2008). Microbial community gene expression in ocean surface waters. Proc. Natl. Acad. Sci. U.S.A. *105*, 3805–3810.

Gerlach, W., Junemann, S., Tille, F., Goesmann, A., and Stoye, J. (2009). WebCARMA: a web application for the functional and taxonomic classification of unassembled metagenomic reads. BMC. Bioinformatics *10*, 430.

Gilbert, J., Field, D., Huang, Y., Edwards, R., Li, W., Gilna, P., and Joint, I. (2008). Detection of large numbers of novel sequences in the metatranscriptomes of complex marine microbial communities. PLoS One. 3, e3042.

Goh, F., Allen, M., Leuko, S., Kawaguchi, T., Decho, A., Burns, B., and Neilan, B. (2009). Determining the specific microbial populations and their spatial distribution within the stromatolite ecosystem of Shark Bay. ISME. J. *3*, 383–396.

He, Z., Xu, M., Deng, Y., Kang, S., Kellogg, L., Wu, L., Nostrand, J.V., Hobbie, S., Reich, P., and Zhou, J. (2010). Metagenomic analysis reveals a marked divergence in the structure of below-ground microbial communities at elevated CO_2. Ecol. Lett. *13*, 564–575.

Henne, A., Schmitz, R.A., Bomeke, M., Gottschalk, G., Daniel, R. (2000). Screening of environmental DNA libraries for the presence of genes conferring lipolytic activity on *Escherichia coli* Appl. Environ. Microbiol. *66*, 3113–3116.

Huang, W., and Marth, G. (2008). EagleView: a genome assembly viewer for next-generation sequencing technologies. Genome Res. *18*, 1538–1543.

Huang, X., and Yang, S. (2005). Generating a genome assembly with PCAP. Curr. Protoc. Bioinformatics *Chapter 11*, Unit11.3.

Human Microbiome Jumpstart Reference Sequence Consortium, Nelson, K.E., Weinstock, G.M., Highlander, S.K., Worley, K.C., Creasy, H.H., Wortman, J.R., Rusch, D.B., Mitreva, M. *et al.*, (2010). A catalog of reference genomes from the human microbiome. Science *328*, 994–999.

Huson, D., Auch, A., Qi, J., and Schuster, S. (2007). MEGAN analysis of metagenomic data. Genome Res. *17*, 377–386.

International Human Genome Sequencing Consortium (2001). Initial sequencing and analysis of the human genome. Nature *409*, 860–921.

Jeck, W.R., Reinhardt, J.A., Baltrus, D.A., Hickenbotham, M.T., Magrini, V., Mardis, E.R., Dangl, J.L., and Jones, C.D. (2007). Extending assembly of short DNA sequences to handle error. Bioinformatics *23*, 2942–2944.

Kaminuma, E., Mashima, J., Kodama, Y., Gojobori, T., Ogasawara, O., Okubo, K., Takagi, T., and Nakamura, Y. (2010). DDBJ launches a new archive database with analytical tools for next-generation sequence data. Nucleic Acids Res. *38*, D33-D38.

Kannan, N., Taylor, S., Zhai, Y., Venter, J., and Manning, G. (2007). Structural and functional diversity of the microbial kinome. PLoS Biol. *5*, e17.

Kao, W., Stevens, K., and Song, Y.S. (2009). BayesCall: A model-based base-calling algorithm for high-throughput short-read sequencing. Genome Res. *19*, 1884–1895.

Kellenberger, E. (2001). Exploring the unknown. The silent revolution of microbiology. EMBO Rep. *2*, 5–7.

Kent, W.J. (2002). BLAT – the BLAST-like alignment tool. Genome Res *12*, 656–664

Kircher, M., Stenzel, U., and Kelso, J. (2009). Improved base calling for the Illumina Genome Analyzer using machine learning strategies. Genome Biol. *10*, R83.

Krause, L., Diaz, N.N., Goesmann, A., Kelley, S., Nattkemper, T.W., Rohwer, F., Edwards, R.A., and Stoye, J. (2008). Phylogenetic classification of short environmental DNA fragments. Nucleic Acids Res. *36*, 2230–2239.

Langmead, B., Trapnell, C., Pop, M., and Salzberg, S.L. (2009). Ultrafast and memory-efficient alignment of short DNA sequences to the human genome. Genome Biol. *10*, R25.

Leininger, S., Urich, T., Schloter, M., Schwark, L., Qi, J., Nicol, G., Prosser, J., Schuster, S., and Schleper, C. (2006). Archaea predominate among ammonia-oxidizing prokaryotes in soils. Nature *442*, 806–809.

Li, H., and Durbin, R. (2009). Fast and accurate short read alignment with Burrows-Wheeler transform. Bioinformatics *25*, 1754–1760.

Li, H., Handsaker, B., Wysoker, A., Fennell, T., Ruan, J., Homer, N., Marth, G., Abecasis, G., and Durbin, R. (2009). The Sequence Alignment/Map format and SAMtools. Bioinformatics *25*, 2078–2079.

Li, R., Zhu, H., Ruan, J., Qian, W., Fang, X., Shi, Z., Li, Y., Li, S., Shan, G. *et al.*, (2010). *de novo* assembly of human genomes with massively parallel short read sequencing. Genome Res. *20*, 265–272.

Li, W., and Godzik, A. (2006). Cd-hit: a fast program for clustering and comparing large sets of protein or nucleotide sequences. Bioinformatics *22*, 1658–1659.

Lumini, E., Orgiazzi, A., Borriello, R., Bonfante, P., and Bianciotto, V. (2009). Disclosing arbuscular mycorrhizal fungal biodiversity in soil through a land-use gradient using a pyrosequencing approach. Environ. Microbiol. *12*, 2165–2179.

Margulies, M., Egholm, M., Altman, W.E., Attiya, S., Bader, J.S., Bemben, L.A., Berka, J., Braverman, M.S., Chen, Y., Dewell, S.B. *et al.* (2005). Genome sequencing in microfabricated high-density picolitre reactors. Nature *437*, 376–380.

Miller, J.R., Koren, S., and Sutton, G. (2010). Assembly algorithms for next-generation sequencing data. Genomics *437*, 315–327.

Moriya, Y., Itoh, M., Okuda, S., Yoshizawa, A.C., and Kanehisa, M. (2007). KAAS: an automatic genome annotation and pathway reconstruction server. Nucleic Acids Res. *35*, W182–5.

Myers, E., Sutton, G., Delcher, A., Dew, I., Fasulo, D., Flanigan, M., Kravitz, S., Mobarry, C., Reinert, K., Remington, K. *et al.* (2000). A whole-genome assembly of Drosophila. Science *287*, 2196–2204.

Ning, Z., Cox, A.J., and Mullikin, J.C. (2001). SSAHA: a fast search method for large DNA databases. Genome Res. *11*, 1725–1729.

Olsson-Francis, K., de la Torre, R., and Cockell, C. (2010). Isolation of novel extreme-tolerant cyanobacteria from a rock-dwelling microbial community by using exposure to low Earth orbit. Appl. Environ. Microbiol. *76*, 2115–2121.

Phillippy, A.M., Schatz, M.C., and Pop, M. (2008). Genome assembly forensics: finding the elusive mis-assembly. Genome Biol. *9*, R55.

Pond, S.K., Wadhawan, S., Chiaromonte, F., Ananda, G., Chung, W., Taylor, J., Nekrutenko, A., and Team, G. (2009). Windshield splatter analysis with the Galaxy metagenomic pipeline. Genome Res. *19*, 2144–2153.

Pruesse, E., Quast, C., Knittel, K., Fuchs, B.M., Ludwig, W., Peplies, J., and Glockner, F.O. (2007). SILVA: a comprehensive online resource for quality checked and aligned ribosomal RNA sequence data compatible with ARB. Nucleic Acids Res. *35*, 7188–7196.

Prufer, K., Stenzel, U., Dannemann, M., Green, R.E., Lachmann, M., and Kelso, J. (2008). PatMaN: rapid alignment of short sequences to large databases. Bioinformatics *24*, 1530–1531.

Qin, J., Li, R., Raes, J., Arumugam, M., Burgdorf, K.S., Manichanh, C., Nielsen, T., Pons, N., Levenez, F., Yamada, T. *et al.* (2010). A human gut microbial gene catalogue established by metagenomic sequencing. Nature *464*, 59–65.

Rondon, M., August, P., Bettermann, A., Brady, S., Grossman, T., Liles, M., Loiacono, K., Lynch, B., MacNeil, I., Minor, C. (2000). Cloning the soil metagenome: a strategy for accessing the genetic and functional diversity of uncultured microorganisms. Appl. Environ. Microbiol. *66*, 2541–2547.

Rougemont, J., Amzallag, A., Iseli, C., Farinelli, L., Xenarios, I., and Naef, F. (2008). Probabilistic base calling of Solexa sequencing data. BMC Bioinformatics *9*, 431.

Rusch, D.B., Halpern, A.L., Sutton, G., Heidelberg, K.B., Williamson, S., Yooseph, S., Wu, D., Eisen, J.A., Hoffman, J.M., Remington, K. *et al.* (2007). The Sorcerer II Global Ocean Sampling expedition: northwest Atlantic through eastern tropical Pacific. PLoS Biol. *5*, e77.

Sayers, E.W., Barrett, T., Benson, D.A., Bolton, E., Bryant, S.H., Canese, K., Chetvernin, V., Church, D.M., Dicuccio, M., Federhen, S. *et al.* (2010). Database resources of the National Center for Biotechnology Information. Nucleic Acids Res. *38*, D5–16.

Schloss, P., and Handelsman, J. (2005). Introducing DOTUR, a computer program for defining operational taxonomic units and estimating species richness. Appl. Environ. Microbiol. *71*, 1501–1506.

Schloss, P., Westcott, S., Ryabin, T., Hall, J., Hartmann, M., Hollister, E., Lesniewski, R., Oakley, B., Parks, D., Robinson, C.J. *et al.* (2009). Introducing mothur: Open-Source, Platform-Independent, Community-Supported Software for Describing and Comparing Microbial Communities. Appl. Environ. Microbiol. *75*, 7537–7541.

Schuster, S.C. (2008). Next-generation sequencing transforms today's biology. Nat. Methods *5*, 16–18.

Seshadri, R., Kravitz, S., Smarr, L., Gilna, P., and Frazier, M. (2007). CAMERA: a community resource for metagenomics. PLoS. Biol. *5*, e75.

Shannon, C.E and Weaver, W. (1949). The mathematical theory of communication. Bell Syst. Tech. J. *27*, 379–423.

Simpson, E. (1949). Measurement of diversity. Nature *163*, 688.

Simpson, J.T., Wong, K., Jackman, S.D., Schein, J.E., Jones, S.J.M., and Birol, I.c. (2009). ABySS: a parallel assembler for short read sequence data. Genome Res. *19*, 1117–1123.

Smith, A.D., Xuan, Z., and Zhang, M.Q. (2008). Using quality scores and longer reads improves accuracy of Solexa read mapping. BMC Bioinformatics *9*, 128.

Smith, E., and Vanbelle, G. (1984). Nonparametric-Estimation of Species Richness. Biometrics *40*, 119–129.

Sogin, M., Morrison, H., Huber, J., Mark, W., Huse, S., Neal, P., Arrieta, J., and Herndl, G. (2006). Microbial diversity in the deep sea and the underexplored 'rare biosphere'. Proc. Natl. Acad. Sci. U.S.A *103*, 12115–12120.

Sun, Y., Cai, Y., Liu, L., Yu, F., Farrell, M.L., McKendree, W., and Farmerie, W. (2009). ESPRIT: estimating species richness using large collections of 16S rRNA pyrosequences. Nucleic Acids Res. *37*, e76.

Tarailo-Graovac, M., and Chen, N. (2009). Using RepeatMasker to identify repetitive elements in genomic sequences. Curr Protoc Bioinformatics Chapter *4*, Unit 4.10.

Tatusov, R.L., Fedorova, N.D., Jackson, J.D., Jacobs, A.R., Kiryutin, B., Koonin, E.V., Krylov, D.M., Mazumder, R., Mekhedov, S.L., Nikolskaya, A.N., *et al.* (2003). The COG database: an updated version includes eukaryotes. BMC Bioinformatics *4*, 41.

Turnbaugh, P.J., Hamady, M., Yatsunenko, T., Cantarel, B.L., Duncan, A., Ley, R.E., Sogin, M.L., Jones, W.J., Roe, B.A., Affourtit, J.P. *et al.* (2009). A core gut microbiome in obese and lean twins. Nature *457*, 480–484.

Tyson, G., Chapman, J., Hugenholtz, P., Allen, E., Ram, R., Richardson, P., Solovyev, V., Rubin, E., Rokhsar, D., and Banfield, J. (2004). Community structure and metabolism through reconstruction of microbial genomes from the environment. Nature *428*, 37–43.

Venter, J.C., Adams, M.D., Myers, E.W., Li, P.W., Mural, R.J., Sutton, G.G., Smith, H.O., Yandell, M., Evans, C.A., Holt, R.A. *et al.* (2001). The sequence of the human genome. Science *291*, 1304–1351.

Venter, J.C., Remington, K., Heidelberg, J.F., Halpern, A.L., Rusch, D., Eisen, J.A., Wu, D., Paulsen, I., Nelson, K.E. *et al.* (2004). Environmental genome shotgun sequencing of the Sargasso Sea. Science *304*, 66–74.

Wang, Q., Garrity, G.M., Tiedje, J.M., and Cole, J.R. (2007). Naive Bayesian classifier for rapid assignment of rRNA sequences into the new bacterial taxonomy. Appl. Environ. Microbiol. *73*, 5261–5267.

Warnecke, F., Luginbuhl, P., Ivanova, N., Ghassemian, M., Richardson, T., Stege, J., Cayouette, M., McHardy, A., Djordjevic, G., Aboushadi, N. *et al.* (2007). Metagenomic and functional analysis of hindgut microbiota of a wood-feeding higher termite. Nature *450*, 560–565.

Warren, R.L., Sutton, G.G., Jones, S.J.M., and Holt, R.A. (2007). Assembling millions of short DNA sequences using SSAKE. Bioinformatics *23*, 500–501.

Wegley, L., Edwards, R., Rodriguez-Brito B., Liu, H., and Rohwer, F. (2007). Metagenomic analysis of the microbial community associated with the coral Porites astreoides. Environ. Microbiol. *11*, 2707–2019.

Wu, C.H., Apweiler, R., Bairoch, A., Natale, D.A., Barker, W.C., Boeckmann, B., Ferro, S., Gasteiger, E., Huang, H., Lopez, R., Magrane, M., Martin, M.J., Mazumder, R., O'donovan, C., Redaschi, N., and Suzek, B. (2006). The Universal Protein Resource (UniProt): an expanding universe of protein information. Nucleic Acids Res. *34*, D187–91.

Wu, T.D., and Watanabe, C.K. (2005). GMAP: a genomic mapping and alignment program for mRNA and EST sequences. Bioinformatics *21*, 1859–1875.

Yooseph, S., Sutton, G., Rusch, D., Halpern, A., Williamson, S., Remington, K., Eisen, J., Heidelberg, K., Manning, G.G.W., *et al.* (2007). The Sorcerer II Global Ocean Sampling expedition: expanding the universe of protein families. PLoS. Biol. *5*, e16.

Zerbino, D.R., and Birney, E. (2008). Velvet: algorithms for *de novo* short read assembly using de Bruijn graphs. Genome Res. *18*, 821–829.

Stable Isotope Probing: Uses in Metagenomics

6

Ondrej Uhlik, Lucie Musilova, Katerina Demnerova, Tomas Macek and Martina Mackova

Abstract

Until recently, investigating functions of microbial populations was restricted to thorough studying of pure cultures. Molecular biology tools enabled scientists to take a much deeper insight into the phylogenetic as well as metabolic diversity but hardly allowed for linking the phylogenetic identity with the metabolic activity which the microbes disposed of. Stable isotope probing (SIP) was one of the first microbial ecology tools to enable such a linkage. SIP consists of providing the community with a stable isotope labelled substrate and subsequent extraction and analysis of the labelled biomarkers. This text summarizes why SIP is a technique of an outstanding importance for microbial ecology and related fields and further focuses on SIP using DNA as a biomarker since DNA is one of the few molecules that bears both phylogenetic and functional information. Much deeper insight into stable isotope labelled DNA has been allowed especially due to current advances in high-throughput sequencing technologies. Metagenomic analyses, however, profit from stable isotope probing as well by reduced complexity of studied DNA (only populations actively performing a particular process). Therefore, when DNA-SIP is used in metagenomics, it is much easier to reconstruct individual genomes of key uncultured microbes as well as detect genes of interest.

Introduction

One of the major challenges of microbial ecology and environmental microbiology has always been the thorough investigation of microbial diversity. Characterization of microbes inhabiting individual ecological niches, from soil and water to extreme environments, has been of interest in order to understand the role of each microbe in that particular habitat. Years of microbiological research have shown that microbes play key roles in global element cycling, decomposition of organic matter, production of many unique substances, as well as in the degradation of xenobiotics and pollutants. Understanding the links between community structure and function is therefore a task of high importance. Research in this field has grown from the cultivation-based studies via 16S rRNA gene and functional gene analyses to metagenomics.

Cultivation

Initially, environmental microorganisms were characterized after they had been extracted from the environment and isolated in a pure culture. Such pure cultures are relatively easy to characterize from the point of view of morphology, biochemistry, immunology, or genomics (Sutherland and Rafii, 2006), in other words, cultivated microbes can be identified and their metabolic capabilities associated with their function can be determined. In the 1980s it was described that the number of microbes that can be assessed by microscopy is much higher than that obtained by plate counting (Staley and Konopka, 1985). Later the discrepancy was specified to be on average about 1% of microbes being routinely cultured (Torsvik and Ovreas, 2002; Lozupone and Knight, 2008; Lopez-Garcia and Moreira, 2008; Zhang and Xu, 2008). In the course of time, basic strategies of cultivation have been modified in the attempt to isolate more microbes in the pure culture which in some cases have yielded some novel results (reviewed by Tyson and Banfield, 2005). However, for the large majority of microbes conditions that would allow for their cultivation have not yet been described (Cardenas and Tiedje, 2008). Therefore, current research on environmental microbes is mostly based on molecular biology tools that have progressively evolved from 16S rRNA gene and other conservative genes analyses via functional gene analyses and metagenomics.

16S/18S rRNA gene analyses: what microbes inhabit the site?

By the 1970s, ribosomal RNA (rRNA) analyses was already being used for phylogenetic analyses revealing the primary kingdoms, the eubacteria, the archaea, and the eukaryotes (Woese and Fox, 1977). It became apparent that classifying organisms according to similarities in morphological, physiological, and biochemical characteristics and nutritional needs does not comply with evolutionary relationships. For phylogenetic analyses, comparing sequences of specific macromolecules appeared suitable. Of those macromolecules the small subunit rRNA (16S for prokaryotes and 18S for eukaryotes) had the highest potential (Fox *et al.*, 1980; Lane *et al.*, 1985) due to its universal distribution, sequence conservation, structure with variable evolutionary rates, and the fact it cannot be subjected to horizontal transfer (Tringe and Hugenholtz, 2008). Ever since, analysing 16S rRNA and/or its genes has become the most common tool for phylogenetic studies of prokaryotes – in the early 1990s, universal primers were developed that enabled the amplification of almost the full length of 16S rRNA genes in most bacteria (Weisburg *et al.*, 1991). Although these universal primers were later slightly modified to target even more eubacteria, they have been widely used until today. Nowadays, 16S rRNA gene sequence analyses are used for phylogeny-based classification/identification of populations within communities as well as isolates.

Functional gene analyses: what are the metabolic capabilities of the microbes?

In terms of microbial function, however, 16S rRNA does not have a substantial informative value. In order to study metabolic capabilities of the microbes with the aim to determine their potential function, genes encoding key enzymes involved in a particular geochemical cycling process must be investigated. Although in some cases the number of the genes shows the reaction potential, the actual process is better reflected by studying expressed genes, i.e. mRNA (Lovley, 2003). In complex microbial communities where microbes were not cultivated, however, the problem remained how to link the function or metabolic activity with the phylogenetic identity.

Metagenomics

Metagenomics is currently experiencing a boom especially due to more readily available high-throughput sequencing and improved bioinformatic tools. The aim of the sequencing-based metagenomics is to perform a reconstruction of the metabolic pathways of those microorganisms that compose the community and to identify the role that they play in the ecosystem (Vieites *et al.*, 2009). Since the first experiments in which a whole-community shotgun sequencing approach was used, however, it has become apparent that this approach can yield results only in the sense of revealing the metabolic potential of the community as a whole. Assigning sequences, and hence functions, to individual microbes is possible only in the case of communities with a very low diversity or of very abundant species within the community (Chen and Murrell, 2010). Yet in most cases the low-abundant microbes play key roles for community functioning. As metagenomics is highly biased towards the high-abundant species, it often fails to link microbial populations to their metabolic activity.

Metagenomic analyses can be carried out, except for high-throughput sequencing, by high-throughput microarray technologies. Although these technologies are not commonly mentioned along with metagenomics, they are based on comprehensive analyses of metagenomes. These microarray technologies include in particular functional gene microarrays GeoChip (He *et al.*, 2007; He *et al.*, 2010), which contain almost 28,000 probes, covering about 57,000 genes involved in basic elements (C, N, P, S) cycling, energy processes, metal resistance, organic contaminant degradation, and antibiotic resistance, as well as gyrB as a phylogenetic marker (He *et al.*, 2010).

Stable isotope probing: linking identity to function

Stable isotope probing (SIP) was developed as one of the major tools of microbial ecology research that enabled scientists to directly link the metabolic activity to the phylogenetic identity of the microbes (Radajewski *et al.*, 2000) without the need of isolating them. Therefore, SIP joins the advantages of cultivation-based approaches and molecular biology approaches based on 16S rRNA gene sequencing and eliminates their disadvantages. Small subunit rRNA (16S or 18S rRNA) gene sequencing is a robust technique to describe the phylogenetic diversity of microbial communities but provides few direct links between the identity of microorganisms and their role in the environment. On the other hand, after cultivation, microbial strains can be identified and characterized at the physiological, biochemical and genetic levels but only a small fraction of microorganisms have been isolated and cultivated so far. SIP, in contrast, covers both cultured and as yet uncultured microbes and allows for identifying which microbes support the particular function (Uhlík *et al.*, 2009a).

SIP is based on analysing biomarkers, molecules that bear phylogenetic information. These biomarkers are isolated directly from the environmental sample after it had been incubated with a stable isotope-labelled compound whose metabolism is studied. If the stable isotope labelled compound is utilized, the stable isotopes are incorporated into the cell components including the biomarkers. The biomarkers used for these purposes were introduced sequentially over the last years from polar lipid-derived fatty acids (Boschker *et al.*, 1998), DNA (Radajewski *et al.*, 2000), rRNA (Manefield *et al.*, 2002), and proteins (Jehmlich *et al.*, 2008a,b), to the recent use of mRNA (Huang *et al.*, 2009). Furthermore, only stable isotope-labelled biomarkers originating from the metabolically active populations are analysed whereas the inactive populations are left out (Fig. 6.1). Stable isotopes commonly used include ^{13}C or ^{15}N; less common are ^{2}H or ^{17}O (Radajewski *et al.*, 2003; Dumont and Murrell, 2005; Neufeld *et al.*, 2007a; Uhlík *et al.*, 2009a).

The appropriate constituent biomarkers are used based on the situation since all of them have specific advantages as well as limitations that were discussed elsewhere (Dumont and Murrell, 2005; Neufeld *et al.*, 2007a; Uhlík *et al.*, 2009a). From the point of view of metagenomics, DNA-SIP is of particular interest. A stable isotope labelled metagenome, after its separation from the unlabelled DNA of inactive populations, offers the possibility of studying phylogenetically relevant genes as well as functional genes. The main benefit of DNA-SIP-enabled metagenomics is the ability to reconstruct individual genomes as a result of lower diversity of labelled populations compared to the whole community (Chen and Murrell, 2010). Similarly, a stable isotope-labelled metagenome can be investigated through GeoChip (Leigh *et al.*, 2007).

Although the main focus of this text is the use of SIP in metagenomics, applications of DNA-based SIP before the use of the present metagenomic tools will be discussed as well.

Stable isotope probing in the 'pre-metagenomics' era

Initial DNA-SIP experiments were focused on the metabolism of C_1-compounds, such as methane or methanol. Methylotroph populations were studied in oak forest soil (Radajewski *et al.*, 2000, 2002), peat soil (Morris *et al.*, 2002), Transbaikal soda lake sediments (Lin *et al.*, 2004), cave groundwater and floating microbial mat (Hutchens *et al.*, 2004), or Lake Washington sediments (Nercessian *et al.*, 2005). Phylogenetically, most methylotrophs were members of Proteobacteria. These studies, however, brought some interesting findings, such as a novel methanotroph species related to *Methylocella palustris*, or a

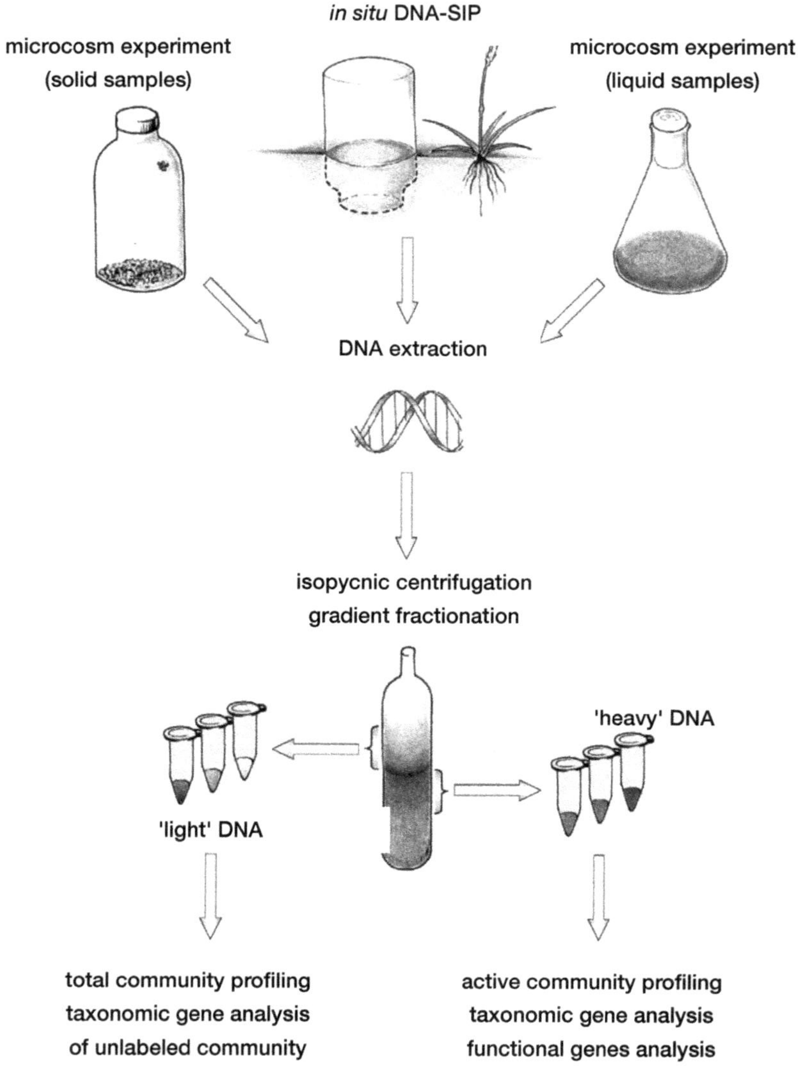

Figure 6.1 Scheme of an experiment employing SIP in metagenomics (reprinted from Science of the Total Environment, 407, Uhlik *et al.*, DNA-based stable isotope probing: a link between community structure and function, 3611–3619, 2009, with permission from Elsevier).

eukaryotic organism related to alga *Ochromonas danica*. Recently, DNA-SIP was used to characterize methylotrophs that are associated with the metabolism of methanol and methylamine in the marine environment (Neufeld *et al.*, 2007b). The authors underlined the importance of performing SIP experiments in marine environments not only because of the significant role of C_1 compounds in the marine ecology but also because of the potential of discovering novel and deeply branching as yet uncultured bacteria. They supplemented this hypothesis with exemplary results – the large portion of *Methylophaga*-related 16S rRNA gene sequences as well as methanol and methylamine dehydrogenase sequences retrieved from ^{13}C-labelled DNA were associated with clades that lacked close affiliations with sequences from cultured bacteria. Moreover, unique populations and functional genes were discovered to be associated with the metabolism of methanol and methylamine, which is an evidence of ecological adaptation of different populations to specific C_1 compounds. More recently, DNA-SIP was used to identify the active microorganisms in wastewater that were involved in methane-dependent denitrification in the presence of oxygen (Osaka *et al.*, 2008). The results showed the association of methanotrophs, such as Methylococcaceae, and methylotrophs, such as Hyphomicrobiaceae or Methylophilaceae, in the methane-dependent denitrification ecosystem. Other processes involved in nitrogen cycling were also studied by means of DNA-SIP, including ^{13}C-acetate stimulated denitrification in sludge which was performed by Comamonadaceae and Rhodocyclaceae members (Ginige *et al.*, 2005), or ammonia oxidation in freshwater sediment (Whitby *et al.*, 2001) and amended sediment from the Ythan estuary, Scotland, UK (Freitag *et al.*, 2006), which were both performed mainly by *Nitrosomonas*-related bacteria. Miller *et al.* (2004) employed DNA-SIP in order to identify populations metabolizing methyl halides in soil where these gases are naturally released. These authors discovered that methyl chloride is metabolized by more diverse groups of microbes compared to methyl bromide whose metabolism was dominated by organisms closely related to *Burkholderia* (Miller *et al.*, 2004).

Similarly, SIP was successfully applied to identify microbes responsible for the metabolism of aromatics, such as benzene, benzoic acid, phenol, polycyclic aromatic hydrocarbons (PAHs), or biphenyl and polychlorinated biphenyls (PCBs). The use of benzene, a proven carcinogen, as a solvent and fuel component has lead to the contamination of surface waters, sediments, and subsurface habitats. Liou and colleagues (2008) identified by DNA-SIP members of a coal-tar waste-contaminated sediment microbial community potentially responsible for the metabolism of benzene. Besides *Ralstonia*, *Pseudomonas* and *Propionibacterium*, they detected a novel benzene-degrading *Pelomonas* population dominating benzene metabolism (Liou *et al.*, 2008). Benzoic acid is a naturally occurring aromatic acid that enters the soil through root exudates or plant decomposition, and a diverse number of bacteria are able to metabolize it. Tracking ^{13}C from ^{13}C-benzoic acid added to agricultural field plots revealed a predominance of *Burkholderia* species within benzoic acid metabolism (Pumphrey and Madsen, 2008). Degradation of phenol was investigated together with the metabolism of glucose, caffeine, and naphthalene in field soil directly *in situ* (Padmanabhan *et al.*, 2003). The populations associated with the degradation of the ^{13}C-labelled glucose, phenol, naphthalene, and caffeine were identified as members of the genera *Arthrobacter*, *Pseudomonas*, *Acinetobacter*, *Massilia*, *Flavobacterium*, and *Pedobacter*; *Pseudomonas*, *Pantoea*, *Acinetobacter*, *Enterobacter*, *Stenotrophomonas*, and *Alcaligenes*; *Pseudomonas*, *Acinetobacter*, and *Variovorax*; and *Acinetobacter*, *Enterobacter*, *Stenotrophomonas*, and *Pantoea*, respectively. Naphthalene utilization was also monitored in contaminated sediment and a previously unknown bacterium *Polaromonas naphthalenivorans* sp. nov. was identified to have dominated this process (Jeon *et al.*, 2003, 2004). PAH metabolism was a subject of research even in a bioreactor treating contaminated soil, where bacteria metabolizing naphthalene, phenanthrene, and salicylate (a stimulator of growth for a range of PAH-degrading bacteria) were studied (Singleton *et al.*, 2005). Salicylate and naphthalene degradation was found to be primarily performed by the populations of *Pseudomonas* spp. and *Ralstonia* spp., whereas phenanthrene degradation was mainly associated with *Acidovorax* spp. Besides PAHs, PCBs also represent a large group of aromatic contaminants present in many sites all over the world. Metabolism of biphenyl, a basic analogue of PCBs, was investigated in long-term PCB- and PAH-contaminated soil, where biphenyl can also be found among contaminants. Contaminated soil transferred to pots planted with horseradish or left unplanted was differentially dominated by biphenyl-utilizing populations identified as *Hydrogenophaga* spp. and *Paenibacillus* spp., respectively (Uhlik *et al.*, 2009b).

Stable isotope probing used in metagenomics

The first two examples of using stable isotope probing in metagenomics were published in 2006. The first one dealt with isolating genes which encode coenzyme B_{12}-dependent glycerol dehydratases from enrichment cultures of glycerol-fermenting microorganisms from a sediment sample (Schwarz *et al.*, 2006). This study showed that metagenomic analysis of ^{13}C-labelled DNA significantly increases the frequency of target genes compared to total community DNA. In the other study (Dumont *et al.*, 2006), the authors identified a complete methane monooxygenase (pmoCAB) operon from soil by combining stable isotope probing and metagenomic analysis. This study was the first one that investigated stable isotope labelled DNA after it had been cloned into a vector. Analysing functional genes this way circumvents the need to isolate them by PCR, which can be very problematic. In many cases, not enough sequences are known to design adequately specific primers. These, when designed incorrectly, can introduce many biases into the analyses. Additionally, Dumont and colleagues (2006) found in the same clone which bore the pmoCAB operon other genes that are involved in the utilization of C_1-compounds.

Leigh and colleagues (2007) employed DNA-SIP to study active biphenyl-metabolizing bacteria in a root zone of pine growing naturally in a long-term PCB-contaminated soil in the Czech Republic. The metagenome of active populations was analysed through GeoChips after multiple displacement amplification of the whole metagenome was performed. Such an analysis resulted in the detection of 30 genes involved in biodegradation of organic contaminants with 90% being involved in the degradation of aromatics, including four bph genes (genes of the upper biphenyl pathway), genes associated with a β-ketoadipate pathway, and several oxygenase genes from the pathways of benzoate, dibenzofuran, phenol, phenylpropionate, and protocatechuate degradation. Additionally, many genera were detected to be involved in biphenyl utilization that had not been associated with this process before, such as *Pseudonocardia* spp., *Nocardiodes* spp., *Variovorax* spp. or *Polaromonas* spp. (Leigh *et al.*, 2007).

Multiple displacement amplification was also used after stable isotope probing to obtain a sufficient amount of DNA for subsequent metagenomic analyses (namely fosmid library construction and screening) by Murrell and colleagues (Chen *et al.*, 2008; Neufeld *et al.*, 2008). Employing multiple displacement amplification allows for the use of environmentally relevant concentrations of the labelled substrate which solves one of the most criticized problems associated with SIP, i.e. using much higher concentrations of the labelled substrate than are present in the environment. In the first of the two studies, *Methylophaga* spp. clade was identified to be involved in marine methanol cycling, and a DNA fragment was found that encoded the enzymes involved in methanol dehydrogenase synthesis, regulation, and assembly (Neufeld *et al.*, 2008). In the second study, 16S rRNA genes of *Methylocella/Methylocapsa*-related species and *Methylocystis* species were detected. The analyses also resulted in the assembly of a gene cluster mxaFJGIRSAC, which contains genes involved in methanol utilization (Chen *et al.*, 2008).

Another proof-of-principle study that used SIP and metagenomics was the analysis of populations involved in C_1 compounds utilization in the water and sediments of Lake Washington, WA, USA (Kalyuzhnaya *et al.*, 2008). Besides the finding that bacteria of the *Verrucomicrobia*, *Nitrospirae* and *Planctomycetes* spp., which had not been traditionally associated with methylotrophy, were associated with the metabolism of methane, methanol, methylamine, formaldehyde, and formate, the authors also successfully reconstructed a nearly complete genome of a novel methylotrophic bacterium *Methylotenera mobilis*. Therefore, this study successfully demonstrated a facilitated genome reconstruction of rare but key microbes of the community when stable isotope probing preceded metagenomic analysis.

Another study that combined DNA-SIP with metagenomics dealt with bacteria capable of growing on biphenyl in the PCB-contaminated sediment of the Raisin River, MI, USA (Sul *et al.*, 2009). After cosmid library construction and screening, one bphA sequence was found which was highly similar to bphA in *Pseudomonas* sp. Cam-1 and bphA1 in *Pseudomonas* pseudoalcaligenes KF707. The same clone also contained a bphE gene (which encodes a small subunit of biphenyl dioxygenase whereas bphA encodes the large one) but not any other genes of the biphenyl upper pathway, suggesting that these genes were transferred by a horizontal gene transfer between bacteria. Expression study employing detected bphAE genes revealed the ability of the enzymes encoded by these genes to metabolize biphenyl and unsubstituted and para-susbtituted biphenyl rings. Overall, this study showed the ability of retrieving key functional genes without the need of isolating them by PCR (Sul *et al.*, 2009).

Discussion and future trends

DNA-SIP as a method to link microbial community structure with ecosystem processes and functions has gained a lot of attention over the last decade. Although DNA-based SIP is the least sensitive of all SIP techniques, especially when combined with metagenomics, its informative value is outstanding. Since the first studies that identified metabolically active methylotrophs, SIP was employed to study the metabolism of several C_1 compounds as well as more complex organic substrates (glucose, salicylate, benzoate) and a wide variety of organic contaminants (benzene, PAHs, biphenyl/PCBs). However, much of the essential information about most of the processes occurring in nature remains unknown. Until now, only a few studies were published that combined SIP with metagenomics, and screening using clone libraries approach is hardly ever capable of covering all the genes present. Therefore, it is supposed that the combination of SIP and metagenomics will fill some of these gaps. However, major problems of metagenomic analyses remain to be overcome, primarily rather incomplete databases for annotations which result in a large portion of obtained sequences not being successfully assigned to a known function (Vieites *et al.*, 2009).

The main merit of DNA-SIP is the informative value of DNA, which covers both phylogenetic and corresponding functional information. In the future, it is very likely that research in this area will head towards actively transcribed genes. Recently, mRNA-based SIP was demonstrated as a possible way of retrieving functional information and in combination with rRNA-SIP also phylogenetic information (Huang *et al.*, 2009). Isolating sufficient amount of mRNA from environmental samples and its further processing, however, is often very challenging. Firstly, in prokaryotes, mRNA usually represents only a minority of the total RNA, and secondly, a universal reverse transcriptase signal, such as poly(A) tail, is lacking which hinders uniform amplification of mRNAs (Vieites *et al.*, 2009). Yet there have been procedures designed to help overcome these problems, i.e. using *E. coli* poly(A) polymerase for separating mRNA from rRNAs for the former (Frias-Lopez *et al.*, 2008), and using whole-community RNA amplification for the latter (Gao *et al.*, 2007). When mRNA-SIP becomes as routinely used as DNA-SIP, it can be combined with a metatranscriptomic approach, i.e. metatranscriptome (the expressed genetic information of an ecosystem) can be reversely transcribed into cDNA and it can be further analysed through high-throughput sequencing or library construction and screening. Metatranscriptomes may alternatively be analysed using GeoChips or similar high-throughput microarray technologies.

DNA-SIP enabled metagenomics is still at the beginning of its application. However, it is clear now that it represents a revolutionary technique to combine approaches for recovering the genes of interest (independently of microbe cultivability). This provides comprehensive information to better understand global environmental processes and microbial ecology interactions, as well as to obtain new proteins and new activities which can be used in biotechnology research and practice.

Acknowledgements

The authors acknowledge the language-related comments of Alice Naomi Mitchell. This work was supported by the Ministry of Education, Youth and Sports of the Czech Republic (grants ME 10041, NPVII 2B06151, and research projects MSM 6046137305 and Z 40550506).

References

Boschker, H.T.S., Nold, S.C., Wellsbury, P., Bos, D., de Graaf, W., Pel, R., Parkes, R.J., and Cappenberg, T.E. (1998). Direct linking of microbial populations to specific biogeochemical processes by ^{13}C-labelling of biomarkers. Nature *392*, 801–805.

Cardenas, E., and Tiedje, J.M. (2008). New tools for discovering and characterizing microbial diversity. Curr. Opin. Biotechnol. *19*, 544–549.

Chen, Y., Dumont, M.G., Neufeld, J.D., Bodrossy, L., Stralis-Pavese, N., McNamara, N.P., Ostle, N., Briones, M.J., and Murrell, J.C. (2008). Revealing the uncultivated majority: combining DNA stable-isotope probing, multiple displacement amplification and metagenomic analyses of uncultivated Methylocystis in acidic peatlands. Environ. Microbiol. *10*, 2609–2622.

Chen, Y., and Murrell, J.C. (2010). When metagenomics meets stable-isotope probing: progress and perspectives. Trends Microbiol. *18*, 157–163.

Dumont, M.G., and Murrell, J.C. (2005). Stable isotope probing – linking microbial identity to function. Nat. Rev. Microbiol. *3*, 499–504.

Dumont, M.G., Radajewski, S.M., Miguez, C.B., McDonald, I.R., and Murrell, J.C. (2006). Identification of a complete methane monooxygenase operon from soil by combining stable isotope probing and metagenomic analysis. Environ. Microbiol. *8*, 1240–1250.

Fox, G.E., Stackebrandt, E., Hespell, R.B., Gibson, J., Maniloff, J., Dyer, T.A., Wolfe, R.S., Balch, W.E., Tanner, R.S., Magrum, L.J. *et al.* (1980). The phylogeny of prokaryotes. Science *209*, 457–463.

Freitag, T.E., Chang, L., and Prosser, J.I. (2006). Changes in the community structure and activity of betaproteobacterial ammonia-oxidizing sediment bacteria along a freshwater-marine gradient. Environ. Microbiol. *8*, 684–696.

Frias-Lopez, J., Shi, Y., Tyson, G.W., Coleman, M.L., Schuster, S.C., Chisholm, S.W., and Delong, E.F. (2008). Microbial community gene expression in ocean surface waters. Proc. Natl. Acad. Sci. U. S. A. *105*, 3805–3810.

Gao, H., Yang, Z.K., Gentry, T.J., Wu, L., Schadt, C.W., and Zhou, J. (2007). Microarray-based analysis of microbial community RNAs by whole-community RNA amplification. Appl. Environ. Microbiol. *73*, 563–571.

Ginige, M.P., Keller, J., and Blackall, L.L. (2005). Investigation of an acetate-fed denitrifying microbial community by stable isotope probing, full-cycle rRNA analysis, and fluorescent *in situ* hybridization-microautoradiography. Appl. Environ. Microbiol. *71*, 8683–8691.

He, Z., Deng, Y., Van Nostrand, J.D., Tu, Q., Xu, M., Hemme, C.L., Li, X., Wu, L., Gentry, T.J., Yin, Y. *et al.* (2010). GeoChip 3.0 as a high-throughput tool for analyzing microbial community composition, structure and functional activity. ISME J. Epub ahead of print.

He, Z., Gentry, T.J., Schadt, C.W., Wu, L., Liebich, J., Chong, S.C., Huang, Z., Wu, W., Gu, B., Jardine, P., Criddle, C., and Zhou, J. (2007). GeoChip: a comprehensive microarray for investigating biogeochemical, ecological and environmental processes. ISME J. *1*, 67–77.

Huang, W.E., Ferguson, A., Singer, A.C., Lawson, K., Thompson, I.P., Kalin, R.M., Larkin, M.J., Bailey, M.J., and Whiteley, A.S. (2009). Resolving genetic functions within microbial populations: *in situ* analyses using rRNA and mRNA stable isotope probing coupled with single-cell raman-fluorescence *in situ* hybridization. Appl. Environ. Microbiol. *75*, 234–241.

Hutchens, E., Radajewski, S., Dumont, M.G., McDonald, I.R., and Murrell, J.C. (2004). Analysis of methanotrophic bacteria in Movile Cave by stable isotope probing. Environ. Microbiol. *6*, 111–120.

Jehmlich, N., Schmidt, F., Hartwich, M., von, B.M., Richnow, H.H., and Vogt, C. (2008a). Incorporation of carbon and nitrogen atoms into proteins measured by protein-based stable isotope probing (Protein-SIP). Rapid Commun. Mass Spectrom. *22*, 2889–2897.

Jehmlich, N., Schmidt, F., von, B.M., Richnow, H.H., and Vogt, C. (2008b). Protein-based stable isotope probing (Protein-SIP) reveals active species within anoxic mixed cultures. ISME. J. *2*, 1122–1133.

Jeon, C.O., Park, W., Ghiorse, W.C., and Madsen, E.L. (2004). Polaromonas naphthalenivorans sp. nov., a naphthalene-degrading bacterium from naphthalene-contaminated sediment. Int. J. Syst. Evol. Microbiol. *54*, 93–97.

Jeon, C.O., Park, W., Padmanabhan, P., DeRito, C., Snape, J.R., and Madsen, E.L. (2003). Discovery of a bacterium, with distinctive dioxygenase, that is responsible for *in situ* biodegradation in contaminated sediment. Proc. Natl. Acad. Sci. U.S.A. *100*, 13591–13596.

Kalyuzhnaya, M.G., Lapidus, A., Ivanova, N., Copeland, A.C., McHardy, A.C., Szeto, E., Salamov, A., Grigoriev, I.V., Suciu, D., Levine, S.R. *et al.* (2008). High-resolution metagenomics targets specific functional types in complex microbial communities. Nat. Biotechnol. *26*, 1029–1034.

Lane, D.J., Pace, B., Olsen, G.J., Stahl, D.A., Sogin, M.L., and Pace, N.R. (1985). Rapid determination of 16S ribosomal RNA sequences for phylogenetic analyses. Proc. Natl. Acad. Sci. U. S. A *82*, 6955–6959.

Leigh, M.B., Pellizari, V.H., Uhlik, O., Sutka, R., Rodrigues, J., Ostrom, N.E., Zhou, J., and Tiedje, J.M. (2007). Biphenyl-utilizing bacteria and their functional genes in a pine root zone contaminated with polychlorinated biphenyls (PCBs). ISME. J. *1*, 134–148.

Lin, J.L., Radajewski, S., Eshinimaev, B.T., Trotsenko, Y.A., McDonald, I.R., and Murrell, J.C. (2004). Molecular diversity of methanotrophs in Transbaikal soda lake sediments and identification of potentially active populations by stable isotope probing. Environ. Microbiol. *6*, 1049–1060.

Liou, J.S., DeRito, C.M., and Madsen, E.L. (2008). Field-based and laboratory stable isotope probing surveys of the identities of both aerobic and anaerobic benzene-metabolizing microorganisms in freshwater sediment. Environ. Microbiol. *10*, 1964–1977.

Lopez-Garcia, P., and Moreira, D. (2008). Tracking microbial biodiversity through molecular and genomic ecology. Res. Microbiol. *159*, 67–73.

Lovley, D.R. (2003). Cleaning up with genomics: applying molecular biology to bioremediation. Nat. Rev. Microbiol. *1*, 35–44.

Lozupone, C.A., and Knight, R. (2008). Species divergence and the measurement of microbial diversity. FEMS Microbiol. Rev. *32*, 557–578.

Manefield, M., Whiteley, A.S., Griffiths, R.I., and Bailey, M.J. (2002). RNA stable isotope probing, a novel means of linking microbial community function to phylogeny. Appl. Environ. Microbiol. *68*, 5367–5373.

Miller, L.G., Warner, K.L., Baesman, S.M., Oremland, R.S., McDonald, I.R., Radajewski, S., and Murrell, J.C. (2004). Degradation of methyl bromide and methyl chloride in soil microcosms: Use of stable C isotope fractionation and stable isotope probing to identify reactions and the responsible microorganisms. Geochim. Cosmochim. Acta *68*, 3271–3283.

Morris, S.A., Radajewski, S., Willison, T.W., and Murrell, J.C. (2002). Identification of the functionally active methanotroph population in a peat soil microcosm by stable-isotope probing. Appl. Environ. Microbiol. *68*, 1446–1453.

Nercessian, O., Noyes, E., Kalyuzhnaya, M.G., Lidstrom, M.E., and Chistoserdova, L. (2005). Bacterial populations active in metabolism of C_1 compounds in the sediment of Lake Washington, a freshwater lake. Appl. Environ. Microbiol. *71*, 6885–6899.

Neufeld, J.D., Chen, Y., Dumont, M.G., and Murrell, J.C. (2008). Marine methylotrophs revealed by stable-isotope probing, multiple displacement amplification and metagenomics. Environ. Microbiol. *10*, 1526–1535.

Neufeld, J.D., Dumont, M.G., Vohra, J., and Murrell, J.C. (2007a). Methodological considerations for the use of stable isotope probing in microbial ecology. Microb. Ecol. *53*, 435–442.

Neufeld, J.D., Schafer, H., Cox, M.J., Boden, R., McDonald, I.R., and Murrell, J.C. (2007b). Stable-isotope probing implicates *Methylophaga* spp and novel Gammaproteobacteria in marine methanol and methylamine metabolism. ISME J. *1*, 480–491.

Osaka, T., Ebie, Y., Tsuneda, S., and Inamori, Y. (2008). Identification of the bacterial community involved in methane-dependent denitrification in activated sludge using DNA stable-isotope probing. FEMS Microbiol. Ecol. *64*, 494–506.

Padmanabhan, P., Padmanabhan, S., DeRito, C., Gray, A., Gannon, D., Snape, J.R., Tsai, C.S., Park, W., Jeon, C., and Madsen, E.L. (2003). Respiration of ^{13}C-labeled substrates added to soil in the field and subsequent 16S rRNA gene analysis of ^{13}C-labeled soil DNA. Appl. Environ. Microbiol. *69*, 1614–1622.

Pumphrey, G.M., and Madsen, E.L. (2008). Field-based stable isotope probing reveals the identities of benzoic acid-metabolizing microorganisms and their *in situ* growth in agricultural soil. Appl. Environ. Microbiol. *74*, 4111–4118.

Radajewski, S., Ineson, P., Parekh, N.R., and Murrell, J.C. (2000). Stable-isotope probing as a tool in microbial ecology. Nature *403*, 646–649.

Radajewski, S., McDonald, I.R., and Murrell, J.C. (2003). Stable-isotope probing of nucleic acids: a window to the function of uncultured microorganisms. Curr. Opin. Biotechnol. *14*, 296–302.

Radajewski, S., Webster, G., Reay, D.S., Morris, S.A., Ineson, P., Nedwell, D.B., Prosser, J.I., and Murrell, J.C. (2002). Identification of active methylotroph populations in an acidic forest soil by stable-isotope probing. Microbiology *148*, 2331–2342.

Schwarz, S., Waschkowitz, T., and Daniel, R. (2006). Enhancement of gene detection frequencies by combining DNA-based stable-isotope probing with the construction of metagenomic DNA libraries. World J. Microbiol. Biotechnol. *22*, 363–367.

Singleton, D.R., Powell, S.N., Sangaiah, R., Gold, A., Ball, L.M., and Aitken, M.D. (2005). Stable-isotope probing of bacteria capable of degrading salicylate, naphthalene, or phenanthrene in a bioreactor treating contaminated soil. Appl. Environ. Microbiol. *71*, 1202–1209.

Staley, J.T., and Konopka, A. (1985). Measurement of *in situ* activities of nonphotosynthetic microorganisms in aquatic and terrestrial habitats. Annu. Rev. Microbiol. *39*, 321–346.

Sul, W.J., Park, J., Quensen, J.F., III, Rodrigues, J.L., Seliger, L., Tsoi, T.V., Zylstra, G.J., and Tiedje, J.M. (2009). DNA-stable isotope probing integrated with metagenomics for retrieval of biphenyl dioxygenase genes from polychlorinated biphenyl-contaminated river sediment. Appl. Environ. Microbiol. *75*, 5501–5506.

Sutherland, J.B., and Rafii, F. (2006). Cultural, Serological, and Genetic Methods for Identification of Bacteria. In Identification of Microorganisms by Mass Spectrometry, Wilkins, C.L., and Lay, J.O. Jr., eds. (Hoboken, NJ: John Wiley & Sons, Inc.), pp. 1–21.

Torsvik, V., and Ovreas, L. (2002). Microbial diversity and function in soil: from genes to ecosystems. Curr. Opin. Microbiol. *5*, 240–245.

Tringe, S.G., and Hugenholtz, P. (2008). A renaissance for the pioneering 16S rRNA gene. Curr. Opin. Microbiol. *11*, 442–446.

Tyson, G.W., and Banfield, J.F. (2005). Cultivating the uncultivated: a community genomics perspective. Trends Microbiol. *13*, 411–415.

Uhlik, O., Jecna, K., Leigh, M.B., Mackova, M., and Macek, T. (2009a). DNA-based stable isotope probing: a link between community structure and function. Sci. Total Environ. *407*, 3611–3619.

Uhlik, O., Jecna, K., Mackova, M., Vlcek, C., Hroudova, M., Demnerova, K., Paces, V., and Macek, T. (2009b). Biphenyl-metabolizing bacteria in the rhizosphere of horseradish and bulk soil contaminated by polychlorinated biphenyls as revealed by stable isotope probing. Appl. Environ. Microbiol. *75*, 6471–6477.

Vieites, J.M., Guazzaroni, M.E., Beloqui, A., Golyshin, P.N., and Ferrer, M. (2009). Metagenomics approaches in systems microbiology. FEMS Microbiol. Rev. *33*, 236–255.

Weisburg, W.G., Barns, S.M., Pelletier, D.A., and Lane, D.J. (1991). 16S ribosomal DNA amplification for phylogenetic study. J. Bacteriol. *173*, 697–703.

Whitby, C.B., Hall, G., Pickup, R., Saunders, J.R., Ineson, P., Parekh, N.R., and McCarthy, A. (2001). ^{13}C incorporation into DNA as a means of identifying the active components of ammonia-oxidizer populations. Lett. Appl. Microbiol. *32*, 398–401.

Woese, C.R., and Fox, G.E. (1977). Phylogenetic structure of prokaryotic domain: The primary kingdoms. Proc. Natl. Acad. Sci. U. S. A. *74*, 5088–5090.

Zhang, L., and Xu, Z. (2008). Assessing bacterial diversity in soil. J. Soil Sediments *8*, 379–388.

DNA Sequencing of Uncultured Microbes from Single Cells

7

Roger S. Lasken, Mary-Jane Lombardo, Mark Novotny, Joyclyn Yee-Greenbaum and Rashel V. Grindberg

Abstract

Development of a method to sequence DNA from a single cell has enabled new strategies to investigate the microbial world. Only a few years ago, sequencing from one cell was not feasible. A bacterium only contains a few femtograms of DNA which is insufficient for current sequencing technologies. This limitation was overcome with the development of a method to amplify DNA called multiple displacement amplification (MDA) which can generate micrograms of genomic sequence from one cell. Improvements have also been made in our ability to isolate cells by flow cytometry, micromanipulation and microfluidics and to lyse the cells to release the single genome copy as a template for MDA. Large portions of the genome can be obtained from each cell and this has opened up a new front in the effort to sequence uncultivated species. Cells can be isolated from an environment or clinical specimen and directly sequenced with no need to develop culture methods. This chapter will review the current methodologies, the strengths and limitations of the single-cell approach and its application to microbial genomics.

Single-cell sequencing

In 2001 The Department of Energy funded development of a method to sequence from a single cell and this was accomplished (Raghunathan, 2005) at Molecular Staging Inc., where MDA was developed. From an *E. coli* cell isolated by flow cytometry, MDA generated about 25 μg of DNA, more than enough for DNA sequencing and genotyping applications (Fig. 7.1).

It was estimated that about 30% of the amplified DNA was the desired *E. coli* sequences and the rest was nonspecific synthesis such as primer derived products or amplification from contaminating DNA. This was still a remarkable technology considering that a >billion-fold amplification had occurred from the 5 fg of DNA template in the cell. MDA performance has improved with more experience in methods to isolate and handle cells, lyse them, and also in avoiding contaminating DNA in the MDA reagents. Care must be taken to avoid DNA contamination in the lab. While contamination must be limited to a level far less than the targeted single copy of genomic DNA template, standard methods developed for procedures such as PCR have proved sufficient including use of dedicated pipettors and careful lab technique in avoiding contamination of reagent stocks (Ishoey *et al.*, 2008).

Combined use of the single-cell and metagenomic approaches

Single-cell sequencing should be complementary to research based on shotgun sequencing of total DNA extracted from environmental samples such as soil or sea water. This method, called metagenomics, provides overall gene frequencies in the community, but rarely reveals the genetic linkage of sequences within individual cells. It can fail to distinguish where highly homologous sequences were actually derived from very divergent species. Conversely, it can fail to distinguish where there are highly variable sequence differences

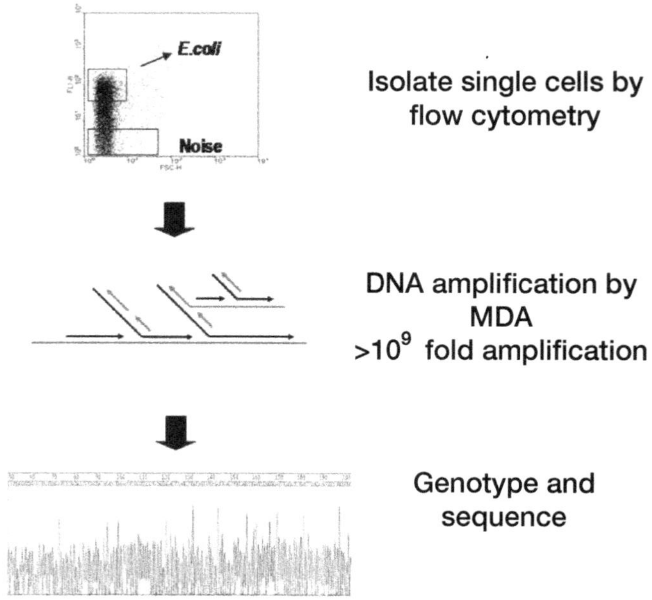

Figure 7.1 Single-cell isolation and DNA amplification by MDA. *E.coli* cells were isolated by flow cytometry. The single cells were lysed and the DNA amplified by MDA giving a $>10^9$-fold amplification (Raghunathan, 2005). The amplified DNA was suitable for use as template for genotyping and sequencing reactions.

between cells of the same species. In contrast, single-cell sequencing resolves the genetic linkage of sequences within a discrete organism. It provides the biological context of observing what is really possible in individuals where genetic linkage has been driven by natural selection. Finally, single-cell sequencing may provide a new means for addressing the problem of genetic diversity even where the traditional terminology of 'species' is problematic. In these cases of extreme variation the concept of a 'pan-genome' (the range of sequences present) and 'core-genome' (the defining sequences essential to the group) have been used instead of relying on the more limiting terminology of 'species'. Single-cell sequencing can contribute a measurement of the diversity between cells and the sequence context and genetic linkage needed to define boundaries of the pan-genome.

Sequencing of uncultivated bacteria

One immediate application with single cells was that bacterial species could be sequenced with no need to develop culture methods (Lasken, 2007). Sequences from a novel Crenarchaeota species were obtained using FISH probes specific for both Archaea and Crenarchaeota to select the individual cells (Kvist *et al.*, 2007). As expected, some of the genomic sequence was lost during the MDA reaction, and amplification bias resulted in underrepresentation of some other regions. Nevertheless, numerous genes were sequenced from this previously inaccessible organism. Even a partial genomic draft can greatly advance biological research. About 70% of the genome was recovered by MDA for a *Beggiatoa* species (Mussmann *et al.*, 2007), a marine bacteria that had not been successfully cultured. Predicted genes were identified for enzymes involved in sulfur oxidation, nitrate and oxygen respiration and CO_2 fixation confirming a putative chemolithoautotrophic physiology. Two studies carried out genomic sequencing from single cells for TM7, a proposed phylum for which no sequenced members had existed. In one, FISH probes were used to isolate cells by flow cytometry (Podar *et al.*, 2007). In the other (Marcy *et al.*, 2007b), the cells were isolated on a microfluidic chip and MDA carried out in a 60nl chamber. DNA sequencing was highly accurate by the 454 method using MDA from single cells whether the amplification was carried out on the microfluidic chip or by conventional MDA in 50 µl reactions (Marcy *et al.*, 2007a). In other studies, about 70–80% of novel bacterial genomes have been obtained starting with single cells (Rodrigue *et al.*, 2009; Woyke *et al.*, 2009; Zhang *et al.*, 2006).

One strategy to reduce amplification bias and loss of sequence is to carry out MDA with multiple cells of the same species. In one study, the complete genome was obtained for an uncultivable bacteria derived from a protist inhabiting the termite gut (Hongoh *et al.*, 2008). The genome was reconstructed by carrying out MDA on pooled bacterial cells derived from a single protist host cell so that individual bacteria would have only slight genome variations (genomovars). While exciting progress on the biology and biochemistry of these bacteria resulted, it will be important to remember that the derived genome was a composite from multiple individuals. Similarly, the first reported human genome was a composite of individuals and even the recently completed genome of a single human (Levy *et al.*, 2007) was a composite from cells differing by somatic mutations. Until improved technologies are available, completing the genomes of uncultured microbes will require reconstruction from highly homologous cells. Another approach for obtaining multiple cells is to use microcolony methods (Abulencia *et al.*, 2006) which require growth from one cell, but only for a low number of cell divisions. The demands for development of media are not as great as for conventional culturing and the few progeny cells should be highly clonal.

A different approach that should be possible is to obtain closely related cells from the environment by fluorescence activated cell sorting (FACS) using hybridization of FISH probes for specific 16S ribosomal RNA (rRNA) sequences. There are sufficient copies of the 16S rRNA molecules to allow hybridization of fluorescently labelled probes and isolation by micromanipulation or FACS (Kvist *et al.*, 2007; Lasken *et al.*, 2005). Actively metabolizing bacteria may be easier to detect because they will have a higher number of ribosomes. Unfortunately, single-copy genes are far more difficult to detect by FISH. 16S rRNA has been the only target reported for FACS isolation of bacterial cells and the probes may identify a fairly broad range of related species. Even cells with identical 16S rRNA gene sequences can have considerable variation in other genes. One way to overcome this is to isolate cells based on 16S rRNA, carry out MDA, and then further characterize their amplified genomes for other PCR targets or by preliminary sequencing to a low depth. In this way, closely related amplified genomes could be identified prior to investing funds in extensive sequencing. As sequencing costs drop in the future, it should even become possible to sequence deeply from many single cell amplifications and then use these initial genome drafts to confirm close homology and complete composite genomes assemblies. Metagenomic databases from environmental shotgun sequencing should also be a useful tool to guide assembly from single-cell data (Lasken, 2007), or at least in defining consensus sequences of the so called core genome for species having high genetic variation between individuals.

The MDA reaction for amplification of DNA

MDA reaction mechanisms

MDA achieves whole genome amplification by use of random hexamer primers (Fig. 7.2A) in a manner similar to older PCR-based methods that used random primers.

However, MDA was the first whole genome amplification method that was carried out isothermally, usually at 30°C (Dean *et al.*, 2001). This enabled MDA to avoid some of the limitations created by thermal cycling such as extreme amplification bias and short DNA products, typically only several hundred base pairs in length (Lasken and Egholm, 2003). MDA has relatively low amplification bias compared to PCR and the DNA products average about 12 kb in length, ranging up to an estimated >100kb. MDA replicates double stranded DNA in a similar manner to the cellular process (which is necessarily isothermal) in which the DNA polymerase invades a replication fork, displacing the template's complementary strand as it progresses. As the random primers are extended in MDA the polymerase invades and displaces downstream 5′-ends (Fig. 7.2B). In

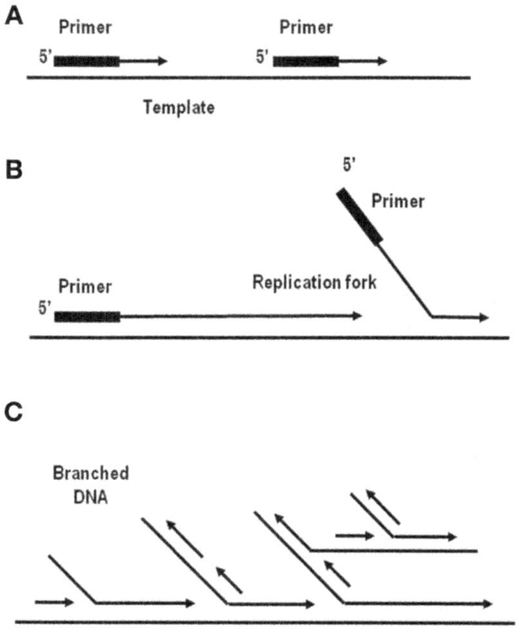

Figure 7.2 Mechanism of the MDA reaction. (A) Two of the random primers are shown annealed to the template. As the phi29 DNA polymerase extends a primer, downstream primers and their elongation products are in its path. (B) The phi29 DNA polymerase is capable of establishing new replication forks when it encounters the 5′ end of these downstream products. (C) As single-stranded DNA is displaced, it in turn becomes available for yet more primers to anneal to. The amplification becomes exponential by a branching mechanism.

the cell an enzyme called a helicase precedes the DNA polymerase and assists it in unwinding the double stranded template at the replication fork (Kornberg and Baker, 1991). In contrast, MDA utilizes DNA polymerases that have strong strand displacing ability facilitating invasion of a replication fork without the aid of a helicase. Usually, MDA uses the DNA polymerase from the *B. subtilis* bacteriophage phi29 (Dean *et al.*, 2001). The phi29 DNA polymerase's extremely tight binding to the template and strong strand displacement activity allows it to advance into a replication fork displacing the complementary strand of the template as it proceeds.

In the initial step of MDA, genomic DNA template is usually denatured by treatment with KOH and then neutralization with Tris-HCl, or by heating at 95°C. When MDA is carried out from whole cells, KOH or heat can often serve to lyse the cell as well as denature the released DNA template. Random hexamer primers and the phi29 DNA polymerase are then added to the single stranded DNA and primer extension begins (Fig. 7.2A). The amplification becomes exponential because of the ability of the phi29 DNA polymerase to establish new replication forks. As the DNA polymerase is extending a primer on a single stranded DNA template, other downstream primers and their extension products are in the polymerase's path. Some DNA polymerases have been shown to pass over a double stranded region and proceed downstream without initiating strand displacement synthesis (Kornberg and Baker, 1991). However, the phi29 DNA polymerase is proficient at invading the duplex at the 5′ end of the downstream primer and establishing a new replication fork (Fig. 7.2B, replication fork). Furthermore, the displaced strand becomes available for annealing of yet more random primers. The DNA polymerases create a branched structure (Lizardi *et al.*, 1998) extending additional primers as they concurrently displace downstream products (Fig. 7.2C). The branching creates a highly exponential amplification. For example, more than a billion-fold amplification was achieved from the genome of a single bacterial cell in an overnight MDA reaction (Raghunathan, 2005). About 5 fg of genomic DNA in the single bacterium yielded 25 μg of amplified DNA.

The phi29 DNA polymerase

For MDA, the most important characteristics of the phi29 DNA polymerase are its exceptionally tight binding to DNA and the related properties of processivity and strand displacement synthesis (Lasken, 2005). The polymerase binds to a primed template forming a complex that has a half-life estimated at between 1 and 13 minutes depending on reaction conditions (k_{off} ~ 0.01–0.001/s). The phi29 DNA polymerase has a processivity of about 70,000 nucleotides (Blanco *et al.*, 1989) meaning that it adds an average of 70,000 nucleotides each time it binds to the primer/template. For comparison, the large fragment of *E. coli* DNA polymerase I (Klenow fragment) has a processivity of only about 10 nucleotides. The high processivity of the phi29

DNA polymerase contributes to the robust strand displacement synthesis and great length of the MDA products. However, the constraints of the branching mechanism (Fig. 7.2C) also presumably determine the average length of the amplified DNA products in MDA. The average length is about 12 kb (Dean *et al.*, 2002) when run on a denaturing agarose gel so that all of the DNA forms present (linear single-stranded and double-stranded DNA and some reaction intermediates still in complex branched forms) are resolved as the single-stranded products. The length would be affected by the spacing of primer annealing. Also, as the amplification proceeds, annealing of new primers internally on previously displaced DNA strands should act to limit the average product length generated so that it will be less than the 70,000 nt processivity observed in single primer extension experiments.

Exonuclease-resistant random primers

Early attempts to develop isothermal methods like MDA achieved only a few fold amplification. In the late 1990s a biotechnology company called Molecular Staging Inc. took on this problem by studying the kinetics of the reaction. It was necessary to maintain very high DNA polymerase and primer concentrations throughout the reaction in order to create the branching structures that could yield exponential amplification (Dean *et al.*, 2001). The phi29 DNA polymerase (Fig. 7.3, X-axis) and the primers (Fig. 7.3, 5 vs. 50 mM hexamer) were both strongly limiting for the reaction. However, the phi29 DNA polymerase rapidly degrades normal primers with its 3′–5′exonuclease proofreading activity. The proofreading activity contributes to high fidelity during DNA replication (see below) by excising nucleotides misincorporated by the DNA polymerase activity. Unfortunately, it also avidly degrades single stranded oligonucleotides like the random hexamer primers used in MDA (Fig. 7.3, exo-sensitive).

The problem was solved by using random hexamers that are protected from degradation (Fig. 7.3, exo-resistant) with two phosphorothioate linkages at the 3′-terminus (5′-NpNpNpNp^{S}Np^{S}N-3′).

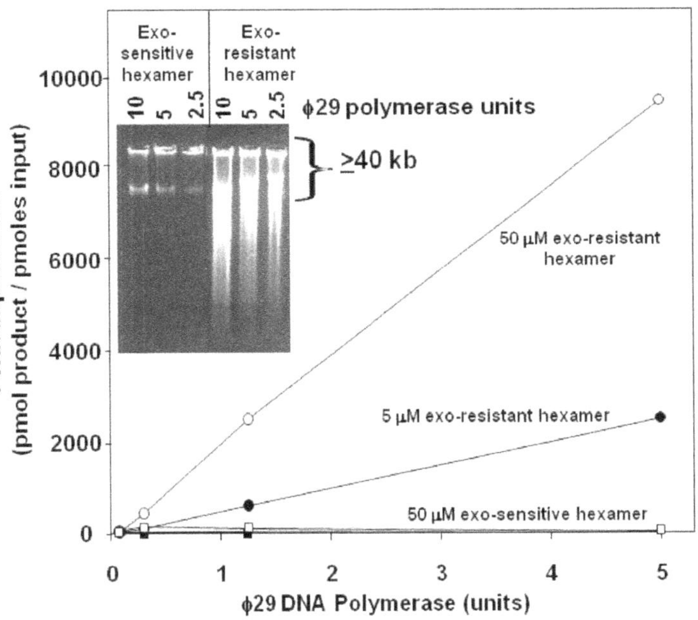

Figure 7.3 MDA with exonuclease sensitive or exonuclease resistant random primers. 20 µl MDA reactions contained 1 ng of M13 DNA as template with exo-resistant or exo-sensitive primers and phi29 DNA polymerase as indicated. DNA synthesis was quantified by the incorporation of radioactive nucleotides. The insert shows MDA reactions products resolved by agarose gel electrophoresis. Reprinted with permission of Genome Research, Cold Spring Harbour Laboratory Press (Dean *et al.*, 2001).

With the primers protected from degradation, the amplification could occur.

DNA yield and fold-amplification

Fold-amplification depends on the starting amount of DNA template since about the same amount of DNA is synthesized no matter how much template is added. MDA tends to be a self-limiting reaction that slows as it approaches a high DNA yield. Typically a concentration of a little less than 1 milligram DNA per ml is produced in one commercial kit tested by the PicoGreen assay (which primarily detects double stranded DNA but not the free nucleotides and random primers in the reaction) (Lasken and Egholm, 2003). Under these conditions, using a 50 μl MDA reaction, one *E. coli* cell would contribute about 5×10^{-15} g of DNA template and yield $30–40 \times 10^{-6}$ g of amplified DNA giving more than a billion-fold amplification. If 1000 cells were used instead (contributing 5 10^{-12} g of DNA for template) about the same yield of 30–40 X 10^{-6} grams of DNA would result but the amplification would only be a million-fold.

Characteristics of amplified DNA and its use in genomic applications

The early published work on the properties of the amplified DNA was carried out with near optimum amounts (>3000 genome copies) of DNA template. These studies established the basic capabilities of MDA where template is not limiting. MDA using a single genome copy from one cell is more challenging and will be discussed later.

Replication fidelity

Fidelity of the DNA polymerase can limit the usefulness of the amplified DNA for accurate genotyping, DNA sequencing, and cloning purposes. The phi29 DNA polymerase has high fidelity with a rate of only 1 error in 10^5–10^6 nucleotides synthesized (Esteban *et al.*, 1993) (or, in an alternative form frequently used to express error rates, 10^{-5}–10^{-6} errors/nucleotide) which is about 100-fold lower (Nelson *et al.*, 2002) than that of Taq DNA polymerase. While the 3′–5′ exonuclease proofreading activity of the phi29 DNA polymerase was problematic because it degraded the random primers, it is necessary to achieve this high fidelity. DNA polymerases lacking a 3′–5′ exonuclease, or phi29 DNA polymerase with this activity removed (unpublished) have not been favoured for use in MDA.

Another critical consideration for DNA amplification is the accumulation of sequence errors throughout the reaction as newly synthesized DNA serves as template for subsequent DNA synthesis. This is extreme for PCR where the DNA products are estimated to have about one mutation per 900 bases after 20 cycles (Saiki *et al.*, 1988); therefore, a minimum of cycles is generally used when PCR products are to be cloned. MDA has an advantage over PCR-based methods because errors accumulate at a far lower rate. This is because the DNA does not double in a cyclic manner in MDA. Rather, many primers are thought to be extended from each template as the branched structures grow (Fig. 7.2C).

The high fidelity of the amplification has been reported in studies of Sanger sequencing from MDA products (Detter *et al.*, 2002; Hawkins *et al.*, 2002). A 500,000 bp region of the amplified DNA yielded the same error rate as unamplified DNA (Paez, 2004). However, it is important to note that the Sanger sequencing method itself typically introduces one error in 100–1000 nucleotides which will mask the less frequent errors introduced by MDA of the template. Therefore, Sanger sequencing is not the most sensitive method to determine the fidelity of MDA. Accuracy in the Sanger method is ultimately obtained by repeatedly sequencing over the same regions at sufficient depth to identify and eliminate the errors, whether introduced by the Sanger method or other sources such as PCR or MDA of the template. At present, the Sanger sequencing data can only demonstrate that MDA did not introduce a significant increase in errors above those inherent to the Sanger method. In the most revealing study of MDA fidelity, the mutation rate generated in an expressed protein was measured and the apparent MDA error rate was derived from this (Nelson *et al.*, 2002). A plasmid containing the lacZ gene (containing 350 detectable nucleotide sites) was used as the MDA template for a 10,000-fold amplification. In a subsequent colony screening about 1.4% were mutated (1488 blue, wild type colonies and 21 white, mutated colonies). It could be calculated that one polymerase error had accumulated per 25,000 bp in the final MDA product (error frequency = 4×10^{-5}).

A far higher number of errors might be expected to accumulate in the billion-fold amplification that occurs for MDA from one cell. Surprisingly, Sanger sequencing of a billion-fold amplification from one *E. coli* cell suggested an error rate similar to that reported for the 10,000-fold amplification used in the lacZ mutation study. 81 errors were found in 59,652 nucleotides of sequence (error rate = 1.36×10^{-3}), however, analysis of each error suggested that all were due to Sanger sequencing and the actual MDA error rate was $<1.7 \times 10^{-5}$ (Zhang *et al.*, 2006). Even without this correction for Sanger-generated errors, MDA is introducing far fewer errors than PCR. Possibly, DNA produced early in the MDA reaction, and having few accumulated errors, provides most of the template for subsequent amplification so that 10^4-fold and 10^9-fold amplifications have a similar error rate. However, more research will be needed to fully understand this. The average number of times that MDA recopies a template (the mechanism by which errors accumulate) has not yet been reported. In any case, the accumulation of errors is many orders of magnitude higher in PCR since all templates are recopied at each cycle resulting in an error rate of about 1×10^{-3} after only a million-fold amplification.

Even though MDA has fewer errors than PCR, in some cases they could still adversely affect a study. When cloning for the purpose of expressing proteins it will be important to carefully consider the mutation rate (Nelson *et al.*, 2002). Another example would be use of MDA to amplify human DNA from a tumour biopsy in order to detect rare somatic mutations. The errors introduced by MDA could be problematic if present at a comparable rate to the cellular mutations under investigation.

In the case of amplification from a single cell, there are additional considerations about the impact of errors that occur early in the reaction. Consider PCR from one bacterial cell. For one DNA polymerase error in the first cycle, three accurate and one mutated strand would result (one product strand with the error, and one product and the two original template strands with no error). Therefore, at the completion of the PCR reaction, about 25% of the DNA obtained would conserve this error. If the starting template had been one single stranded viral genome, 50% of the final PCR product would contain the mutation. Again, MDA has the advantage of many primer extensions from the original template early in the reaction (Fig. 7.2). This is important for genomic sequencing from single cells. A mutation that did occur early in the reaction would still only be present in a small percentage of the final amplified copies and these errors could be eliminated by sequencing to a sufficient depth.

The high fidelity of MDA also allows its use in single nucleotide polymorphism (SNP) assays. SNPs are single base pairs in the genome that vary between individuals and are useful in many genetic studies. The available data is for the human genome and for the case where optimum amounts of >3000 DNA template copies were added to the MDA. When human DNA was amplified by MDA and used to determine genotypes of 2320 SNPs located throughout the genome (Barker *et al.*, 2004), >99.9% of SNP calls were accurate. Other studies reported >99.8% concordance to unamplified genomic DNA (Paez, 2004) and 99.95% accuracy for 3,814 genotyping calls (Tranah *et al.*, 2003). These error rates, on the order of one in a thousand, are higher than the rates found by genomic sequencing or mutation assays because they are only partly due to DNA polymerase misincorporation errors. A more important factor is amplification bias which can cause the drop out of an under-representation allele and consequent miscalling of a heterozygote as homozygous (see below, Allele amplification bias in diploid genomes). MDA accuracy is still high enough to allow use of MDA for many genetic studies where DNA is in short supply (Barker *et al.*, 2004). In one of the most thorough studies to date (Lovmar *et al.*, 2003), MDA was preferred to the PCR-based amplification method 'Improved PEP' (iPEP) based on genotyping success, signal-to-noise ratio, power of discrimination between homozygous and heterozygous SNP genotypes, yield, and authenticity of allele representation.

Amplification bias

Amplification bias refers to the over- and under-representation of different sequences in the amplified DNA. Local sequence effects can reproducibly cause some DNA regions to be amplified better than others. Also, early random events in the reaction can produce an uneven outcome in the final amplified DNA. This results in a stochastic effect in which a given region is over-represented in some MDA reactions and under-represented in others. Up until the late 1990s, the only methods for whole genome amplification were based on PCR, for example, using random primers [primer extension preamplification (PEP) or degenerate primers (degenerate oligonucleotide primed PCR (DOP)]. These methods produce large amplification bias, possibly from thermal cycling due to the wide range of melting temperatures (T_m) for both template denaturation and primer annealing. DOP and PEP caused amplification bias several orders of magnitude higher than for MDA (Fig. 7.4) for eight human loci measured by quantitative PCR (qPCR) (Dean *et al.*, 2002).

Bias was the lowest when the MDA reaction was designed to amplify the template by 10^2-, 10^3- or 10^4-fold. For 10^5-fold amplification, MDA began to create strong bias in one locus. In another study 47 single copy loci, one on the p and q arm of each human chromosome, were tested for bias (Hosono *et al.*, 2003). MDA was carried out with the optimum 10 ng template (≥ 3000 genome copies) to generate 40 µg of amplified DNA, a 4,000-fold amplification. All of the loci, which are single copy in the genome, were still represented at between about 0.5 and 3 copies per genome in the amplified DNA (Fig. 7.5).

The pattern of bias that did occur was reproducible with some loci over- and some under-represented supporting the idea that there are sequence effects on the efficiency of MDA priming. The error bars for each locus are for the 47 different patient blood samples used in MDA and give a good idea of the random bias that occurs between reactions.

In the extreme case of single bacterial cells (which result in more than billion-fold amplification) the random type of bias introduced early in the reaction is far greater and the representation of different sequences will vary over a large range (Raghunathan, 2005). Amplification bias could cause inaccurate results where the goal is to quantify the relative numbers of different species present in a microbial community (Abulencia *et al.*, 2006). For example, one might want to determine how the ratio between two species changes over the seasons of the year. However,

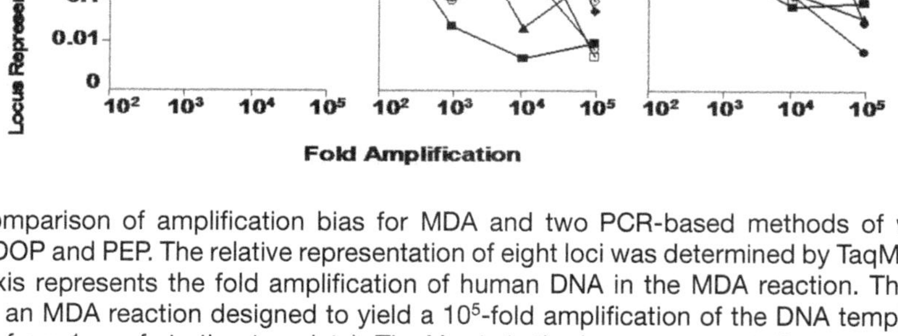

Figure 7.4 Comparison of amplification bias for MDA and two PCR-based methods of whole genome amplification, DOP and PEP. The relative representation of eight loci was determined by TaqMan quantitative PCR. The X-axis represents the fold amplification of human DNA in the MDA reaction. The highest level tested was for an MDA reaction designed to yield a 10^5-fold amplification of the DNA template (100 mg of DNA amplified from 1 ng of starting template). The Y-axis is the locus representation expressed relative to the starting genomic DNA template. A value of 100% indicates that the locus is present in the same copy number in the amplified DNA as in the original DNA. Reprinted with permission from Proc. Natl. Acad. of Sci. U.S.A. (Dean *et al.*, 2002).

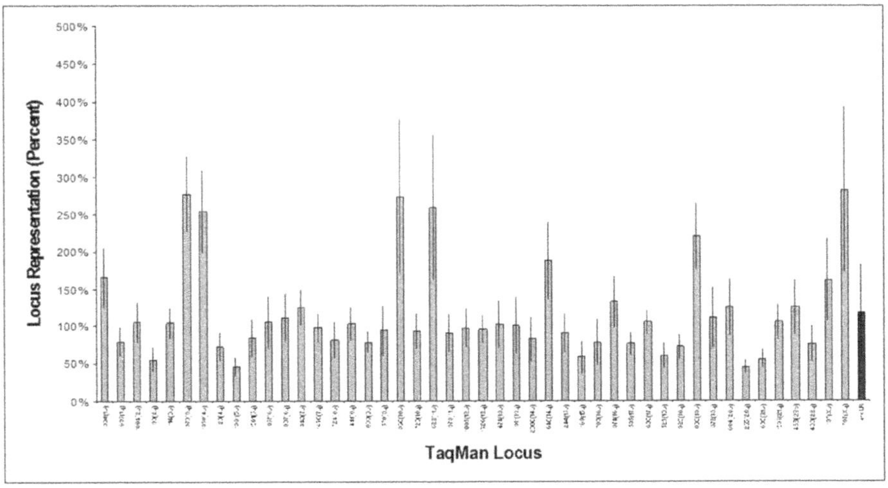

Figure 7.5 Amplification bias analysis by the quantitative TaqMan PCR assay for 47 human loci. Locus representation expressed as the per cent of locus copies in the amplified DNA relative to the starting DNA template. Each bar represents the average of amplifications of 44 different individual's DNA. The average of all of the 47 loci was 117% (far right bar). Reprinted with permission of Genome Research, Cold Spring Harbour Laboratory Press (Hosono *et al.*, 2003).

MDA can give accurate results when the goal is only to detect the presence or absence of a microbe (as in medical or biodefence applications) or to obtain the genomic sequence or the genotype of particular genes of interest. Many detection and sequencing methods are based on use of specific oligonucleotides, used as hybridization probes or primers for PCR or sequencing, and can be highly tolerant of amplification bias in the DNA.

Allele amplification bias in diploid genomes

The genotyping studies with human DNA confirm reliable representation in the amplified DNA of both alleles (one from the maternal and one from the paternal chromosome) in the diploid human genome. Poor amplification of either parental chromosome relative to the other can result in allele drop out artefacts in which a heterozygote is incorrectly scored as a homozygote. Therefore, accurate SNP genotyping requires that the phi29 DNA polymerase copy the template with high sequence fidelity and also that both parental alleles are well represented. Presumably, the alleles are about equally amplified because they are surrounded by identical flanking sequences in the two chromosomes that would contribute the same priming efficiencies in MDA. An exception would be a heterozygous haplotype difference where the alleles had different flanking sequences (R. Lasken, unpublished).

These genotyping studies used an optimum concentration of input human DNA template in the MDA reaction. However, allele amplification bias depends on the starting number of template copies. When less than optimal amounts of DNA template are used, or when the DNA is of poor quality, then amplification bias between alleles results in a higher rate of genotyping miscalls. Use of MDA to generate sufficient template for SNP genotyping from single human blastomeres was only partially successful and will require improved methods to reduce amplification bias between alleles (Handyside, 2004). In contrast, bacteria have haploid genomes. While amplification bias can result in complete loss of sequence, it does not have the disadvantage of causing genotyping miscalls where a heterozygote is mistaken for a homozygote. Single-cell MDA should be useful in demonstrating the presence of SNP genotypes that correlate to important phenotypes such as the virulence of pathogens. However, the absence of a particular genetic marker in the sequence cannot be easily interpreted since it may have been lost in the MDA reaction.

Cell lysis and the effect of DNA template quality on MDA performance

For individual cells, any DNA loss or damage to the single template copy will negatively impact the amplification. The options for single-cell lysis are limited to procedures that avoid harsh chemicals, such as SDS, that would inhibit any enzymatic method like MDA or PCR. Cell lysis methods that are compatible with MDA include heating the cell to 95°C (Dean *et al.*, 2001), and treatment with KOH and neutralization with buffer (Dean *et al.*, 2002). Both methods also serve to denature the DNA providing single stranded template for the MDA. Sonication and pressure are other potential methods to lyse single cells (unpublished) although KOH or heat might still be needed to denature the template. Lysozyme and related proteins might be used to digest bacterial cell walls if they do not inhibit downstream applications and if they are free of DNA contamination. This can be tested in negative controls lacking a DNA template.

Harsh enough treatment is needed to effectively lyse the cell, but this must be balanced against the need to minimize DNA damage which will result from high temperature, exposure to high pH, or sonication. Exposure time to KOH must be long enough to lyse the cell, depending on its particular morphology (and potentially its physiological state); however, damage to the DNA will occur when it is released. Similar considerations are needed for heat or sonication which lyse cells but also damage DNA. For an environmental sample with mixed species, a generic lysis protocol should be selected that maximizes results. Or, increasingly harsh conditions might be used on different sample aliquots in order to expand the range of species obtained ranging from easily lysed to difficult cells or spores.

Early studies on the effect of template length on MDA were also carried out with optimal amounts of >3000 template copies. For sonicated DNA, samples of at least 2–4 kb average length were most efficiently amplified based on qPCR for the representation of a genetic marker (Fig. 7.6, Locus Representation on the Y axis).

There was limited amplification for template of 1–2 kb length. A consistent finding was also reported for partially degraded DNA archived in long-term storage (Lasken, 2005). However, it has not been determined whether the limited amplification resulted from the 1–2 kb template still remaining or, alternatively, from trace amounts of high molecular weight DNA still present in these sonicated or stored samples. MDA fails from cDNA template suggesting that more study is needed to determine the minimal template length required for MDA. One solution is a kit (Qiagen Inc.) that ligates cDNAs into concatemers or circles that effectively extend the length of the template. The highly processive phi29 DNA polymerase is expected to utilize longer DNA templates most efficiently. The MDA branching mechanism (Fig. 7.2C) might also require sufficient length template to occur efficiently.

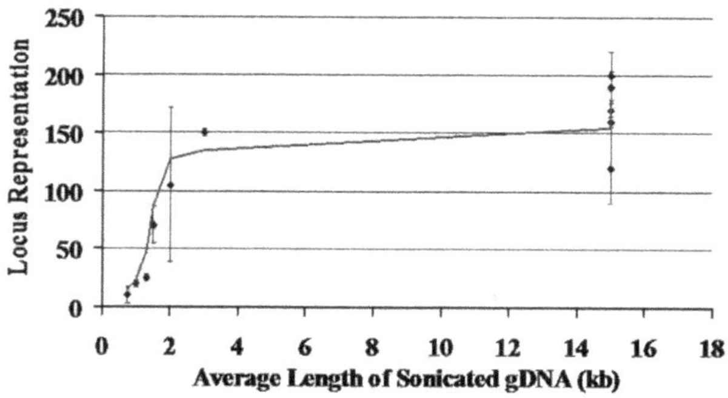

Figure 7.6 The effect of DNA template length on MDA performance. Human DNA was sonicated to produce samples of different average length. The samples were then used as template for MDA reactions. Quantitative PCR by the TaqMan assay was carried out for a test locus to determine its representation in the amplified DNA. Use of DNA templates shorter than 2–4 kb in the MDA reactions resulted in a loss of locus representation in the amplified DNA. Reprinted with permission from Scion Press (Lasken, 2005).

Damage to both DNA strands of the single genome copy of a bacterium will necessarily result in loss of sequence. Double-stranded breaks in the DNA will cause under-representation of the ends which are known to be poorly amplified by MDA. Furthermore, during the assembly of sequences into a complete genome it will not be possible to form contigs across a site where there was a double-stranded DNA break in the template since the phi29 DNA polymerase will not generate amplification products spanning this region. Other types of DNA template damage such as chemical modification or loss of the base result in replication errors if not a complete block of the DNA polymerase in MDA. One of the challenges of discovering novel microbes will be in optimizing cell lysis. Microbes requiring harsher treatments will suffer greater insults to the DNA template. High-temperature KOH treatment and sonication are known to damage DNA and need to be used only to the extent required (Lasken *et al.*, 2005). If the lysed cell is not immediately used in MDA, the DNA quality can also depend on how long it is stored and storage conditions, and the number of times frozen and thawed. Finally, loss of DNA that is not released from the cell after the lysis step, or that is released but becomes bound to material used in the procedure will also contribute to incomplete sequencing. Low DNA binding tubes and microtitre wells such as those appropriate for PCR should used in all MDA steps.

Effect of DNA template concentration on the MDA reaction

Total copy number of template affected an MDA's quality when tested in SNP genotyping assays (Fig. 7.7).

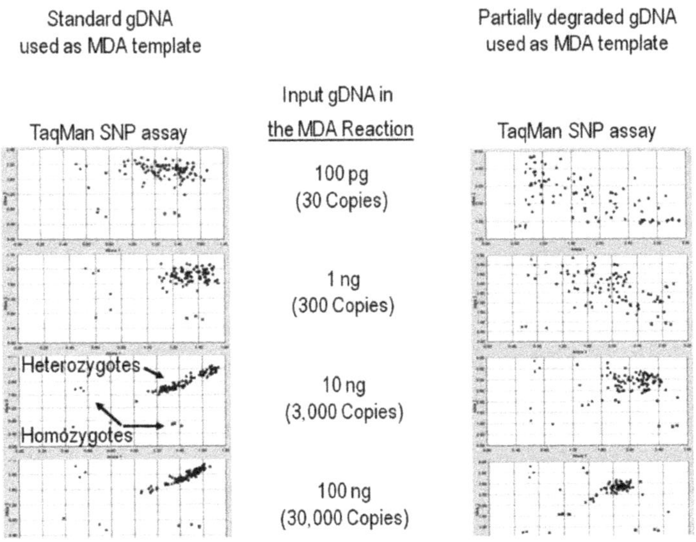

Figure 7.7 The effect of DNA template concentration and quality on MDA. Purified genomic DNA (standard gDNA) of an individual heterozygous for the test locus was amplified in 84 replicate MDA reactions. DNA that had been in long term storage and was known to be of poor quality (Partially degraded gDNA) from a second heterozygous individual was also amplified in 84 replicate MDA reactions. The amplified DNA was then tested for genotyping by the TaqMan SNP assay (locus CYP2C19*2, ABI TaqMan Pre-Developed Allelic Discrimination Assay). A TaqMan SNP assay uses fluorescent detection of a PCR specific for one allele of the SNP (X axis) and another PCR specific for the other allele (Y axis). Homozygotes will be near either the X or Y axis (homozygote control data points indicated by arrows). Heterozygous individuals (giving PCR detection of both alleles) should appear as a tight population along the diagonal of the graph (Heterozygotes arrow). Scattering of the data points from the diagonal axis indicates unequal amplification of the two alleles during the MDA reaction. The amplification bias increased when the MDA reaction had less input gDNA template or when the template was partially degraded. DNA was added as template for the MDA reactions in amounts indicated. More of the degraded gDNA template was required (100 ng) than standard gDNA (10 ng) to generate amplified DNA that was optimal in the TaqMan assay. Reprinted with permission of Scion Press (Lasken, 2005).

As human DNA template decreased in the MDA, amplification bias between the alleles increased. This is reflected as scattering of the heterozygous data points which should cluster in a narrow band along the diagonal of the graph (Fig. 7.7, 100 ng, 30,000 copies). Presumably, early in the MDA reaction, favoured amplification of one allele over the other occurs on a random basis when there are very few template molecules present. As expected from this kind of stochastic process, the two loci had about the same likelihood of being over- or under-represented (demonstrated by the scattering of the data points to either side of the diagonal line). It is instructive that DNA damage had the same effect as limiting input template copy number. In effect, a larger amount of partially degraded DNA was needed to achieve the same MDA performance as high molecular weight template. 100 ng of partially degraded DNA was needed as MDA template to get about the same results as 10 ng of standard gDNA template based on the diagonal distribution of heterozygote data points.

Chimeric DNA rearrangements caused by the MDA reaction

Chimeras are DNA rearrangements such as the joining together of two segments that are separated in the original genomic DNA template. They can be formed as an artefact of DNA amplification methods and are of concern for MDA when used for genomic sequencing. Sanger sequencing of cloned MDA products revealed up to 50% chimeric inserts (3 kb average clone insert length) (Zhang *et al.*, 2006). Chimeras were reduced by about 80% (to a final rate of 6% of inserts screened) by three enzymatic treatments carried out after the MDA reaction was completed: (1) removal of random primers and addition of more phi29 DNA polymerase and dNTPs to complete extension of growing 3′-termini (thereby 'de-branching' the DNA), (2) S1 nuclease to digest remaining single stranded DNA, and 3) flow shearing and size selection on agarose gels followed by nick translation with *E. coli* DNA polymerase I. It was postulated that chimeras were generated by the cloning step, with *E. coli* somehow rearranging branched DNA forms present in the library inserts. However, pyrosequencing of MDA products by the 454 Life Sciences method, a cloning-independent method, revealed a similar chimera frequency (one chimeric junction per ~20 kb) indicating that the MDA reaction itself, rather than cloning, generated the chimeras (Lasken and Stockwell, 2007).

The principal mechanism of chimera formation results from the nature of MDA in which 5′-ends are displaced by the DNA polymerase (Fig. 7.8). These single strands provide alternative templates for any nearby 3′-primer terminus. Following and initial extension of a random primer (Fig. 7.7A), this growing strand can switch to a new template (Fig. 7.8B–D). Growing 3′-ends are often facilitated in switching by annealing to any short region of complementarity in the new template (Fig. 7.8D).

The complementarity ranged from 2 to 21 base pairs (Fig. 7.8E). The MDA reaction model revealed why the majority of chimeras (85%) were inversions. Growing 3′ termini are exposed primarily to nearby 5′ strands and, by annealing to these new templates of opposite polarity, the fully extended product will consist of two sequences joined in inverted orientation. This study suggests an alternative explanation for how post-MDA treatment with S1 nuclease reduces chimeras. Digestion of the single stranded regions postulated as MDA reaction intermediates (Fig. 7.8D) would result in fewer joined segments both in Sanger sequencing and the clone-free sequencing methods.

Fortunately, chimeras tend to occur randomly and so they can be resolved by sequencing a genome to a sufficient depth (Ishoey *et al.*, 2008; Zhang *et al.*, 2006). For example, by sequencing to a depth of 10×, there will be an average of nine correct sequences to compare to any chimeric sequence. However, chimeras will be a greater difficulty in instances were low sequencing depth is obtained including environmental shotgun sequencing where only low coverage is possible for most of the genomes represented.

Genome coverage by MDA using optimum amounts of DNA template

The percent of the genome still present after amplification reflects many factors including

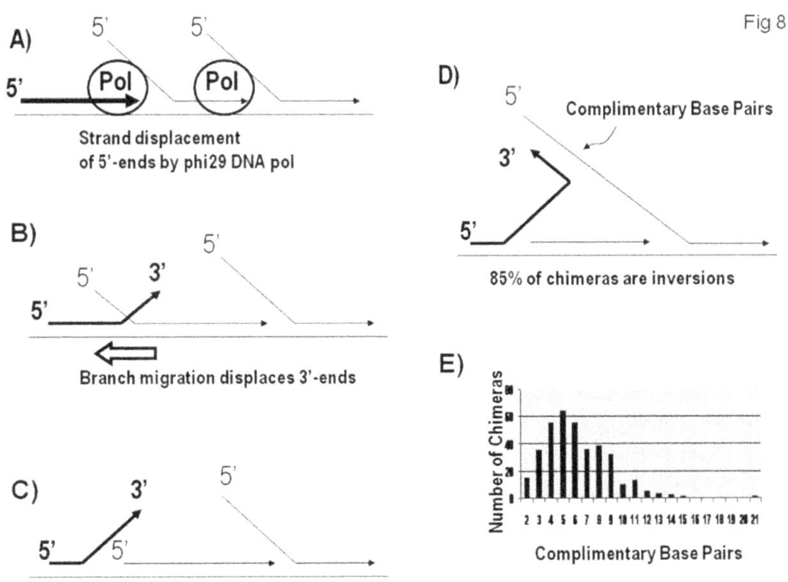

Figure 7.8 Mechanism creating chimeric DNA inversions in MDA. In MDA, as the DNA polymerase (Pol) extends the 3′-ends of primers (arrows), it displaces downstream 5-ends. If these 5′-ends reanneal via branch migration, the partially extended 3′-ends will be displaced. Re-annealing of these 3′-ends to other nearby 5-ends results in extension on a second template of opposite polarity and, consequently, formation of chimeric inversions. Re-annealing of 3′-ends was facilitated at short complementary sequences on the new templates of about 2 to 21 nucleotides in length (bar graph insert). Reprinted with permission of BioMed Central (Lasken and Stockwell, 2007).

(1) certain sequences that are apparently not amplified well by MDA such as human telomeres (Dean *et al.*, 2002), (2) amplification bias which causes drop out of some sequences on a random basis (Ishoey *et al.*, 2008), and (3) DNA damage, particularly in the case where very low amounts of template are present such as single-cell MDAs. With optimum levels of template in the MDA reaction, most of the human genome is covered. The studies of SNP genotyping accuracy provide some information on the per cent of the genome conserved. Of 2,320 SNPs assayed throughout the human genome (Barker *et al.*, 2004), all were present in the amplified DNA with no significant difference in call rate between amplified and unamplified DNA. Another study estimated that 99.82% of the human genome was present in amplified DNA based on array probing of >10,000 SNP alleles (Paez, 2004). When low levels of template must be used, DNA damage and amplification bias affect the ultimate coverage of the genome in the amplified DNA.

Amplification bias from single cells and genome coverage

An important factor affecting the outcome of a genomic sequencing project is the completeness of genome coverage from the MDA reaction and the degree of amplification bias. While MDA has relatively low amplification bias for optimum amounts of template, it is no simple matter to amplify the genome of one cell a billion-fold. Priming and the start of exponentially growing branched structures (Fig. 7.1) would not be synchronous across the genome and early variations occurring on the single template copy will result in bias in the final MDA product. On a random basis, some sequences are amplified more than others ranging over thousands of fold (Ishoey *et al.*, 2008; Lasken, 2007; Raghunathan, 2005). In general, a greater sequencing depth (repeated sequencing over the same DNA) will be required to maximize genome assemblies from one cell and some genomic regions will be lost. Typically, about 50–80% of genomes have been obtained using enough depth to recover the underrepresented sequences (Ishoey *et al.*, 2008). Some strategies to reduce amplification bias and loss of sequence are covered in later sections.

MDA quality control assays for improved genomic sequencing

One approach to improving success in single-cell studies is to develop quality control assays that identify the most specific and unbiased MDA reactions. A DNA quantification assay (for example, PicoGreen) will confirm that typical DNA yields were achieved in the MDA and provide concentrations for use in designing sequencing procedures. Low MDA yields can indicate the presence of reaction inhibitors contaminating the DNA template, such as humic acid from soil samples, or nucleases introduced by biological specimens or as cross contamination during experimental procedures, that can degrade MDA primers and reaction products. Where the DNA sequence of the target microbe is known, PCR can be used to test MDA reactions and the most complete amplification chosen for sequencing (Marcy *et al.*, 2007a,b). For example, this strategy should be useful for re-sequencing of known pathogens in studies of the genetic diversity between individual cells. Quantitative PCR methods reveal the presence of desired sequences and also give an indication of the amplification bias. By carrying out qPCR at several different loci it is also possible to get an indication of the specificity of the amplification for the template. Some loci will be overrepresented and others underrepresented, however, if the average level of all loci is low, this will indicate that much of the amplified DNA (quantified by PicoGreen) is nonspecific such as primer derived artefacts or products from contaminating DNAs. This type of quantitative quality control assay was used to identify amplified human DNA suitable for use in genotyping (Hosono *et al.*, 2003; Yan *et al.*, 2004). Analysis of two test loci by real time qPCR TaqMan assays was fairly diagnostic for the entire genome. When these loci were represented at normal levels, it was generally correlated to comprehensive representation of the entire genome.

For discovery of novel microbes, PCR might also be used to test MDA reactions for some sequences normally present such as the 16S rRNA gene (Lasken, 2007). Other sequences that are postulated to be present based on related organisms might also be used to design PCR primers. It is convenient that PCR products derived from a single-cell MDA reaction can be directly analysed by cycle sequencing. There is no need to clone the PCR products since they are all derived from the same organism. This will enable a low cost strategy of (1) enriching for a desired microbe (for example, by cell sorting or micromanipulation using FISH or other labelling techniques), (2) carrying out many single-cell MDA reactions, (3) screening the amplified DNAs by PCR for the 16S rRNA gene to identify the desired taxa, and (4) screening the amplified DNAs for other potential PCR targets to confirm the MDAs of greatest interest. If the PCR reactions are quantitative (i.e. qPCR) they will also provide an indication of per cent coverage and amplification bias prior to sequencing.

Strategies for genomic sequencing projects

Pyrosequencing directly from MDA reactions

The performance of amplified DNA as a template for pyrosequencing (454 Life Sciences) was evaluated using MDA from single *E. coli* cells (Marcy *et al.*, 2007a). MDA was carried out in standard 50 μl reactions (Repli-g kit, Qiagen) and also in 60 nl reactions using a microfluidic chip technology. For a quality control assay of the MDA reactions, prior to use in pyrosequencing, qPCR was carried out for 10 *E. coli* genomic loci in order to assess the genome coverage and amplification bias and to allow selection of the best MDA reactions for sequencing. The amplification bias was generally greater in the 50 μl MDAs than in the microfluidic reactions. The specificity of MDA for a single plasmid DNA template copy was also reported to be greater with reduced reaction volume (Hutchison *et al.*, 2005). Specificity refers to the ability of the reaction to amplify the intended template as apposed to generating artefact sequences (for example, primer dimers and related non-specific products). Based on qPCR for 10 loci, an average of about 43% of the amplified DNA in the microfluidic MDA reactions was derived from *E. coli* and 26% was specific for the standard 50 μl reactions (Marcy *et al.*, 2007a).

The best 50 μl and 60 nl MDA reactions (with > 80% *E. coli* sequence) were selected for 454 pyrosequencing (Table 7.1). Genome assembly was similar for the two samples: 82% of the reads assembled, 10% of the reads were identified as likely repeats, and less than 0.4% of the reads were identified as problematic by the assembler, possibly due to being chimeric sequence, or for some other reason.

For both samples, > 95% of all reads mapped fully or partially to the *E. coli* reference genome, a level that is reasonably consistent with the qPCR analysis which estimated that > 80% was specific. Therefore, the qPCR reactions provided a useful quality control assay prior to sequencing. Even for novel microbes, qPCR quality control assays based on 16 rRNA gene sequences or other identified loci could be used to assure a high level of success in subsequent sequencing efforts. One possibility that remains to be tested is that the 454 sequencing protocol itself may improve specificity if small nonspecific DNAs generated in the MDA reaction, such as primer dimers, are lost during the 454 steps of PCR or DNA clean-up. The genome assembly process also improved specificity for *E. coli* sequence with over 99% of the contigs mapping to the reference genome. The high level of specificity bodes well for use of MDA in sequencing novel organisms where there are no previously obtained sequences available to guide the genome assembly.

Given the number of nucleotides sequenced and the known size of the *E. coli* genome, it is possible to estimate the per cent of the genome that should be covered using an approach called the Lander–Waterman model. The MDA reactions resulted in less coverage (about 60–70% of the genome) than predicted (~85–95%) from the ~10 million nucleotides of sequence obtained (Table 7.1). The amplification bias created by MDA can account for this discrepancy. Over- and under-representation of sequences should result in the coverage being lower than the ideal value. Nevertheless, the ability to sequence unculturable microbes, even with the requirement of sequencing to greater depth, promises to be a major advancement in microbiology.

Table 7.1 Analysis of 454 sequencing from single *E. coli* cell MDA reactions. About 100,000 reads (10 million nucleotides) were obtained for each MDA reaction

Measurement	Microfluidic, 60 nl MDA	50 μl MDA
Number of 454 sequences	97,470	114,551
% reads assembled	82%	82%
% reads no overlap	7–8	7–8
% likely repeats	10	10
Identified as problematic sequences by 454 assembler	< 0.4%	< 0.4%
% contigs mapping to *E. Coli*	> 99%	> 99%
% Genome coverage by contigs	32.57	38.66
% reads fully or partially mapping to *E. Coli*	> 95%	> 95%
% reads partially mapping to *E. coli*	2.16	4.12
% reads unmapped to *E. coli*	0.36	1.36
Inferred read errors	0.55	1.12
% of K12 genome covered Measured/predicted by Lander Waterman model	60.75/86.84	67.37/90.41
Coverage measured/Coverage predicted	0.70	0.75
Chimera rates	1/11,880 bp	1/21,977 bp

Sequencing of uncultured microbes from single-cell MDA reactions and genome assembly

Sequencing of a novel uncultured microbe was not possible until the advent of MDA and metagenomic technologies. Some practical strategies have been suggested for sequencing from single-cell MDA reactions and there is growing experience in the level of success that is typically obtained (Ishoey *et al.*, 2008). The amplified DNA is most suitable for pyrosequencing (454 Life Sciences) and Sanger sequencing at this time. The newer technologies of Illumina sequencing (Illumina Inc.) and SOLiD sequencing (Invitrogen/Applied Biosystems) have extremely high throughput and lower cost, but they generate very short sequence reads that make assembly of the genome more challenging. However, these new sequencing platforms are very valuable for 're-sequencing' projects where a known reference genome is available to guide the assembly, as well as for genome polishing (the final steps of resolving any gaps or uncertainties in the completed genome). Re-sequencing of known genomes of microbial pathogens, for example, would allow analysis of genetic diversity of virulence factors between individual cells. There is also great interest in sequencing novel organisms with these technologies; however, work is still in progress to enable genome assembly from the many short sequencing reads in the absence of a known reference sequence. Improved versions of the methods are also being introduced that will yield longer reads as well as paired end reads (short reads derived from both ends of a longer sequence) which may soon enable sequencing of novel genomes at greatly reduced cost.

The Sanger and 454 methods have worked well for single-cell sequencing although draft assemblies have tended to be more fragmented compared to assemblies from unamplified DNA due to DNA template damage or loss, and amplification bias from MDA. The amplification bias in the single-cell MDA reactions can be seen in a depth distribution graph of the sequence reads (Fig. 7.9A) (Ishoey *et al.*, 2008).

For single-cell MDA of Flavobacteria and TM7a (Fig. 7.9A, top and middle panels) the representation of different genomic sequences varies over a wide range compared to a typical example with unamplified genomic DNA (Fig. 7.9A, bottom panel, *Xylanimonas cellulosilytica*). Greater sequencing depth is required to complete such assemblies with added cost and labour. The presence of about one chimeric DNA rearrangement per 10–20 kb in the MDA reaction is also a problem for large insert clone libraries like BACs, used in Sanger sequencing, since a majority of inserts would contain at least one rearrangement. The inversions (85% of all MDA chimeras) would disrupt paired end reads from these long inserts which are very valuable in the assembly, in particular for genomes with long repeat sequences. However, the less frequent chimeras in small to mid-size insert clone libraries have not prevented sequence assembly when sufficient sequencing coverage was used. Chimeras occur randomly throughout the genome (Lasken and Stockwell, 2007) so that most reads across a given sequence will be correct and a rearranged sequence read will be eliminated.

An integrated approach to genomic sequencing from single cells was evaluated at the DOE Joint Genome Institute. Pyrosequencing was chosen because of lower labour and cost compared to Sanger sequencing. It also avoided use of clone libraries which create cloning bias in Sanger sequencing. However, the addition of Sanger sequences provided high-quality coverage of homopolymeric regions and improved the assembly by bridging regions through paired end information of larger insert clones (Ishoey *et al.*, 2008). Some of the remaining gaps could be closed by primer walking (using sequences already obtained to design new PCR primers spanning a gap) on medium and small insert size clones. Also, PCR/adapter PCR could be used from the original MDA reaction. Apparently, these sequences that were not originally recovered by 454 pyrosequencing were only underrepresented in the MDA product and could be recovered by PCR (Ishoey *et al.*, 2008; Zhang *et al.*, 2006). In spite of the limitations, current MDA protocols enable the reconstruction of major portions of the genome from one cell of uncultured microorganisms. MDA from single cells of *Prochlorococcus*, *E. coli*, and TM7 yielded up to 70% of their genomes. For a marine Flavobacterium approximately 80%

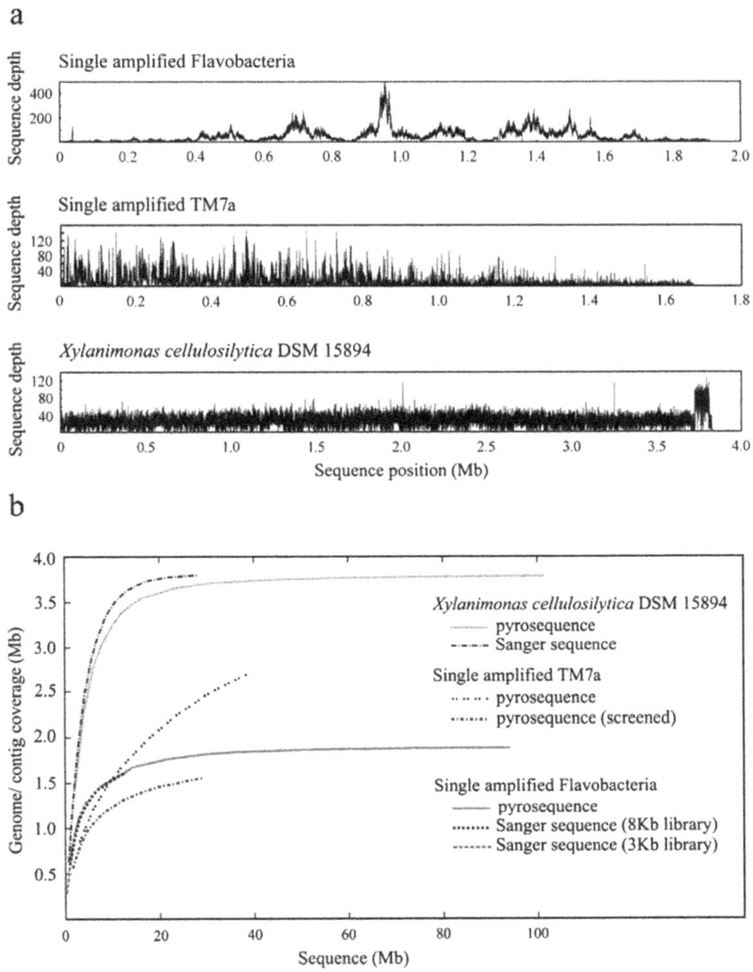

Figure 7.9 Characterization of single-cell MDA DNA sequence. (A) 454 Pyrosequence depth distribution for a single amplified Flavobacterium (Top panel) and a single amplified TM7a cell (middle panel) demonstrate that MDA resulted in high regional fluctuations in sequence coverage, while the sequence depth for a typical unamplified sample, *Xylanimonas cellulosilytica* DSM 15894 (bottom panel), exhibits a low degree of variation. Sequence positions were determined via arranging contigs by length. TM7a pyrosequence was screened to exclude reads contributing to contigs <300 bp and human reads. The mean sequence depth is 85.7 (± 73.7) for the Flavobacterium, 14.4 (± 16.2) for TM7a, and 31.9 (± 13.9) for *X. cellulosilytica*. A certain degree of sequence depth fluctuation may be attributed to repeats and mis-assemblies. (B) Genome coverage as a function of the genome sequencing effort. Additional shotgun sequencing would not be effective in recovering the underamplified genomic regions in the single amplified Flavobacterium (estimated genome size of ~2.3 Mb) or TM7a (no genome estimate available). The *X. cellulosilytica* genome is estimated to be ~3.8 Mb. Reprinted with permission of Current Opinion in Microbiology (Ishoey *et al.*, 2008).

of the genome was recovered (Ishoey *et al.*, 2008). A graph of the genome recovery as a function of the sequencing effort indicates near-saturation (Fig. 7.9B), that is, additional shotgun sequencing would mostly result in repeated sampling of the overamplified genomic regions. Recovery of the remaining 20% of the genome and genome closure may require a combination of approaches including use of multiple cells and integration of single-cell and metagenomic data. Another approach may be to use information in the sequence drafts to guide successful development of culture methods, still a critical goal of the biological research. At present, partial genomes and completed composite genomes are sufficient for many exciting applications.

Future directions

MDA is enabling new progress in the discovery of uncultured environmental microbes. Single-cell genomics also promises to be useful in the study of the human microbiome and in infectious disease research. It will be possible to discover bacteria living in the human body by sequencing directly from individual cells. Even for pathogens that can be cultured, it will be possible to analyse genetic diversity between individual cells without the selective loss of genotypes that results from growth in culture (Fig. 7.10).

Improvements to the MDA reaction reducing amplification bias will be needed to reduce genotyping errors when amplifying from single human cells and loss of genomic sequence when amplifying to sequence novel microbes. However, the future will bring new advancements in the enzymology of DNA amplification methods. Perhaps even the ultimate goal in the biotechnology of DNA amplification, complete chromosomal replication *in vitro,* will be achieved. Efforts to improve the MDA reaction and optimize it for new applications are under way in many labs. The growing demand for DNA substrate for use in genomic and detection applications underscores the continued need for innovation in the field of DNA amplification.

Another area of rapid advancement is transcriptional analysis from single cells. By taking advantage of next-generation sequencing methods, such as the SOLiD sequencing platform, single eukaryotic cells have been investigated for gene expression and regulation (Lao *et al.,* 2009). These researchers have succeeded in synthesizing cDNA from the RNA contained in one cell

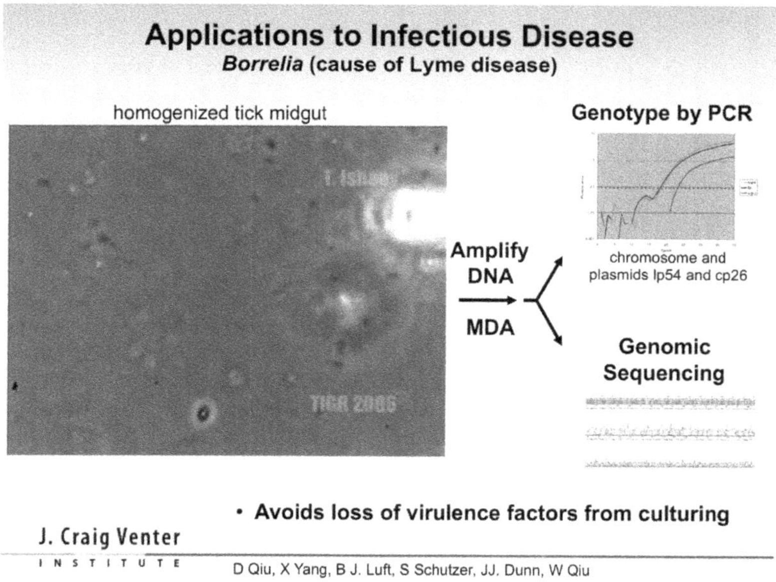

Figure 7.10 Use of single-cell MDA to analyse microbial pathogens. Single cells of *Borrelia* (arrow), the cause of Lyme disease, were isolated directly from tick midgut tissue by micromanipulation. The individual cells were amplified in 5 μl MDA reactions yielding approximately 4.5μg DNA product which was used as template in PCR reactions with primers targeting *Borrelia* genes. The PCR products were sequenced and confirmed the genotypes of outer surface proteins (ospA, ospB, osp C). In older studies using PCR from total DNA extracted from the tissue, it was not possible to determine if genotypes of different loci were genetically linked within individual cells since the data was derived from all of the bacterial cells present. While culturing isolates is an alternative method that does provide linkage information within single cells, the organisms for which adequate culture conditions are available are very limited and there is often strong selection for fast growing strains biasing the results. Moreover, virulence factors associated with plasmids can be rapidly lost in culture. Reprinted with permission from Curr. Opin. Microbiol. (Ishoey *et al.*, 2008).

and sequencing this by the SOLiD method. The method has also been demonstrated for single embryonic cells (Tang *et al.*, 2010a), oocytes and blastomeres from mice (Lao *et al.*, 2009; Tang *et al.*, 2010b). These massively parallel sequencing efforts yield enormous amounts of data, on the order of 10^7–10^8 total reads per run (Lao *et al.*, 2009). Sequencing of cDNA has not been achieved yet from single bacterial cells. Single-cell mRNA sequencing methods make use of the 3′-poly(A) tail of eukaryotic transcripts to initiate reverse transcriptase mediated cDNA synthesis (Tang *et al.*, 2010a). In combination with poly(A)-targeted PCR amplification they demonstrate ultra high throughput sequencing, digital gene expression profiling, and transcript variant discovery from a single cell. Since a 3′-poly(A) sequence is absent in bacterial mRNA, cDNA synthesis necessitates alternative methods to prime the reaction. Methods such as the addition of poly A tails with poly(A) terminal transferase (Fu *et al.*, 2005) or adapter ligation (Meiri *et al.*, 2010) are currently being explored. We anticipate that these kinds of obstacles may soon be overcome and that transcriptomics from single bacterial cells is an attainable goal in the near future.

References

Abulencia, C.B., Wyborski, D.L., Garcia, J.A., Podar, M., Chen, W., Chang, S.H., Chang, H.W., Watson, D., Brodie, E.L., Hazen, T.C. *et al.* (2006). Environmental whole-genome amplification to access microbial populations in contaminated sediments. Appl. Environ. Microbiol. *72*, 3291–3301.

Barker, D.L., Hansen, M.S., Faruqi, A.F., Giannola, D., Irsula, O.R., Lasken, R.S., Latterich, M., Makarov, V., Oliphant, A.; Pinter, J.H., *et al.* (2004). Two methods of whole-genome amplification enable accurate genotyping across a 2320-SNP linkage panel. Genome Res. *14*, 901–907.

Blanco, L., Bernad, A., Lazaro, J.M., Martin, G., Garmendia, C., and Salas, M. (1989). Highly efficient DNA synthesis by the phage phi 29 DNA polymerase. Symmetrical mode of DNA replication. J. Biol. Chem. *264*, 8935–8940.

Dean, F.B., Nelson, J.R., Giesler, T.L., and Lasken, R.S. (2001). Rapid amplification of plasmid and phage DNA using Phi 29 DNA polymerase and multiply-primed rolling circle amplification. Genome Res. *11*, 1095–1099.

Dean, F.B., Hosono, S., Fang, L., Wu, X., Faruqi, A.F., Bray-Ward, P., Sun, Z., Zong, Q., Du, Y., Du, J., *et al.* (2002). Comprehensive human genome amplification using multiple displacement amplification. Proc. Natl. Acad. Sci. U.S.A. *99*, 5261–5266.

Detter, J.C., Jett, J.M., Lucas, S.M., Dalin, E., Arellano, A.R., Wang, M., Nelson, J.R., Chapman, J., Lou, Y., Rokhsar, D. *et al.* (2002). Isothermal strand-displacement amplification applications for high- throughput genomics. Genomics *80*, 691–698.

Esteban, J.A., Salas, M., and Blanco, L. (1993). Fidelity of phi 29 DNA polymerase. Comparison between protein-primed initiation and DNA polymerization. J. Biol. Chem. *268*, 2719–2726.

Fu, H., Tie, Y., Xu, C., Zhang, Z., Zhu, J., Shi, Y., Jiang, H., Sun, Z., and Zheng, X. (2005). Identification of human fetal liver miRNAs by a novel method. FEBS Lett. *579*, 3849–3854.

Handyside, A.H., Robinson, M.D., Simpson, R.J., Omar, M., Shaw, M.A., Grudzinskas, J.G., and Rutherford, A. (2004). Isothermal whole genome amplification from single and small numbers of cells: a new era for preimplatation genetic diagnsosis of inherited disease. Mol. Hum. Reprod. *10*, 767–772.

Hawkins, T.L., Detter, J.C., and Richardson, P.M. (2002). Whole genome amplification – applications and advances. Curr. Opin. Biotechnol. *13*, 65–67.

Hongoh, Y., Sharma, V.K., Prakash, T., Noda, S., Taylor, T.D., Kudo, T., Sakaki, Y., Toyoda, A., Hattori, M., and Ohkuma, M. (2008). Complete genome of the uncultured Termite Group 1 bacteria in a single host protist cell. Proc. Natl. Acad. Sci. U.S.A. *105*, 5555–5560.

Hosono, S., Faruqi, A.F., Dean, F.B., Du, Y., Sun, Z., Wu, X., Du, J., Kingsmore, S.F., Egholm, M., and Lasken, R.S. (2003). Unbiased whole-genome amplification directly from clinical samples. Genome Res. *13*, 954–964.

Hutchison, C.A., III, Smith, H.O., Pfannkoch, C., and Venter, J.C. (2005). Cell-free cloning using ϕ29 DNA polymerase. Proc. Natl. Acad. Sci. U.S.A. *102*, 17332–17336.

Ishoey, T., Woyke, T., Stepanauskas, R., Novotny, M., and Lasken, R.S. (2008). Genomic sequencing of single microbial cells from environmental samples. Curr. Op. Microbiol. *11*, 198–204.

Kornberg, A., and Baker, T.A. (1991). DNA Replication: W.H. Freeman and Company.

Kvist, T., Ahring, B.K., Lasken, R.S., and Westermann, P. (2007). Specific single-cell isolation and genomic amplification of uncultured microorganisms. Appl. Microbiol. Biotechnol. *74*, 926–935.

Lao, K.Q., Tang, F., Barbacioru, C., Wang, Y., Nordman, E., Lee, C., Xu, N., Wang, X., Tuch, B., Bodeau, J. *et al.* (2009). mRNA-sequencing whole transcriptome analysis of a single cell on the SOLiD system. J. Biomol. Tech. *20*, 266–271.

Lasken, R.S. (2005). Multiple displacement amplification of genomic DNA: In Whole Genome Amplification, Hughes, S., and Lasken, R., eds. (Bloxham, UK: Scion Publishing), pp. 99–118.

Lasken, R.S. (2007). Single cell genomic sequencing using Multiple Displacement Amplification. Curr. Op. Microbiol. *10*, 1–7.

Lasken, R.S., and Egholm, M. (2003). Whole genome amplification: abundant supplies of DNA from precious samples or clinical specimens. Trends Biotechnol. *21*, 531–535.

Lasken, R.S., Raghunathan, A., Kvist, T., Ishøy, T., Westermann, P., Ahring, B.K., and Boissy, R. (2005). Multiple displacement amplification from single bacterial cells. In Whole Genome Amplification, Hughes, S., and Lasken, R., eds. (Bloxham, UK: Scion Publishing), pp. 119–147.

Lasken, R.S., and Stockwell, T.B. (2007). Mechanism of chimera formation during the Multiple Displacement Amplification reaction. BMC Biotechnol. 7, 19.

Levy, S., Sutton, G., Ng, P.C., Feuk, L., Halpern, A.L., Walenz, B.P., Axelrod, N., Huang, J., Kirkness, E.F., Denisov, G. *et al.* (2007). The diploid genome seqeunce of an individual human. PLoS Biol 5:e254.

Lizardi, P.M., Huang, X., Zhu, Z., Bray-Ward, P., Thomas, D.C., and Ward, D.C. (1998). Mutation detection and single-molecule counting using isothermal rolling-circle amplification. Nat. Genet. *19*, 225–232.

Lovmar, L., Fredriksson, M., Liljedahl, U., Sigurdsson, S., and Syvanen, A.C. (2003). Quantitative evaluation by minisequencing and microarrays reveals accurate multiplexed SNP genotyping of whole genome amplified DNA. Nucleic Acids Res. *31*, e129.

Marcy, Y., Ishoey, T., Lasken, R.S., Stockwell, T.B., Walenz, B.P., Halpern, A.L., Beeson, K.Y., Goldberg, S.M.D., and Quake, S.R. (2007a). Nanoliter Reactors Improve Multiple Displacement Amplification of Genomes from Single Cells. PLOS 3, 1702–1708.

Marcy, Y., Ouverney, C., Bik, E.M., Lösekann, T., Ivanova, N., Martiny, H.G., Szeto, E., Platt, D., Hugenholtz, P., Relman, D.A. *et al.* (2007b). Dissecting biological 'dark matter' with single-cell genetic analysis of rare and uncultivated TM7 microbes from the human mouth. Proc. Natl. Acad. Sci. U.S.A. *104*,11889–11894.

Meiri, E., Levy, A., Benjamin, H., Ben-David, M., Cohen, L., Dov, A., Dromi, N., Elyakim, E., Yerushalmi, N., Zion, O. *et al.* (2010). Discovery of microRNAs and other small RNAs in solid tumors. Nucleic Acids Res. *38*, 6234–6246.

Mussmann, M., Hu, F.Z., Richter, M., de Beer, D., Preisler, A., Jorgensen, B.B., Huntemann, M., Glöckner, F.O., Amann, R., Koopman, W.J.H. *et al.* (2007). Insights into the genome of large sulphur bacteria revealed by analysis of single filaments. PLOS Biol, 5,1923–1937 (e230).

Nelson, J.R., Cai, Y.C., Giesler, T.L., Farchaus, J.W., Sundaram, S.T., Ortiz-Rivera, M., Hosta, L.P., Hewitt, P.L., Mamone, J.A., Palaniappan, C. *et al.* (2002). TempliPhi, phi29 DNA polymerase based rolling circle amplification of templates for DNA sequencing. Biotechniques Suppl, 44–47.

Paez, J.G., Lin, M., Beroukhim, R., Lee, J.C., Zhao, X., Richter, D.J., Gabrial, S., Herman, P., Sasaki, H., Altshuler, D. *et al.* (2004). Genome coverage and sequence fidelity of phi29 polymerase-based multiple-strand displacement whole-genome amplification. Nucleic Acids Res. *32*, e71.

Podar, M., Abulencia, C.B., Walcher, M., Hutchison, D., Zengler, K., Garcia, J.A., Holland, T., Cotton, D., Hauser, L., and Keller, M. (2007). Targeted Access to the Genomes of Low Abundance Organisms in Complex Microbial Communities. Appl. Environ. Microbiol. *73*, 3205–3214.

Raghunathan, A., Ferguson, H.R., Bornarth, C.J., Driscoll. M., and Lasken, R.S. (2005). Genomic DNA amplification from a single bacterium. Appl. Environ. Microbiol. *71*, 3342–3347.

Rodrigue, S., Malmstrom, R.R., Berlin, A.M., Birren, B.W., Henn, M.R., and Chisholm, S.W. (2009). Whole genome amplification and *de novo* assembly of single bacterial cells. PLoS One *4*, e6864

Saiki, R.K., Gelfand, D.H., Stoffel, S., Scharf, S.J., Higuchi, R., Horn, G.T., Mullis, K.B., and Erlich, H.A. (1988). Primer-directed enzymatic amplification of DNA with a thermostable DNA polymerase. Science *239*, 487–491.

Tang, F., Barbacioru, C., Bao, S., Lee, C., Nordman, E., Wang, X., Lao, K., and Surani, M.A. (2010a). Tracing the derivation of embryonic stem cells from the inner cell mass by single-cell RNA-Seq analysis. Cell Stem Cell *6*, 468–478.

Tang, F., Barbacioru, C., Nordman, E., Li, B., Xu, N., Bashkirov, V.I., Lao, K., and Surani, M.A. (2010b). RNA-Seq analysis to capture the transcriptome landscape of a single cell. Nat. Protoc. *5*, 516–535.

Tranah, G.J., Lescault, P.J., Hunter, D.J., and De, V.I. (2003). Multiple displacement amplification prior to single nucleotide polymorphism genotyping in epidemiologic studies. Biotechnol. Lett. *25*, 1031–1036.

Woyke, T., Xie, G., Copeland, A., Gonzalez, J.M., Han, C., Kiss, H., Saw, J.H., Senin, P., Yang, Ch., Chatterji, S. *et al.* (2009). Assembling the marine metagenome, one cell at a time. PLoS One *4*, e5299.

Yan, J., Feng, J., Hosono, S., and Sommer, S.S. (2004). Assessment of multiple displacement amplification in molecular epidemiology. BioTechniques *37*, 136–143.

Zhang, K., Martiny, A.C., Reppas, N.B., Barry, K.W., Malek, J., Chisholm, S.W., and Church, G.M. (2006). Sequencing genomes from single cells by polymerase cloning. Nat. Biotechnol. *24*, 680–686.

8 A Community Cyberinfrastructure Resource for Metagenomics Research: CAMERA 2.0

Jing Chen, Shulei Sun, Weizhong Li and John C. Wooley

Abstract

The sustained deluge of complex experimental data in metagenomics opens extraordinary opportunities for this new science, but also requires an advanced, cyberinfrastructure-based knowledge resource that provides a platform for rigorous analysis of the data sets and in turn, of the microbial communities. In response, we established the Community Cyberinfrastructure for Advanced Microbial Ecology Research and Analysis or CAMERA (http://camera.calit2.net/) and recently released a very major extension to this resource, termed CAMERA 2.0.

In this chapter, we describe the essential attributes of CAMERA 2.0 and their relationship to expectations of the experimental and computational researchers in metagenomics. An analysis portal provides access to a semantically aware database along with computational infrastructure. Meticulous attention is paid to annotating the sequence data with metadata, the associated contextual information (environmental parameters). Using the standards established by the Genomics Standard Consortium, CAMERA 2.0 enables ready depositing, locating, analysing, visualizing and sharing data and the use of a range of actual analytical processes or workflows.

These workflows, most notably, are user-selected, allowing a flexible approach in which an investigator can set their own requirements for a given sequential analysis. The analytical and computational history is archived, allowing accurate reproducibility, detailed further analysis, and validation.

Introduction

Research on microbes has undergone a scientific revolution in the past decade, building from early work using molecular probes for phylogeny and population monitoring and then empowered by ever cheaper, faster sequencing technology and advanced computational methods. Responding and exploiting the resulting transformation to a data rich science, the convergence of numerous different subdisciplines gave rise to metagenomics as a newly emerging scientific discipline (NRC Metagenomics Committee, 2007). The general biological details, the state of the art discoveries, and the potential for the new science are extensively and rigorously reviewed in the other chapters in this volume, along with the earlier book in the series (Marco, 2010), as well in earlier reviews, such as Arima and Wooley (2008), Handelsman (2004) and Riesenfeld *et al.* (2004). Correspondingly, we will not describe the biological methods and research findings in this article, but will focus on a response to the especially complex and data-rich nature of experimental observations in metagenomics. First and foremost, ever larger, more disparate and more rapidly acquired data sets are intrinsic features for modern biology. In particular, the now established science of genomics, with its high throughput technologies, already has brought an expanded scale and complexity to extant biological data, and metagenomics will build upon this status. The deluge of complex data has brought new opportunities in bioinformatics and extensive requirements for data or knowledge resources to sustain experimental work and the growth of theoretical or *in silico* analysis.

Contemporary commercial instruments and methods, the so-called next generation sequencing technologies, such as 454 pyrosequencing, Illumina, and SOLiD, generate large amounts of sequencing data in a short time with affordable costs, and there clearly will be continuing 'next generation' technologies driving exponential increases in data at a system levels, such as genomics or transcriptomics. However, the sequencing reads are sometimes only low quality reads and represent only partial fragments of the genomes sampled (Wooley *et al.*, 2010). Natural or *in situ* populations of microbes can contain even more than 10,000 species, leading to very complex data sets. At the same time, there remain many computational challenges to remove artefacts, annotate, and analyse the primary or sequence data (Wooley and Ye, 2010; Wooley *et al.*, 2010), which must be linked to environmental context information. The essential data resource for sequence-related biology research is provided by the comprehensive sequence information available within the International Nucleotide Sequence Database or INSD (Cochrane *et al.*, 2010), an integrated, shared dataset, established by the curators of DDBJ, ENA and GenBank [which is the INSD Collaboration or INSDC; Cochrane *et al.*, 2010]. Most notably during this explosive growth phase of the new science of metagenomics, specialized community information resources are also needed to address the complexity of the microbial communities, their genomes, and their natural (non-laboratory/non-cultured) conditions including population data and environmental context. At the same time, we emphasize that it is important to sustain the high quality of the central community resource, INSD, for many reasons, including that so many boutique databases obtain their fundamental primary input from one or another collaborator, and that these sites serve as the ultimate archive for nucleic acid sequence data.

An added impetus for a specialized metagenomics resource is the complexity of the requisite contextual data or metadata, which provides further information about the primary or sequence data. Genomic data per se have had a policy of stable annotation from the initial investigator, and while in 2002, a secondary resource was created to include third party genomic annotation, at the present, the need for archival fidelity has meant usage in the sequence database itself only of the annotation originally contributed by the sequence team (Cochrane *et al.*, 2006; NRC Metagenomics Committee, 2007). The very process and details of metadata annotation for metagenomics will differ for different environments (NRC Metagenomics Committee, 2007) and will change over time, in part since best practices and researcher participation in community standards for collecting and annotating is actively being established at this time (Field *et al.*, 2008). Most notably, annotating the sequences from microbial populations from vastly different environments will necessitate some unique features for the diverse categories, and the majority of the metadata is likely to be generally added or updated over time by curators and members of the research community, not by the researchers doing the initial sequencing (NRC Metagenomics Committee, 2007). In sum, the rapid growth of metagenomics research and the unparalleled growth in sequence data as well as the complexity of the population data creates challenges and expectations for contemporary information technology and computing tools, and a requirement for a new specialized resource to empower further research and insight at the scale of the experimental data.

The Community Cyberinfrastructure for Advanced Microbial Ecology Research and Analysis or CAMERA (http://camera.calit2.net/) was established in response to these novel features, challenges and opportunities in metagenomics research, with a strong mandate to provide an information framework for advanced insight into microbial ecology. The earliest data ingestion involved largely data from marine and soil metagenomic samples (Seshadri *et al.*, 2007), but the incorporation of all categories of metagenomic data and reference genomes or microbial data sets is underway. Thus, the requisite data for comparative studies and for an in depth analysis of the world of microbes will be included to allow researchers to use all of relevant data for their analyses.

The CAMERA team has recently released a very major extension of the resource, termed version 2.0 (Sun *et al.*, 2010). An advanced web-based analysis portal provides the community of users with access to a semantically aware database along with a suite of associated computational infrastructure. Overall, as an information resource, the CAMERA project has created a single system for microbial biology that enables ready depositing, locating, analysing, visualizing and the sharing of data and actual analytical processes (workflows: see below). In addition, CAMERA follows the standards established by the Genomics Standard Consortium (GSC) and works within the GSC in advancing the standards and notably, to enhance the potential value of the metadata. Meticulous attention is paid to annotating the primary or sequence data with metadata, that is, the associated contextual information (environmental parameters). At the same time, by providing a computational infrastructure at the same time, the resource can provide a wide and growing collection of established and newly developed bioinformatics tools. Most notably, the resource includes a flexible or user-selected work flows to allow researchers to work within their own requirements for a given analysis in order to characterize better the functional implications of their experimental data.

We introduce, immediately below, the four most essential features of CAMERA 2.0 (below termed simply as CAMERA) in terms of its primary data and metadata; computational and bioinformatics tools; novel software and process pipelines or scientific workflows; provenance (i.e. the recording of all steps and features associated with the processing of data); our continuing, essential prioritization in a dialogue with the broad science community; and the provision of options (by way of workflows) for users. Next, we describe the essential attributes of this current release of CAMERA with regard to value for users, indicate its unique features and underlying IT considerations, and outline ongoing development and future plans and their relationship to expectations of the experimental and computational researchers in metagenomics. We also include examples of research accomplishments using CAMERA and thus, provide suggestions on how new users can use CAMERA effectively to explore the microbial world, reflect on central decisions for overall implementation and describe our ongoing development efforts and expectations for the future.

Metagenomic sequence, metadata, and annotation ingestions and queries

CAMERA has integrated 103 billion bps of metagenomic sequencing data, along with available metadata. [The specific breakdown in base pairs as of 10 November 2010: Sequencing Reads, 58 billion (B); Assembly, 14.8B; ORFs, 21B; Proteins, 7.8B; and Non-Coding RNA, 0.8B.] It covers environmental metagenomic projects deposited in NCBI (Sayers *et al.*, 2010), and includes additional, still ongoing projects. Researchers are able to query on metadata categories such as habitat, sample collection location, time, salinity, pH, and other environmental physicochemical measurements to match their own needs. Importantly, CAMERA is compliant with the standards promulgated by the Genome Standards Consortium (GSC) (Field *et al.*, 2008), and sustains a role within GSC for extending standards for content and format of the metagenomic data and metadata submissions. In addition, for its metagenomics data sets, CAMERA incorporates pre-computed annotation results using reference data resources such as PFAM (Finn *et al.*, 2010), COG (Tatusov *et al.*, 1997), KEGG (Kanehisa *et al.*, 2010), Geno Ontology (GO-EBI and EMBL-EBI, 2010).

High-performance computational resources and bioinformatics tools

CAMERA has developed and/or integrated tools for sequence quality control, assembly, gene prediction, annotation, and functional and comparative genomics applications. Via a standards compliant, centralized, open access Web Portal interface, CAMERA also provides high-performance networking access and grid-based computing, which provides storage and visualization capacity as well. Through the portal users can customize and execute their chosen workflows, run sophisticated queries, and submit data. The data server complex contains a large amount of storage and large computational clusters. The tools and resources make more rapid, comprehensive metagenomic analysis possible. The compute cluster currently consists of 103 Dell 1950s with two dual-core 2.33 GHz Intel Xeon processors and between 4 and 16 Gb of

RAM and 8 Dell 1935s with two quad-core AMD Opteron 2356 processors with 16 gigabytes of ram. This cluster is backed by 18 terabytes of storage on a network file system hosted by a Sun X4500 server. The compute cluster uses ROCKS to manage the cluster, ganglia for statistics, and Intermapper for monitoring. Sun Grid Engine installation is used to run jobs on the cluster. The system can, when needed, access additional, shared, large computational resources, such as UC San Diego's Triton (http://tritonresource.sdsc.edu).

Scientific workflows, the provenance browser and sharing of data within research groups

CAMERA organizes tools in the Kepler (http://kepler-project.org) scientific workflow system (Altintas *et al.*, 2010b; Deelman *et al.*, 2009; Ludaescher *et al.*, 2006). The Kepler system provides a graphical user interface for users to manage and in batch mode run a suite of serial computational analyses that they have selected; it also includes a framework for provenance, the archive of all workflow execution and thus, the capacity to reproduce and validate. In supporting interaction of computational tools and manual analysis, the system allows users not only to view and download results, but also, by using the provenance information, to browse the step by step details on how the results have been generated. To increase efficiency in analysis and avoid redundancy, users are able to create research teams or collaborating groups that share results and data.

Interaction with research communities

CAMERA provides a platform for users to upload their own workflows, which thereby support tool integration from the broader research communities. (As of November, 2010, this cyber resource has 4050 users from 93 nations.) CAMERA also provides data submission tool to allow researchers to share and forward data to other metagenomics sites and to other community archives such as GenBank, to ensure wide access to data and annotation. We hold workshops on specific use cases where there is a major effort under way, hold tutorials at workshops and meetings, are extending web-based tutorials and other user information, and encourage interactions on the part of users including visits to optimize the value of the resource.

CAMERA database requirements, architecture and management

Functional requirements for the CAMERA database

In essence, the CAMERA project involves a very interdisciplinary 'journey' with molecular microbial ecology at its core and the use of state of the art cyberinfrastructure to open new windows to knowledge. The computing and information technologies need to serve many functions to support diverse experimental and theoretical/modelling research in metagenomics. As such, the backend database system, the corner stone for CAMERA, has been constructed to meet the wide range of requirements that come from the broad, interdisciplinary community. We describe the key features below.

Data repository

Across the biological sciences the rapid growth of data, often complex data, has increased the need for computational and bioinformatics approaches. For metagenomics (as well as genomics, of course) sequencing projects (ranging from those of major centres to individual researchers) generate an increasing number of data daily since the new sequencing technologies routinely lower the cost of sequencing. Many metagenomics research projects need to cross-reference a range of sequence data beyond their own experimental data, i.e. data sets likely to be derived from other group projects. Individual groups do not have the facility to acquire, host and maintain such large amounts of data. Thus, the database needs to provide storage for a large collection of sequence data. At the same time, the data stored in the database, despite its volume, need to be easily and quickly accessed by researchers, matters we address below.

An integrated data source

In metagenome analysis, researchers often need to compare sequences from different metagenomic samples in order to find meaningful relationships among them in the ecosystem. For example, an ecologist might want to compare sequences from many samples based on different spatial, temporal or some biochemical parameters to find certain patterns. Many metagenomic samples have been sequenced; these are stored in different labs, sequencing centres, and data repositories. All these data sources might have different schemas and data policies, and provide different query interfaces to the users. In order to query multiple data sources under these conditions, it would be time-consuming, if not difficult, to come to understand the data schema from the disparate data resources and also, to learn how to use different query tools. To alleviate this, CAMERA has been committed to providing a unified data source that integrates all metagenomic samples and allows users to query across them, and to download data they are interested in from a single interface. By integrating these data into one database, most of the metagenomic sequences that currently exist in isolated reports have been collected, curated and managed in one place, making the data far more accessible and doing so especially for comparative or integrative studies. Among the fundamental questions researches want to answer for the microbes in populations are 'Who they are, what do they do and how they do it?' As a consequence, in addition to analysing metagenomic data, an essential operation involves comparing metagenome sequences to reference genome sequences, which will provide information such as taxonomy, protein family, gene and pathway information. The CAMERA database , thus, provides updated reference genome sequences from the organisms that have been fully sequenced.

Managing metadata for sequence data

The sequence metadata can be critical to certain metagenomic studies, since they provide contextual information to the primary information that carried by the sequence data, and the metadata itself might represent the key differences between two environments. Such metadata includes temporal data such as when the sample was collected (winter or summer, day time or at night), spatial data such as geo-coordinates, depth or height, and the chemical environmental data, such as temperature, pH and salinity. In order to allow users to cross-referencing such types of metadata, the Database needs to provide a consistent method to manage metadata and to provide query interfaces to users so that they can easily find out environment samples that are interest to them, based on a requested/specified search criteria.

Readily used cyberinfrastructure

Another role of the CAMERA database, besides as a data repository, is to provide backend data management for various data analysis applications (or workflows) in CAMERA's computational environment. This includes analytical results, workflow parameters and provenance data, and requires the Database be able to handle a high amount of online transactions.

Sustained inclusion of leading edge technology

Being able to support semantic queries is another requirement for the Database. Semantic queries can help users to obtain or manipulate data in different data resources or computer-based databases without knowing their detailed syntactic structures. Users just need to focus on constructing the logic of their queries (i.e. what information is needed) without having to know advanced computer science techniques (such as joins across a database). More importantly, semantic queries enable retrieval of both explicitly and implicitly derived information. For example, if a user wants to find all aquatic metagenomic samples, a semantic query will not only find the samples whose habitat types are explicitly identified or tagged as 'aquatic', it can also find sample that are tagged as 'open ocean', 'river', 'lagoon', etc. On the contrary, a simple text-based query will not retrieve the samples that don't mention 'aquatic'. This simple example reflects the more complex queries that we need as researchers in today's molecular-based microbial ecology in order to probe the complexities of the natural microbial world and the impact of microbes on the environment and their response and influence, for example, on climate, pollution and its remediation, and carbon cycling.

CAMERA database architecture

To enhance effective use of the CAMERA database as a resource for the community, essential technical details about the components in the database are summarized in the next few sections. For users with experience in bioinformatics queries, these technical details may help them obtain an in depth analysis of the data stored in SDB and discover the range of questions that can be answered from the database, especially by the SDB. In addition, CAMERA's Database schema can be used as a reference implementation to help others in implementing similar sequence data management systems. For a more technical, detailed description on the advanced web-based portal, developed using GridSphere (http://www.gridsphere.org), and the CAMERA 2.0 user services, based on REST architectural principles, see Lin *et al.* (2011). GridSphere is a standards-compliant portal framework and the REST (Representative State Transfer) technology is an alternative, data-centric approach to building Web Services and describes a loosely coupled and lightweight means for implementing conventional Service-Oriented Architecture; using such an approach allows CAMERA Web Services to operate similarly to the Web itself. In addition, for an in depth description of the use in CAMERA 2.0 of scientific workflows and their provenance, and a comprehensive review of the database architecture and related cyberinfrastructure technologies of CAMERA 2.0, see Altintas *et al.* (2010b,c). For more background on the open source Kepler (http://kepler-project.org) community development of scientific workflows, see Altintas *et al.* (2010b), Deelman *et al.* (2009) and Ludaescher *et al.* (2006) and references therein. The importance and use of provenance for scientific data, and the more extensive requirements than for other forms of provenance, are active research topics in computer science; its status and attributes have been analysed and reviewed extensively (Altintas *et al.*, 2010a; Davidson and Freire, 2008; Freire *et al.*, 2008; Serra da Cruz *et al.*, 2009).

In order to meet the requirements described in the previous (Requirements) section, we divided the Database into four sections or schemas (see Fig. 8.1) based on their functionality:

1. Data submission: If a user submits metagenomic data to CAMERA for publishing, the data will be managed in this schema. Accession numbers for projects, samples, libraries will be assigned in this schema. Once these data pass the QA (or curation) process of CAMERA, they will be transformed to the semantically aware database of CAMERA database.
2. Semantically aware database (SDB): This schema supports different primary sequence and metadata queries, which includes advanced sequence metadata searches, exporting sequence data for BLAST (creating various BLAST databases for the CAMERA portal) and other workflow applications.
3. User data: This schema manages users' own private data, such as BLAST results generated from portal applications.
4. Provenance data: This schema stores workflow provenance data, i.e. ownership and process, computer state and computational history.

The basis for such divisions is that certain types of functionality might have different impacts on implementation. For example, the schema that supports user applications needs to be able to handle a higher transaction rate, whereas the schema that supports query is typically only a read-only data warehouse for users. By dividing them into four schemas, it is easier to implement security policy, provide hardware storage, establish maintenance routines and optimize the performance. In order to provide high availability for the applications, we use an Oracle cluster instead of a single instance of a database server. Given that the size of the database is already quite large and growing very rapidly, this architecture provides better load balancing and scalability. On the application level, besides allowing CAMERA applications to make direct connections to the Database, a service layer provides essential queries to applications via web Services.

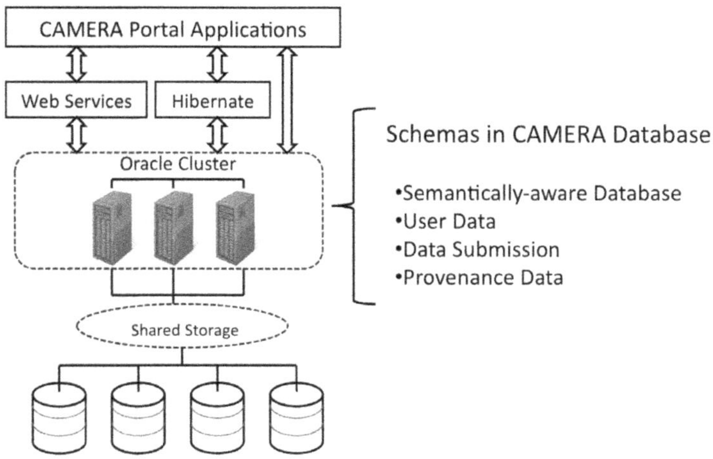

Figure 8.1 Database Architecture.

Metadata management

Metagenomic researchers need to access not just large amount of primary sequence data, but also the complex metadata that provides the context for sequence data. Meeting metadata requirements is one important aspect of the overall CAMERA database and is one of the key components in the SDB. Metadata for genome/metagenome samples is difficult to manage due to the lack of consistency in how much metadata need to be collected, how they need to be called, what unit need to be used, and so on. The Genomic Standards Consortium (http://www.gensc.org) has made a significant contribution in helping the research community have and use standards for the metadata collection. Of especial relevance to metagenomics, their MIGS/MIMS/MIENS (Field *et al.*, 2008) check list (http://www.gsc.org) has an extensive list of metadata. This list provides guideline on what types of metadata need to be collected for different environmental conditions. GSC has also recommended canonical names for these metadata types. Doing so provides a solid foundation for metadata management, integration and exchange.

The MIxS check list established by the GSC has broad coverage of metadata, but there remain large amounts of data that are not defined in the standard. The GSC has standardized metadata names, but has not yet standardized units for the environmental measurements. Of course, the GSC efforts to expand and extend standards are an ongoing process. Without comprehensively standardized units, it will be hard to implement queries interfaces that can support complex queries on numeric values, such as using the potential information about physical environments believed to play important roles in community ecology. For an example of the difficulty, one study might record sample depths in feet, whereas other studies might record sample depth in metres, or in inches. Without active (and sometimes, impossible) editing of input data by curators or manual intervention by users, this inconsistency will make it harder to answer integrative queries, such as in this simple example, 'Find me all the samples whose sample depth is between 0 – 10 metres'.

In order to make our data model fully compliant with the MIMS/MIGS/MIENS (MIxS) and still have the capability to handle any non standard data, we created a data dictionary table called CAMERA_DATATYPES in our data model to store information such as the metadata name, description, type, and unit. Both standard and non-standard metadata are managed and every metadata type is registered in this metadata dictionary. If a metadata type is already defined by GSC standard, we use the standard name. If not, we create a CAMERA standardized name for it. The units for numeric values are also defined in this data dictionary. An advantage of this design is that it can be easily updated whenever the GSC releases new standards or adds or updates arbitrary metadata types. Metadata that are compliant with MIxS can be easily loaded into the SDB, which also includes the documentation for the information about each kind of metadata, such as, its description, data type, category and value range.

The SDB organizes metadata with an entity relationship (ER) model, as shown in Fig. 8.2.

The entity PROJECTS includes project type, description, publications, principal investigators and a number of cross database links related to a metagenome or genome project. Within PROJECTS, data related to the raw material collected from the environment is stored in the entity termed BIO_MATERIALS. Temporal data, such as when the sample was obtained, are stored in BIO_MATERIALS. Geospatial data is stored in the COLLECTION_SITE entity and referenced by the BIO_MATERIALS entity. The data include geographical location, sample depth, region name and additional attributes such as habitat types. Besides storing geographical coordinates as floating point numbers in the traditional way, we created an Oracle geometry type field to store the location of the site. The GEOMETRY DATA TYPE allows the storage of sample sites that are not just a simple point and provides an interface for advanced GIS queries with Oracle's geospatial support. For the cases that the raw material is collected from a host organism, metadata related to the host, such as the species' name and health status, is stored in the COLLECTION_HOST entity. The metadata

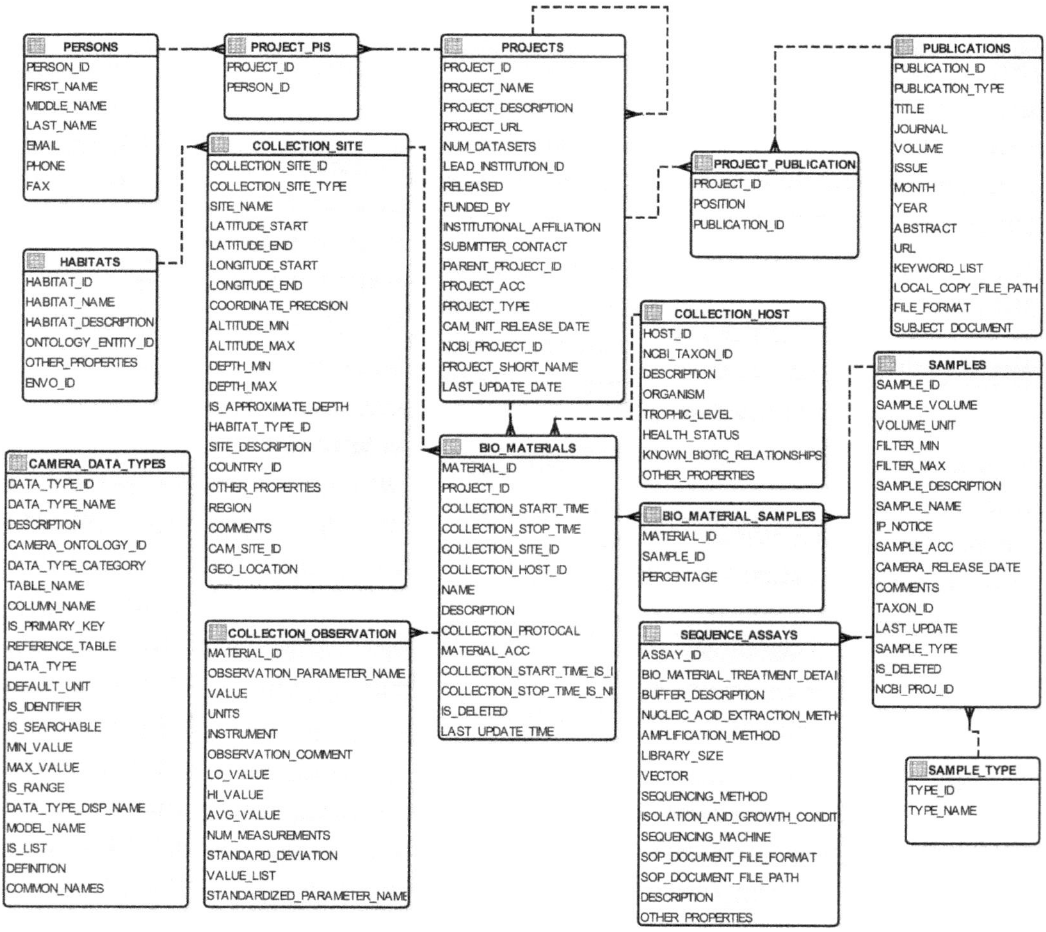

Figure 8.2 The Entity-Relationship (ER) diagram for Metadata Management in SDB.

required by the MIxS specification is normalized and stored in various entities. Most of the metadata in the MIxS checklist are stored in the COLLECTION_OBSERVATION table, which include metadata such as temperature, salinity and chlorophyll density. The schema of the COLLECTION_OBSERVATION table is able to store any metadata related to the raw material, not just the metadata defined by the GSC. The PUBLICATIONS entity includes the information about the publications, and has a many-to-many relationship with the PROJECTS entity. The HABITATS table includes a collection of defined habitat types. At the moment, these are all from the community development resource known as EnvO, or Environment Ontology (http://www.environmentontology.org/), but if or when needed, this table can also store habitats that are not (yet) in EnvO.

Samples collected from the environment are initially prepared for processing and this preparation process and the prepared sample are defined as the SAMPLES entity. It includes attributes such as sample volume, maximum filter size (and pre-filter size), minimum filter size, description and the NCBI project ID (if there is any). The sequencing process of a sample is defined as the SEQUENCE_ASSAYS entity, which stores information including the sequencing method, the sequencing machine and DNA type. The SAMPLE_TYPE entity defines the investigation type of a DNA sample, as specified by the MIMS check list: eukaryotes, bacteria and archaea, plasmid, virus, organelle, metagenome.

The use of semantic applications to search existing databases to derive extensive and even unexpected knowledge has increased rapidly. Bioscience researchers recognize the value of access to semantic databases that provide computer-mediated content rather than exclusively human-oriented content. The advantage of a semantic database is that data are defined and linked in a way that enables more effective discovery, automation, integration, and the re-use across various applications. However, existing technologies to allow users to create data repositories using semantic models are not mature enough to handle a very large database like the CAMERA database . In order to support semantic queries that are implemented in the application layer, we added semantic information on top of our existing relational data model through a metadata dictionary table; namely, CAMERA_DATATYPES. This table is also a semantic data dictionary in which metadata in the SDB can be mapped to external ontology terms. For example, the metadata type 'habitat' is mapped to Environment Ontology term 'habitat' in this data dictionary. By looking up the CAMERA_DATATYPES table, applications can also map the values of metadata type 'habitat' to EnvO terms. Currently, inference in semantic queries of CAMERA relies exclusively on external ontologies, but we are implementing a 'CAMERA Ontology,' which in turn will be treated as another 'external ontology' by the system. Entities, attributes and relations in the CAMERA database are also stored in CAMERA_DATATYPES table with their related ontology links. This semantic data dictionary table not only allows us to integrate camera data with external ontologies, e.g. Gene Ontology (GO) and EnvO, but also is essential for integrating analytical results from scientific workflows into the CAMERA database.

Managing metagenomics sequencing reads, assemblies and annotations

The Metagenomic Sequence Data is the core module in the CAMERA database. Integrated sequencing reads, assembled sequences, open reading frames (ORFs) and hypothetical proteins and annotations of other attributes are combined into one system (Fig. 8.3).

The different types of sequence data in the CAMERA database are represented using several entities based on their semantic meaning. The reads that come directly from sequencing machines are stored in the RAW_SEQUENCES table. The assembled sequences such as contigs and scaffolds are stored in the CONSTRUCTED_SEQUENCE_REGISTRY table. The mapping between assembled sequences and reads is preserved in the CONSTRUCTED_SEQ_RAW_SEQ table. This allows researchers to track back the abundance information from reads even if their analysis is done on assembled

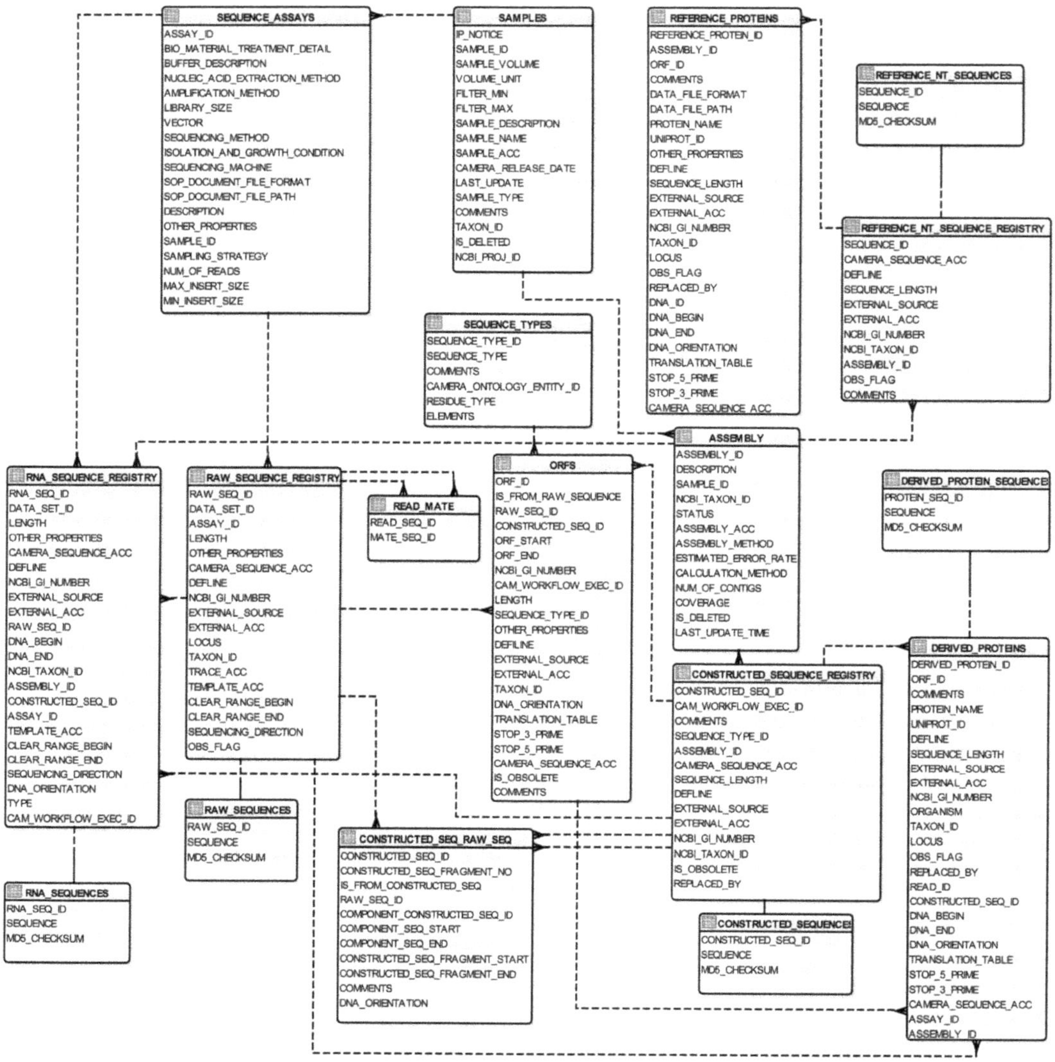

Figure 8.3 Diagram of the Sequence Store in the CAMERA database.

sequences, and could be very useful for analysing metagenomic short reads from next generation sequencers. Mate pair information is stored in the READ_MATE table.

ORFs called from reads or assembled sequences are stored in the ORFs table, and translated peptide sequences from ORFs are stored in DERIVED_PROTEINS. CAMERA also does automatic annotations on the reads and/or assemblies stored in CAMERA database. A PROTEIN_SEQUENCE_ANNONTIONS table is defined to store the heterogeneous annotation information in a generalized format. Doing so makes the data model flexible enough to store almost any kind of annotation, while still achieving a reasonable query performance using XML search functions in Oracle. In addition, a customized annotation module for cluster data provides fast querying of protein clustering results.

Integrating workflow analytical data with the SDB

In order to integrate workflow data into the CAMERA database infrastructure, we extended CAMERA's data model to manage workflows and workflow runs. In this extension, we defined a generalized data model to store arbitrary data for workflow input and output. In this generalized data model, workflow inputs and outputs are stored as objects. Each object has an object identifier (OID) assigned by the database, and the OID can have an unlimited number of attributes. When objects are imported from workflows, the workflow run ID, which identifies a workflow run, will be carried by those objects. These IDs connect the user data schema with objects in the workflow provenance database. Each object, or its attributes, can optionally have a link to a field or a row in the SDB. This link establishes the connection between workflow results and the SDB. On top of this data model, a set of software was implemented to extract, transform and load the workflow input and output to the database. With this software, workflow authors define how to convert the results of workflow runs results into an object or a group of objects and how to map them with objects in the SDB. These mappings add an important semantic meaning to the objects created from workflow runs and are used to analyse how data flows in different workflow runs, how workflows interact with each other, and how new user collaborations can be formed from common workflow or data usage patterns.

Integrating metagenomic, genomic and reference data

In additional to metagenomic sequences and their metadata, researchers also need data from reference genomes, their annotations, and other databases, such as taxonomy, Pfam, TIGRFam (Haft *et al.*, 2003), Gene Ontology, KEGG, COG, in the course of metagenomic data analysis. The data sources mentioned above have correspondingly been integrated into the SDB. At the same time, data analysis algorithms like RAMMCAP (Haft *et al.*, 2003; Li, 2009) can easily generate integrated annotation reports by querying these databases. RAMMCAP and its implementation in a workflow are described in more detail below.

Only the most up-to-date version of these data sources can provide the requisite values for users. To maintain timely updates of such a large data warehouse is always a challenge, because large amounts of data need to be processed, the data types and sources are highly diverse, and the range of data sources change rapidly, not only the amount of data growth, there are also many information been changed or deleted. For example, many sequences are added into each release of GenBank, and many sequences are also removed from GenBank for reasons such as sequencing contamination. In CAMERA, we designed and implemented a series of tools to update our reference database automatically and rapidly, so that in updating the SDB, we can readily conform to the release cycle of the external data sources. For example, CAMERA's GenBank and RefSeq database is update after every GenBank and RefSeq release by NCBI. Blast nodes for various reference sequences are also updated accordingly. These update information are also available to users so that they can easily find out when the Database was refreshed and what data are in each dataset.

The four paths we created in CAMERA to update both the genome and metagenome sequence database are described in Fig. 8.4. The reference genome data can be updated in two ways depending on the data sources: (1) reference sequences from NCBI are updated through our standard NCBI updated pipeline; (2) reference sequences directly from sequencing centres or other non-NCBI data sources are updated through CAMERA's in house command line tools. Metagenomic sequence data are updated through two similar paths that are described in Fig. 8.4.

Information life cycle management

The nature of metagenomic research requires paying attention to complete information life cycle management, i.e. information too has its own life cycle. A typical life cycle is considered to have five phases, namely creation and receipt, distribution, use, maintenance, and disposition.

Developers and managers routinely focus on the first four phases of this cycle, but the last phase is often ignored. Deleting information before an appropriate/carefully assigned retention period can cause valuable data to be lost. On the other hand, keeping information that is less frequently accessed or has met its useful retention period online can slow down access to current data and increase the cost of routine maintenances such as backup and archiving. This will over time add significant system performance and operating costs.

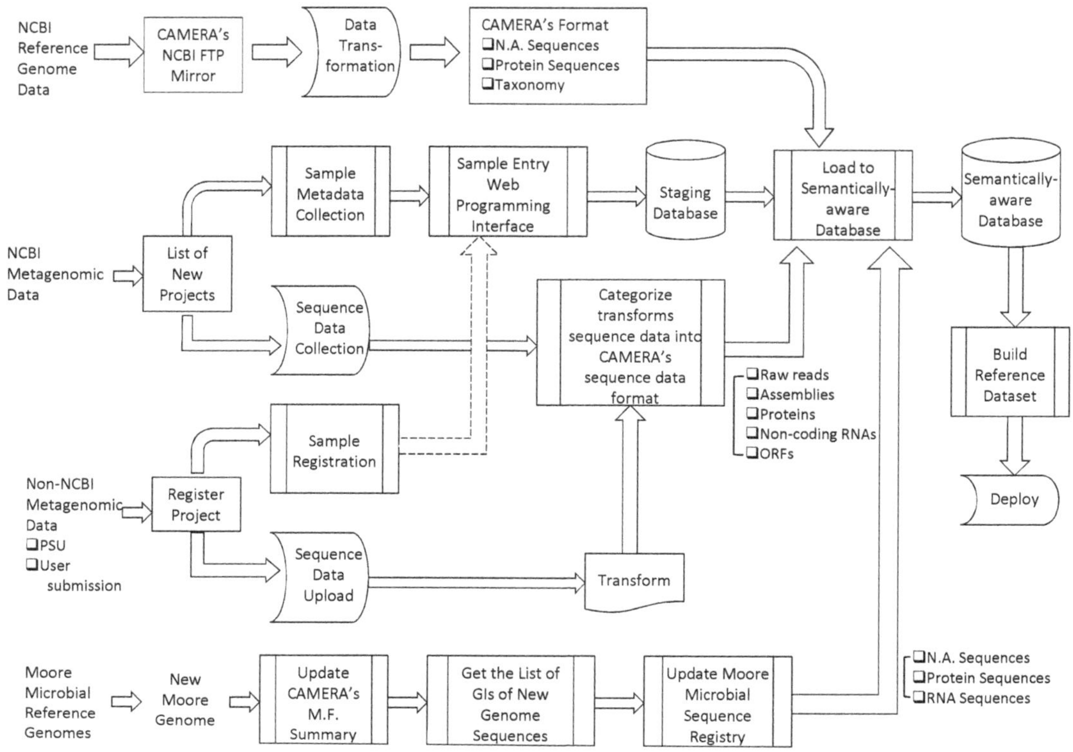

Figure 8.4 SDB update path.

How CAMERA manages this potential impact is described below.

The specific complication for delivering the best user services is that the sources of CAMERA's reference data change routinely. For example, each NCBI GenBank release might have some sequence entries deleted or updated. In order to have proper support for provenance data, the SDB needs to have the ability to reconstruct a reference dataset back to a certain point of time. In our reference database, all the deletions are done at the logical level; they will not be included in a conventional search result, but they are still in the database. Users always have the ability to refer to a logically deleted sequence in the database. All the updates are also tracked, so we know when and what was changed. The CAMERA database is partitioned in the way that logically deleted information can be managed in different storage from the more 'active' information. This implementation allows us to keep all the relevant information online with relatively low cost and minimum impact to the system performance. A similar strategy is also applied to user data management. User result data are kept for six months in CAMERA. If not touched again by users, they will be deleted from CAMERA database after this retention period. The system will reclaim the resources after the deletion.

Metagenomic data analysis

CAMERA organizes data analysis tools through collaborative, user selected scientific workflows. At the core of this environment is the Kepler scientific workflow platform that supports the integration of data and provides capabilities for provenance. Using this platform, CAMERA's workflow system currently makes the following metagenomic analyses available to researchers: a QC filter for raw 454 sequencing reads; the identification of duplicates from pyrosequencing reads; the assembly of pyrosequencing sequencing reads; gene prediction, clustering and functional annotation (including many tools); BLAST searches; KEGG pathway analyses; community diversity analyses.

QC filtering for raw reads

The second-generation sequencing technologies can provide a large amount of sequence at very low cost; however, they produce higher error rates than a traditional Sanger Sequencer. The QC filter is to remove the low quality reads prior to other data analyses. The QC Filter Workflow takes FASTA and QUAL files or a FASTQ file as input, calculates the average score for each read, then fetches high quality reads, filters out reads shorter than a user-defined minimum read length, and generates a statistical analysis.

Identification of duplicates from pyrosequencing reads

The 454 sequencers regularly produce artificially duplicated reads, which can lead to misleading conclusions. Exact duplicates sometimes have been removed before data analyses (Dinsdale *et al.*, 2008; Haft *et al.*, 2003). Recently, nearly identical duplicates – i.e. those reads that begin at the same position but may vary in length or bear mismatches – have also been classified as artefacts. Together, these exact and nearly identical duplicates can make up 11–35% of the raw reads (Gomez-Alvarez *et al.*, 2009).

The Duplicate Identification Workflow (Niu *et al.*, 2010) was implemented based on the use of the sequence clustering algorithm CD-HIT(http://cd-hit.org) (Li and Godzik, 2006; Li *et al.*, 2001a, 2002) to identify duplicated reads. Duplicates are either exactly identical, or otherwise meet these criteria: (1) they start at the same position; (2) their lengths can be different, but the shorter one must be fully aligned with the longer one (the seed); (3) they can only have up to 4% mismatches (insertion, deletion, and substitution); and (4) only 1 base is allowed per insertion or deletion. Criteria settings for (3) and (4) can be adjusted by users. This workflow allows mismatches in order to tolerate sequencing errors and takes a FASTA file as input, and outputs two files. One file lists the clusters of duplicates with similarity information. The second file in FASTA format is the reads sequences with duplicates removed.

In metagenomics, the amount of reads is used as an abundance measure, so artificial duplicates will introduce an overestimate of the abundances of taxa, genes, and functions. Duplicated reads observed in a pyrosequencing run include not only artificial duplicates, but also natural duplicates, i.e. reads from the same fragments that start at the same genomic position by chance. Therefore, simply removing all duplicates might also cause an underestimation of abundance associated with naturally duplicated reads. The occurrence of natural duplicates can be very low for metagenomic samples lacking dominant species; in this case, duplicates from those samples should be removed. However, our analysis shows that transcriptomic samples contain more natural duplicates than artificial duplicates, so removal of the duplicates from transcriptomic reads is not useful (Niu *et al.*, 2010).

Assembly of pyrosequencing reads

CAMERA's Assembly Workflow is a consensus-based meta-assembler designed for assembly of 454 reads. The method is called MetaRAC, i.e. Metagenomic Reads Assembly by Consensus. First, seven assembly programs [Velvet (Zerbino and Birney, 2008), ABySS (Simpson *et al.*, 2009), SSAKE (Warren *et al.*, 2007), Taipan (Schmidt *et al.*, 2009), Newbler [http://www.roche.com], Celera (Myers *et al.*, 2000) and SOAPdenovo (Li *et al.*, 2010)] are used to generate a pool of contigs. Then, each contig is evaluated by how many times this contig is supported by the seven assembly programs. Based on this analysis, we assign each base a quality score. Finally, the contigs are re-assembled using Phrap (http://www.phrap.org). Our analysis shows that MetaRAC is better than any one of the seven individual assembly programs. For example, it offers better coverage and accuracy in gene recovery. The Assembly Workflow takes a FASTA file as input and gives two output files. One output has the contigs in FASTA format. The second output lists the mapping between reads and contigs.

Gene prediction, clustering and functional annotation

A primary approach for these aspects is to use the Rapid Analysis of Multiple Metagenomes with a

Cluster and Annotation Pipeline, or RAMMCAP (see Li, 2009), which is a workflow for rapid and comprehensive annotation. RAMMCAP was implemented using an ultra-fast sequence clustering algorithm, fast protein family annotation tools, and a list of tools that cover rRNA identification, tRNA identification, ORF call, and other standard methods used in metagenomics analysis. Through the CAMERA web portal, users can run RAMMCAP jobs, visualize the results, download and export the results, and manage the jobs. The current (updated) version as described in Fig. 8.5.

The input to RAMMCAP is a FASTA file of reads from one or more metagenomes. The reads are pre-processed to remove artificial duplicates (454 reads) and low quality reads. (This is accomplished by the use of the QC Filter Workflow and the Duplicate Identification Workflow.) The output files include statistics of input file, predicted genes (rRNAs, tRNAs and proteins), read clusters, protein clusters, and functional annotations. RAMMCAP can cluster and annotate one million metagenomic reads in only hundreds of CPU hours.

Identification of rRNA

The identification of genes coding for ribosomal RNA (rRNA) is an important goal in the analysis of data from metagenomics projects. For datasets such as metatranscriptomes, the first step in rigorously identifying specific rRNAs usually is to filter out other un-wanted rRNAs. We developed a program called Meta_RNA, designed for the identification of rRNA genes from metagenomic fragments based on Hidden Markov models (HMMs). This program provides rRNA gene predictions with high sensitivity and specificity on artificially fragmented genomic DNAs (Huang *et al.*, 2009). Alternatively, rRNA can be identified by BLASTN-based rRNA finders we developed. The detailed comparison between these two methods can be found in Huang *et al.* (2009) and from http://tools.camera.calit2.net/camera/meta_rna/.

Identification of tRNA

After the rRNA identification, the rRNAs are masked from the input reads, and tRNAs are predicted with tRNA-scan (Lowe and Eddy, 1997). The tRNAs are also masked from the input reads. Masked reads are used in further analyses.

Clustering sequencing reads

Cluster analysis is an important tool for exploring large, complex datasets such as metagenomic sequences. CAMERA uses the fast clustering program CD-HIT, available as open source software (http://cd-hit.org) for clustering and comparing large sets of protein or nucleotide sequences.

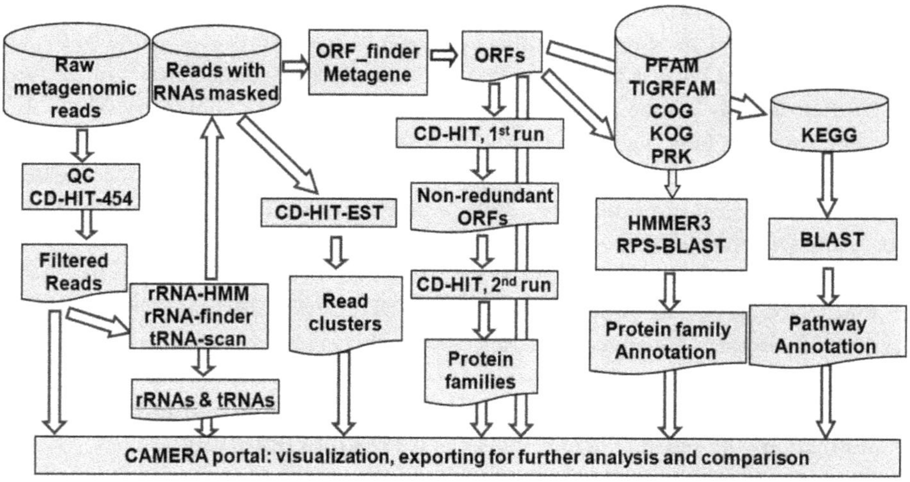

Figure 8.5 Metagenomic data analysis pipeline RAMMCAP.

It identifies similar sequences from a dataset or between two datasets. CD-HIT helps to significantly reduce the computational and manual efforts in database search, family classification, phylogenic analysis, and so on. Clustering also helps to understand and correct the bias within a dataset that commonly contains unevenly distributed sequences. CD-HIT can be 2 to 3 orders of magnitude faster than other methods like BLASTCLUST, has the capacity to process hundreds of millions of sequences, and processes millions of sequences in hours (Li and Godzik, 2006; Li *et al.*, 2001; Li *et al.*, 2002). As a result, besides use in other community databases, it has been widely applied in metagenomic studies (Turnbaugh *et al.*, 2009; Yooseph *et al.*, 2007).

In RAMMCAP, after the identification of rRNA and tRNA, the masked reads are clustered with CD-HIT-EST (a specific DNA version of CD-HIT) at 95% sequence identity over 80% of the length of shorter sequences (clustering parameters can be adjusted by users) to identify clusters of unique read sequences, referred to as read clusters. It takes ~1 hour to cluster a million 200 base reads.

Clustering analysis offers novel ways to view and analyse large metagenomic datasets.

a It provides an overview of a dataset's internal redundancy structure and gives the distribution of clusters.
b It identifies gene families.
c It identifies novel gene families that usually cannot be found by homology search.
d It saves resources in computational analysis by removing redundant sequences.
e It reduces efforts in sequence assembly by filtering extra redundant sequences.
f When sequence reads from multiple samples are clustered together, the clusters can be used to compare samples based on how the sequences from different samples occur across the clusters. Fig. 8.6 shows an example in which sequencing reads from multiple samples are clustered and compared together.

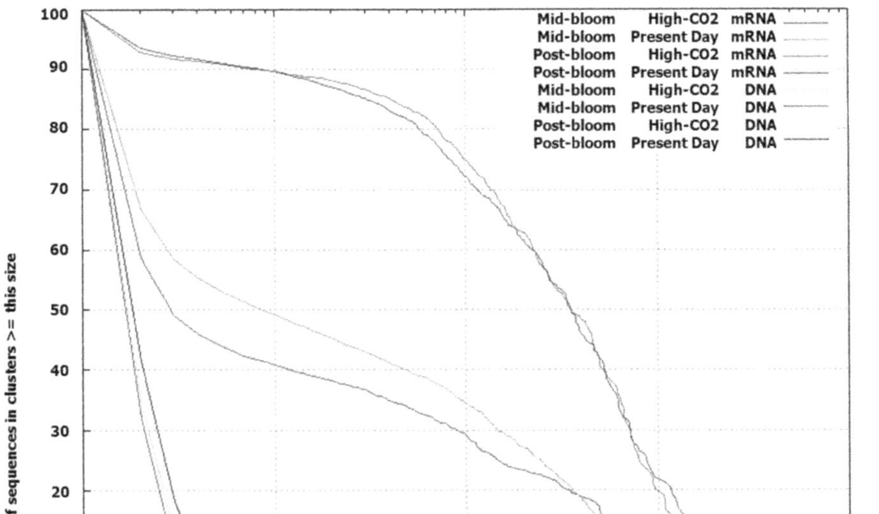

Figure 8.6 Distribution of clusters from eight samples. The *x*-axis is cluster size. The *y*-axis is the percentage of total sequences within clusters ≥ this size. The clustering analysis was made for four metagenomic and four metatranscriptomic samples from two time points and two conditions (high CO_2 and normal CO_2) (Gilbert *et al.*, 2008). Clear patterns are observed for different type of samples. A colour version of this figure is located in the plate section at the back of the book.

Predicting open reading frames (ORFs)

ORFs are collected from sequencing reads with ORF_finder, an ORF calling program implemented here by six reading frame translations. Within each reading frame, an ORF starts at the beginning of a read or the first ATG after a previous stop codon; it ends at the first stop codon or the end of that read. The minimal length of ORFs can be specified by users.

ORFs can also be called from sequencing reads using the program Metagene (Noguchi *et al.*, 2006). Since these sequence reads are short, a predicted ORF may be a portion of a complete ORF. An ORF may also be a translation from a non-coding frame: such an ORF is called a spurious ORF, as defined in the original GOS study (Yooseph *et al.*, 2007). The GOS study also introduced a spurious ORF detection method using a non-synonymous to synonymous substitution test (Yooseph *et al.*, 2008), which can be used within the CAMERA environment. While this method is not integrated within RAMMCAP, it can be used independently to detect the spurious ORFs.

Because metagenomic sequences are short, both ORF_finder and Metagene may miss real ORFs and may produce wrong predictions. Table 8.1 shows the sensitivity and specificity of ORF_finder and Metagene based on 4 simulated datasets of 100, 200, 400, and 800 bp.

Clustering ORFs

Similar to read clusters, ORFs are also clustered using CD-HIT. ORFs are first clustered at 90–95% identity to identify the non-redundant sequences, which are further clustered into families (ORF clusters) at a conservative threshold, so that each cluster contains sequences of the same or similar function. A 30% sequence identity indicates significant similarities for full-length proteins. Since ORFs from Sanger reads are long enough, they are clustered at 30% identity over 80% of the ORF length. ORFs from 454 reads are shorter, clustered at 60% identity over 80% of the ORF length. ORF clusters are used for functional studies. The size of an ORF cluster is the number of its non-redundant sequences. For one million ORFs, it takes a few CPU hours to cluster at 60% identity and ~100 CPU hours at 30% identity. This type of multiple-step clustering has been used in independent analysis of GOS ORFs (Li *et al.*, 2008). It should be noted that the current online RAMMCAP service from the CAMERA portal does not support clustering at 30% identity, which needs instead to be performed with the standalone CD-HIT program.

ORF annotation

ORFs are annotated from PFAM and TIGRFAM with HMMER (Eddy, 1998) accelerated with Hammerhead (Portugaly *et al.*, 2007), and from COG and KOG with RPS-BLAST (Altschul *et al.*, 1997). Hits must be with e-values ≤ 0.001, and meet the significant scores in case of HMMER searches. This annotation process only takes ~100 CPU hours for one million ORFs. Through these function annotations from PFAM, TIGRFAM, COG and KOG, additional cross-linked annotations are also generated including GO and EC number.

Table 8.1 Sensitivity and specificity of ORF_finder and metagene

	Sensitivity/specificity (%)			
Datasets Method	100 bp reads	200 bp reads	400 bp reads	800 bp reads
Metagene	59.07/92.58	94.31/92.32	93.84/91.91	94.02/93.25
ORF_finder, cut-off = 30 aa	99.93/36.56	99.74/36.73	99.78/34.49	99.92/32.83
ORF_finder, cut-off = 40 aa		96.74/46.65	97.89/43.81	99.13/41.54
ORF_finder, cut-off = 50 aa		91.65/54.82	94.04/52.51	96.88/50.21
ORF_finder, cut-off = 60 aa		73.66/61.27	88.94/59.84	93.34/57.98

Summarizing annotation

RAMMCAP summarizes the clustering information, functional annotations, and other integrative attributes into a large spreadsheet that users can download for further analysis. In addition, RAMMCAP provide statistical summaries for input reads and summaries on predicted ORFs like GC content and length distributions.

BLAST workflow

The CAMERA BLAST Workflow offers a powerful engine to search reference databases and other metagenomic databases (http://legacy.camera.calit2.net/camera2/dataset-information). Six BLAST programs (BLASTN, BLASTP, TBLASTN, BLASTX, TBLASTX, MEGABLAST) are included in this workflow. A user can input as many as several hundred thousand sequences as query to search against a list of compiled metagenomic datasets or reference databases. When the BLAST job is finished, the CAMERA web portal provides a graphic interface to view, browse, sort, and export the results.

KEGG pathway analysis

The KEGG Annotation Workflow takes a FASTA file of protein sequences. It searches a user selected KEGG sequence database with BLASTP, and assigns the KEGG information to the proteins.

Analysis of community diversity

Phages, the viruses that infect bacteria, are the most abundant biological entities in the biosphere, with an estimated 10^{31} particles on the planet. A metagenomic approach allows the study of viral communities and provides an estimate of their diversity. CAMERA offers two diversity workflows: a-diversity and γ-diversity. The α-diversity Workflow assesses the biodiversity of a single viral community while the γ-diversity predicts the total biodiversity over two or more viral communities. These workflows are derived from PHACCS (Angly *et al.*, 2005), which uses six rank-abundance distributions (broken stick, exponential, logarithmic, log normal, niche pre-emption, and power law) to predict viral diversities from metagenomic sequences.

Examples of CAMERA usage

Another important aspect of the overall workflow environment is that these workflows are organized into a systematic network, in which the output for one functional unit can be used as an input for the next workflow. This allows researchers to build a complete end-to-end analysis stream by choosing to use different combinations of workflows based on their specific needs for a given data analysis in order to solve their scientific questions. The questions can include the basic environmental metagenomic questions such as 'Who are they?', 'What are they doing?', and 'How do they do that?', and more complex questions like 'Which are high expressed or down expressed under certain of environmental physicochemical measurements and/or treatments?' Below are some scientific examples or use cases to show how CAMERA has helped researchers address their bioinformatics challenges; these are meant to be representative of the nature of the research already empowered by CAMERA and from which potential users can extend to their own interests.

CASE 1: genomic diversity in surface Ocean waters

The GOS sequencing effort has vastly expanded the landscape of metagenomics, providing an opportunity to study the genetic potential of surface ocean water bacterioplankton on a global scale. Some questions are the habitat-based diversity, in terms of both taxon evenness and taxon richness, for each GOS site and to estimate genome characteristics of a typical free-living, surface ocean water bacterium. Other challenges are to answer who dominates the 0.1 to 0.8 μm size fraction of surface ocean water bacteria, the proportions of other taxa varied with ocean habitat type, which metadata affect the abundance, and so on. This use case summarizes a study (Biers *et al.*, 2009) that provides a quantitative, categorical assessment of microbial diversity – both the operational taxonomic unit (OTU) evenness/distribution and OTU richness/abundance in surface waters of the northwest Atlantic Ocean through the eastern tropical Pacific Ocean by analysing 16S rRNA gene homologues in the GOS metagenomic library. For a full description of the methods used, the results, and the discussion, see the original publication (Biers *et al.*, 2009). What follows is an abbreviated outline of the methods on how to use CAMERA sequence datasets and associated metadata to retrieve gene homologues. All Basic Local Alignment Search Tool (BLAST) searches for gene homologues in the unassembled GOS metagenome were performed through the CAMERA BLASTN Workflow. To maximize homologue retrieval, the entire 1,542-bp sequence of the 16S rRNA gene (rrsE, locus tag b4007) from the *E. coli* K-12 genome (RefSeq accession no. NC_000913) was set as the query sequence. Nucleotide sequences for 16S rRNA homologues were identified with BLASTN against 'GOS: all metagenomic sequence reads (*N*)' [with cut-off expect (*E*) value of 1E–5; overlap length of ≥ 65 bp; similarity of ≥ 85%]. Before taxonomic assignment of 16S rRNA gene homologues, the retrieved GOS sequences were trimmed to exclude any non-16S rRNA gene portion of the sequence read. Paired-read sequences, two sequence reads originating from the same insert (one forward read, one reverse read), within the 16S rRNA gene homologues were identified and then screened to verify that they were assigned into the same taxonomic bin. Then, one of the paired reads was removed from further analysis. Single-copy gene hits, i.e. hits to genes found once per prokaryotic genome, were retrieved by using CAMERA BLASTP Workflow. Briefly, amino acid sequences for single-copy genes (atpD, gyrB, dnaK, rpoB, tufA, and recA) were identified with BLASTP against 'GOS: all ORF peptides (P)' (cutoff E-value of 1e-20). Hits for each single copy gene, excluding paired reads, were normalized to the relative length of the gene compared to the length of recA, and the average number of recA-normalized single-copy gene hits was used in this analysis.

CASE 2: Expression study on transporter genes

Coastal ocean bacterioplankton controls the flow of dissolved organic carbon (DOC) from terrestrial and oceanic sources into the marine food web, and regulate the release of inorganic carbon to atmospheric and offshore reservoirs. While the fate of the chemically complex coastal DOC reservoir has long been recognized as a critical feature of the global carbon budget, it has been problematic to identify both the compounds that serve as major conduits for carbon flux and the roles of individual bacterioplankton taxa in mediating that flux (Poretsky *et al.*, 2010). In this case, we may ask questions such as: Which transporter genes are over expressed? How many and which components are the key components of the biologically label DOC pool in coastal waters? Which taxa dominate the community? For a full description of the methods used the results, and the discussions, see the original publication (Poretsky *et al.*, 2010). What follows is an abbreviated outline of the methods used in the bacterioplankton transporter genes expression study, which can serve as general instructions on how to use the CAMERA sequence datasets and associated metadata to address questions like those considered in this study.

A first step involved quality control (QC) for the sequences evaluated. The low-quality reads were removed by the QC Filter Workflow; specifically, any reads with average quality score < 21, or read length < 25 bp were removed. The exact duplicate and nearly duplicate reads were removed by the 454 Duplicate Clustering Workflow. Ribosomal RNA sequences were eliminated following identification by the BLASTN Workflow queries against the GenBank nucleotide database (nt).

A key next step involved cDNA annotation. The annotation of proteins using unassembled, potential protein encoding sequences was accomplished by running the BLASTX Workflow analysis against the NCBI non-redundant reference sequence database (RefSeq). COG annotation was done using Metagenomic Annotation and Clustering Workflow. Taxonomic binning of sequences was carried out using MEGAN. The further assignment into higher-resolution marine taxa (e.g. Roseobacter and SAR11) was based on the NCBI taxonomy of closest BLAST hits. The Taxonomic Binning Workflow is able to determine 'who they are' and 'who dominates the community'; BLASTX Workflow helps identify the levels of gene expression, and the key components.

Using workflows to direct experimental design

CAMERA workflows not only play a key role in analysing data, but are also able to direct some experimental designs, such as the design of a microarray experiment or of another probe of an environmental population. To illustrate this specific case of environmental microarray probe design and indicate the potential, Fig. 8.7 provides step by step instructions for using a CAMERA workflow to find the expression level of genes of interest to a user. Identification of the genes, in turn, can direct many subsequent experiments, such as (a) researchers would be able to design their primers and probes for the genes of interest to study the abundance and diversity of one genes or group of genes and (b) researchers could design microarray probes using existing sequencing data.

Retrospectives and perspectives

Retrospection is generally easy, but as is attributed to Yogi Berra: prediction is hard, especially about the future. In this chapter, we have provided the technical basis for our chosen path for erecting the cyberinfrastructure of CAMERA. Before venturing into the hard problem, we first provide comments in a historical perspective of the overall philosophy, which underlies the approaches within CAMERA including its trajectory in the future. It is obvious that building an easy to use, effective resource for metagenomics, one that reflects the range of expectations in providing and sustaining a knowledge framework for this new science, requires considerable use of advanced cyberinfrastructure. To being with, setting aside even the frequently fragmentary characteristic of the primary sequence data, an exceptional range of sciences and subdisciplines have contributed to the development of metagenomics, ranging from biogeochemistry to ecology. Scientists from these diverse disciplines will have correspondingly diverse interests and goals in probing the data. For example, in ecology, the so-called secondary, environmental contextual data might well often be the primary key for analysis. Moving beyond studying the diversity and distribution of DNA (qua genes) and other macromolecules to explore the implications of the geochemical environment directly

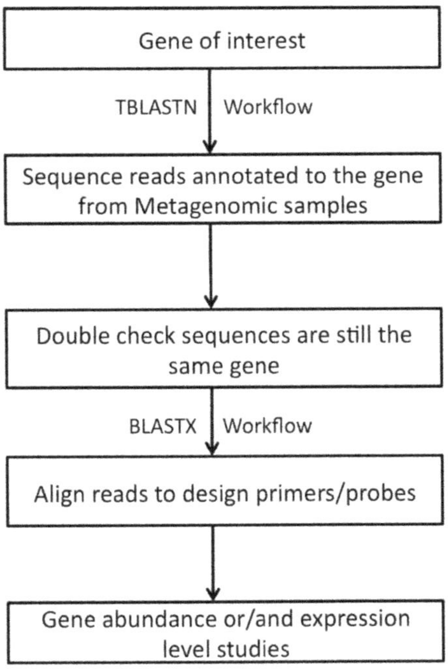

Figure 8.7 A flow chart on how to design experimental probes using metagenomic data.

creates a need for rich contextual metadata on which vigorous and rigorous search and analysis tools can be employed. To insert a perspective note: as long as cyberinfrastructure enabled knowledge resources provide an adequate framework, the wide range of research domains that converged in establishing metagenomics will themselves be profoundly changed. The study that enabled the new science by reaching the funding sources, i.e. the study by the National Research Council (NRC-Metagenomics-Committee, 2007), noted that not just disciplines but the fields themselves of earth, life and biomedical sciences would be transformed, as would biotechnology, agriculture, biodefense, forensics, bioenergy and bioremediation.

Data standardization for both primary and associated contextual data, i.e. metadata, is essential for meeting such an obligation. Thus, the CAMERA team set out from the beginning to work with the Genome Standards Consortium (SGC), in order to include a collection of metadata that is as rich as possible. CAMERA, of course, has also complied with the standards already established, and engaged directly with the GSC's efforts to achieve acceptance and use of the minimum standards, while the GSC aims to expand those and progress towards a set of optimized standards accepted by the community. As one aspect of a set of International Congresses of Metagenomics hosted by CAMERA, we have also attempted to highlight the value of the GSC efforts and engage the experimental community to value capturing and submitting the requisite information for inclusive, reliable metadata.

To the benefit of the community, all of the major metagenomics information resources have both focused on their own set of deliverables for user support and in particular, have chosen to be compliant with the Genomic Standards Consortium minimal standards, and each continues to interact with the rest of the GSC to sustain progress, as do the members of the INSDC. Along with CAMERA, these include the IMG/M (Markowitz *et al.*, 2008), MG-RAST (Meyer *et al.*, 2008), the Megx (Kottmann *et al.*, 2010), and the NIH Human Microbiome Project (HMP) DACC or Data Analysis and Coordination Centre (http://www.hmpdacc.org/).

While considering how we move forward with data exchange standards and metadata standards, it is important to keep in mind that the rapid research advances are, after all, going on in a new science. Thus, while many science questions for research in metagenomics can be answered or at least, initially probed by traditional or conventional (widely used and well established) search algorithms, for a metagenomics knowledge resource to provide only precomputed data sets and the means for conventional analysis paths might restrict the potential innovation and vast exploration of the unknown world of microbes. For example, the study of microbes, among other recently highlighted considerations, needs to deal with the size of pan-genomes and to evaluate the implications of supra-organisms (with regard to both the eukaryotic host and their microbes). Thus, the key word in providing a resource seemed to us to be flexibility, being able to respond to the pace of the science and most notably, putting options in the hands of the users. To create a flexible workflow environment, we set out, correspondingly, to draw from the cyberinfrastructure (CI) used successfully in neuroinformatics (most notably, the BIRN project; Grethe *et al.*, 2005) and other successful physical and life sciences CI projects. For efficiency, we also sought to provide ready access to the breadth of community knowledge/community databases as well as analysis tools, and thus to enable 'one stop shopping.' The range of ongoing research and discoveries will certainly grow as we provide the tools for users to introduce, integrate and implement their own analysis software into workflows. The details of the Community CI and Database have been described above, and references have been given to more technical descriptions.

The other articles in this and the preceding volume of the book, the rich literature of reviews and special issues of leading journals, and the 'canonical' NRC study itself provide a glimpse of the challenges, opportunities, and applications. Our role has been, and will continue to be, providing the cyberinfrastructure on which the questions can be answered. Like any community research resource, our regular users are our 'first resource' and 'first obligation' as we look to the future. A number of projects are under way to enhance CAMERA and some are in beta test already. Below, we provide an overview of ongoing projects, near term plans, and implementation efforts for CAMERA. The key features of our efforts will be to continue to seek guidance from the experimental community, to interact extensively with regular users of the database and with the directions the community takes, so we can remain flexible, incorporate and/or repurpose software as needed, and find and encourage the community to incorporate newly established software tools that expand the analytical potential. We will also sustain our dialogue with the Genomic Standards Consortium since community participation in this effort, especially around metadata, is critical for progress.

CAMERA will extend our semantic database (SDB) features to enhance the options for user queries. Much software and many frameworks have been recently created to support semantic technology. In the open source domain, Jena and Pellet are two popular semantic reasoning engines (software able to infer logical consequences from a set of facts provided). Commercial reasoning engines such as OWLIM and OntoBroker are also available. Oracle has also implemented a Semantic Support Module in its 11g database. A linkage between metagenomic metadata and annotation with ontology terms has already been created in CAMERA's database. CAMERA is continuing to evaluate extant software and frameworks for ontology, and will use an existing software package to manage multiple ontologies; CAMERA will also create an application layer to integrate these software tools with the CAMERA database . This layer will allow users to run semantic queries and thus, formally, the CAMERA database would be said to be a semantic database (rather than one that is simply semantically aware).

To extend the SDB further, we plan a visualized metadata search interface in CAMERA using 3rd party data visualization packages, such as the Silverlight PivotViewer from Microsoft. This will provide users with a more interactive web front-end to search (or explore) metagenomic samples. Such an interface can visualize thousands of related samples at once, so that users can see trends and patterns related to certain environmental parameters that would be hidden when looking at one sample at a time.

CAMERA is actively adding new analysis workflows and allows users to contribute community tools as new workflows. In the very near future, our plans are to include a Fragment Recruitment Viewer, which is a workflow to align or recruit raw metagenomic reads to reference genomes or homologue genomes using fast alignment programs such as BLAT (Kent, 2002), and to add a VIROME Workflow designed to explore metagenome sequence data collected from viral assemblages.

The Fragment Recruitment Viewer (FRV) is a user interface that can visualize the fragments (reads) that are recruited to a reference genome. Fragments are aligned to reference with a searching or mapping program such as BLASTN or BLAT. An alignment (or a hit) is defined by the position on the reference genome, the identity, the e-value, the bit-score, and other parameters. Hits are pre-calculated and stored in CAMERA database. With the 2D scatter plot, researchers are able to view, compare, download images, and export text data.

VIROME is a web-application designed for scientific exploration of metagenome sequence data collected from viral assemblages occurring within a number of different environmental contexts. The pipeline focuses on the classification of predicted open-reading frames (ORFs) from viral metagenomes. It processes sequences directly from a sequencing centre, step by step: (a) identification and removal of all reads containing 16S ribosomal gene homologous sequence; (b) identification of tRNA encoding regions within sequence reads; (c) prediction of open-reading frames (ORFs) within viral metagenome sequences; (d) BLAST analysis of ORFs against two subject databases for the identification of peptide homologues; (e) classification of ORFs into one of six VIROME categories; and (f) Assignment of possible taxonomic origin based on the best BLAST hit.

Among other tools to be included soon are gene prediction tools newly developed by the community; one promising example is FragGeneScan (Rho *et al.*, 2010). In addition, a Taxonomy Binning Workflow is under development. This workflow will assign a taxonomy path to a read using tools like MEGAN (Huson *et al.*, 2007), RDP Classifier (Wang *et al.*, 2007), and the Silva database (Pruesse *et al.*, 2007). An especially important development, in beta test now, is a computational tool kit enabling the algorithmic members of the user community to incorporate (i.e. introduce, integrate and implement) their own novel software into a scientific workflow in CAMERA, thus extending their own analyses and providing a path for dissemination and validation. This involvement of the entire community of software developers, offering enhanced research options as well as public usage and evaluation, is an important element of our considerations on the future of data retrieval.

Another feature that will remain not just important but vital for future considerations for metagenomics resources is, as we have emphasized above, the continuing effort to develop and promote data exchange standards and also, standards for metadata. Our continuing involvement in the dialogue hosted by the Genomic Standards Consortium represents, obviously, an important element of future considerations in that regard. The emphasis that the key study (NRC Metagenomics Committee, 2007) on the absence of any one simple approach to metadata and their standardization, while noting their importance, is well described in the study's Table 7.1 on the challenges ahead for the new science and in their Box 5.2 directly illustrating the advantages of robust, inclusive metadata.

Above we have discussed trends in data standardization and retrieval. Another concern for metagenomics is the growing size of any experimental project, let alone that for the cumulative data sets. The prospects for continued exponential growth of highly complex data from research in the biological and biomedical sciences are significant enough to be daunting to all of us. The phenomena of rapidly growing data sets is general to the sciences, of course, and are well known to computer experts in academic and commercial settings. Given the scale of the longer term challenges in effectively managing, storing, querying, and transmitting ultra-large scientific datasets, the trends over the longer term in expanding data storage while lowering costs, and in the transmission of petabyte scale datasets, will largely depend on considerable progress in computer engineering. It appears likely that a new view of community cyber (for science in general) will develop, through academic and commercial engineering efforts connected to scientific computing, to manage the expanding impact of high throughput and high resolution data for all of the sciences, with technology advances ranging from data compression algorithms to advances in data transmission.

A range of near term responses are ongoing within the biological sciences community. A detailed description (written from the ENU perspective) of the recent, well planned and orchestrated responses to and management of petabyte datasets for the INSD has been published (Cochrane *et al.*, 2009). The Genomic Standards Consortium (GSC) has also established a working committee that takes into account the exceptional needs for computing resources for metagenomics and is exploring the potential of cloud computing and other contemporary technologies. This effort is termed M5 for the intersection of 'Metagenomics, Metadata, MetaAnalysis, Models and MetaInfrastructure.' As one of the first steps, CAMERA is working with other members of the GSC subcommittee, including colleagues from MG-RAST, IMG/M and DACC, to develop a data exchange format for transfer of sequences, annotations, and metadata. The subcommittee will design and implement software that is compliant with M5 standards. (As an immediate consequence, CAMERA will be able to share annotation results with other resources or for downloads by the community and will be able to provide an integrated suite of annotations, including those that originated with a range of other annotation pipelines.) With the overarching aim of improving data-sharing and interoperability in metagenomics studies, the M5 Platform is viewed within the GSC as a future solution for the computational needs of the metagenomics (and genomics) science community and the effort; their vision, progress to date, and further details are described on their web site (http://gensc.org/gc_wiki/index.php/M5). M5 discussions have also been archived in several videos taken from the 9th meeting of the GSC and are available on SciVeeTV (http://www.

scivee.tv/). We will sustain our contributions to these efforts and look for other opportunities to facilitate the response of bioscientists to the continuing exponential growth of the data. Overall, as a resource with a deep grounding in computer science and engineering and information technology and their extension to cyberinfrastructure, we believe CAMERA is well positioned to incorporate technology advances as they occur.

Conclusions

In this chapter, we have reviewed a resource using state of the art computer science technologies to respond to the data challenges of the research in metagenomics. Given the challenges with respect to the data and the pace of increases in sequencing technology, as well as the exceptional opportunities enabled by the research, a clear requirement for such a resource arose. In response, we established and are continuing to advance the CAMERA project and its Database. As the first full implementation of our original goals, we have released a major extension, termed CAMERA 2.0, which adds considerable flexibility for users, taking into account that the researchers have come to metagenomics from many disciplines (from biogeochemistry to microbial ecology). Along with providing an advanced analysis portal for access to a semantically aware database and computational resources, CAMERA provides a rich description of the contextual data (environmental parameters) or metadata for the primary sequence data.

In order to optimize archiving and access to the explosion of metagenomics data, we consider the use of the standards established by the Genomics Standard Consortium (GSC) to be very important. CAMERA correspondingly follows those standards and works within the GSC in advancing the standards and notably, to enhance the potential value of the metadata. Overall, CAMERA 2.0 enables ready depositing, locating, analysing, visualizing and sharing data and the use of a range of actual analytical processes or workflows. An important feature that enables user choices consistent with their research objectives and provides for computational efficiency and analytical reproducibility is the use of a scientific workflow, Kepler, an open source community product. We have provided examples of research accomplishments using CAMERA, in order to suggest how new users can use CAMERA effectively to explore the microbial world. We have also described our ongoing efforts in improve the environment for users further and our anticipations for our metagenomics resource in light of the continuing revolutions in microbiology and in sequencing technology.

Acknowledgements

The authors wish to recognize the contributions of the entire CAMERA team and the members of pioneering UCSD cyberinfrastructure efforts such as BIRN, the grand launching vision of David Kingsbury and Larry Smarr for CAMERA, and the advice from the CAMERA's Science Advisory Board and the Science Program Officers of the Gordon and Betty Moore Foundation. We also especially note Mark Ellisman's key insights across every aspect of expectations on CAMERA and his overall direction of the project. The infrastructure implementation for CAMERA 2.0 is supported by the Gordon and Betty More Foundation, and the algorithmic development research by the NIH National Centre for Research Resources (NCRR) (grant R01RR025030).

References

Altintas, I., Anand, M.K., Crawl, D., Belloum, A., Missier, P., Goble, C.A., and Sloot, P.M.A. (2010a). Understanding Collaborative Studies Through Interoperable Workflow Provenance. In The third International Provenance and Annotation Workshop (IPAW2010) (Troy, NY).

Altintas, I., Chen, J., Sedova, M., Gupta, A., Sun, S., Lin, A.W., Gujral, M., Anand, M.K., Li, W., Grethe, J.S. *et al.* (2010b). Extending the Data Model for Data-Centric Metagenomics Analysis using Scientific Workflows in CAMERA. 2010 Sixth IEEE International Conference on e-Science Workshops.

Altintas, I., Lin, A.W., Chen, J., Churas, C., Gujral, M., Sun, S., Li, W., Manansala, R., Sedove, M., Grethe, J.S. *et al.* (2010c). CAMERA 2.0: A Data-Centric Metagenomics Community Infrastructure Driven by Scientific Workflows. IEEE 2010 Fourth International Workshop on Scientific Workflows.

Altschul, S.F., Madden, T.L., Schaffer, A.A., Zhang, J., Zhang, Z., Miller, W., and Lipman, D.J. (1997). Gapped BLAST and PSI-BLAST: a new generation of protein database search programs. Nucleic Acids Res. 25, 3389–3402.

Angly, F., Rodriguez-Brito, B., Bangor, D., McNairnie, P., Breitbart, M., Salamon, P., Felts, B., Nulton, J., Mahaffy, J., and Rohwer, F. (2005). PHACCS, an online tool for estimating the structure and diversity of uncultured viral communities using metagenomic information. BMC Bioinformatics *6*, 41.

Arima, K., and Wooley, J. (2008). METAGENOMICS 2007 Snapshot. In Computational Methods for Understanding Archaeal and Bacterial Genomics, Xu, Y., and Gogarten, P., eds (London, UK: Imperial College Press), pp. 345–396.

Biers, E.J., Sun, S., and Howard, E.C. (2009). Prokaryotic genomes and diversity in surface ocean waters: interrogating the Global Ocean Sampling metagenome. Appl. Environ. Microbiol. *75*, 2221–2229.

Cochrane, G., Bates, K., Apweiler, R., Tateno, Y., Mashima, J., Kosuge, T., Mizrachi, I.K., Schafer, S., and Fetchko, M. (2006). Evidence standards in experimental and inferential INSDC Third Party Annotation data. OMICS *10*, 105–113.

Cochrane, G., Akhtar, R., Bonfield, J., Bower, L., Demiralp, F., Faruque, N., Gibson, R., Hoad, G., Hubbard, T., Hunter, C. *et al.* (2009). Petabyte-scale innovations at the European Nucleotide Archive. Nucleic Acids. Res. *37*, D19–25.

Cochrane, G., Martin, M., and Apweiler, R. (2010). In Metagenomics: Theory, Methods and Applications, Marco, D.E., ed. (Norfolk, UK: Caister Academic Press), pp. 79–105.

Davidson, S.B., and Freire, J. (2008). Provenance and scientific workflows:challenges and opportunities. In Proceedings of the ACM SIGMOD international conference on Management of data (New York, USA: ACM). pp. 1345–1350.

Deelman, E., D., G., M., S., and I., T. (2009). Workflows and e-science: an overview of workflow system features and capabilities. FGCS *25*, 528–540.

Dinsdale, E.A., Edwards, R.A., Hall, D., Angly, F., Breitbart, M., Brulc, J.M., Furlan, M., Desnues, C., Haynes, M., Li, L. *et al.* (2008). Functional metagenomic profiling of nine biomes. Nature *452*, 629–632.

Eddy, S.R. (1998). Profile hidden Markov models. Bioinformatics *14*, 755–763.

Field, D., Garrity, G., Gray, T., and Morrison, T. (2008). The minimum information about a genome sequence (MIGS) specification. Nat. Biotechnol. *18*, 541–547.

Finn, R.D., Mistry, J., Tate, J., Coggill, P., Heger, A., Pollington, J., Gavin, O., Gunasekaran, P., Cerick, G., and Forslund, K. (2010). The Pfam protein families database. Nucleic Acids Res. *38*, D211–D222.

Freire, J., Koop, D., Santos, E., and Silva, C.T. (2008). Provenance for computational tasks: A survey. Comput. Sci. Eng. *10*, 11–21.

Gilbert, J.A., Field, D., Huang, Y., Edwards, R., Li, W., Gilna, P., and Joint, I. (2008). Detection of large numbers of novel sequences in the metatranscriptomes of complex marine microbial communities. PLoS ONE 3, e3042.

GO-EBI, and EMBL-EBI (2010). The gene ontology in 2010: extensions and refinements. Nucleic Acids Res. *38*, D331–D335.

Gomez-Alvarez, V., Teal, T.K., and Schmidt, T.M. (2009). Systematic artifacts in metagenomes from complex microbial communities. Isme J. 3, 1314–1317.

Grethe, J.S., Baru, C., Gupta, A., James, M., Ludaescher, B., Martone, M., Papadopoulos, P.M., Peltier, S.T., Rajasekar, A., Santini, S. *et al.* (2005). Biomedical informatics research network: building a national collaboratory to hasten the derivation of new understanding and treatment of disease. Biomedical Informatics Research Network. In From Grid to Healthgrid: Proceedings of Healthgrid *2005*, Solomonides, T., McClatchey, R., Breton, V., Legré, Y., and Nørager, S., eds (Amsterdam: IOS Press).

Haft, D.H., J.D., S., and O., W. (2003). The TIGRFAMs database of protein families. Nucleic Acids Res *31*, 371–373.

Handelsman, J. (2004). Metagenomics: genomic analysis of microbial communities. Microbiol. Mol. Biol. Rev. *68*, 669–685.

Huang, Y., Gilna, P., and Li, W.Z. (2009). Identification of ribosomal RNA genes in metagenomic fragments. Bioinformatics 25, 1338–1340.

Huson, D.H., Auch, A.F., Qi, J., and Schuster, S.C. (2007). MEGAN Analysis of Metagenomic Data. Genome Res. *17*, 377–386.

Kanehisa, M., Goto, S., Furumichi, M., Tanabe, M., and Hirakawa, M. (2010). KEGG for representation and analysis of molecular networks involving diseases and drugs. Nucleic Acids Res. *38*, D355–D360.

Kent, W.J. (2002). BLAT – the BLAST-like alignment tool. Genome Res. *12*, 656–664.

Kottmann, R., Kostadinov, I., Duhaime, M., Buttigieg, P., Yilmaz, P., Hankeln, W., Waldmann, J., and Glockner, F. (2010). Megx.net: Integrated database resource for marine ecological genomics. Nucleic Acids Res. *38*, D391–D395.

Li, R., Zhu, H., Ruan, J., Qian, W., Fang, X., Shi, Z., Li, Y., Li, S., Shan, G., Kristiansen, K., *et al.* (2010). De novo assembly of human genomes with massively parallel short read sequencing. Genome Res. *20*, 265–272.

Li, W. (2009). Analysis and comparison of very large metagenomes with fast clustering and functional annotation. BMC Bioinformatics *10*, 359.

Li, W., and Godzik, A. (2006). Cd-hit: a fast program for clustering and comparing large sets of protein or nucleotide sequences. Bioinformatics *22*, 1658–1659.

Li, W., Jaroszewski, L., and Godzik, A. (2001). Clustering of highly homologous sequences to reduce the size of large protein databases. Bioinformatics *17*, 282–283.

Li, W., Jaroszewski, L., and Godzik, A. (2002). Tolerating some redundancy significantly speeds up clustering of large protein databases. Bioinformatics *18*, 77–82.

Li, W., Wooley, J.C., and Godzik, A. (2008). Probing metagenomics by rapid cluster analysis of very large datasets. PLoS ONE 3, e3375.

Lin, A.W., Altintas, I., Churas, C., Gujral, M., Grethe, J., and Ellisman, M. (2011). Case Study on the Use of REST Architectural Principles for Scientific Analysis: CAMERA – Community Cyberinfrastructure for Advanced Microbial Ecology Research and Analysis.

In REST: From Research to Practice, Cesare, E.W. , and Pautasso, ed. (New Yor, USA: Springer).

Lowe, T.M., and Eddy, S.R. (1997). tRNAscan-SE: a program for improved detection of transfer RNA genes in genomic sequence. Nucleic Acids Res. *25*, 955–964.

Ludaescher, B., Altintas, I., Berkley, C., Higgins, D., Jaeger, E., Jones, M., Lee, E., Tao, J., and Zhao, Y. (2006). Scientific workflow management and the Kepler system. Concurrency Comput. Pract. Exp., 18: 1039–1065.

Marco, D. (ed.) (2010). Metagenomics: Theory, Methods and Applications. (Norfolk: Caister Academic Press).

Markowitz, V.M., Ivanova, N., Szeto, E., Palaniappan, K., Chu, K., Dalevi, D., Chen, I., Grechkin, Y., Dubchak, I., and Anderson, I. (2008). IMG/M: A data management and analysis system for metagenomes. Nucleic Acids Res. *36*, D534-D538.

Meyer, F., Paarmann, D., d'Souza, M., Olson, R., Glass, E.M., Kubal, M., Paczian, T., Rodriguez, A., Stevens, R., Wilke, A., *et al.* (2008). The metagenomics RAST server – a public resource for the automatic phylogenetic and functional analysis of metagenomes. BMC Bioinformatics 9, 286.

Myers, E.W., Sutton, G., Delcher, A., Dew, I., Fasulo, D., Flanigan, M., Kravitz, S., Mobarry, C., Reinert, K., and Remington, K. (2000). A whole-genome assembly of Drosophila. Science *287*, 2196–2204.

Niu, B., Fu, L., Sun, S., and Li, W. (2010). Artificial and natural duplicates in pyrosequencing reads of metagenomic data. BMC Bioinformatics *11*, 187.

Noguchi, H., Park, J., and Takagi, T. (2006). MetaGene: prokaryotic gene finding from environmental genome shotgun sequences. Nucleic Acids Res. *34*, 5623–5630.

NRC-Metagenomics-Committee (2007). The New Science of METAGENOMICS

In Revealing the Secrets of Our Microbial Planet. (Washington, DC: The National Academies Press).

Poretsky, R.S., Sun, S., Mou, X., and Moran, M.A. (2010). Transporter genes expressed by coastal bacterioplankton in response to dissolved organic carbon. Environ. Microbiol. *12*, 616–627.

Portugaly, E., Johnson, S., Ninio, M., and Eddy, S. (2007). Improved HMMERHEAD for Better Sensitivity. RECOMB 07 Poster.

Pruesse, E., Quast, C., Knittel, K., Fuchs, B.M., Ludwig, W., Peplies, J., and Glöckner, F.O. (2007). SILVA: a comprehensive online resource for quality checked and aligned ribosomal RNA sequence data compatible with ARB. Nucleic Acids Res. *35*, 7188–7196.

Rho, M., Tang, H., and Ye, Y. (2010). FragGeneScan: predicting genes in short and error-prone reads. Nucleic Acids Res. *38*, e191.

Riesenfeld, C.R., Schloss, P.D., and Handelsman, J. (2004). Metagenomics: genomic analysis of microbial communities. Annu. Rev. Genetics *38*.

Sayers, E.W., Barrett, T., Benson, D., Bryant, S., Canese, K., Chetvernin, V., Church, D., Dicuccio, M., Edgar, R., and Federhen, S. (2010). Database resources of the National Center for Biotechnology Information. Nucleic Acids Res. *38*, D5-D16.

Schmidt, B., Sinha, R., Beresford-Smith, B., and Puglisi, S.J. (2009). A fast hybrid short read fragment assembly algorithm. Bioinformatics *25*, 2279–2280.

Serra da Cruz, S.M., Campos, M.L.M., and Mattoso, M. (2009). Towards a taxonomy of provenance in scientific workflow management systems. IEEE Congress on Services, 259–266.

Seshadri, R., Kravitz, S.A., Smarr, L., Gilna, P., and Frazier, M. (2007). CAMERA: a community resource for metagenomics. PLoS Biol. *5*, e75.

Simpson, J.T., Wong, K., Jackman, S.D., Schein, J.E., Jones, S.J.M., and Birol, I. (2009). ABySS: A parallel assembler for short read sequence data. Genome Res. *19*, 1117–1123.

Sun, S., Chen, J., Li, W., Altinatas, I., Lin, A., Peltier, S., Stocks, K., Allen, E.E., Ellisman, M., Grethe, J. *et al.* (2010). Community cyberinfrastructure for Advanced Microbial Ecology Research and Analysis: the CAMERA resource. Nucleic Acids Res. 1–6.

Tatusov, R.L., Koonin, E., and Lipman, D. (1997). A Genomic perspective on protein families. Science *278*, 631–637.

Turnbaugh, P.J., Hamady, M., Yatsunenko, T., Cantarel, B.L., Duncan, A., Ley, R.E., Sogin, M.L., Jones, W.J., Roe, B.A., Affourtit, J.P. *et al.* (2009). A core gut microbiome in obese and lean twins. Nature *457*, 480-U487.

Wang, Q., Garrity, G.M., Tiedje, J.M., and Cole, J.R. (2007). Bayesian classifier for rapid assignment of rRNA sequences into the new bacterial taxonomy. Appl. Environ. Microbiol. 73, 5261–5267.

Warren, R.L., Sutton, G.G., Jones, S.J.M., and Holt, R.A. (2007). Assembling millions of short DNA sequences using SSAKE. Bioinformatics 23, 500–501.

Wooley, J.C., Godzik, A., and Friedberg, I. (2010). A Primer on Metagenomics. PLoS Comput. Biol. *6*, e1000667.

Wooley, J.C., and Ye, Y. (2010). Metagenomics: Facts and artifacts, and computational challenges. J. Comput. Sci. Technol. *25*, 71–81.

Yooseph, S., Li, W., and Sutton, G. (2008). Gene identification and protein classification in microbial metagenomic sequence data via incremental clustering. BMC Bioinformatics 9, 182.

Yooseph, S., Sutton, G., Rusch, D.B., Halpern, A.L., Williamson, S.J., Remington, K., Eisen, J.A., Heidelberg, K.B., Manning, G., Li, W. *et al.* (2007). The Sorcerer II Global Ocean Sampling Expedition: Expanding the Universe of Protein Families. PLoS Biol. 5, e16.

Zerbino, D.R., and Birney, E. (2008). Velvet: Algorithms for *de novo* short read assembly using de Bruijn graphs. Genome Res. *18*, 821–829.

Metagenomics for the Identification of Novel Viruses

9

Vincent Montoya, Charles Y. Chiu and Patrick Tang

Abstract

Viruses are the most abundant and genetically diverse biological entities on Earth and the vast majority is yet to be discovered. Therefore, systematic surveillance for viruses requires techniques that have both broad specificity and high sensitivity. Conventional laboratory techniques in virology often fail to detect a specific aetiology in many syndromes that are thought be caused by viruses. Metagenomics-based tools such as pan-viral microarrays and ultra-high-throughput sequencing have significantly improved our ability to detect and characterize divergent as well as novel viruses. Some of these methods rely on the fact that any one virus will possess some degree of conservation within its genomic sequence with other members of the same family. Thus, nucleic acid amplification tests targeting conserved regions in the viral families associated with a particular disease can often lead to a successful diagnosis. However, metagenomics-based techniques such as pan-viral microarrays are able to transcend our predetermined lists of viruses associated with each syndrome and allow for the simultaneous interrogation of thousands of conserved and specific genetic regions within all taxa of known virus families. Second generation high-throughput sequencing offers the unique opportunity to discover novel pathogens with no a priori sequence information with sensitivities comparable to that of PCR. As the costs for these techniques continue to decrease and the technology becomes more widely available, they will have the potential to revolutionize our approach to detecting viruses and diagnosing viral diseases.

Introduction

In a recent survey of emerging infectious diseases between 1940 and 2004, there have been an estimated 335 pathogens introduced into the human population (Jones *et al.*, 2008). In addition to improvements in laboratory detection methods, other anthropogenic trends such as immunosuppression, globalization, and environmental changes have collectively fuelled the emergence of new pathogens. Whether drug, disease, or poverty induced, the increase in immunocompromised individuals provides new opportunities for otherwise harmless viruses to infect human hosts. Given that intercontinental travel is shorter than the incubation period for many infectious diseases, the effects of globalization can facilitate the efficient spread of novel virus infections. As humans encroach into previously uninhabited regions of the planet, novel human–animal interactions will likely result in cross-species exchange of viruses. Climate change also compounds this situation as changes in weather patterns and temperature lead to the redistribution of animal vectors and the microorganisms that they harbour (Brodie *et al.*, 2007). Indeed, it has been estimated that over 60% of emerging infectious diseases are caused by zoonotic pathogens, the majority of which derive from wildlife populations (Jones *et al.*, 2008). Recent examples include HIV (Chen *et al.*, 1996; Keele *et al.*, 2006; Marx *et al.*, 1991), SARS (Drosten *et al.*, 2003; Ksiazek *et al.*, 2003; Peiris *et al.*, 2003), and the latest pandemic virus, 2009 swine-origin H1N1 influenza (CDC, 2009; Dawood *et al.*, 2009; Fraser *et al.*, 2009). A recent study in wild non-human primates demonstrated that co-infections with simian immunodeficiency virus (SIV), simian T-cell lymphotropic virus type 1 (STLV-1), and simian foamy virus (SFV) were as high as 45% (Leendertz *et al.*, 2010). In addition, central African bush hunters have been found to be infected with SFV as well as HTLV-3 and -4 (Wolfe *et al.*, 2005), suggesting that cross-species viral transmission is occurring and the potential for novel retrovirus recombination is real.

Cell culture, electron microscopy, serology, and PCR can be powerful methods for virus identification, but only if the respective cell lines, sera, and primers for downstream detection are compatible with the respective viral target. The traditional paradigm in diagnostic microbiology relies upon tests tailored for the already known aetiological agents associated with the observed clinical syndrome. Using conventional laboratory assays, an estimated 40% of gastrointestinal (Finkbeiner *et al.*, 2008) and 15% of respiratory illnesses (Juven *et al.*, 2000) for which the causative agents are believed to be of mostly viral in origin, fail to be diagnosed. With advances in molecular methods, namely the introduction of microarrays and ultra-high-throughput sequencing, the field of viral metagenomics has emerged as a powerful and unbiased collection of techniques for the detection and discovery of viruses. Viral metagenomics, or comprehensive nucleic acid sequence-based characterization of a given sample, promises to revolutionize the way we approach the field of virology much akin to how sequence-based bacterial classification has opened new realms in bacteriology. For instance, in two metagenomic analyses of human stool, roughly 28%–59% of the sequences were shown to have no significant matches in GenBank (E value > 0.001) (Breitbart *et al.*, 2003; Finkbeiner *et al.*, 2008). In contrast to detection strategies for cellular organisms which target universally conserved genes such as ribosomal DNA, viruses do not possess such conserved genes and therefore more advanced metagenomic techniques are essential for unbiased characterizations of any given virus community.

Metagenomic strategies for virus discovery typically employ an algorithmic approach as shown in Fig. 9.1.

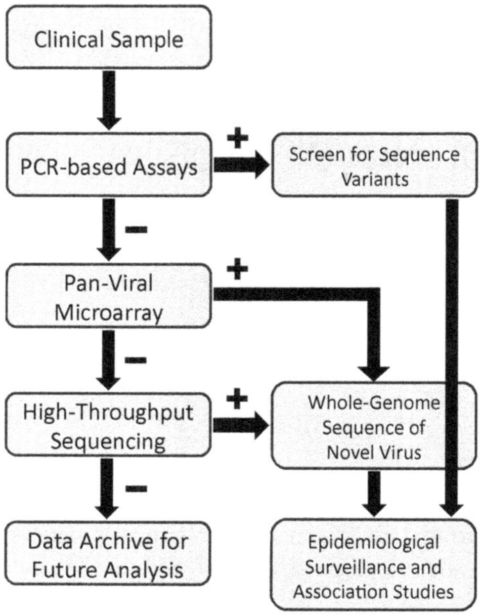

Figure 9.1 Algorithm for the identification of novel and uncommon viruses.

In a clinical laboratory, when conventional tests fail to identify viral pathogens associated with a particular syndrome, single or multiplexed PCR assays for conserved regions within known virus families (Mahony *et al.*, 2009; Rohayem *et al.*, 2004), differentially labelled nucleic acid probes, as well as direct sequencing can be employed (Molenkamp *et al.*, 2007; Raymond, 2009; Roh *et al.*, 2008). Samples that remain undiagnosed are then subjected to metagenomic analyses, such as pan-viral microarrays, simultaneously testing for all known, divergent, as well as novel viruses (Palacios *et al.*, 2007; Wang *et al.*, 2002). Positive signals on the microarrays can then be confirmed by sequence-based identification of these viruses. If the aetiological agent is undetectable by microarray, the samples can then be subjected to high-throughput sequencing to detect the potentially novel virus. Based on the relative costs of PCR, microarrays and high-throughput sequencing, this algorithm is the most cost-effective yet comprehensive strategy currently available for virus discovery. For roughly two decades, this powerful collection of molecular methods has been utilized in the discovery of multiple viruses from various families (Table 9.1).

Table 9.1 Viruses discovered using metagenomics techniques

Virus	Method	Family	Year	Reference
KSHV	RDA*1	Herpesviridae	1994	Chang *et al.* (1994)
GB viruses A and B	RDA	Flaviviridae	1995	Simons *et al.* (1995)
GB viruses C	SISPA*2	Flaviviridae	1996	Linnen *et al.* (1996)
TTV	RDA	Anelloviridae	1997	Nishizawa *et al.* (1997)
Bovine Parvovirus	SISPA	Parvoviridae	2001	Allander *et al.* (2001)
Coronavirus, HCoV-NL63	Cell culture and VIDISCA*3	Coronaviridae	2004	van der Hoek *et al.* (2004)
Bocavirus	Shotgun sequencing	Parvoviridae	2005	Allander *et al.* (2005)
Parvovirus 4/TT-like viruses	SISPA	Parvoviridae	2005	Jones *et al.* (2005)
XMRV	Microarray	Retroviridae	2006	Urisman *et al.* (2006)
Israeli acute paralysis virus	SISPA, high-throughput sequencing	Dicistroviridae	2007	Cox-Foster *et al.* (2007)
WU polyomavirus	Shotgun sequencing	Polyomaviridae	2007	Gaynor *et al.* (2007)
KI polyomavirus	Shotgun sequencing	Polyomaviridae	2007	Allander *et al.* (2007)
Human TMEV-like cardiovirus	Microarray	Picornaviridae	2008	Chiu *et al.* (2008a)
Human cosavirus E1	SISPA	Picornaviridae	2008	Holtz *et al.* (2008)
Unnamed arenavirus	High-throughput sequencing	Arenaviridae	2008	Palacios *et al.* (2008)
Merkel cell carcinoma	High-throughput sequencing	Polyomaviridae	2008	Feng *et al.* (2008)
Coronavirus SW1	Microarray	Coronaviridae	2008	Mihindukulasuriya *et al.* (2008)
Human cosavirus A1	Shotgun sequencing	Picornaviridae	2008	Kapoor *et al.* (2008)
Sea turtle tornovirus 1	SIA*4, shotgun sequencing	Unclassified ssDNA virus	2009	Ng *et al.* (2009)
Klassevirus	High-throughput sequencing	Picornaviridae	2009	Greninger *et al.* (2009)
TSV	High-throughput sequencing	Polyomaviridae	2010	van der Meijden *et al.* (2010)
Cyclovirus	SISPA, high-throughput sequencing	Circoviridae	2010	Li *et al.* (2010a)

*1Representational differential analysis.
*2Sequence-independent single primer amplification.
*3Virus-discovery-cDNA-AFLP.
*4Sequence-independent amplification.

Traditional strategies

Conventional laboratory methods in diagnostic virology, including cell culture, electron microscopy, and serology, are proven tools in the identification and characterization of viruses (Landry and Hsiung, 2000). Cell culture for viruses is an immeasurably important tool as it is essential for addressing fundamental questions regarding pathogenesis, host–virus interactions, and drug susceptibility (Leland and Ginocchio, 2007). In particular, it is a method for the amplification of viruses that facilitates all downstream diagnostic and characterization assays. Cell culturing coupled with electron microscopy is a methodology that has been used to identify many clinically important pathogens such as adenoviruses, enteroviruses, paramyxoviruses, reoviruses; and more recently the SARS coronavirus (Biel, 1999; Curry *et al.*, 2006; Peiris *et al.*, 2003). Electron microscopy can also be used independently from virus culture, providing an avenue for virus detection directly from clinical samples (Bishop *et al.*, 1973; Cossart *et al.*, 1975; Kapikian *et al.*, 1972; Morris *et al.*, 1975). Electron microscopic detection can be enhanced with sera from recently infected hosts to improve the visualization of the viral particles, as illustrated by the discoveries of hepatitis A and B viruses (Blumberg *et al.*, 1965; Dane *et al.*, 1970; Feinstone *et al.*, 1973). In solid-phase immunoelectron microscopy, antibodies can be attached by their Fc portion to the microscopic grid, leaving the antigen specific portion available to concentrate the corresponding virus for inspection (Gerna *et al.*, 1985). Conjugating metals such as gold or ferritin to the antibodies has also been successfully utilized for virus subtype differentiation (Hopley and Doane, 1985). Upon visualization, structural information about the virus can be obtained providing directional guidance in downstream experiments when more definitive identification tests are employed (Chua *et al.*, 1999; Nichol *et al.*, 1993). Immune serum can also be used to screen peptide libraries obtained from infected hosts such as that used for the discovery of hepatitis C virus (Choo *et al.*, 1989).

These strategies however, like all others, have their limitations. The most obvious limitation to virus culture systems is the availability of an appropriate cell line that can efficiently amplify the suspected virus. Although the majority of viruses pathogenic to humans can be grown, cell lines for many common viruses (e.g. Norwalk virus, bocavirus, BK virus, JC virus, human herpesviruses 6, 7, and 8) have yet to be established despite some of their identifications decades ago (Landry and Hsiung, 2000; Leland and Ginocchio, 2007). Immuno-based detection systems also have their limitations as sera derived from previously infected hosts can be difficult to obtain and antibodies may not be present in sufficient concentrations (Doane, 1987; Roingeard, 2008). Furthermore, electron microscopy is relatively insensitive as typically 10^5 to 10^6 particles/ml are needed for viral detection (Roingeard, 2008).

The human virome

Globally, it is estimated there are over 10^{30} virus-like particles in existence and that 10^{27} are being generated every minute (Breitbart and Rohwer, 2005; Suttle, 2005). Although the vast majority infect bacteria, their effects on the human population can be significant given that prophage insertions make up the majority of the genetic differences between a bacterial pathogen and its commensal relatives. A metagenomic analysis of the human gastrointestinal tract has estimated that there are over 1000 distinct types of viruses present (Breitbart *et al.*, 2003). A recent study compared the gastrointestinal viromes in four pairs of monozygotic twins and their respective mothers (Reyes *et al.*, 2010). The virus communities were unique to each individual and relatively stable, as >95% of the identified viruses were recovered in three time points over a 1-year period. Interestingly, about 81% of the sequences in this study had no significant matches to known viruses in a non-redundant viral sequence database. Various independent studies have identified a diverse set of viruses which persist with their human hosts for great lengths of time. We currently know of at least 8–12 viruses that are presumed to be present in at least 90% of the human population, many of which have not been positively associated with any known diseases (Virgin *et al.*, 2009). The significance of the human virome is still far from understood, but it is intriguing to consider the existence of commensal viruses and whether or not they have significant effects, both beneficial and deleterious, on our health (Barin, 2000; Barton *et al.*, 2007).

Identifying new viruses

Clues to the presence of a new virus typically depend upon the detection of unusual clinical symptoms and epidemiological patterns or cases for which routine laboratory tests fail to diagnose a known pathogen (Fredricks, 2004). Such unknown clinical specimens are often banked and amenable to retrospective testing using more advanced techniques such as metagenomics (Svraka *et al.*, 2010).

Screening individuals with an elevated risk of cross-species transfer (e.g. African bushmeat hunters and animal workers) (Myers *et al.*, 2006; Wolfe *et al.*, 2005) as well as the most common vectors responsible for facilitating these transfers (e.g. wild and domesticated animals, blood-feeding insects, rodents and bats) is a promising approach to discovering novel viruses (Dinsdale *et al.*, 2008; Engels *et al.*, 2002; Hall-Mendelin *et al.*, 2010; Heeney, 2006; Kuiken *et al.*, 2005; Wolfe *et al.*, 2007). One recent study identified viruses from multiple families in the guano of insectivorous bats (Li *et al.*, 2010b). In addition to plant- and insect-specific viruses, relatives of known mammalian viral families such as Parvoviridae, Circoviridae, Adenoviridae, Poxviridae, Picornaviridae, Astroviridae, and Coronaviridae were found. Notably, a hybrid sequence with homology between Picornaviridae and Dicistroviridae was also found, possibly indicating a new family of viruses. As these studies continue, coupled with new bioinformatics methods which can predict the potential hosts from viral sequence data (Kapoor *et al.*, 2010), we may be able to identify pathogens much earlier in an outbreak, and potentially even before they enter the human population.

Although detecting novel viruses in acute infections can be difficult, pathogens associated with neoplastic, inflammatory, autoimmune and other chronic diseases are even more elusive (Fredricks, 2004; Fredricks and Relman, 1998). In these cases, the association between a pathogen and the clinical syndrome is more difficult to establish as the disease may have multiple variables involved in its pathology, including host genetics and environmental factors (Virgin *et al.*, 2009). In a recent example illustrating the complexities of an inflammatory disease, only mice containing a susceptibility allele infected by a specific virus produce symptoms of Crohn's disease (Cadwell *et al.*, 2010). As metagenomic techniques expand to explore chronic diseases, we will likely discover new virus associations and ultimately better strategies for treatment and prevention of these diseases.

Metagenomics

Preanalytical preparations

Perhaps the most important and often neglected step in virus discovery using molecular methods is the extraction of viral nucleic acid (Delwart, 2007; Petrich *et al.*, 2006; Wiedbrauk *et al.*, 1995). The absolute and relative concentrations of viral to host nucleic acids will heavily influence the success of all subsequent tests. Various methods have been used to increase the ratio of viral nucleic acids, and each of these depends on the sample matrix. Clinical specimens derived from solid tissue have high concentrations of host genetic and structural material, and therefore require homogenization or enzymatic digestion. Given that the genomes of viral particles are protected by capsid proteins and/or envelope proteins, exonucleases such as DNase, can be added prior to extraction to reduce non-viral nucleic acid interference. This method was first employed by Allander *et al.* for the discovery of two novel parvoviruses in bovine serum albumin (Allander *et al.*, 2001). However, this method is limited to those viruses which do not exist in episomal forms and are not integrated into the host genome. Given that ribosomal nucleic acids predominate in RNA extractions, another method of reducing nucleic acid complexity is to enrich for polyadenylated transcripts (Hu and Hirshfield, 2005).

Samples that contain particulate matter require filtration and concentration (Loens *et al.*, 2002; Spackman and Suarez, 2008; Thurber *et al.*, 2009). Tangential flow filtration has been used to concentrate virus particles from multiple sample types (Bench *et al.*, 2007; Breitbart *et al.*, 2003). In standard impact filtration, the flow of material is perpendicular to the filter and therefore the flow-through of the desired substrate may be obstructed by the accumulation of unwanted material (Schoenfeld *et al.*, 2008). In tangential flow filtration, the flow is cycled through the system tangential to the filter, where a back flow, generated by clamping the exiting sample, provides an adequate pressure differential to drive particles through the filter. This tangential flow system effectively minimizes the accumulation of material along the filter. Viral particles larger than the pores in the filter repeatedly cycle through the system reaching the desired concentration for downstream detection. For samples with extremely high concentrations of particulate matter, pre-filtering with a coarse pore filter prior to tangential flow filtration is recommended. Ultracentrifugation is typically the next step in virus particle concentration where virus particles of the desired density gradient can readily be removed for downstream processes. Detailed procedures on these methods have been described elsewhere (Stang *et al.*, 2005; Thurber *et al.*, 2009).

Subtraction and differential display

Two related methods to identify large differences in the nucleic acid content between two samples are subtraction and differential display (Sagerstrom *et al.*, 1997; Stein and Liang, 2002). These strategies require a high percentage of genetic similarity between the samples being tested and although samples from the same patient are preferred, the same tissue type from a control subject may also be useful (Ambrose and Clewley, 2006). For samples with a relatively high concentration of virus, differential display requires a nonspecific amplification step where upon gel electrophoresis, unique amplicons relative to the uninfected samples are subsequently cloned and sequenced (Lu *et al.*, 2004). Subtraction or representational differential analysis (RDA) is an alternative method that can address the challenge of viral infections with low nucleic acid content (Muerhoff *et al.*, 1997). This method consists of adding the uninfected sample (driver) in excess to that of the test sample with common sequences removed via subtractive hybridization (Sagerstrom *et al.*, 1997). This process may be performed in single or multiple rounds until the resolution of unique single stranded DNA sequences are sufficient for isolation and subsequent characterization. As stated above, episomal or integrated viruses cannot be concentrated using basic techniques such as centrifugation or in some cases enzymatic digestion. Subtractive hybridizations have frequently been used as a solution to this problem in the past. Historically, Bishop and Varmus used this method for the characterization of the first oncogene in an avian sarcoma virus (Stehelin *et al.*, 1976). More recently a variation of this method has been used to discover human herpesvirus 8 from Kaposi sarcoma lesions (Chang *et al.*, 1994). Despite these successes, this somewhat laborious method lacks the sensitivity of current high-throughput sequencing methods.

Recently, modifications of these subtractive hybridizations have been specifically designed to complement second generation sequencing (Adams *et al.*, 2009). Here, both the healthy and affected samples are first amplified, where biotinylated dUTP in addition to dTTP are incorporated into the healthy sample. Streptavidin beads are then used to subtract the common amplicons between the healthy and affected samples. Through this method, a novel cucumovirus was discovered in infected *Gomphrena globosa* plants.

Sequence-independent amplification

The most successful methods for identifying novel viruses that are too divergent to be detected by conventional assays rely on sequence-independent amplification. The core algorithm for all of these methods is similar and begins with nucleic acid amplification followed by subcloning and sequencing of amplified fragments. A reverse transcription step prior to amplification to convert RNA to complementary DNA (cDNA) can be used to detect viral RNA. Homology searches are performed in GenBank to identify sequenced fragments that possess similarity to known viruses. Novel viruses can be identified by fragments that are distantly related to known viral sequences.

Many different techniques for sequence-independent amplification have been described, and can be divided roughly into three groups. The first group, including sequence-independent single primer amplication (SISPA) or linked amplified shotgun libraries (LASL), is based on restriction enzyme digestion or shearing of viral DNA/cDNA and ligation of adapters, followed by PCR amplification using the adapter sequence (Reyes and Kim, 1991). These methods have been successful in identifying among others, new parvoviruses in human plasma (Jones *et al.*, 2005). The second group includes PCR amplification methods using arbitrary primers at low annealing temperatures or random primers, which have been successful in detecting numerous viruses in respiratory secretions and stool, including human metapneumoviruses, bocaviruses, polyomaviruses, cosaviruses, and kobuviruses (Allander *et al.*, 2005, 2007; Gaynor *et al.*, 2007; Greninger *et al.*, 2009; Holtz *et al.*, 2009; Kapoor *et al.*, 2008; Li *et al.*, 2009; van den Hoogen *et al.*, 2001). The final group consists of methods which amplify DNA by displacement, including PhiX29 DNA polymerase-based amplification to amplify genomic DNA in an unbiased fashion, and rolling-circle amplification to amplify circular DNA viral genomes (Rector *et al.*, 2004). A rolling-circle amplification strategy was recently used to identify a novel polyomavirus associated with a rare skin disease, trichodysplasia spinulosa, in an immunocompromised patient (van der Meijden *et al.*, 2010). Since nearly all sequence-independent amplification methods result in unbiased amplification of DNA, it is critical that background nucleic acids, such as from the human host or bacteria, are first removed to enrich samples for viral nucleic acids and thus maximize the probability of discovering sequences corresponding to novel viruses.

Microarrays

Microarrays have proven to be a powerful metagenomic tool to quantify and characterize environments (Brodie *et al.*, 2007), transcriptomes (Schena *et al.*, 1995; Welsh *et al.*, 2001), and clinical specimens (Chiu *et al.*, 2008b; Palacios *et al.*, 2007; Wang *et al.*, 2002). Diagnostic microarrays hold great promise for metagenomics as they can be simultaneously utilized for detecting and resequencing known, divergent, and novel viruses. For example, shortly after the swine flu pandemic was declared, an influenza microarray platform was utilized to obtain genome-wide sequence information about the emerging reassortant influenza virus (Berthet *et al.*, 2010). In contrast to these targeted viral microarrays which detect a narrower range of viruses, the detection of a broad range of viruses requires a different type of microarray design.

The most widely used pan-viral microarrays, the Virochip and the Greenechip, have tens of thousands of probes, each with a collection accounting for all known virus families known to infect humans (Palacios *et al.*, 2007; Wang *et al.*, 2002). In reference to quantitative microarrays, the probes on these arrays are elongated to increase their tolerance for mismatches between each probe and its complementary sequence thereby increasing the ability to detect novel viruses. These arrays need to be updated regularly as new viruses are discovered.

In 2002, the Virochip was first introduced and shortly afterwards, it demonstrated its potential in the initial identification and sequence characterization of the SARS coronavirus (Wang *et al.*, 2003). Probe selection for the original Virochip is genome based: all sequenced viral genomes are aligned and a 70-nucleotide window is used to select the five best sequences which contain at least 20 bases in common. Additional probes were chosen for certain genera that are genetically similar but have clinically diverse outcomes so as to further assist in their differentiation (Wang *et al.*, 2002). The current version of the Virochip, updated in October of 2009, contains ~36,000 probes. The sensitivity of the Virochip is superior to direct fluorescent antibody detection and comparable to PCR for diagnosis of viral respiratory infections (Chiu *et al.*, 2008b; Kistler *et al.*, 2007), with detection levels as low as 100 viral genome copies based on testing for rhinoviruses (Wang *et al.*, 2002). The Virochip is currently being used as a diagnostic tool at the British Columbia Centre for Disease Control as well as University of California, San Francisco, where it has successfully identified uncommon, divergent, as well as novel viruses. For example in 2006, it identified a novel murine retrovirus in prostate cancer patients with mutations in RNase L, a known ribonuclease involved in type 1 interferon-mediated antiviral responses (Urisman *et al.*, 2006). Other significant findings include a new clade of rhinoviruses (Kistler *et al.*, 2007), a novel human cardiovirus (Chiu *et al.*, 2008a), as well as an uncommon human parainfluenza virus type 4 and a divergent human metapneumovirus infection in patients with severe respiratory infection (Chiu *et al.*, 2007; Chiu *et al.*, 2006). More recently, the Virochip was used to rapidly identify 2009 pandemic influenza as a novel swine influenza variant even though the viral probes were designed prior to the emergence of the pandemic (Greninger *et al.*, 2010).

The diagnostic microarray with the widest range in microorganism identification is the GreeneChip. This array is a pan-microbial microarray containing 29,495 probes specific for pathogens ranging from viruses to eukaryotic parasites (Palacios *et al.*, 2007). In contrast to the Virochip, this array design is protein-based where three nucleic acid probes were chosen for each family based on alignments of amino acid sequences. These three regions when possible correspond to one non-structural and two structural regions. For cellular parasites, specific portions of highly conserved ribosomal DNA sequences were selected. Collectively the chip contains 9477 oligonucleotides for all vertebrate viruses (as of 2007, totalling 1710 species), 11,479 16S rRNA probes for 135 bacteria, 1120 18S rRNA probes for 73 fungi, and 848 18S rRNA probes for 63 parasite genera. In addition to these probes, 300 host-immune response genes were included for additional information. Based on real-time PCR analyses using viral RNA isolated from cultured cells, the sensitivity of this array platform reached 1900 RNA copies for RNA viruses and 10,000 RNA copies for DNA viruses. Illustrating the potential of this array, a patient with symptoms presumably indicative of viral haemorrhagic fever was shown to be positive for Plasmodium falciparum. Extensive subtyping of influenza viruses using a set of specific primers prior to hybridization was also demonstrated (Quan *et al.*, 2007).

The crux to all microarrays is their sensitivity as well as their reproducibility. With reference to diagnostic arrays, this is greatly determined by the viral to host ratio of nucleic acids prior to amplification. Variables such as the type of sample and the site of infection greatly influence this ratio. For instance, there is a relatively high success rate of virus identification in acute respiratory and gastrointestinal infections which tend to have higher virus titres compared to other infections (Chiu *et al.*, 2008b; Kistler *et al.*, 2007). Acellular samples such as those derived from cell culture supernatant or blood serum improve this ratio due to minimal host nucleic acid content (Leski *et al.*, 2010). In tissue samples that have less desirable virus to host nucleic acid ratios, methods such as DNase treatment can be useful prior to reverse transcription as stated in the previous section.

Various methods in reverse transcription and subsequent amplifications also influence the success of virus identification (Leski *et al.*, 2010). The standardized method for amplification using the Virochip consists of a modified sequence independent amplification (SIA) strategy that incorporates a reverse transcription step. Reverse transcription is directed by random nonamer primers flanked by a unique 5′ adapter sequence. Second strand synthesis is carried out, much like SIA (Bohlander *et al.*, 1992), with Sequenase, a T7 DNA polymerase. The cDNA is subsequently amplified prior to enzymatic amino-allyl nucleotide incorporation for downstream dye conjugation, both using a Klenow fragment DNA polymerase. The above procedure is analogous to that used for the GreeneChip except that dendrimers are employed for fluorescent dye molecule incorporation.

Each virus genome will bind in a unique pattern as dictated by the probes available as well as the hybridization conditions used. Given that there are thousands of probes, computer algorithms and statistical methods are essential to ascertain whether an observed pattern is indicative of a true virus genome. One such strategy designed for this purpose is E-Predict, an algorithm that compares an observed hybridization pattern to a pre-constructed theoretical pattern (Urisman *et al.*, 2005). These theoretical hybridization patterns are generated *in silico* where the free energy of hybridization is calculated for each virus genome when compared to all probes on the microarray. Furthermore, by comparing related microarray data this calculated value can be adjusted in a Bayesian fashion. With reference to novel viruses for which a theoretical pattern is non-existent, the intensity of each individual probe can also be compared to its intensity in controls as well as related samples hybridized in the past. When taxonomically arranged, these statistically significant oligonucleotides can provide sufficient data to determine the identity of a given virus (Chiu *et al.*, 2006). Other algorithms assessing the significance of a given pathogen in microarray data have also been employed such as PhyloDetect (Rehrauer *et al.*, 2008), and DetectiV (Watson *et al.*, 2007).

Deep sequencing

The development of second generation high-throughput or 'deep' sequencing technology for metagenomics has rapidly impacted the field of virus discovery by improving the detection of viruses that are highly divergent and which may be present at very low titres. These platforms have the capacity to generate hundreds of thousands to millions of sequences per run (Shendure and Ji, 2008). Currently, the two most widely used deep sequencing platforms for virus discovery are the pyrosequencing system of 454 Life Sciences/ Roche (CT, USA) and the 'sequencing-by-synthesis' system of Solexa/Illumina (CA, USA), although several third generation sequencing systems are in development.

The utility of a deep sequencing platform for virus discovery relies on both the average read length and total number of sequence reads generated. The sequences must be long enough to permit unambiguous identification of sequences corresponding to novel viral genomes. This is especially true for clinical samples that may have a complex nucleic acid background from environmental sources or as part of the microbial flora, such as respiratory secretions and stool; the problem still exists, however, for body sites that are considered normally 'sterile', including sera, tissue, and cerebrospinal fluid. One potential solution is to assemble short sequence reads that overlap to generate longer contiguous sequences (contigs) which can in turn be more readily identified. This method was recently used to detect a novel arenavirus related to lymphocytic choriomeningitis virus in a cluster of fatal cases in transplant recipients (Palacios *et al.*, 2008). In addition to read length, there must be enough reads generated for adequate sequencing depth. Methods described above to enrich viral nucleic acids in samples still do not negate the need to obtain as many sequences as possible to maximize detection sensitivity for viruses that may be present at exceedingly low titres. One example is the recent discovery of a novel polyomavirus associated with Merkel cell carcinoma by pyrosequencing (Feng *et al.*, 2008). After selection for polyadenylated RNA and analysis of over 400,000 sequence reads, only two sequences were found to correspond to the novel polyomavirus. The Roche Genome Sequencer FLX™ currently produces up to 1 million sequence reads per run with an average read length ranging from 200 to 450 bp, while the Illumina Genome Analyzer IIx can generate up to 100 million sequencer reads with fixed paired-end read lengths of 100–200 bp (paired-end reads are sequences generated from the opposite ends of the same individual DNA molecule). Barcoding of samples permits analysis of multiple samples per run at the cost of proportionally fewer sequences generated per specimen.

Analysis of deep sequencing data

The analysis of deep sequencing data for novel viral sequences in clinical samples poses a major computational challenge. We and others have developed automated bioinformatics pipelines to carry out the analysis, as shown in Fig. 9.2. First, low-complexity and duplicate reads are removed and sequence reads are classified by sample barcode. Overlapping reads can be computationally assembled at this stage to generate contigs. The sequence reads or assembled contigs are then subjected to similarity searches to non-redundant organism sequence databases. The criterion for classifying sequences is arbitrary, although an E-score of $<10^{-5}$ is considered to be fairly stringent (Allander *et al.*, 2005). To avoid misclassifications,

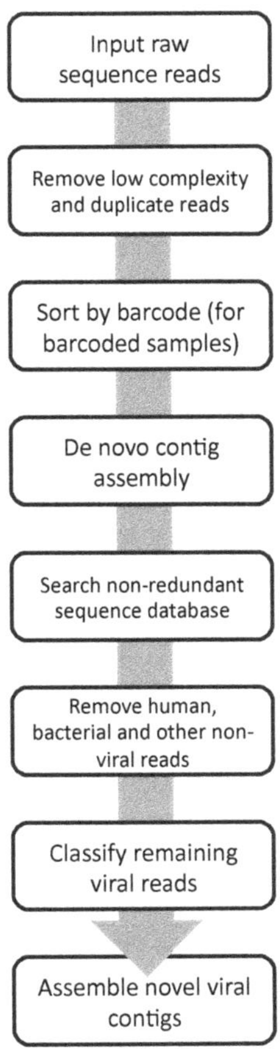

Figure 9.2 Analysis of deep sequencing data for the discovery of novel viruses.

similarity searches can be performed initially at a very high stringency and then repeated at gradually decreasing levels of stringency. Sequence reads corresponding to human, bacterial, and other non-viral organisms are removed from the dataset. However, this procedure will remove viral sequence that may have previously integrated in the sequenced human genome (e.g. human endogenous retroviruses and bornaviruses) (Horie *et al.*, 2010; Martin *et al.*, 1981). The remaining sequence reads are then compared with viral databases or the NCBI non-redundant (nr) database using programs that examine homology first at the nucleotide (BLASTN) and then at the amino acid levels (BLASTX and TBLASTX) (Altschul *et al.*, 1990). Detected viral sequences corresponding, albeit distantly, to known viruses are taxonomically classified, and attempts are made to assemble larger viral contigs or genomes from sequences corresponding to these viruses. If there are matches to multiple regions of the viral genome, the mapped sequences can be used to generate primers for subsequent PCR and Sanger sequencing to fill in the missing gaps in the sequence.

A key problem in viral discovery is how to identify viruses that are so highly divergent that they elude detection even by TBLASTX. In some cases, a remote homology program such as PHI-BLAST can be used for detection of these extremely divergent viruses (Altschul *et al.*, 1997). Direct searches for highly conserved viral protein motifs or hallmark genes may be helpful. In addition, the detection sensitivity may be improved by use of substitution matrices or hidden Markov models to quantify sequence similarities (Altschul *et al.*, 2010). Other BLAST-independent methods to discriminate sequences corresponding to novel viruses include searching for sequence 'markers' that may not bear any homology to known sequences but may be common to samples from specific clinical cohorts (e.g. aseptic meningitis, idiopathic hepatitis). These markers may correspond to a novel virus, and the viral genome may be recovered by primer walking or further sequencing. Finally, with the current attainable sequencing read depths, it is now feasible to sequence entire viral genomes *de novo* without the use of any reference alignment. For example, by pooling 17 samples from patients from influenza-like illness at the onset of the 2009 H1N1 pandemic, 90% of the genome of 2009 H1N1 influenza was assembled *de novo* without the availability of any sequences from the Orthomyxoviridae (influenza) family to use as a reference (Greninger *et al.*, 2010).

Many steps in the pipeline, particularly the BLASTN to the nr database, TBLASTX, PHI-BLAST, and *de novo* assembly steps, can be computationally time-consuming. The required processing time will only increase in the future as publicly available sequence databases are expected to increase in size at an exponential rate (Stratton *et al.*, 2009). Thus, it is preferable for software implementing a deep sequencing analysis pipeline to be scalable and thus suitable for use on a desktop workstation, small server cluster, large computational core, or even on a web-based cloud computing system (Schadt *et al.*, 2010).

To date, deep sequencing methods have been used to identify several novel viruses, including a haemorrhagic fever virus from South Africa (Briese *et al.*, 2009), a human kobuvirus (Greninger *et al.*, 2009), animal circoviruses found in faecal specimens (Li *et al.*, 2010a), an arenavirus in transplant recipients (Palacios *et al.*, 2008), and Merkel cell polyomavirus (Feng *et al.*, 2008). Currently, the high costs, lengthy times for sample preparation, and need for sophisticated bioinformatics expertise prevent deep sequencing from being applied on a routine basis for viral discovery. Nevertheless, in the future, we envisage that this technology will revolutionize not only viral discovery, but also clinical diagnostics in laboratory and public health settings.

Evidence is not causality

Discovering a novel virus within a clinical sample is merely the first step in the meticulous process of determining associations of the virus with the presented disease, much less causality. Viruses often fail to meet the demands of Koch's postulates for causality due to the difficulty in obtaining models for the infection both *in vitro* and *in vivo* (Fredricks and Relman, 1998). As stated above, a sufficient culture system as well as a suitable host that can serve as a model for those pathologies found in humans can be arduous tasks. The absence of these luxuries severely hampers downstream analyses as they can provide valuable information regarding the cell, tissue, and ultimately host tropisms. These systems also allow one to ascertain the complexities of host–virus interactions at the molecular level laying the foundation for the elucidation of possible mechanisms of pathogenesis. Finally, concentrated titres of virus can be obtained and utilized for ultrastructural studies or to produce antigens in order to assess its seroprevalence in the population in question.

Confronted with these difficulties, other alternative epidemiological parameters such as serological and epidemiological analyses have been used to supplement the difficulties of Koch's postulates (Falkow, 1988; Inglis, 2007; Rivers, 1937). Epidemiological data such as the prevalence of nucleic acid and sera in the infected population relative to the healthy population can solitarily or collectively provide robust and unambiguous lines of evidence (Dong *et al.*, 2007). Identifying sera from infected patients that are specific to antigens of the novel virus may solidify a causal relationship as well. For example, in the 1980s an infectious pathogen associated with non-A, non-B hepatitis was presumed to be of viral origin, however no available cell lines were successful in culturing any virus. A breakthrough was made in 1989 when a lambda phage cDNA expression library was shown to react with the serum from a patient with non-A, non-B hepatitis (Choo *et al.*, 1989). This novel antigen was found to be of viral origin and was adopted in a powerful assay which was utilized in the establishment of the virus, named hepatitis C, in various liver pathologies. Initially, encouraging statistics were obtained when it was found in 65% of patients with hepatocellular carcinomas (HCC) and 74% of non-A, non-B chronic hepatitis and cirrhosis (risk factors for HCC), and only 1–2% of healthy patients without risk factors for hepatitis (Colombo *et al.*, 1989; Esteban *et al.*, 1989). Thus, these studies provided encouraging lines of evidence for the subsequent solidification of hepatitis C as a major risk factor in developing HCC without fulfilling Koch's postulates for causality. However, for ubiquitous viruses (e.g. Epstein–Barr virus, human herpesvirus 6 and torque teno virus) or diseases in which additional variables such as host genetics and environmental factors play significant roles in the development of disease, proof of viral causation may be exceedingly difficult (Gewurz *et al.*, 2008; Okamoto, 2009; Salvetti *et al.*, 2009).

Future perspective

Addressing the demands of assays with broad specificity and high sensitivity has been a problem since the inception of diagnostic microbiology. Owing to the abundance of viral genome sequences and advances in molecular biology, these demands can both be met simultaneously by simply analysing all of the nucleic acid material in the specimen. As these problems continue to be addressed in previously unimaginable ways, current strategies in virus identification should be updated accordingly (Svraka *et al.*, 2010). In the field of molecular biology, changes in virtually every step of nucleic acid and protein analyses are accelerating at such staggering rates that our ability to store and process the information is now the largest obstacle. Each step can be specialized to solve the individual problems unique to each sample source. Recent analyses into sequencing the genomes of extinct animals such as mammoths or Neanderthals highlight these specialized techniques (Dalton, 2006; Poinar *et al.*, 2006). Furthermore, as the automation and miniaturization of nucleic acid amplification and detection platforms, incorporation of microfluidics, and the development of more sensitive hybridization methods continue to advance, the chances of adapting them into the clinical setting becomes more promising (Bergeron, 2008; Holland and Kiechle, 2005; Jain, 2005). The rapid pace at which next-generation sequencing is evolving will soon provide more affordable, higher throughput, single molecule sequencing, such that it may even replace standard nucleic acid amplification and microarray hybridization techniques (Harris *et al.*, 2008). Our enhanced ability to identify and characterize more viruses will likely reveal a significant fraction of novel viruses infecting any species under study, allowing us to appreciate the extent of their interactions with their hosts and ultimately the global ecosystem (Kristensen *et al.*, 2010; Relman, 1999; Wolfe *et al.*, 2007). Akin to the elegant viral metagenomic studies characterizing various environments, the human body must also be studied similarly where all microbial species of both cellular and viral origins, should be tabulated and characterized (Anderson *et al.*, 2003). Given that the human genome has viral sequences derived from families such as Retroviridae and Bornaviridae, viruses have likely made significant contributions towards the evolution of our genome (Horie *et al.*, 2010; Martin *et al.*, 1981). Although not formally represented in the three domains of life, viruses should at the least be considered an essential part of the human metagenome, a collective network of genomes consisting of multiple players, all influencing the health and disease of the ecosystem that is the human body (Virgin *et al.*, 2009).

References

Adams, I.P., Glover, R.H., Monger, W.A., Mumford, R., Jackeviciene, E., Navalinskiene, M., Samuitiene, M., and Boonham, N. (2009). Next-generation sequencing and metagenomic analysis: a universal diagnostic tool in plant virology. Mol. Plant. Pathol. *10*, 537–545.

Allander, T., Andreasson, K., Gupta, S., Bjerkner, A., Bogdanovic, G., Persson, M.A., Dalianis, T., Ramqvist, T., and Andersson, B. (2007). Identification of a third human polyomavirus. J. Virol. *81*, 4130–4136.

Allander, T., Emerson, S.U., Engle, R.E., Purcell, R.H., and Bukh, J. (2001). A virus discovery method incorporating DNase treatment and its application to the identification of two bovine parvovirus species. Proc. Natl. Acad. Sci. U.S.A. *98*, 11609–11614.

Allander, T., Tammi, M.T., Eriksson, M., Bjerkner, A., Tiveljung-Lindell, A., and Andersson, B. (2005). Cloning of a human parvovirus by molecular screening of respiratory tract samples. Proc. Natl. Acad. Sci. U.S.A. *102*, 12891–12896.

Altschul, S.F., Gish, W., Miller, W., Myers, E.W., and Lipman, D.J. (1990). Basic local alignment search tool. J. Mol. Biol. *215*, 403–410.

Altschul, S.F., Madden, T.L., Schaffer, A.A., Zhang, J., Zhang, Z., Miller, W., and Lipman, D.J. (1997). Gapped BLAST and PSI-BLAST: a new generation of protein database search programs. Nucleic Acids Res. *25*, 3389–3402.

Altschul, S.F., Wootton, J.C., Zaslavsky, E., and Yu, Y.K. (2010). The construction and use of log-odds substitution scores for multiple sequence alignment. PLoS Comput. Biol. *6*, e1000852.

Ambrose, H.E., and Clewley, J.P. (2006). Virus discovery by sequence-independent genome amplification. Rev. Med. Virol. *16*, 365–383.

Barin, F. (2000). The virus isolated from patient TT (TTV): still an orphan 2 years after its discovery. Transfus. Clin. Biol. *7*, 79–83.

Barton, E.S., White, D.W., Cathelyn, J.S., Brett-McClellan, K.A., Engle, M., Diamond, M.S., Miller, V.L., and Virgin, H.W. (2007). Herpesvirus latency confers symbiotic protection from bacterial infection. Nature *447*, 326–329.

Bench, S.R., Hanson, T.E., Williamson, K.E., Ghosh, D., Radosovich, M., Wang, K., and Wommack, K.E. (2007). Metagenomic characterization of Chesapeake Bay virioplankton. Appl. Environ. Microbiol. *73*, 7629–7641.

Berthet, N., Leclercq, I., Dublineau, A., Shigematsu, S., Burguiere, A.M., Filippone, C., Gessain, A., and Manuguerra, J.C. (2010). High-density resequencing DNA microarrays in public health emergencies. Nat. Biotechnol. *28*, 25–27.

Biel S.S., G.H. (1999). Diagnostic electron microscopy is still a timely and rewarding method. J. Clin. Virol. *13*, 105–119.

Bishop, R.F., Davidson, G.P., Holmes, I.H., and Ruck, B.J. (1973). Virus particles in epithelial cells of duodenal mucosa from children with acute non-bacterial gastroenteritis. Lancet *2*, 1281–1283.

Blumberg, B.S., Alter, H.J., and Visnich, S. (1965). A 'new' antigen in leukemia sera. JAMA *191*, 541–546.

Bohlander, S.K., Espinosa, R., 3rd, Le Beau, M.M., Rowley, J.D., and Diaz, M.O. (1992). A method for the rapid sequence-independent amplification of microdissected chromosomal material. Genomics *13*, 1322–1324.

Breitbart, M., Hewson, I., Felts, B., Mahaffy, J.M., Nulton, J., Salamon, P., and Rohwer, F. (2003). Metagenomic analyses of an uncultured viral community from human feces. J. Bacteriol. *185*, 6220–6223.

Breitbart, M., and Rohwer, F. (2005). Here a virus, there a virus, everywhere the same virus? Trends Microbiol. *13*, 278–284.

Briese, T., Paweska, J.T., McMullan, L.K., Hutchison, S.K., Street, C., Palacios, G., Khristova, M.L., Weyer, J., Swanepoel, R., Egholm, M. *et al.* (2009). Genetic detection and characterization of Lujo virus, a new hemorrhagic fever-associated arenavirus from southern Africa. PLoS Pathog. *5*, e1000455.

Brodie, E.L., DeSantis, T.Z., Parker, J.P., Zubietta, I.X., Piceno, Y.M., and Andersen, G.L. (2007). Urban aerosols harbor diverse and dynamic bacterial populations. Proc. Natl. Acad. Sci. U.S.A. *104*, 299–304.

Cadwell, K., Patel, K.K., Maloney, N.S., Liu, T.C., Ng, A.C., Storer, C.E., Head, R.D., Xavier, R., Stappenbeck, T.S., and Virgin, H.W. (2010). Virus-plus-susceptibility gene interaction determines Crohn's disease gene Atg16L1 phenotypes in intestine. Cell *141*, 1135–1145.

CDC (2009). Swine influenza A (H1N1) infection in two children – Southern California, March-April 2009. MMWR Morb. Mortal. Wkly. Rep. *58*, 400–402.

Chang, Y., Cesarman, E., Pessin, M.S., Lee, F., Culpepper, J., Knowles, D.M., and Moore, P.S. (1994). Identification of herpesvirus-like DNA sequences in AIDS-associated Kaposi's sarcoma. Science *266*, 1865–1869.

Chen, Z., Telfier, P., Gettie, A., Reed, P., Zhang, L., Ho, D.D., and Marx, P.A. (1996). Genetic characterization of new West African simian immunodeficiency virus SIVsm: geographic clustering of household-derived SIV strains with human immunodeficiency virus type 2 subtypes and genetically diverse viruses from a single feral sooty mangabey troop. J. Virol. *70*, 3617–3627.

Chiu, C.Y., Alizadeh, A.A., Rouskin, S., Merker, J.D., Yeh, E., Yagi, S., Schnurr, D., Patterson, B.K., Ganem, D., and DeRisi, J.L. (2007). Diagnosis of a critical respiratory illness caused by human metapneumovirus by use of a pan-virus microarray. J. Clin. Microbiol. *45*, 2340–2343.

Chiu, C.Y., Greninger, A.L., Kanada, K., Kwok, T., Fischer, K.F., Runckel, C., Louie, J.K., Glaser, C.A., Yagi, S., Schnurr, D.P. *et al.* (2008a). Identification of cardioviruses related to Theiler's murine encephalomyelitis virus in human infections. Proc. Natl. Acad. Sci. U.S.A. *105*, 14124–14129.

Chiu, C.Y., Rouskin, S., Koshy, A., Urisman, A., Fischer, K., Yagi, S., Schnurr, D., Eckburg, P.B., Tompkins, L.S., Blackburn, B.G. *et al.* (2006). Microarray detection of human parainfluenzavirus 4 infection associated with respiratory failure in an immunocompetent adult. Clin. Infect. Dis. *43*, e71–76.

Chiu, C.Y., Urisman, A., Greenhow, T.L., Rouskin, S., Yagi, S., Schnurr, D., Wright, C., Drew, W.L., Wang, D., Weintrub, P.S. *et al.* (2008b). Utility of DNA microarrays for detection of viruses in acute respiratory tract infections in children. J. Pediatr. *153*, 76–83.

Choo, Q.L., Kuo, G., Weiner, A.J., Overby, L.R., Bradley, D.W., and Houghton, M. (1989). Isolation of a cDNA clone derived from a blood-borne non-A, non-B viral hepatitis genome. Science *244*, 359–362.

Chua, K.B., Goh, K.J., Wong, K.T., Kamarulzaman, A., Tan, P.S., Ksiazek, T.G., Zaki, S.R., Paul, G., Lam, S.K., and Tan, C.T. (1999). Fatal encephalitis due to Nipah virus among pig-farmers in Malaysia. Lancet *354*, 1257–1259.

Colombo, M., Kuo, G., Choo, Q.L., Donato, M.F., Del Ninno, E., Tommasini, M.A., Dioguardi, N., and Houghton, M. (1989). Prevalence of antibodies to hepatitis C virus in Italian patients with hepatocellular carcinoma. Lancet *2*, 1006–1008.

Cossart, Y.E., Field, A.M., Cant, B., and Widdows, D. (1975). Parvovirus-like particles in human sera. Lancet *1*, 72–73.

Cox-Foster, D.L., Conlan, S., Holmes, E.C., Palacios, G., Evans, J.D., Moran, N.A., Quan, P.L., Briese, T., Hornig, M., Geiser, D.M. *et al.* (2007). A metagenomic survey of microbes in honey bee colony collapse disorder. Science *318*, 283–287.

Curry, A., Appleton, H., and Dowsett, B. (2006). Application of transmission electron microscopy to the clinical study of viral and bacterial infections: present and future. Micron *37*, 91–106.

Dane, D.S., Cameron, C.H., and Briggs, M. (1970). Virus-like particles in serum of patients with Australia-antigen-associated hepatitis. Lancet *1*, 695–698.

Dawood, F.S., Jain, S., Finelli, L., Shaw, M.W., Lindstrom, S., Garten, R.J., Gubareva, L.V., Xu, X., Bridges, C.B., and Uyeki, T.M. (2009). Emergence of a novel swine-origin influenza A (H1N1) virus in humans. N. Engl. J. Med. *360*, 2605–2615.

Delwart, E.L. (2007). Viral metagenomics. Rev. Med. Virol. *17*, 115–131.

Dinsdale, E.A., Edwards, R.A., Hall, D., Angly, F., Breitbart, M., Brulc, J.M., Furlan, M., Desnues, C., Haynes, M., Li, L., McDaniel, L. *et al.* (2008). Functional metagenomic profiling of nine biomes. Nature *452*, 629–632.

Doane, F.W. (1987). Immunoelectron microscopy in diagnostic virology. Ultrastruct. Pathol. *11*, 681–685.

Dong, B., Kim, S., Hong, S., Das Gupta, J., Malathi, K., Klein, E.A., Ganem, D., Derisi, J.L., Chow, S.A., and Silverman, R.H. (2007). An infectious retrovirus susceptible to an IFN antiviral pathway from human prostate tumors. Proc. Natl. Acad. Sci. U.S.A. *104*, 1655–1660.

Drosten, C., Gunther, S., Preiser, W., van der Werf, S., Brodt, H.R., Becker, S., Rabenau, H., Panning, M., Kolesnikova, L., Fouchier, R.A. *et al.* (2003). Identification of a novel coronavirus in patients with severe acute respiratory syndrome. N. Engl. J. Med. *348*, 1967–1976.

Engels, E.A., Frisch, M., Goedert, J.J., Biggar, R.J., and Miller, R.W. (2002). Merkel cell carcinoma and HIV infection. Lancet *359*, 497–498.

Esteban, J.I., Esteban, R., Viladomiu, L., Lopez-Talavera, J.C., Gonzalez, A., Hernandez, J.M., Roget, M., Vargas, V., Genesca, J., Buti, M. *et al.* (1989). Hepatitis C virus antibodies among risk groups in Spain. Lancet *2*, 294–297.

Falkow, S. (1988). Molecular Koch's postulates applied to microbial pathogenicity. Rev. Infect. Dis. 10 Suppl 2, S274–276.

Feinstone, S.M., Kapikian, A.Z., and Purceli, R.H. (1973). Hepatitis A: detection by immune electron microscopy of a viruslike antigen associated with acute illness. Science *182*, 1026–1028.

Feng, H., Shuda, M., Chang, Y., and Moore, P.S. (2008). Clonal integration of a polyomavirus in human Merkel cell carcinoma. Science *319*, 1096–1100.

Finkbeiner, S.R., Allred, A.F., Tarr, P.I., Klein, E.J., Kirkwood, C.D., and Wang, D. (2008). Metagenomic analysis of human diarrhea: viral detection and discovery. PLoS Pathog. *4*, e1000011.

Fraser, C., Donnelly, C.A., Cauchemez, S., Hanage, W.P., Van Kerkhove, M.D., Hollingsworth, T.D., Griffin, J., Baggaley, R.F., Jenkins, H.E., Lyons, E.J. *et al.* (2009). Pandemic potential of a strain of influenza A (H1N1): early findings. Science *324*, 1557–1561.

Fredricks, D.N. (2004). Diseases of unknown etiology: the role of infectious agents (London, Elsevier).

Fredricks, D.N., and Relman, D.A. (1998). Infectious agents and the etiology of chronic idiopathic diseases. Curr. Clin. Top. Infect. Dis. *18*, 180–200.

Gaynor, A.M., Nissen, M.D., Whiley, D.M., Mackay, I.M., Lambert, S.B., Wu, G., Brennan, D.C., Storch, G.A., Sloots, T.P., and Wang, D. (2007). Identification of a novel polyomavirus from patients with acute respiratory tract infections. PLoS Pathog. *3*, e64.

Gerna, G., Passarani, N., Sarasini, A., and Battaglia, M. (1985). Characterization of serotypes of human rotavirus strains by solid-phase immune electron microscopy. J. Infect. Dis. *152*, 1143–1151.

Gewurz, B.E., Marty, F.M., Baden, L.R., and Katz, J.T. (2008). Human herpesvirus 6 encephalitis. Curr. Infect. Dis. Rep. *10*, 292–299.

Greninger, A.L., Chen, E.C., Sittler, T., Schneierman, A., Roubinian, N., Yu, G., Kim, E., Pillai, D., Guyard, C., Mazzulli, T. *et al.* (2010). A metagenomic analysis of pandemic influenza A (2009 H1N1) infection in patients from North America. PLoS One (in press).

Greninger, A.L., Runckel, C., Chiu, C.Y., Haggerty, T., Parsonnet, J., Ganem, D., and DeRisi, J.L. (2009). The complete genome of klassevirus – a novel picornavirus in pediatric stool. Virol J *6*, 82.

Hall-Mendelin, S., Ritchie, S.A., Johansen, C.A., Zborowski, P., Cortis, G., Dandridge, S., Hall, R.A., and van den Hurk, A.F. (2010). Exploiting mosquito sugar feeding to detect mosquito-borne pathogens. Proc. Natl. Acad. Sci. U.S.A. *107*, 11255–11259.

Heeney, J.L. (2006). Zoonotic viral diseases and the frontier of early diagnosis, control and prevention. J. Intern. Med. *260*, 399–408.

Holtz, L.R., Finkbeiner, S.R., Kirkwood, C.D., and Wang, D. (2008). Identification of a novel picornavirus related to cosaviruses in a child with acute diarrhea. Virol. J. *5*, 159.

Holtz, L.R., Finkbeiner, S.R., Zhao, G., Kirkwood, C.D., Girones, R., Pipas, J.M., and Wang, D. (2009). Klassevirus *1*, a previously undescribed member of the family Picornaviridae, is globally widespread. Virol. J. *6*, 86.

Hopley, J.F., and Doane, F.W. (1985). Development of a sensitive protein A-gold immunoelectron microscopy method for detecting viral antigens in fluid specimens. J. Virol. Methods *12*, 135–147.

Horie, M., Honda, T., Suzuki, Y., Kobayashi, Y., Daito, T., Oshida, T., Ikuta, K., Jern, P., Gojobori, T., Coffin, J.M. *et al.* (2010). Endogenous non-retroviral RNA virus elements in mammalian genomes. Nature *463*, 84–87.

Hu, Y., and Hirshfield, I. (2005). Rapid approach to identify an unrecognized viral agent. J. Virol. Methods *127*, 80–86.

Inglis, T.J. (2007). Principia aetiologica: taking causality beyond Koch's postulates. J. Med. Microbiol. *56*, 1419–1422.

Jones, K.E., Patel, N.G., Levy, M.A., Storeygard, A., Balk, D., Gittleman, J.L., and Daszak, P. (2008). Global trends in emerging infectious diseases. Nature *451*, 990–993.

Jones, M.S., Kapoor, A., Lukashov, V.V., Simmonds, P., Hecht, F., and Delwart, E. (2005). New DNA viruses identified in patients with acute viral infection syndrome. J. Virol. *79*, 8230–8236.

Juven, T., Mertsola, J., Waris, M., Leinonen, M., Meurman, O., Roivainen, M., Eskola, J., Saikku, P., and Ruuskanen, O. (2000). Etiology of community-acquired pneumonia in 254 hospitalized children. Pediatr. Infect. Dis. J. *19*, 293–298.

Kapikian, A.Z., Wyatt, R.G., Dolin, R., Thornhill, T.S., Kalica, A.R., and Chanock, R.M. (1972). Visualization by immune electron microscopy of a 27-nm particle associated with acute infectious nonbacterial gastroenteritis. J. Virol. *10*, 1075–1081.

Kapoor, A., Simmonds, P., Lipkin, W., Zaidi, S., and Delwart, E. (2010). Use of nucleotide composition analysis to infer hosts for three novel picorna-like viruses. J. Virol. *84*, 10322–10328.

Kapoor, A., Victoria, J., Simmonds, P., Slikas, E., Chieochansin, T., Naeem, A., Shaukat, S., Sharif, S., Alam, M.M., Angez, M. *et al.* (2008). A highly prevalent and genetically diversified Picornaviridae genus in South Asian children. Proc. Natl. Acad. Sci. U.S.A. *105*, 20482–20487.

Keele, B.F., Van Heuverswyn, F., Li, Y., Bailes, E., Takehisa, J., Santiago, M.L., Bibollet-Ruche, F., Chen, Y., Wain, L.V., Liegeois, F. *et al.* (2006). Chimpanzee reservoirs of pandemic and nonpandemic HIV-1. Science *313*, 523–526.

Kistler, A., Avila, P.C., Rouskin, S., Wang, D., Ward, T., Yagi, S., Schnurr, D., Ganem, D., DeRisi, J.L., and Boushey, H.A. (2007). Pan-viral screening of respiratory tract infections in adults with and without asthma reveals unexpected human coronavirus and human rhinovirus diversity. J. Infect. Dis. *196*, 817–825.

Ksiazek, T.G., Erdman, D., Goldsmith, C.S., Zaki, S.R., Peret, T., Emery, S., Tong, S., Urbani, C., Comer, J.A., Lim, W. *et al.* (2003). A novel coronavirus associated with severe acute respiratory syndrome. N. Engl. J. Med. *348*, 1953–1966.

Kuiken, T., Leighton, F.A., Fouchier, R.A., LeDuc, J.W., Peiris, J.S., Schudel, A., Stohr, K., and Osterhaus, A.D. (2005). Public health. Pathogen surveillance in animals. Science *309*, 1680–1681.

Landry, M.L., and Hsiung, G.D. (2000). Primary isolation of viruses. In Clinical Virology Manual, Specter, S., Hodinka, R.L., and Young, S.A., eds (Washington, D.C., ASM Press), pp. 27–42.

Leendertz, S.A., Junglen, S., Hedemann, C., Goffe, A., Calvignac, S., Boesch, C., and Leendertz, F.H. (2010). High prevalence, coinfection rate, and genetic diversity of retroviruses in wild red colobus monkeys (*Piliocolobus badius badius*) in Tai National Park, Cote d'Ivoire. J. Virol. *84*, 7427–7436.

Leland, D.S., and Ginocchio, C.C. (2007). Role of cell culture for virus detection in the age of technology. Clin. Microbiol. Rev. *20*, 49–78.

Leski, T.A., Malanoski, A.P., Stenger, D.A., and Lin, B. (2010). Target amplification for broad spectrum microbial diagnostics and detection. Future Microbiol. *5*, 191–203.

Li, L., Kapoor, A., Slikas, B., Bamidele, O.S., Wang, C., Shaukat, S., Masroor, M.A., Wilson, M.L., Ndjango, J.B., Peeters, M. *et al.* (2010a). Multiple diverse circoviruses infect farm animals and are commonly found in human and chimpanzee feces. J. Virol. *84*, 1674–1682.

Li, L., Victoria, J., Kapoor, A., Blinkova, O., Wang, C., Babrzadeh, F., Mason, C.J., Pandey, P., Triki, H., Bahri, O. *et al.* (2009). A novel picornavirus associated with gastroenteritis. J. Virol. *83*, 12002–12006.

Li, L., Victoria, J.G., Wang, C., Jones, M., Fellers, G.M., Kunz, T.H., and Delwart, E. (2010b). Bat guano virome: predominance of dietary viruses from insects and plants plus novel mammalian viruses. J. Virol. *84*, 6955–6965.

Linnen, J., Wages, J., Jr., Zhang-Keck, Z.Y., Fry, K.E., Krawczynski, K.Z., Alter, H., Koonin, E., Gallagher, M., Alter, M., Hadziyannis, S. *et al.* (1996). Molecular cloning and disease association of hepatitis G virus: a transfusion-transmissible agent. Science *271*, 505–508.

Loens, K., Ursi, D., Ieven, M., van Aarle, P., Sillekens, P., Oudshoorn, P., and Goossens, H. (2002). Detection of Mycoplasma pneumoniae in spiked clinical samples by nucleic acid sequence-based amplification. J. Clin. Microbiol. *40*, 1339–1345.

Lu, Y., Wang, S.Y., and Lotz, J.M. (2004). The use of differential display to isolate viral genomic sequence for rapid development of PCR-based detection methods. A test case using Taura syndrome virus. J. Virol. Methods *121*, 107–114.

Mahony, J.B., Hatchette, T., Ojkic, D., Drews, S.J., Gubbay, J., Low, D.E., Petric, M., Tang, P., Chong, S., Luinstra, K., Petrich, A., and Smieja, M. (2009). Multiplex PCR tests sentinel the appearance of pandemic influenza viruses including H1N1 swine influenza. J. Clin. Virol. *45*, 200–202.

Martin, M.A., Bryan, T., Rasheed, S., and Khan, A.S. (1981). Identification and cloning of endogenous retroviral sequences present in human DNA. Proc. Natl. Acad. Sci. U.S.A. *78*, 4892–4896.

Marx, P.A., Li, Y., Lerche, N.W., Sutjipto, S., Gettie, A., Yee, J.A., Brotman, B.H., Prince, A.M., Hanson, A., Webster, R.G. *et al.* (1991). Isolation of a simian immunodeficiency virus related to human immunodeficiency virus type 2 from a west African pet sooty mangabey. J. Virol. *65*, 4480–4485.

Mihindukulasuriya, K.A., Wu, G., St Leger, J., Nordhausen, R.W., and Wang, D. (2008). Identification of a novel coronavirus from a beluga whale by using a panviral microarray. J. Virol. *82*, 5084–5088.

Molenkamp, R., van der Ham, A., Schinkel, J., and Beld, M. (2007). Simultaneous detection of five different DNA targets by real-time Taqman PCR using the Roche LightCycler480: Application in viral molecular diagnostics. J. Virol. Methods *141*, 205–211.

Morris, C.A., Flewett, T.H., Bryden, A.S., and Davies, H. (1975). Epidemic viral enteritis in a long-stay children's ward. Lancet *1*, 4–5.

Muerhoff, A.S., Leary, T.P., Desai, S.M., and Mushahwar, I.K. (1997). Amplification and subtraction methods and their application to the discovery of novel human viruses. J. Med. Virol. *53*, 96–103.

Myers, K.P., Olsen, C.W., Setterquist, S.F., Capuano, A.W., Donham, K.J., Thacker, E.L., Merchant, J.A., and Gray, G.C. (2006). Are swine workers in the United States at increased risk of infection with zoonotic influenza virus? Clin. Infect. Dis. *42*, 14–20.

Ng, T.F., Manire, C., Borrowman, K., Langer, T., Ehrhart, L., and Breitbart, M. (2009). Discovery of a novel single-stranded DNA virus from a sea turtle fibropapilloma by using viral metagenomics. J. Virol. *83*, 2500–2509.

Nichol, S.T., Spiropoulou, C.F., Morzunov, S., Rollin, P.E., Ksiazek, T.G., Feldmann, H., Sanchez, A., Childs, J., Zaki, S., and Peters, C.J. (1993). Genetic identification of a hantavirus associated with an outbreak of acute respiratory illness. Science *262*, 914–917.

Nishizawa, T., Okamoto, H., Konishi, K., Yoshizawa, H., Miyakawa, Y., and Mayumi, M. (1997). A novel DNA virus (TTV) associated with elevated transaminase levels in posttransfusion hepatitis of unknown etiology. Biochem. Biophys. Res. Commun. *241*, 92–97.

Okamoto, H. (2009). History of discoveries and pathogenicity of TT viruses. Curr. Top. Microbiol. Immunol. *331*, 1–20.

Palacios, G., Druce, J., Du, L., Tran, T., Birch, C., Briese, T., Conlan, S., Quan, P.L., Hui, J., Marshall, J. *et al.* (2008). A new arenavirus in a cluster of fatal transplant-associated diseases. N. Engl. J. Med. *358*, 991–998.

Palacios, G., Quan, P.L., Jabado, O.J., Conlan, S., Hirschberg, D.L., Liu, Y., Zhai, J., Renwick, N., Hui, J., Hegyi, H. *et al.* (2007). Panmicrobial oligonucleotide array for diagnosis of infectious diseases. Emerg. Infect. Dis. *13*, 73–81.

Peiris, J.S., Lai, S.T., Poon, L.L., Guan, Y., Yam, L.Y., Lim, W., Nicholls, J., Yee, W.K., Yan, W.W., Cheung, M.T. *et al.* (2003). Coronavirus as a possible cause of severe acute respiratory syndrome. Lancet *361*, 1319–1325.

Petrich, A., Mahony, J., Chong, S., Broukhanski, G., Gharabaghi, F., Johnson, G., Louie, L., Luinstra, K., Willey, B., Akhaven, P. *et al.* (2006). Multicenter comparison of nucleic acid extraction methods for detection of severe acute respiratory syndrome coronavirus RNA in stool specimens. J. Clin. Microbiol. *44*, 2681–2688.

Quan, P.L., Palacios, G., Jabado, O.J., Conlan, S., Hirschberg, D.L., Pozo, F., Jack, P.J., Cisterna, D., Renwick, N., Hui, J. *et al.* (2007). Detection of respiratory viruses and subtype identification of influenza A viruses by GreeneChipResp oligonucleotide microarray. J. Clin. Microbiol. *45*, 2359–2364.

Raymond, F. (2009). Comparison of automated microarray detection with real-time PCR assays for detection of respiratory specimens obtained from children. J. Clin. Microbiol. *47*, 743–750.

Rector, A., Bossart, G.D., Ghim, S.J., Sundberg, J.P., Jenson, A.B., and Van Ranst, M. (2004). Characterization of a novel close-to-root papillomavirus from a Florida manatee by using multiply primed rolling-circle amplification: *Trichechus manatus latirostris* papillomavirus type 1. J. Virol. *78*, 12698–12702.

Rehrauer, H., Schonmann, S., Eberl, L., and Schlapbach, R. (2008). PhyloDetect: a likelihood-based strategy for detecting microorganisms with diagnostic microarrays. Bioinformatics *24*, i83–89.

Reyes, A., Haynes, M., Hanson, N., Angly, F.E., Heath, A.C., Rohwer, F., and Gordon, J.I. (2010). Viruses in the faecal microbiota of monozygotic twins and their mothers. Nature *466*, 334–338.

Reyes, G.R., and Kim, J.P. (1991). Sequence-independent, single-primer amplification (SISPA) of complex DNA populations. Mol. Cell. Probes *5*, 473–481.

Rivers, T.M. (1937). Viruses and Koch's Postulates. J. Bacteriol. 33, 1–12.

Roh, K.H., Kim, J., Nam, M.H., Yoon, S., Lee, C.K., Lee, K., Yoo, Y., Kim, M.J., and Cho, Y. (2008). Comparison of the Seeplex reverse transcription PCR assay with the R-mix viral culture and immunofluorescence techniques for detection of eight respiratory viruses. Ann. Clin. Lab. Sci. *38*, 41–46.

Rohayem, J., Berger, S., Juretzek, T., Herchenroder, O., Mogel, M., Poppe, M., Henker, J., and Rethwilm, A. (2004). A simple and rapid single-step multiplex RT-PCR to detect Norovirus, Astrovirus and Adenovirus in clinical stool samples. J. Virol. Methods *118*, 49–59.

Roingeard, P. (2008). Viral detection by electron microscopy: past, present and future. Biol. Cell. *100*, 491–501.

Sagerstrom, C.G., Sun, B.I., and Sive, H.L. (1997). Subtractive cloning: past, present, and future. Annu. Rev. Biochem. *66*, 751–783.

Salvetti, M., Giovannoni, G., and Aloisi, F. (2009). Epstein–Barr virus and multiple sclerosis. Curr. Opin. Neurol. *22*, 201–206.

Schadt, E.E., Linderman, M.D., Sorenson, J., Lee, L., and Nolan, G.P. (2010). Computational solutions to large-scale data management and analysis. Nat. Rev. Genet. *11*, 647–657.

Schena, M., Shalon, D., Davis, R.W., and Brown, P.O. (1995). Quantitative monitoring of gene expression patterns with a complementary DNA microarray. Science *270*, 467–470.

Schoenfeld, T., Patterson, M., Richardson, P.M., Wommack, K.E., Young, M., and Mead, D. (2008). Assembly of viral metagenomes from yellowstone hot springs. Appl. Environ. Microbiol. *74*, 4164–4174.

Shendure, J., and Ji, H. (2008). Next-generation DNA sequencing. Nat. Biotechnol. *26*, 1135–1145.

Simons, J.N., Pilot-Matias, T.J., Leary, T.P., Dawson, G.J., Desai, S.M., Schlauder, G.G., Muerhoff, A.S., Erker, J.C., Buijk, S.L., Chalmers, M.L. *et al.* (1995). Identification of two flavivirus-like genomes in the GB hepatitis agent. Proc. Natl. Acad. Sci. U.S.A. *92*, 3401–3405.

Spackman, E., and Suarez, D.L. (2008). Avian influenza virus RNA extraction from tissue and swab material. Methods Mol. Biol. *436*, 13–18.

Stang, A., Korn, K., Wildner, O., and Uberla, K. (2005). Characterization of virus isolates by particle-associated nucleic acid PCR. J. Clin. Microbiol. *43*, 716–720.

Stehelin, D., Varmus, H.E., Bishop, J.M., and Vogt, P.K. (1976). DNA related to the transforming gene(s) of avian sarcoma viruses is present in normal avian DNA. Nature *260*, 170–173.

Stein, J., and Liang, P. (2002). Differential display technology: a general guide. Cell. Mol. Life Sci. *59*, 1235–1240.

Stratton, M.R., Campbell, P.J., and Futreal, P.A. (2009). The cancer genome. Nature *458*, 719–724.

Suttle, C.A. (2005). Viruses in the sea. Nature *437*, 356–361.

Svraka, S., Rosario, K., Duizer, E., van der Avoort, H., Breitbart, M., and Koopmans, M.P.G. (2010). Metagenomic sequencing for virus identification in a public health setting. J. Gen. Virol. E pub: doi:10.1099/vir.0.024612-0.

Thurber, R.V., Haynes, M., Breitbart, M., Wegley, L., and Rohwer, F. (2009). Laboratory procedures to generate viral metagenomes. Nat. Protoc. *4*, 470–483.

Urisman, A., Fischer, K.F., Chiu, C.Y., Kistler, A.L., Beck, S., Wang, D., and DeRisi, J.L. (2005). E-Predict: a computational strategy for species identification based on observed DNA microarray hybridization patterns. Genome Biol. *6*, R78.

Urisman, A., Molinaro, R.J., Fischer, N., Plummer, S.J., Casey, G., Klein, E.A., Malathi, K., Magi-Galluzzi, C., Tubbs, R.R., Ganem, D. *et al.* (2006). Identification of a novel Gammaretrovirus in prostate tumors of patients homozygous for R462Q RNASEL variant. PLoS Pathog. *2*, e25.

van den Hoogen, B.G., de Jong, J.C., Groen, J., Kuiken, T., de Groot, R., Fouchier, R.A., and Osterhaus, A.D. (2001). A newly discovered human pneumovirus isolated from young children with respiratory tract disease. Nat. Med. *7*, 719–724.

van der Hoek, L., Pyrc, K., Jebbink, M.F., Vermeulen-Oost, W., Berkhout, R.J., Wolthers, K.C., Wertheim-van Dillen, P.M., Kaandorp, J., Spaargaren, J., and Berkhout, B. (2004). Identification of a new human coronavirus. Nat. Med. *10*, 368–373.

van der Meijden, E., Janssens, R.W., Lauber, C., Bouwes Bavinck, J.N., Gorbalenya, A.E., and Feltkamp, M.C. (2010). Discovery of a new human polyomavirus associated with trichodysplasia spinulosa in an immunocompromized patient. PLoS Pathog. *6*, e1001024.

Virgin, H.W., Wherry, E.J., and Ahmed, R. (2009). Redefining chronic viral infection. Cell *138*, 30–50.

Wang, D., Coscoy, L., Zylberberg, M., Avila, P.C., Boushey, H.A., Ganem, D., and DeRisi, J.L. (2002). Microarray-based detection and genotyping of viral pathogens. Proc. Natl. Acad. Sci. U.S.A. *99*, 15687–15692.

Wang, D., Urisman, A., Liu, Y.T., Springer, M., Ksiazek, T.G., Erdman, D.D., Mardis, E.R., Hickenbotham, M., Magrini, V., Eldred, J., Latreille, J.P., Wilson, R.K., Ganem, D., and DeRisi, J.L. (2003). Viral discovery and sequence recovery using DNA microarrays. PLoS Biol. *1*, E2.

Watson, M., Dukes, J., Abu-Median, A.B., King, D.P., and Britton, P. (2007). DetectiV: visualization, normalization and significance testing for pathogen-detection microarray data. Genome Biol. *8*, R190.

Welsh, J.B., Zarrinkar, P.P., Sapinoso, L.M., Kern, S.G., Behling, C.A., Monk, B.J., Lockhart, D.J., Burger, R.A., and Hampton, G.M. (2001). Analysis of gene expression profiles in normal and neoplastic ovarian tissue samples identifies candidate molecular markers of epithelial ovarian cancer. Proc. Natl. Acad. Sci. U.S.A. *98*, 1176–1181.

Wiedbrauk, D.L., Werner, J.C., and Drevon, A.M. (1995). Inhibition of PCR by aqueous and vitreous fluids. J. Clin. Microbiol. *33*, 2643–2646.

Wolfe, N.D., Dunavan, C.P., and Diamond, J. (2007). Origins of major human infectious diseases. Nature *447*, 279–283.

Wolfe, N.D., Heneine, W., Carr, J.K., Garcia, A.D., Shanmugam, V., Tamoufe, U., Torimiro, J.N., Prosser, A.T., Lebreton, M., Mpoudi-Ngole, E. *et al.* (2005). Emergence of unique primate T-lymphotropic viruses among central African bushmeat hunters. Proc. Natl. Acad. Sci. U.S.A. *102*, 7994–7999.

10 Metagenomics Applied to Arbuscular Mycorrhizal Fungal Communities

Valeria Bianciotto, Erica Lumini, Alberto Orgiazzi, Roberto Borriello and Paola Bonfante

Abstract

Metagenomics studies have recently offered new approaches that shed light on microbial communities in a variety of environments. In this context, DNA pyrosequencing is being used more and more to investigate prokaryotic assemblages in soil environment. Fungi, which are crucial components of soil microbial communities, functioning as decomposers, pathogens and mycorrhizal symbionts, have instead been largely neglected. However, the last year has been characterized by an explosion of metagenomic studies applied to fungal communities, based on the pyrosequencing technology. The aim of this chapter is to focus on arbuscular mycorrhizal fungi (AMF), the most widespread symbionts in many ecosystems, and to demonstrate how metagenomics may help us to understand the composition and dynamics of AMF communities. At the moment, only two studies have investigated AMF biodiversity using the pyrosequencing approach and SSU rDNA as the target gene. Although both studies targeted the same group of fungi, they focused on different habitats. Compared to similar studies, carried out using a cloning–sequencing approach or DNA barcoding, the main outcome that emerged from pyrosequencing analyses applied to AMF and to other fungal communities is the unexpected fungal biodiversity observed in the analysed environments. Interestingly, the pyrosequencing approach applied to isolated spores of AMF has demonstrated that they are a niche for highly polymorphic endobacterial communities. The data confirm the powerfulness of the pyrosequencing approach, which represents a promising new tool to better understand the natural distribution of an essential group of soil microorganisms, such as fungi. The large number of reads that have been obtained increases the likelihood of capturing sequences from rare organisms, which would instead remain undetected with the cloning–sequencing approach.

Introduction

One of the major goals of microbial ecology is to link specific microbial populations to environmental processes. The first step to achieve this aim is the characterization of the microbial diversity present in a certain ecosystem. Studying biodiversity means studying the environmental genome, the so-called 'metagenome'. The term 'metagenomics' was first used over a decade ago and it refers to the possibility of analysing microbial community DNA extracted directly from environmental samples (Chen *et al.*, 2005). A microbial community is a complex assemblage of different organisms living together and belonging to both prokaryotic and eukaryotic genera (Hugenholtz and Tyson, 2008). The most important eukaryotic component is represented by fungi.

Because of their filamentous organization, fungi exploit very diverse substrates on the basis of their nutritional strategy. Saprobes thrive in soil, water and on decaying animal and plant tissues, while a smaller group of fungi, such as parasitic and mutualistic symbionts, feed on living organisms. Such a classification cannot easily be applied to mycorrhizal fungi, a heterogeneous group of species spread over diverse fungal taxa (Girlanda *et al.*, 2007; Bonfante and Genre, 2010). Although they can spend part of their life cycle as free-living organisms, mycorrhizal fungi always associate with the roots of higher plants: with over 90% of plant species, including forest trees, wild grasses and many crops. Their name derives from the Greek words for 'fungus' and 'root'. Mycorrhizal fungi develop an extensive hyphal network in the soil, the aptly named wood-wide web, which can connect whole plant communities and offer an efficient horizontal transfer of nutrients. Together with their plant hosts, they develop specialized areas, called symbiotic interfaces. Thanks to this interaction, both partners benefit from the relationship: mycorrhizal fungi improve the nutrient status of their host plants, thus influencing mineral nutrition, water absorption, growth and disease resistance, whereas in exchange, the host plant is necessary for fungal growth and reproduction. Mycorrhizal fungi colonize environments such as alpine and boreal zones, tropical forests, grasslands and croplands. They play a major role in nutrient cycling, through the specific activity of their mycelium, in absorbing soil nutrients and supplying them to the plant, although their role in carbon flux is less well defined (Smith and Read, 2008; Balestrini *et al.*, 2010).

Mycorrhizal fungi can be divided into two major groups: aseptate endophytes such as Glomeromycota, or septate Asco- and Basidiomycota. Mycorrhiza classifications usually reflect anatomical aspects and identify two broad categories, referred to as ectomycorrhizas (EMs) and endomycorrhizas, depending on whether the fungus colonizes the root intercellular spaces or develops inside cells. Endomycorrhizas are further divided into orchid, ericoid and arbuscular mycorrhizas (AMs) (Balestrini *et al.*, 2010).

The variety of mycorrhizal associations established between plants and fungi has been divided into seven categories (arbuscular mycorrhiza, ectomycorrhiza, ectendomycorrhiza, arbutoid mycorrhiza, monotropoid mycorrhiza, ericoid mycorrhiza and orchid mycorrhiza), primarily on the basis of the structural characteristics of the symbiotic interfaces and the taxonomic identity of the symbionts (Smith and Read, 2008).

The aim of this chapter is to focus on arbuscular mycorrhizal fungi (AMF) in order to provide a short summary of their main biological features and an update on the techniques applied to investigate the biodiversity of this still enigmatic group of soil fungi.

An overview of the biology of arbuscular mycorrhizal fungi

The symbiotic fungi that form arbuscular mycorrhizae are among the most important soil organisms (Bonfante and Genre, 2008). They have been separated from all other major fungal groups and grouped in a monophyletic clade, named Glomeromycota (Schüßler *et al.*, 2001), within which four orders (Glomerales, Diversisporales, Archaeosporales and Paraglomerales) and approximately 200 AMF species have been described and grouped into 14 genera: *Acaulospora, Ambispora, Archaeospora, Diversispora, Entrophospora, Geosiphon, Gigaspora, Glomus, Intraspora, Otospora, Pacispora, Paraglomus, Racocetra* and *Scutellospora* (www.amf-phylogeny.com).

AMF are thought to be the oldest group of organisms living in symbiosis with land plants. The first fossil evidence of glomeromycota-like fungi dates back to the Ordovician, 460 million years ago (Redecker *et al.*, 2000). Estimates, based on molecular clock assumptions, vary from 353 to 462 Mya (Simon *et al.*, 1993a) to 1200 Mya (Heckman *et al.*, 2001). These ancient fungi have coevolved with plants for the last 400 million years and the uniformity of plant cell responses to AM colonization in haploid gametophytes and diploid sporophytes, in non-root organs, and throughout many seed plant clades highlights the ancient origin of the interaction and suggests the existence of common molecular and cellular processes (Bonfante and Genre, 2008). The associations of these fungi with some bryophyte precursors suggest that they largely anticipate the emergence of vascular plants and many studies support the hypothesis that AM fungi carried out a critical role in the colonization of dry land by plants (Simon *et al.*, 1993a; Taylor *et al.*, 1995; Phipps and Taylor, 1996). Owing to the low nutrient mobility in the soil, in particular that of phosphate, the area around plant absorption organs soon becomes poor in nutrients, and creates a so-called 'depletion zone'. The first plant that colonized dry land had in fact to face this critical problem. The AMF ancestor before and the AM fungi today, help plants overcome this zone by making new nutrients, that are out of range of the plant, available. Therefore, symbiosis was a key factor in the colonization of land by plants (Nicolson, 1967; Pyrozynski and Malloch, 1975).

Today, AM fungi are recognized to be symbiotically associated to 70–90% of land plant species (Smith and Read, 2008). A characteristic of this association is a structure called arbuscule (Latin: *arbusculum* = shrub or small tree) which is formed by the symbiotic fungus inside the colonized plant cell. This apparatus connects plant roots to the arbuscular mycorrhizal fungus in an intimate way and becomes a bridge between the plant and the soil (Parniske, 2008). AM fungi (AMF) colonize the roots of most land plants, where they facilitate mineral nutrient uptake from the soil in exchange for plant-assimilated carbon (Smith and Read, 2008). Although their spores can germinate in the absence of host plants, AMF are obligate biotrophs and they depend on a living photoautotrophic partner to complete their life cycle. When not in association with a plant, AM fungi exist in the soil as resting spores that are asexual, multinucleate and produced directly by the mycelium, either inside or outside the root. Following germination, the hyphal germ tube, protruding from the spores, grows through the soil in search of a host root, and produces a pre-symbiotic mycelium. The mycelium, usually aseptate and coenocytic, can grow either outside (extraradical) or inside the roots (intraradical). Fungal hyphae within the root are connected to the extraradical mycelium and form a single continuum. Extraradical hyphae acquire phosphate, initiate the colonization of other roots, and, in most species, are also the site of sporulation (Parniske, 2008). Owing to the complexity of their life cycle it is difficult to study AM fungi in either soil or roots (Fig. 10.1).

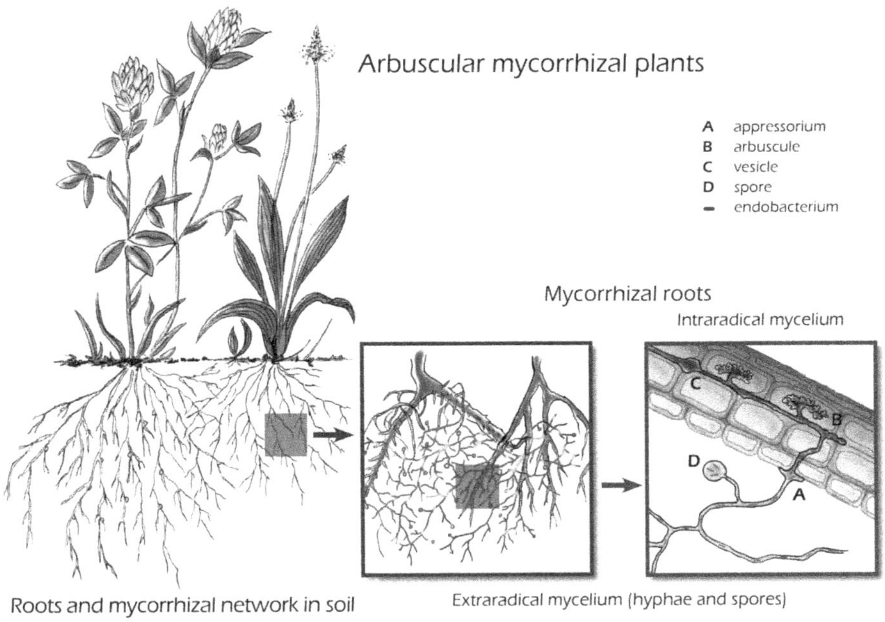

Figure 10.1 Schematic view of arbuscular mycorrhizal (AM) endosymbiosis in two plant species. The figure shows the development of the AM fungal partner in the soil and host plant compartments. The soil extraradical phase is represented by highly branched hyphal networks (central box) that explore soil and connect different host plants to form a single continuum. AM fungi exist in the soil as resting spores (D) and as extraradical hyphae. Hyphae develop from a globose spore (D) and produce an appressorium (A) (also called hyphopodium) on the root epidermis. Intraradical colonization proceeds both intra- and intercellularly and culminates with the formation of arbuscules (B), little fungal trees, inside inner cortical cells. In some AMF species the intraradical mycelium could form vesicles (C) and endobacteria could be present inside their cytoplasm.

Strategies to study the biodiversity of arbuscular mycorrhizal fungal assemblages

Although arbuscular mycorrhizas are geographically ubiquitous (Reddy *et al.*, 2005), the diversity and composition of AM fungal communities vary among habitats (Öpik *et al.*, 2010). In fact, arbuscular mycorrhizal fungal taxa have a specific multidimensional niche that is determined by the plant species that are present at a site and by edaphic factors such as pH, moisture content, phosphorus (P) and nitrogen (N) availability (Helgason and Fitter, 2009). As a result, large between- and within-site variations in the composition of AMF communities have been described (Burrows and Pfleger, 2002; Klironomos and Hart, 2002; Dumbrell *et al.*, 2010).

Field studies on arbuscular mycorrhizal communities have employed a range of approaches, either using indirect methods such as isolation of spores from the soil or from trap cultures, or direct methods via isolation of DNA from mycorrhized roots and from soils.

Indirect methods

Spore isolation from soil as a method to investigate AMF diversity can be criticized as it reflects the sporulation stage of the AMF life cycle (a dormant phase) rather than the activity that occurs within a symbiosis. However, the method is rapid, and it may reveal taxa that are not easily detected by molecular methods. On the other hand, such a method requires taxonomic expertise. Intensive soil sampling, including the use of successive trap culturing, takes much longer, but could reveal significantly greater diversity than a 'one-off' spore extraction (Oehl *et al.*, 2009). Most studies addressing AMF diversity rely on morphological identification of AMF spores extracted directly from the field (Douds and Millner, 1999; Landis *et al.*, 2004; Gai *et al.*, 2006) or from so-called 'trap cultures' in which soil samples from the field sites are used as inocula to propagate the indigenous AMF populations on the roots of appropriate host plants in pot cultures (Jansa *et al.*, 2002; Oehl *et al.*, 2003, 2005; Mathimaran *et al.*, 2005). However, these techniques have their problems as the spore density assessed in field samples does not necessarily reflect the AMF population actually colonizing the roots (Sýkorová *et al.*, 2007); moreover, distinction between freshly formed spores and spores formed during an earlier season is almost impossible (Lee and Koske, 1994). Some AMF species might sporulate only infrequently, or not at all, in the field as well as under trap culture conditions (Clapp *et al.*, 1995; Oehl *et al.*, 2004).

Gai *et al.* (2006) reviewed 20 years of work in China and reported a total of 104 species from nine genera from spore isolations or trap cultures from many soils. These results reflect the world-wide situation in that the frequency of occurrence and abundance of *Glomus* spores were much higher than other genera, with *Glomus mosseae* being reported in over 90% of the sites investigated. Studies based on indirect methods have shown that AMF spore species distribution and abundance is affected by different farming systems. Agricultural soils subjected to conventional farming exhibit a dominant presence of Glomus species, with respect to soil subjected to low input/organic agriculture where it is also possible to detect *Acaulospora*, *Scutellospora* and *Paraglomus* spores (Jansa *et al.*, 2002; Oehl *et al.*, 2003, 2004, 2009; Hijri *et al.*, 2006). Oehl *et al.* (2009), in a 3-year survey on three different agroecosystems (grassland, crop rotation, maize monocropping), found that AMF consortia from all the field sites exhibited strong temporal sporulation dynamics.

Direct methods

The advent of molecular biology has revolutionized the analysis of fungal populations colonizing roots and forming mycorrhizas (Clapp *et al.*, 2002). DNA sequence-based methods are very useful for detecting organisms at different community levels, but they depend on reliable baseline databases and tools for ecological work. Suitable molecular methods are crucial to overcome the limitations of morphological identification (Walker and Schüßler, 2004; Walker *et al.*, 2007; Stockinger *et al.*, 2009; Gamper *et al.*, 2010). The molecular characterization of AMF is in most cases achieved by sequencing DNA amplified from the roots of host plants, spores or soil samples. The main aim of these studies is the characterization of the AMF assemblages present in environments of interest. These kinds of studies, which focus on a specific component of the whole microbial community, has been recently labelled as DNA barcoding (metabarcoding) instead of metagenomic studies (Martin *et al.*, 2011). This is owing to the use of a barcode DNA fragment (e.g. small subunit ribosomal RNA – SSU rRNA or Internal Transcribed Spacer – ITS) in order to identify a specific group of organisms. Several primers that target the rDNA regions as molecular markers have been claimed to be AMF specific, but most of them only amplify a restricted number of glomeromycotan taxa or DNA of non-target organisms. The most comprehensive taxa sampling of Glomeromycota covers the small subunit (SSU) rDNA region (Schüßler *et al.*, 2001), for which some AMF specific primer pairs have been published: NS31 and AM1 (Helgason *et al.*, 1998), AM2 and AM3 (Santos-Gonzalez *et al.*, 2007) and AMV4.5NF and AMV4.5NR (Saito *et al.*, 2004). None of these couples easily amplify Archaeosporaceae and Paraglomeraceae members and they suffer from nonspecific amplification and host-dependent performance (Douhan *et al.*, 2005; Alguacil *et al.*, 2008). A new primer set, AML1 and AML2 (Lee *et al.*, 2008), unlike the previous primers, is perhaps suitable for amplifying sequences from all AMF taxa. The small subunit ribosomal RNA gene is the most used target region, since most data on Glomeromycota natural diversity are obtained using the region between primers NS31 and AM1, which provides the largest comparative DNA sequence data set of those currently available for other genomic regions (Öpik *et al.*, 2009, 2010).

The use of other highly variable regions, such as the ITS or the LSU rDNA regions (or, even better, a combination of both; Gamper *et al.*, 2009) is advisable for the molecular characterization of AMF communities. Glomeromycota specific primers have recently been developed for the amplification of the ITS region flanked by parts of the SSU and LSU regions (Krüger *et al.*, 2009). Inclusion of the internal transcribed spacer (ITS) and the large subunit (LSU) rDNA region leads to both robust phylogenetic analyses and species level resolution (Gamper *et al.*, 2009; Stockinger *et al.*, 2009). However, so far, the number of LSU and ITS sequences in public databases is lower than that of the SSU region.

Field molecular studies of AM communities have been carried out considering all these target regions in many different environments: temperate (Merryweather and Fitter, 1998; Helgason *et al.*, 2002), boreal (Öpik *et al.*, 2003, 2008; DeBellis and Widden, 2006), tropical African (Wubet *et al.*, 2004, 2006), and tropical American woodlands (Husband *et al.*, 2002a,b; Aldrich-Wolfe, 2007); grasslands (Vandenkoornhuyse *et al.*, 2003; Stukenbrock and Rosendahl, 2005; Santos-González *et al.*, 2007; Sýkorová *et al.*, 2007); and in both wet (Wirsel, 2004; Nielsen *et al.*, 2004) and dry (Sun *et al.*, 2008; Alguacil *et al.*, 2009; Al-Yahya'ei *et al.*, 2011) habitats.

Such sequence-based studies of DNA obtained directly from environmental samples involved the use of PCR to amplify the target region of the DNA extracted from the mycorrhized roots. The amplicons were then cloned into bacterial cells (*E. coli*) to isolate and sequence individual fragments. Clone-library analyses, applied to a wide range of environmental samples, have resulted in the identification of a large number of AM fungal taxa, which had never been found as spores in soils using direct strategies, being associated to several plant species. These sequence-based studies could be considered as the first metabarcoding works on AMF. At present, clone-library analysis is the main molecular strategy used to assess AMF fungal diversity in both roots and soil. Since the first works, dating back to the 1990s (Simon *et al.*, 1992, 1993b), this technique can be coupled with RFLP typing in order to reduce the number of sequenced clones. However, thanks to the continuous reduction in sequencing costs and the shorter analysis times, researchers are more and more inclined to sequence all the available clones, in order to have a better understanding of the considered fungal assemblages.

An overview of AM fungal community surveys from different ecosystems is presented in Table 10.1.

The increasing number of AMF biodiversity studies based on cloning and sequencing and, consequently, the higher and higher number of sequences retrieved from many environments has led to the surfacing of a relevant problem: the

Table 10.1 Overview of arbuscular mycorrhizal (AM) fungal community surveys from different ecosystems in relation to sampling, sample screening and marker region

Reference	Year	Strategy	Ecosystem	Environment	Marker region	No. of AMF taxa	No. of AMF sequenced
Lumini *et al.*	2010	Pyroseq	Mediterranean soils (Italy)	Soil	SSU	19/80[a]	2815
Balestrini *et al.*	2010	Clon and seq	Vineyard (Italy)	Soil and plant root	SSU	37	681
Dumbrell *et al.*	2010	Clon and seq	Woodland (UK)	Plant root	SSU	33/37[b]	617
Alguacil *et al.*	2009	Clon and seq	Mediterranean semi-arid soils (Spain)	Plant root	SSU	21	1443
Öpik *et al.*	2009	Pyroseq	Boreal forest (Estonia)	Plant root	SSU	47	111580
Cesaro *et al.*	2008	Clon and seq	Arable field (Italy)	Soil and plant root	LSU	8	183
Öpik *et al.*	2008	Clon and seq	Boreal forest (Estonia)	Plant root	SSU	26/27[c]	911
Alguacil *et al.*	2008	Clon and seq	Arable field (Mexico)	Plant root	SSU	16	213
Schechter and Bruns	2008	Clon and seq	Serpentine soils (USA)	Plant root	SSU	19	1249
Toljander *et al.*	2008	Clon and seq	Arable field (Sweden)	Plant root	SSU	8	115
Hempel *et al.*	2007	Clon and seq	Meadow (Germany)	Soil, plant root and spores	ITS	>18	180
Ligrone *et al.*	2007	Clon and seq	Liverworts (world-wide)	Rhizoids	SSU	10	150
Pivato *et al.*	2007	Clon and seq	Arable field (France)	Soil and plant root	LSU	12	246
Santos-Gonzalez *et al.*	2007	Clon and seq	Grassland (Sweden)	Plant root	SSU	19	185
Wu *et al.*	2007	Clon and seq	Volcanic desert (Japan)	Plant root	ITS	11	205
Vallino *et al.*	2006	Clon and seq	Pollute soils (Italy)	Plant root	SSU	12	115
Yamato and Iwase	2005	Clon and seq	Warm-temperate deciduous forest (Japan)	Plant root	SSU	5	394
Wirsel	2004	Clon and seq	Wetland (Germany)	Plant root	SSU	35	546

Rosendahl and Stukenbrock	2004	Clon and seq	Grassland (Denmark)	Plant root	ITS	11	158
Gollotte *et al.*	2004	Clon and seq	Pasture (UK)	Plant root	ITS	10	30
Saito *et al.*	2004	Clon and seq	Grassland (Japan)	Plant root	SSU	8	200
Wubet *et al.*	2003	Clon and seq	Afromontane forests (Ethiopia)	Plant root	ITS	20	92
Heinemeyer *et al.*	2003	Clon and seq	Grassland (UK)	Plant root	SSU	9	606
Öpik *et al.*	2003	Clon and seq	Boreal forest (Estonia)	Plant root	SSU	6	16
Vandenkoornhuyse *et al.*	2002	Clon and seq	Seminatural grassland (UK)	Plant root	SSU	24	88
Helgason *et al.*	2002	Clon and seq	Woodland (UK)	Plant root	SSU	13	232
Husband *et al.*	2002A	Clon and seq	Tropical forest (Republic of Panama)	Plant root	SSU	18/23[d]	1536
Husband *et al.*	2002B	Clon and seq	Tropical forest (Republic of Panama)	Plant root	SSU	18	558
Daniell *et al.*	2001	Clon and seq	Arable field (UK)	Plant root	SSU	8	303
Helgason *et al.*	1999	Clon and seq	Seminatural woodland (UK)	Plant root	SSU	6/8[e]	141
Helgason *et al.*	1998	Clon and seq	Woodland (UK)	Plant root	SSU	6/10[b]	253

[a]Taxa obtained with different primer sets.

[b]Taxa obtained in different study sites.

[c]Taxa obtained from forest ecosystems of different age and management intensity.

[d]Taxa obtained from roots of different plant species.

[e]Taxa obtained in different sampling time.

majority of AMF sequences deposited in databases are labelled as 'uncultured' Glomeromycota (12,274 sequences up to September 2010), since they are obtained from molecular analysis of plant roots, and have no corresponding morphospecies description. However, more matches with known AM fungal species could be expected when more effort is directed towards sequencing DNA from morphologically characterized AMF isolates. In the last few years, several groups have dedicated a great deal of effort to describing new taxa using spore morphology coupled with molecular analyses (Blaszkowski *et al.*, 2008; Stockinger *et al.*, 2009).

Until 2008, the highest taxon richness in a single site had been reported in an herb-rich boreal coniferous forest in Panama – 30 taxa in the roots of three host species (Husband *et al.*, 2002b; 29 according to the taxon synonymy used by Öpik *et al.*, 2006). Taxon-rich fungal communities (over 20 taxa) are also known in temperate grassland and forest locations (Vandenkoornhuyse *et al.*, 2002; Saito *et al.*, 2004; Helgason *et al.*, 2007). Tropical forests may exhibit a significantly higher average fungal richness than other ecosystems (expressed as the number of fungal taxa per plant species: 18; Öpik *et al.*, 2006). On the other hand, human-impacted habitats, such as arable fields (Helgason *et al.*, 1998; Daniell *et al.*, 2001) or sites polluted with heavy metals (Whitfield *et al.*, 2004), may exhibit low AM fungal taxon diversity. However, evidence of higher richness in such habitats (Hijri *et al.*, 2006; Vallino *et al.*, 2006) suggests that the relationship with management is more complex than expected. AM fungal communities in deciduous forests in temperate zones in the UK have been described by Helgason *et al.* (1998, 1999, 2002, 2007). Most surveys on AMF assemblies have been based on the identification of spore or intra-radical mycelium, but the overlap between the species detected in roots and the spore population is remarkably low (Clapp *et al.*, 1995; Merryweather and Fitter, 1998; Renker *et al.*, 2005). On the other hand, only a few articles have taken into account the AM extraradical phase to study fungal biodiversity. Two studies, Ma *et al.* (2005) and Liang *et al.* (2008), extracted DNA directly from soil to characterize AMF populations via the PCR-DGGE technique. In order to assess the reliability of the method, DGGE profiles obtained from bulk soil were then compared with those obtained from spores maintained in pot cultures. A step forward has been made by Hempel *et al.* (2007), who analysed and sequenced DNA extracted from different compartments: spores collected from soil, mycorrhizal roots and, for the first time, the AMF extraradical phase (e.g. extraradical mycelium and spores). This study, based on cloning and sequencing SSU rDNA, has revealed remarkable differences in the composition of AMF taxa associated to the different compartments. Other researchers (Pivato *et al.*, 2007; Cesaro *et al.*, 2008) have analysed AMF biodiversity associated to both bulk soil and plant host roots using the LSU region. In spite of the absence of clear host specificity in mycorrhizal symbiosis, the results of the latter studies have shown the occurrence of preferential associations between some AM fungal and plant genotypes.

In conclusion, an overview of these data, obtained by focusing on a specific DNA region and on a limited number of sequences, has demonstrated that AM fungal biodiversity is characterized by a core of AM fungi that can be considered as 'generalist', since they are globally distributed (e.g. *Glomus intrardices* group), and by a number of AM fungal taxa that are specific for each environment, which can be considered 'specialist' and as sorts of typical markers of their habitat (Dumbrell *et al.*, 2010).

From traditional to new approaches

In 2009, a new era in large-scale DNA barcoding analysis of AMF was born thanks to new high-throughput technologies (e.g. 454 pyrosequencing), which allows cloning steps, that were previously essential, to be bypassed and a much higher number of sequences to be obtained than those available with cloning and sequencing approaches. Furthermore, the large number of obtained reads increases the likelihood of capturing sequences from rare organisms, which might remain undetected when classic approaches are used. Pyrosequencing is a sequencing-by-synthesis strategy which was developed by Pål Nyrén, professor of biochemistry at the Royal Institute of Technology in Stockholm (Nyrén, 2007). This new way of sequencing has been exploited and patented by the 454 Life Science company to develop a platform, the Genome Sequencer FLX System, that is able to generate 400 million high quality bases per 10 hour instrument run.

So far, just a few research groups have used this new approach for the metabarcoding analysis of the overall fungal community present in different environments: Buée *et al.* (2009) investigated French forest soils, Jumpponen *et al.* (2009, 2010) the phyllosphere and rhizosphere of *Quercus* spp., Ghannoum *et al.* (2010) the oral apparatus of healthy individuals, Lim *et al.* (2010) soils from three islands in the Yellow Sea in Korea, Rousk *et al.* (2010) different pH spots of an arable soil, Tedersoo *et al.* (2010) tropical soils from the Korup National Park, south-west Cameroon, and Wallander *et al.* (2010) managed Norway spruce (*Picea abies*) stands in south-western Sweden.

Only two research groups have focused on the AM fungal component and have published their results to date (Öpik *et al.*, 2009; Lumini *et al.*, 2010). Öpik *et al.* (2009) used this large-scale approach to analyse AMF communities in plant roots in a forest in central Estonia. This study has increased the number of AM fungal taxa retrieved from this forest site from the previous 31 taxa discovered using the cloning and sequencing approach (34 taxa according to the former classification, Öpik *et al.*, 2008) to 51 taxa using 454 pyrosequencing, and an unexpected high diversity has been highlighted. Lumini *et al.* (2010) characterized and compared AMF communities in five Italian soils along a land-use gradient: this study is the first and the largest AMF-soil-based sequence dataset so far obtained (Table 10.1).

A particular aspect of AMF symbiosis is the presence of bacterial communities associated to AMF structures such as hyphae or spores (Bonfante and Anca, 2009). It is known that AMF harbour intracellular bacteria that live in their cytoplasm. The presence of endobacteria inside AMF cytoplasms has long been documented by means of electron microscopy, which has distinguished two bacterial morphotypes. The first, restricted to a phylogenetically relatively young AMF family (Gigasporaceae), is rod shaped, related to *Burkholderia* and has been described as an uncultured taxon, *Candidatus* Glomeribacter gigasporarum (Bianciotto *et al.*, 1996, 2000, 2003). The other bacterial type has been detected inside AMF spores and hyphae colonizing plant roots sampled in the field. It is coccoid in shape and has been called 'bacterium-like organism' (BLO), as its identity is still obscure (MacDonald *et al.*, 1982; Scannerini and Bonfante, 1991).

Naumann *et al.* (2010) have recently demonstrated that this population of coccoid endobacteria is harboured in the cytoplasm of other AMF families, including several *Glomus* isolates. These bacteria are phylogenetically related to *Mollicutes*. A pyrosequencing approach was used to confirm that *Mollicutes*-related BLOs were exclusively present inside the cytoplasm and not as spore surface associated bacteria. The metagenomic DNA was derived from either (1) washed spores, (2) sonicated and washed spores or (3) the water in which the spores were held during sonication. While sequences detected from the sonicated water were considered to be derived from spore surface-associated bacteria, BLO sequences were exclusively obtained from washed and sonicated spores, but never from sonicated water. The pyrosequencing analysis demonstrated that the BLO sequences represent the vast majority (89.8%) of bacterial sequences recovered from cleaned spores.

Furthermore, the pyrosequencing approach allowed another aspect to be clarified concerning a previous clone based analysis of 16S rDNA which revealed an unexpected diversity of BLO sequences associated to the same spore. In order to understand whether this was due to an insufficient sampling density, spores of a single species (*G. etunicatum*), originating from diverse lineages, were 454 GS-FLX pyrosequenced. This approach has demonstrated that different spores of the same species contain different BLO phylotype compositions (Naumann *et al.*, 2010).

AMF biodiversity studies based on pyrosequencing

Although the two studies mentioned above targeted the same fungal group (AM fungi) using the same technique (pyrosequencing), they are not easy to compare because the aims were different.

Lumini *et al.* (2010) verified the effects of different human management systems and ecological traits on soil AMF populations, while Öpik *et al.* (2009) tried to understand whether there was a significant difference between the AMF communities present in the roots: forest specialist and habitat generalist plant species. Therefore, the main difference between the two studies was the considered compartment: bulk soils or plant roots, representing the AM fungal extraradical phase (e.g. mycelium and spores) for Lumini *et al.* (2010) and the intraradical (e.g. mycelium, arbuscules and vescicules) phase for Öpik *et al.* (2009).

Even though both studies targeted the same rDNA gene (18S SSU), they used different primer sets: AMV4.5NF/AMDGR and NS31/AMmix (AM1, AM2 and AM3) and NS31-AM1. Öpik and colleagues used the bar-code strategy for their pyrosequencing study in order to have the highest possible number of reads. 454 pyrosequencing allows 6- to 10-bp-long bar-code sequences, inserted before the chosen primer sequences, to be used. Since each sample is amplified with a known tagged primer, it is possible to sequence an equimolar mixture of PCR-amplified DNA from each sample and assign sequences to samples only on the basis of the barcodes. The other possible strategy consists in keeping the different PCR products physically separated in the plate during the pyrosequencing run, as Lumini and her colleagues did (2010).

One of the most critical aspects of a biodiversity study is the identification of the retrieved sequences. Identification using a BLAST algorithm in particular requires a database of already described sequences. A 'homemade' database, called MaarjAM (Öpik *et al.*, 2010) containing already published Glomeromycota SSU rRNA gene sequences, was used for the analysis of the AMF communities in forests and generalist plant species. The GenBank non-redundant nucleotide database was instead used for the analysis of the soil AMF assemblages to query the obtained reads. These different approaches mean the studied AMF assemblages were described at different taxonomic levels: in the first work, the possibility of having a groomed database made possible to analyse the AMF populations at a 'species level', defined as a group of sequences with ≥ 97% of similarity ('virtual taxa' – VT), while in the second study the authors were more prudent and preferred to classify their assemblages at the order level.

Despite the difference between the two works, they have some points in common. For example, both studies have underlined the current lack of AMF specific primer pairs which are totally suitable for a 454 pyrosequencing study. This is due to different reasons: the difficulty of discriminating at a species level, the excessive fragment lengths, the relatively low number of well characterized and deposited AMF sequences, the primer specificity limited to a single or a few AMF groups and the frequent amplifications of 'contaminant' sequences. Even though Öpik *et al.* (2009) supported the possibility of using a portion of the NS31-AM1 fragment to discriminate AM fungi at a species level, they were forced to discard a significant number of sequences (40,360, corresponding to more than 22% of the total) because they did not belong to the Glomeromycota group. The same happened to Lumini *et al.* (2010), where the NS31-AMmix targeted an even higher number of sequences, corresponding to 62% of the retrieved reads, that did not belong to AMF. This is probably due to the greater complexity of the analysed matrices: mycorrhized plant roots versus bulk soil.

From an ecological point of view, the main outcome that has emerged from both AM fungal biodiversity studies is the unexpected and unprecedented fungal biodiversity observed in the analysed environments. Öpik *et al.* (2008) detected 20 new AMF virtual taxa, which had not been retrieved in a previous study based on a traditional cloning-sequencing approach conducted in the same boreonemoral forest. Taxa belonging to *Glomus* group A were dominant and, in particular, *Glomus fasciculatum* was the most abundant taxon in all the plant samples. The forest plant species hosted more AM fungal taxa than the generalist plant species: 22 taxa were exclusive to the forest plant roots, while just one was typical of the generalist roots. The results obtained by Öpik *et al.* showed, for the first time, that there can be a clear distinction between AM fungal communities associated with host species that differ in habitat preference: the assemblages of the forest plant species were similar to each other, but were significantly different when compared to those of the generalist plant species.

Lumini *et al.* (2010) confirmed the tendency of unexpected and unprecedented biodiversity, and found up to 74 virtual taxa in a single soil. This result represents the highest number so far retrieved in soil. Taxa belonging to Glomerales were dominant in all the five investigated soils (tilled vineyard, covered vineyard, managed meadow, pasture and cork-oak formation). The greatest biodiversity was observed in the pasture soil, while the lowest was in the cork-oak formation. In general, the obtained results showed the influence of two different factors on AM fungal biodiversity: human input and ecological traits. A high human input (e.g. the managed meadow) means low AMF biodiversity and vice versa. In the same way, the presence of forest traits (e.g. the cork-oak formation) determines a reduction in AMF diversity.

Despite the different results, due to the different aims, both studies confirm the powerfulness of the 454 pyrosequencing approach, which represents a promising and exciting new tool to better understand the natural distribution and role of an essential group of soil microorganisms, such as AM fungi.

Problems in AMF metabarcoding studies

Considering the probable diffusion of metabarcoding studies, based on both traditional and new sequencing technology, in fungal biodiversity studies, some general considerations on the unavoidable problems, that will need to be faced, are obligatory. It is very likely that the use of a second-generation sequencing technology will become common not only in prokaryotic (bacterial), but also in fungal ecology, since this technology will surely allow fungal communities to be described in greater detail than ever before.

The first problem that needs to be faced in a metabarcoding study is the sampling strategy: researchers should think carefully about the questions they are addressing when planning experiments. They should critically assess the techniques and experimental design they intend using to encounter the fungal species in the samples. They should consider, in advance, how they will present and analyse their data (Prosser, 2010). However, in complex microbial communities, such as those present in soil, a large fraction of fungal species is present in low amounts and even with the recent advances in sequencing technologies, a complete census of all fungal species in complex ecosystems remains a daunting task. Despite this assumption, microbial ecologists agree that a strategy involving statistical replicates of the samples is obligatory to obtain the best coverage of the studied biodiversity and to compare it in different environments (Prosser, 2010).

Once the experimental design is ready, the second problem that needs to be overcome concerns the optimization of the DNA extraction. This step is particularly difficult when the examined sample is a complex matrix, such as soil. Extracting DNA from environmental samples is possible using commercial kits or 'homemade' protocols. However, in some cases, such as soils rich in clay and/or humic substances, it is necessary to modify the available protocols in order to obtain a good quality DNA. Several articles have reported bias associated with DNA extraction from soil and possible solutions to circumvent this problem (Feinstein *et al.*, 2009).

Once metagenomic DNA is extracted, the third problem that needs to be faced by a researcher interested in AM fungal assemblages, concerns the choice of a suitable primer pair to analyse the overall AM fungal biodiversity. In addition to the lack of a single primer set that is able to cover all the currently described Glomeromycota taxa, the opportunity to use pyrosequencing has led to evaluating the problem of the amplicon length. The available pyrosequencing platform (GS FLX Titanium Series) allows one to obtain reads of an average length of 400 bp, which might not be informative enough to discriminate at low taxonomic levels or to make phylogenetic inferences. This problem is particularly relevant if a well-populated backbone database, such as the MaarjAM database, is not available. Such database can in fact facilitate the detection of biological and technical artefacts, which are very frequent, as has been reported in some recent studies using the pyrosequencing approach (Quinlan *et al.*, 2008; Kunin *et al.*, 2010; Medinger *et al.*, 2010). Furthermore, interest in developing new algorithms that are able to manage large datasets and eliminate sequencing mistakes is growing with the aim of overcoming this problem (Gomez-Alvarez *et al.*, 2009; Quince *et al.*, 2009).

With a pyrosequencing run, it is possible to obtain up to 1,200,000 sequences. The large number of data obtained with this methodology in a relatively short time leads to many questions about their management. This leads to the fourth problem: species identification. Ecologists and molecular biologists usually identify OTUs (operational taxonomic units) at an established similarity level, as a group of sequences belonging to an organism of a certain species. However, the level of similarity chosen to separate OTUs as unique species is arbitrary and controversial. Both of the aforementioned pyrosequencing studies on AMF assemblages used 97% as the cut-off level to discriminate at a 'species level'. This, of course, is a very crucial topic because it directly influences the level of detected biodiversity and it is not easy to solve since a global standard to define fungal species does not exist and genetic diversity varies from taxon to taxon.

Another problem concerns by the correct identification of the sequences. Many of the obtained reads do not in fact match any sequence deposited in the databases. This is due to the very low percentage (0.6–1.3%, according to whether you consider 1.5 or 3.5 million as the number of estimated fungal species) of fungal sequences represented in databases (Hibbett *et al.*, 2009). There is a risk that the increasing number of studies based on the new technology could make this situation worse and block databases with sequences named 'environmental samples' or 'unknown organisms'. For this reason, collaboration among taxonomists, ecologists and bioinformaticians is necessary to establish accepted standards in order to identify, describe and register new fungal species, not only using traditional methodologies but also the new molecular resources.

A recent research (Tedersoo *et al.*, 2010) has compared the performance of the pyrosequencing and traditional sequencing approaches for the species recovery of ectomycorrhizal fungi. Even though pyrosequencing and the traditional method yielded qualitatively similar results, there were slight, but significant, differences that affected the taxonomic view of the fungal community, particularly in the rare taxa that were retrieved (the so-called 'singletons', that are taxa represented by only one sequence). Compared with Sanger sequencing-based methods, pyrosequencing provides more data of some orders of magnitude on the diversity of organisms in their natural habitat. In conclusion, it is widely recognized that pyrosequencing offers a powerful alternative for the characterization of fungal assemblages, but it still requires careful development and selection of the molecular tools (e.g. a well-populated backbone database) in order to understand and reduce its technological biases.

From metagenomics to metatranscriptomics

Since the introduction of molecular methods, our appreciation of natural AMF ribotype richness (Hijri *et al.*, 2006; Hempel *et al.*, 2007; Öpik *et al.*, 2008, 2009) has increased greatly, and fungal distribution and diversity patterns, in relation to biotic and abiotic environmental factors, are emerging (Helgason *et al.*, 1998; Johnson *et al.*, 2004; Öpik *et al.*, 2006). At the moment, ribosomal genes remain the only target with sufficient database depth to conduct discriminative studies at a taxon-specific level. However, the discrimination between different AMF ribotypes is hampered by the high level of intra-isolate and inter-isolate variation in the nuclear encoded ribosomal subunits (Sanders *et al.*, 1995; Lanfranco *et al.*, 1999; Corradi *et al.*, 2007). This heterogeneity may be linked to the proposed polygenomic structure of the glomeromycotan genome (Kuhn *et al.*, 2001; Hijri and Sanders, 2005), but findings on such a phenomenon appear to be in conflict (Pawlowska and Taylor, 2005). Ecological interpretation of nuclear rDNA community data for AMF is seriously impaired by the frequent overlap of sequence haplotypes among related species, making it difficult or even impossible to define species boundaries (Clapp *et al.*, 2002; de Souza *et al.*, 2004; Rosendahl, 2008).

Therefore, even though molecular markers based on the nuclear encoded ribosomal subunit are still the most widely used in ecological and phylogenetic studies, thanks to the large dataset available, the scientific community is looking for possible new molecular markers. Recently developed quantitative (Alkan *et al.*, 2006; Pivato *et al.*, 2007; Jansa *et al.*, 2008; Salvioli *et al.*, 2008; Gamper *et al.*, 2008) and activity-related molecular genetic approaches (Vandenkoornhuyse *et al.*, 2007) only partially satisfy this request through the use of some alternative gene markers. However, it is still difficult to determine the functional significance of the presence and abundance of mycorrhizal fungi in different ecosystems (Gamper *et al.*, 2010).

Although direct DNA extraction and the associated analysis methods have yielded unprecedented insight into soil fungal communities, the persistence of rDNA in metabolically inactive fungi means that the analysed communities comprise both active and inactive fungi (Anderson and Parkin, 2007). These studies in fact represent the first step towards understanding ecosystem functions, since knowing what organisms are present is important to predict ecosystem processes, but understanding whether the most frequently represented organisms are also the most functionally active could be even more interesting.

Gamper *et al.* (2010) suggested a shift towards plant and fungal protein-encoding genes (functional genes) as more immediate indicators of mycorrhizal contributions to ecological processes. Studying the functional gene expressed in an environment means studying the so-called 'metatranscriptome' of that habitat. Gamper and his colleagues also indicate, as possible candidates, a number of genes involved in mineral nutrient uptake (e.g. uptake of phosphorus and nitrogen), carbon cycling and overall metabolic activity. Detection of such genes (DNA) in an environment shows that the encoded gene products could potentially be present, whereas transcript (mRNA) detection indicates that particular mycorrhizal traits are actually expressed. Before it is possible to carry out a metatranscriptomic study, necessary steps should be taken: (1) preliminary nucleotide sequence databases of functional indicator genes should be established over a broad taxonomic range, (2) robust AMF- and gene-specific PCR amplification systems should be developed to examine sequence diversity in environmental samples, (3) reliable mRNA extraction and work-up procedures should be optimized to analyse root and soil samples, and (4) appropriate high-throughput tools should be developed to detect and quantify specific mRNAs.

Functional gene-based approaches may offer several advantages, such as the extension of traditional phylogenetic and rRNA gene-based approaches to AM fungal ecology. First, molecular genetic assays, targeted to functional genes, will directly measure the expression of particular AM traits that are important for ecological processes. For instance, analyses of the diversity and activities of functional genes will allow differences in the role of AMF to be compared from site to site and in time. Such information could subsequently be used to estimate the contribution and importance of arbuscular mycorrhizae, to specific ecosystem processes and eventually to ecosystem functioning.

Metatranscriptomic studies on arbuscular mycorrhizal fungi are not yet available. However, the possibility of combining traditional approaches (e.g. cDNA library construction) with the emerging sequencing technologies (e.g. pyrosequencing) represents a crucial point in achieving the goal of the functional characterization of AMF communities in different ecosystems.

Acknowledgements

Contributions to this chapter have in part been funded by the National Project Soil-Sink-FISR, the ENDURE Project 'European Network of Excellence', Compagnia di San Paolo – Torino, IPP-CNR (AG-P02-006) and ECOFLOR Project (Regione Piemonte). We are grateful to E.L. for the freehand drawing and Giuseppe Torrisi for graphic processing. R.B. was supported by Progetto Lagrange and Tecnoverde SPA grant.

References

Aldrich-Wolfe, L. (2007). Distinct mycorrhizal communities on new and established hosts in a transitional tropical plant community. Ecology *88*, 559–566.

Alguacil, M.M., Lumini, E., Roldán, A., Salinas-Garcia, J.R., Bonfante, P., and Bianciotto, V. (2008). The impact of tillage practices on arbuscular mycorrhizal fungal diversity in subtropical crops. Ecol. Appl. *18*, 527–536.

Alguacil, M.M., Roldán, A., and Torres, M.P. (2009). Complexity of semiarid gypsophilous shrub communities mediates the AMF biodiversity at the plant species level. Microbial Ecol. 57, 718–727.

Alkan, N., Gadkar, V., Yarden, O., and Kapulnik, Y. (2006). Analysis of quantitative interactions between two species of arbuscular mycorrhizal gungi, Glomus mosseae and G. intraradices, by Real-Time PCR. Appl. Environ. Microbiol. 72, 4192–4199.

Al-Yahya'ei, M.N., Oehl, F., Vallino, M., Lumini, E., Redecker, D., Wiemken, A., and Bonfante, P. (2011). Unique arbuscular mycorrhizal fungal communities uncovered in date palm plantations and surrounding desert habitats of Southern Arabia. Mycorrhiza 21, 195–209.

Anderson, I.C., and Parkin, P.I. (2007). Detection of active soil fungi by RT-PCR amplification of precursor rRNA molecules. J. Microbiol. Meth. 68, 248–253.

Balestrini, R., Magurno, F., Walker, C., Lumini, E., and Bianciotto, V. (2010). Cohorts of arbuscular mycorrhizal fungi (AMF) in Vitis vinifera, a typical Mediterranean fruit crop. Environ. Microbiol. Rep. 2,594–604.

Bianciotto, V., Bandi, C., Minerdi, D., Sironi, M., Tichy, H.V., and Bonfante, P. (1996). An obligately endosymbiotic fungus itself harbors obligately intracellular bacteria. Appl. Environ. Microbiol. 62, 3005–3010.

Bianciotto, V., Lumini, E., Bonfante, P., and Vandamme, P. (2003). '*Candidatus* Glomeribacter gigasporarum' gen. nov. sp. nov., an endosymbiont of arbuscular mycorrhizal fungi. Int. J. Syst. Evol. Microbiol. 53, 121–124.

Bianciotto, V., Lumini, E., Lanfranco, L., Minerdi, D., Bonfante, P., and Perotto, S. (2000). Detection and identification of bacterial endosymbionts in arbuscular mycorrhizal fungi belonging to the family Gigasporaceae. Appl. Environ. Microbiol. 66, 4503–4509.

Blaszkowski, J., Czerniawska, B., Wubet, T., Schäfer, T., Buscot, F., and Renker, C. (2008). *Glomus irregulare*, a new arbuscular mycorrhizal fungus in the Glomeromycota. Mycotaxon 106, 247–267.

Bonfante, P., and Anca, I.A. (2009). Plants, mycorrhizal fungi, and bacteria: a network of interactions. Annu. Rev. Microbiol. 63, 363–383.

Bonfante, P., and Genre, A. (2008). Plants and arbuscular mycorrhizal fungi: an evolutionary-developmental perspective. Trends Plant Sci. 13, 492–498.

Bonfante, P., and Genre, A. (2010). Mechanisms underlying beneficial plant – fungus interactions in mycorrhizal symbiosis. Nat. Com. 1, 48.

Buée, M., Reich, M., Murat, C., Morin, E., Nilsson, R.H., Uroz, S., and Martin, F. (2009). 454 pyrosequencing analyses of forest soils reveal an unexpected high fungal diversity. New Phytol. 184, 449–456.

Burrows, R.L., and Pfleger, F.L. (2002). Host responses to AMF from plots differing in plant diversity. Plant Soil 240, 169–180.

Cesaro, P., van Tuinen, D., Copetta, A., Chatagnier, O., Berta, G., Gianinazzi, S., and Lingua, G. (2008). Preferential colonization of Solanum tuberosum L. roots by the fungus Glomus intraradices in arable soil of a potato farming area. Appl. Environ. Microbiol. 74, 5776–5783.

Chen K., and Pachter, L. (2005). Bioinformatics for whole-genome shotgun sequencing of microbial communities. PLoS Comp. Biol. 1, 24.

Clapp, J., Daniell, T., and Young, J. (2002). Genetic studies of the structure and diversity of arbuscular mycorrhizal fungal communities. In Mycorrhizal ecology, van der Heijden, M.G.A., and Sanders, I.R., eds (Berlin: Springer-Verlag), pp. 201–224.

Clapp, J.P., Young, J.P.W., Merryweather, J.W., and Fitter, A.H. (1995). Diversity of fungal symbionts in arbuscular mycorrhizas from a natural community. New Phytol. 130, 259–265.

Corradi, N., Croll, D., Colard, A., Kuhn, G., Ehinger, M., and Sanders, I.R. (2007). Gene copy number polymorphisms in an arbuscular mycorrhizal fungal population. App. Environ. Microbiol. 73, 366–369.

Daniell, T.J., Husband, R., Fitter, A.H., and Young, J.P.W. (2001). Molecular diversity of arbuscular mycorrhizal fungi colonising arable crops. FEMS Microb. Ecol. 36, 203–209.

de Souza, F.A., Kowalchuk, G.A., Leeflang, P., van Veen, J.A., and Smit, E. (2004). PCR-denaturing gradient gel electrophoresis profiling of inter- and intraspecies 18S rRNA gene sequence heterogeneity is an accurate and sensitive method to assess species diversity of arbuscular mycorrhizal fungi of the genus Gigaspora. App. Environ. Microbiol. 70, 1413–1424.

DeBellis, T., and Widden, P. (2006). Diversity of the small subunit ribosomal RNA gene of the arbuscular mycorrhizal fungi colonizing Clintonia borealis from a mixed-wood boreal forest. FEMS Microb. Ecol. 58, 225–235.

Douds, D.D., and Millner, P.D. (1999). Biodiversity of arbuscular mycorrhizal fungi in agroecosystems. Agr. Ecosyst. Environ. 74, 77–93.

Douhan, G.W., Petersen, C., Bledsoe, C.S., and Rizzo, D.M. (2005). Contrasting root associated fungi of three common oak-woodland plant species based on molecular identification: host specificity or non-specific amplification? Mycorrhiza 15, 365–372.

Dumbrell, A.J., Nelson, M., Helgason, T., Dytham, C., and Fitter, A.H. (2010) Relative roles of niche and neutral processes in structuring a soil microbial community. ISME J. 4, 337–345.

Feinstein, L.M., Sul, W.J., and Blackwood, C.B. (2009). Assessment of bias associated with incomplete extraction of microbial DNA from soil. Appl. Environ. Microbiol. 75, 5428–5433.

Gai, J., Christie, P., Feng, G., and Li, X. (2006). Twenty years of research on community composition and species distribution of arbuscular mycorrhizal fungi in China: a review. Mycorrhiza 16, 229–239.

Gamper, H.A., van der Heijden, M.G.A., and Kowalchuk, G.A. (2010). Molecular trait indicators: moving beyond phylogeny in arbuscular mycorrhizal ecology. New Phytol. 185, 67–82.

Gamper, H.A., Walker, C., Schüßler, A. (2009). Diversispora celata sp. nov: molecular ecology and phylotaxonomy of an inconspicuous arbuscular mycorrhizal fungus. New Phytol. *182*, 495–506.

Gamper, H.A., Young, J.P.W., Jones, D.L., and Hodge, A. (2008). Real-time PCR and microscopy: Are the two methods measuring the same unit of arbuscular mycorrhizal fungal abundance? Fungal Genet. Biol. *45*, 581–596.

Ghannoum, M.A., Jurevic, R.J., Mukherjee, P.K., Cui, F., Sikaroodi, M., Naqvi, A., and Gillevet, P.M. (2010). Characterization of the oral fungal microbiome (mycobiome) in healthy individuals. PLoS Pathog. *6*, e1000713.

Girlanda, M., Perotto, S., and Bonfante, P. (2007). Mycorrhizal Fungi: Their Habitats and Nutritional Strategies. In The Mycota. IV. Environmental and Microbial Relationships, 2nd edn. Kubicek, C.P., Druzhinina, I.S., eds (Berlin: Springer-Verlag), pp. 229–256.

Gollotte, A., van Tuinen, D., and Atkinson, D. (2004). Diversity of arbuscular mycorrhizal fungi colonising roots of the grass species Agrostis capillaris and Lolium perenne in a field experiment. Mycorrhiza *14*, 111–117.

Gomez-Alvarez, V., Teal, T.K., and Schmidt, T.M. (2009). Systematic artifacts in metagenomes from complex microbial communities. ISME J. *3*, 1314–1417.

Heckman, D.S., Geiser, D.M., Eidell, B.R., Stauffer, R.L., Kardos, N.L., and Hedges, S.B. (2001). Molecular evidence for the early colonization of land by fungi and plants. Science *293*, 1129–1133.

Heinemeyer, A., Ridgway, K.P., Edwards, E.J., Benham, D.G., Young, J.P.W., and Fitter, A.H. (2003). Impact of soil warming and shading on colonization and community structure of arbuscular mycorrhizal fungi in roots of a native grassland community. Glob. Change Biol. *10*, 52–64.

Helgason, T., and Fitter, A.H. (2009). Natural selection and the evolutionary ecology of the arbuscular mycorrhizal fungi (Phylum Glomeromycota). J. Exp. Bot. *60*, 2465–2480.

Helgason, T., Daniell, T., Husband, R., Fitter, A., and Young, J. (1998). Ploughing up the wood-wide web? Nature *394*, 431.

Helgason, T., Fitter, A.H., and Young, J.P.W. (1999). Molecular diversity of arbuscular mycorrhizal fungi colonising Hyacinthoides non-scripta (bluebell) in a seminatural woodland. Molec. Ecol. *8*, 659–666.

Helgason, T., Merryweather, J., Cole, J., Wilson, P., Young, J., and Fitter, A. (2002). Selectivity and functional diversity in arbuscular mycorrhizas of co-occurring fungi and plants from a temperate deciduous woodland. J. Ecol. *90*, 371–384.

Helgason, T., Merryweather, J., Young, J., and Fitter, A. (2007). Specificity and resilience in the arbuscular mycorrhizal fungi of a natural woodland community. J. Ecol. *95*, 623–630.

Hempel, S., Renker, C., and Buscot, F. (2007). Differences in the species composition of arbuscular mycorrhizal fungi in spore, root and soil communities in a grassland ecosystem. Environ. Microbiol. *9*, 1930–1938.

Hibbett, D.S., Ohman, A., and Kirk, P.M. (2009). Fungal ecology catches fire. New Phytol. *184*, 279–282.

Hijri, I., Sýkorová, Z., Oehl, F., Ineichen, K., Mader, P., Wiemken, A., and Redecker, D. (2006). Communities of arbuscular mycorrhizal fungi in arable soils are not necessarily low in diversity. Mol. Ecol. *15*, 2277–2289.

Hijri, M., and Sanders, I.R. (2005). Low gene copy number shows that arbuscular mycorrhizal fungi inherit genetically different nuclei. Nature *433*, 160–163.

Hugenholtz, P., and Tyson, G.W. (2008). Microbiology: metagenomics. Nature 455: 481–483.

Husband, R., Herre, E.A., and Young, J.P.W. (2002a) Temporal variation in the arbuscular mycorrhizal communities colonising seedlings in a tropical forest. FEMS Microbiol. Ecol. *42*, 131–136.

Husband, R., Herre, E.A., Turner, S.L., Gallery. R., and Young, J.P.W. (2002b). Molecular diversity of arbuscular mycorrhizal fungi and patterns of host association over time and space in a tropical forest. Mol. Ecol. *11*, 2669–2678.

Jansa, J., Mozafar, A., Anken, T., Ruh, R., Sanders, I., and Frossard, E. (2002). Diversity and structure of AMF communities as affected by tillage in a temperate soil. Mycorrhiza *12*, 225–234.

Jansa, J., Smith, F.A., and Smith, S.E. (2008). Are there benefits of simultaneous root colonization by different arbuscular mycorrhizal fungi? New Phytol.*177*, 779–789.

Johnson, D., Vandenkoornhuyse, P.J., Leake, J.R., Gilbert, L., Booth, R.E., Grime, J.P., Young, J.P.W., and Read, D.J. (2004). Plant communities affect arbuscular mycorrhizal fungal diversity and community composition in grassland microcosms. New Phytol. *161*, 503–515.

Jumpponen, A., and Jones, K.L. (2009). Massively parallel 454-sequencing indicates hyperdiverse fungal communities in temperate Quercus macrocarpa phyllosphere. New Phytol. *184*, 438–448.

Jumpponen, A., Jones, K.L., Mattox, J., and Yaege, C. (2010). Massively parallel 454-sequencing of fungal communities in Quercus spp. ectomycorrhizas indicates seasonal dynamics in urban and rural sites. Mol. Ecol. *19*, 41–53.

Klironomos, J., and Hart, M. (2002). Colonization of roots by arbuscular mycorrhizal fungi using different sources of inoculum. Mycorrhiza *12*, 181–184.

Krüger, M., Stockinger, H., Krüger, C., and Schüßler, A. (2009). DNA-based species level detection of Glomeromycota: one PCR primer set for all arbuscular mycorrhizal fungi. New Phytol. *183*, 212–223.

Kuhn, G., Hijri, M., and Sanders, I.R. (2001). Evidence for the evolution of multiple genomes in arbuscular mycorrhizal fungi. Nature *414*, 745–748.

Kunin, V., Engelbrektson, A., Ochman, H., and Hugenholtz, P. (2010). Wrinkles in the rare biosphere: pyrosequencing errors can lead to artificial inflation of diversity estimates. Environ. Microbiol. *12*, 118–123.

Landis, F.C., Gargas, A., and Givnish, T.J. (2004). Relationships among arbuscular mycorrhizal fungi, vascular plants and environmental conditions in oak savannas. New Phytol. *164*, 493–504.

Lanfranco, L., Delpero, M., and Bonfante, P. (1999). Intrasporal variability of ribosomal sequences in the endomycorrhizal fungus *Gigaspora margarita*. Molec. Ecol. *8*, 37–45.

Lee, J., Lee, S., and Young, J.P.W. (2008). Improved PCR primers for the detection and identification of arbuscular mycorrhizal fungi. FEMS Microbiol. Ecol. *65*, 339–349.

Lee, P.J., and Koske, R.E. (1994). Gigaspora gigantea: seasonal abundance and ageing of spores in a sand dune. Mycol. Res. *98*, 453–457.

Liang, Z., Drijber, R.A., Lee, D.J., Dwiekat, I.M., Harris, S.D., and Wedin, D.A. (2008). A DGGE-cloning method to characterize arbuscular mycorrhizal community structure in soil. Soil Biol. Biochem. *40*, 956–966.

Ligrone, R., Carafa, A., Lumini, E., Bianciotto, V., Bonfante, P., and Duckett, J.G. (2007). Glomeromycotean associations in liverworts: a molecular cellular and taxonomic analysis. Am. J. Bot. *94*, 1756–1777.

Lim, Y.W., Kim, B.K., Kim, C., Jung, H.S., Kim, B.S., Lee, J.H., and Chun, J. (2010). Assessment of soil fungal communities using pyrosequencing. J. Microbiol. *48*, 284–289.

Lumini, E., Orgiazzi, A., Borriello, R., Bonfante, P., and Bianciotto, V. (2010). Disclosing arbuscular mycorrhizal fungal biodiversity in soil through a land-use gradient using a pyrosequencing approach. Environ. Microbiol. *12*, 2165–2179.

Ma, W., Siciliano, S., and Germida, J. (2005). A PCR-DGGE method for detecting arbuscular mycorrhizal fungi in cultivated soils. Soil Biol. Biochem. *37*, 1589–1597.

MacDonald, R.M., Chandler, M., and Mosse, B. (1982). The occurrence of bacterium-like organelles in vesicular-arbuscular mycorrhizal fungi. New Phytol. *90*, 659–663.

Martin, F., Cullen, D., Hibbett, D., Pisabarro, A., Spatafora, J.W., Baker, S.E., and Grigoriev, I.V. (2011). Sequencing the fungal tree of life. New Phytol. doi: 10.1111/j.1469-8137.2011.03688.x.

Mathimaran, N., Ruh, R., Vullioud, P., Frossard, E., and Jansa, J. (2005). Glomus intraradices dominates arbuscular mycorrhizal communities in a heavy textured agricultural soil. Mycorrhiza *16*, 61–66.

Medinger, R., Nolte, V., Pandey, R.M., Jost, S., Ottenwälder, B., Schlötterer, C., and Boenigk, J. (2010). Diversity in a hidden world: potential and limitation of next-generation sequencing for surveys of molecular diversity of eukaryotic microorganisms. Molec. Ecol. *19*, 32–40.

Merryweather, J., and Fitter, A. (1998). The arbuscular mycorrhizal fungi of *Hyacinthoides non-scripta*. I. Diversity of fungal taxa. New Phytol. *138*, 117–129.

Naumann, M., Schüßler, A., and Bonfante, P. (2010). The obligate endobacteria of arbuscular mycorrhizal fungi are ancient heritable components related to the Mollicutes. ISME J. *4*, 862–871.

Nicolson, T.H. (1967). Vesicular-arbuscular mycorrhiza – a universal plant symbiosis. Sci. Prog. *55*, 561–581.

Nielsen, K.B., Kjøller, R., Olsson, P.A., Schweiger, P.F., Andersen, F.O., and Rosendahl, S. (2004). Colonisation and molecular diversity of arbuscular mycorrhizal fungi in the aquatic plants Littorella uniflora and Lobelia dortmanna in southern Sweden. Mycol. Res. *108*, 616–625.

Nyrén, P. (2007). The history of pyrosequencing. Methods Mol. Biol. *373*, 1–14.

Oehl, F., Sieverding, E., Ineichen, K., Mader, P., Boller, T., and Wiemken, A. (2003). Impact of land use intensity on the species diversity of arbuscular mycorrhizal fungi in agroecosystems of central Europe. Appl. Environ. Microbiol. *69*, 2816–2824.

Oehl, F., Sieverding, E., Ineichen, K., Mäder, P., Wiemken, A., and Boller, T. (2009). Distinct sporulation dynamics of arbuscular mycorrhizal fungal communities from different agroecosystems in long-term microcosms. Agr. Ecosyst. Environ. *134*, 257–268.

Oehl, F., Sieverding, E., Ineichen, K., Ris, E., Boller, T., and Wiemken, A. (2005). Community structure of arbuscular mycorrhizal fungi at different soil depths in extensively and intensively managed agroecosystems. New Phytol. *165*, 273–283.

Oehl, F., Sieverding, E., Mäder, P., Dubois, D., Ineichen, K., Boller, T., and Wiemken, A. (2004). Impact of long-term conventional and organic farming on the diversity of arbuscular mycorrhizal fungi. Oecologia *138*, 574–583.

Öpik, M., Moora, M., Liira, J., Kõljalg, U., Zobel, M., and Sen, R. (2003). Divergent arbuscular mycorrhizal fungal communities colonize roots of Pulsatilla spp. in boreal Scots pine forest and grassland soils. New Phytol. *160*, 581–593.

Öpik, M., Moora, M., Liira, J., and Zobel, M. (2006). Composition of root-colonizing arbuscular mycorrhizal fungal communities in different ecosystems around the globe. J. Ecol. *94*, 778–790.

Öpik, M., Moora, M., Zobel, M., Saks, Ü., Wheatley, R., Wright, F., and Daniell, T. (2008). High diversity of arbuscular mycorrhizal fungi in a boreal herb-rich coniferous forest. New Phytol. *179*, 867–876.

Öpik, M., Metsis, M., Daniell, T.J., Zobel, M., and Moora, M. (2009). Large-scale parallel 454 sequencing reveals host ecological group specificity of arbuscular mycorrhizal fungi in a boreonemoral forest. New Phytol. *184*, 424–437.

Öpik, M., Vanatoa, A., Vanatoa, E., Moora, M., Davison, J., Kalwij, J.M., Reier, Ü., and Zobel, M. (2010). The online database MaarjAM reveals global and ecosystemic distribution patterns in arbuscular mycorrhizal fungi (Glomeromycota). New Phytol. *188*, 223–241.

Parniske, M. (2008). Arbuscular mycorrhiza: the mother of plant root endosymbioses. Nat. Rev. Micro. *6*, 763–775.

Pawlowska, T.E., and Taylor, J.W. (2005). Arbuscular mycorrhizal fungi: Hyphal fusion and multigenomic structure (reply). Nature *433*, E4. doi:10.1038/nature03295

Phipps, C.J., and Taylor, T.N. (1996). Mixed Arbuscular Mycorrhizae from the Triassic of Antarctica. Mycologia *88*, 707–714.

Pivato, B., Mazurier, S., Lemanceau, P., Siblot, S., Berta, G., Mougel, C., and Tuinen, D.V. (2007). Medicago species affect the community composition of arbuscular mycorrhizal fungi associated with roots. New Phytol. *176*, 197–210.

Prosser, J.I. (2010). Replicate or lie. Environ. Microbiol. 12: 1806–1810.

Pyrozynski, K.A., and Malloch, D.W. (1975). The origin of land plants: a matter of mycotrophism. Biosystems *6*, 153–164.

Quince, C., Lanzén, A., Curtis, T.P., Davenport, R.J., Hall, N., Head, I.M., Read, L.F., Sloan, W.T. (2009). Accurate determination of microbial diversity from 454 pyrosequencing data. Nat. Methods *6*, 639–641.

Quinlan, A.R., Stewart, D.A., Strömberg, M.P., and Marth, G.T. (2008). Pyrobayes: an improved base caller for SNP discovery in pyrosequences. Nat. Methods *5*, 179–181.

Reddy, S., Pindi, P., Reddy, S. (2005). Molecular methods for research on arbuscular mycorrhizal fungi in India: problems and prospects. Curr. Sci. *89*, 1699–1709.

Redecker, D., Kodner, R., Graham, L.E. (2000). Glomalean Fungi from the Ordovician. Science *289*, 1920–1921.

Renker, C., Blanke, V., and Buscot, F. (2005). Diversity of arbuscular mycorrhizal fungi in grassland spontaneously developed on area polluted by a fertilizer plant. Environ. Pollut. *135*, 255–266.

Rosendahl, S. (2008). Communities, populations and individuals of arbuscular mycorrhizal fungi. New Phytol. *178*, 253–266.

Rosendahl, S., and Stukenbrock, E.H. (2004). Community structure of arbuscular mycorrhizal fungi in undisturbed vegetation revealed by analyses of LSU rDNA sequences. Mol. Ecol. *13*, 3179–3186.

Rousk, J., Baath, E., Bookes, P.C., Lauber, C.L., Lozupone, C., Caporaso, J.G., Knight, R., and Fierer, N. (2010). Soil bacterial and fungal communities across a pH gradient in an arable soil. ISME J. *10*, 1–12.

Saito, K., Suyama, Y., Sato, S., and Sugawara, K. (2004). Defoliation effects on the community structure of arbuscular mycorrhizal fungi based on 18S rDNA sequences. Mycorrhiza *14*, 363–373.

Salvioli, A., Lumini, E., Anca, I.A., Bianciotto, V., and Bonfante, P. (2008). Simultaneous detection and quantification of the unculturable microbe Candidatus Glomeribacter gigasporarum inside its fungal host *Gigaspora margarita*. New Phytol. *180*, 248–257.

Sanders, I.R., Alt, M., Groppe, K., Anken, T., and Wiemken, A. (1995). Identification of ribosomal DNA polymorphisms among and within spores of the Glomales: application to studies on the genetic diversity of arbuscular mycorrhizal fungal communities. New Phytol. *130*, 419–427.

Santos-González, J.C., Finlay, R.D., and Tehler, A. (2007). Seasonal dynamics of arbuscular mycorrhizal fungal communities in roots in a seminatural grassland. Appl. Environ. Microbiol. *73*, 5613–5623.

Scannerini, S., and Bonfante, P. (1991). Bacteria and bacterialike objects in endomycorrhizal fungi. In Symbiosis as a Source of Evolutionary Innovation: Speciation and Morphogenesis, Margulis, L., and Fester, R., eds. (Cambridge, MA: MIT Press), pp. 273–287.

Schechter, S.P., and Bruns, T.D. (2008). Serpentine and non-serpentine ecotypes of *Collinsia sparsiflora* associate with distinct arbuscular mycorrhizal fungal assemblages. Molec. Ecol. *17*, 3198–3210.

Schüßler, A., Schwarzott, D., and Walker, C. (2001). A new fungal phylum, the Glomeromycota: phylogeny and evolution. Mycol. Res. *105*, 1413–1421.

Simon, L., Bousquet, J., Levesque, R.C., and Lalonde, M. (1993a). Origin and diversification of endomycorrhizal fungi and coincidence with vascular land plants. Nature *363*, 67–69.

Simon, L., Lalonde, M., and Bruns, T. (1992). Specific amplification of 18S fungal ribosomal genes from vesicular-arbuscular endomycorrhizal fungi colonizing roots. Appl. Environ. Microbiol. *58*, 291–295.

Simon, L., Vesque, R., and Lalonde, M. (1993b). Identification of endomycorrhizal fungi colonizing roots by fluorescent single-strand conformation polymorphism-polymerase chain reaction. Appl. Environ. Microbiol. *59*, 4211–4215.

Smith, S.E., and Read, D.J. (2008). Mycorrhizal symbiosis. 3rd edn. (Academic Press).

Stockinger, H., Krüger, M., and Schüßler, A. (2010), DNA barcoding of arbuscular mycorrhizal fungi. New Phytol. *187*, 461–474.

Stockinger, H., Walker, C., and Schüßler, A. (2009). 'Glomus intraradices DAOM197198', a model fungus in arbuscular mycorrhiza research, is not Glomus intraradices. New Phytol. *183*, 1176–1187.

Stukenbrock, E., and Rosendahl, S. (2005). Distribution of dominant arbuscular mycorrhizal fungi among five plant species in undisturbed vegetation of a coastal grassland. Mycorrhiza *15*, 497–503.

Sun, Y., Li, X., and Feng, G. (2008). Effect of arbuscular mycorrhizal colonization on ecological functional traits of ephemerals in the Gurbantonggut desert. Symbiosis *46*, 121–127.

Sýkorová, Z., Wiemken, A., and Redecker, D. (2007). Co-occurring Gentiana verna and Gentiana acaulis and their neighboring plants in two swiss upper montane meadows harbor distinct arbuscular mycorrhizal fungal communities. App. Environ. Microb. 73, 5426–5434.

Taylor, T.N., Remy, W., Hass, H., and Kerp, H. (1995). Fossil Arbuscular Mycorrhizae from the Early Devonian. Mycologia *87*, 560–573.

Tedersoo, L., Nilsson, R.H., Abarenkov, K., Jairus, T., Sadam, A., Saar, I., Bahram, M., Bechem, E., Chuyong, G., and Kõljalg, U. (2010). 454 Pyrosequencing and Sanger sequencing of tropical mycorrhizal fungi provide similar results but reveal substantial methodological biases. New Phytol. *188*, 291–301.

Toljander, J.F., Santos-González, J.C., Tehler, A., and Finlay, R.D. (2008). Community analysis of arbuscular mycorrhizal fungi and bacteria in the maize

mycorrhizosphere in a long-term fertilization trial. FEMS Microbiol. Ecol. *65*, 323–338.
Vallino, M., Massa, N., Lumini, E., Bianciotto, V., Berta, G., and Bonfante, P. (2006). Assessment of arbuscular mycorrhizal fungal diversity in roots of Solidago gigantea growing in a polluted soil in Northern Italy. Environ. Microb. *8*, 971–983.
Vandenkoornhuyse, P., Baldauf, S.L., Leyval, C., Straczek, J., and Young, J.P.W. (2002). Evolution – extensive fungal diversity in plant roots. Science *295*, 2051.
Vandenkoornhuyse, P., Mahé, S., Ineson, P., Staddon, P., Ostle, N., Cliquet, J., Francez, A., Fitter, A.H., and Young, J.P.W. (2007). Active root-inhabiting microbes identified by rapid incorporation of plant-derived carbon into RNA. P. Nat. Acad. Sci. *104*, 16970–16975.
Vandenkoornhuyse, P., Ridgway, K.P., Watson, I.J., Fitter, A.H., and Young, J.P.W. (2003). Co-existing grass species have distinctive arbuscular mycorrhizal communities. Mol. Ecol. *12*, 3085–3095.
Walker, C., and Schüßler, A. (2004). Nomenclatural clarifications and new taxa in the Glomeromycota Pacispora. Mycol. Res. *108*, 981–982.
Walker, C., Vestberg, M., Demircik, F., Stockinger, H., Saito, M., Sawaki, H., Nishmura, I., and Schüßler, A. (2007). Molecular phylogeny and new taxa in the Archaeosporales (Glomeromycota): Ambispora fennica gen. sp. nov., Ambisporaceae fam. nov., and emendation of Archaeospora and Archaeosporaceae. Mycol. Res. *111*, 137–153.
Wallander, H., Johansson, U., Sterkenburg, E., Brandström Durling, M., and Lindahl, B.D. (2010). Production of ectomycorrhizal mycelium peaks during canopy closure in Norway spruce forests. New Phytol. *187*, 1124–1134.
Whitfield, L., Richards, A.J., and Rimmer, D.L. (2004). Relationships between soil heavy metal concentration and mycorrhizal colonisation in Thymus polytrichus in northern England. Mycorrhiza *14*, 55–62.
Wirsel, S.G.R. (2004). Homogenous stands of a wetland grass harbour diverse consortia of arbuscular mycorrhizal fungi. FEMS Microb. Ecol. *48*, 129–138.
Wu, B.Y., Hogetsu, T., Isobe, K., and Ishii, R. (2007). Community structure of arbuscular mycorrhizal fungi in a primary successional volcanic desert on the southeast slope of Mount Fuji. Mycorrhiza *17*, 495–506.
Wubet, T., Kottke, I., Teketay, D., and Oberwinkler, F. (2003). Mycorrhizal status of indigenous trees in dry Afromontane forests of Ethiopia. Forest Ecol. Manag. *179*, 387–399.
Wubet, T., Weiß, M., Kottke, I., and Oberwinkler, F. (2006). Two threatened coexisting indigenous conifer species in the dry Afromontane forests of Ethiopia are associated with distinct arbuscular mycorrhizal fungal communities. Can. J. Bot. *84*, 1617–1627.
Wubet, T., Weiß, M., Kottke, I., Teketay, D., Oberwinkler, F. (2004). Molecular diversity of arbuscular mycorrhizal fungi in Prunus africana, an endangered medicinal tree species in dry Afromontane forests of Ethiopia. New Phytol. *161*, 517–528.
Yamato, M., and Iwase, K. (2005). Community analysis of arbuscular mycorrhizal fungi in a warm-temperate deciduous broad-leaved forest and introduction to the fungal community into seedlings of indigenous woody plants. Mycoscience *46*, 334–342.

The Human Microbiome: Exploring and Manipulating our Microbial Selves

11

Corinne F. Maurice and Peter J. Turnbaugh

Abstract

The human body is home to roughly 10 times more microbial cells than human cells, containing a vast array of genes referred to in aggregate as the human microbiome. Metagenomics has recently enabled an initial map of the microbial diversity found in different body habitats, individuals, and populations. These developments include an extensive catalogue of genes and genomes, characterization of the human gut virome, a description of the patterns of succession of the gut microbiota during development, and links between obesity and the gut microbiome. The application of principles from macro-ecology, in addition to studies of defined communities in model organisms, have begun to reveal the basic operating principles that govern community assembly, stability, and function. These studies are beginning to move beyond the simple characterization of the organisms and genes found in a given habitat at a single timepoint, to a systems biology approach that allows the characterization of this complex microbial community at a variety of spatial and temporal scales. In the foreseeable future, studies of the human microbiome promise to reveal new biomarkers for disease, novel strategies for manipulation, and a more comprehensive view of human physiology.

Introduction

The human body is constituted of ~10^{13} cells, and although this number might seem astonishingly high, it only represents about 10% of the total cells within or on the surface of the human body (Savage, 1977). Indeed, our body is home to an extraordinary diversity of microbial cells from all three domains of life: bacteria, archaea, and eukarya, as well as viruses (together referred to as the human microbiota). Although the majority of these microbes are found in the gastrointestinal tract, they are also widely distributed along the skin surface, oral and nasal cavities, and urogenital tracts, among other body habitats.

A prescient manuscript written more than a century ago highlighted the fact that most of the bacteria associated with the human body do not cause disease, making it critical to distinguish the commensal (see Box 11.1 for a definition of terms) from the pathogenic bacteria in the gastrointestinal tract (Kendall, 1909).

Arthur Kendall also argued that research should focus on both the ecology and function of this microbiota, and that diet could play a major role in shaping its structure. These initial observations led to a variety of questions; for example, which microbes comprise the human microbiota? How do these communities vary between different body habitats, over time, and between individuals? What factors affect the establishment and maintenance of these communities? What are the principal microbial metabolic pathways? Which microbes are involved in maintaining health or contributing to disease? Is there a core group of abundant bacterial lineages found across all humans? These questions, although central to human biology, would remain largely unanswered until the recent development of 'metagenomic' methods to study the genomes of these microbes, their gene expression, and their metabolic activities.

Box 11.1 Definition of important terms

Bacteriotherapy. The treatment of disease through the use of bacteria or their products. This has recently been used to describe the transplantation of a complex microbial consortia from a human donor to a patient (Khoruts *et al.*, 2010).

Commensalism. A relationship between two different organisms (e.g. humans and a member of their microbiota) where one member derives a benefit while the other is unaffected.

'Core' human gut microbiome. A set of features shared across all or the vast majority of gut microbiomes (e.g. genes and/or metabolic capabilities).

Functional metagenomics. A term commonly used to refer to the analysis of gene expression, enzymatic activity, or other functional properties of a metagenomic library constructed in a host organism, e.g. *E. coli*.

Metagenomics. A rapidly evolving field emerging from advances in DNA sequencing methods, with a focus on the use of culture-independent methods to study the structures, functions and dynamic operations of microbial communities. Frequently used to refer to the technique of isolating and sequencing community DNA directly from an environment, but now extends to the profiling of gene expression at the level of RNA (metatranscriptomics), protein (metaproteomics), and community metabolism (metabolomics).

Microbiome. The aggregate genomes found in the members of a microbiota.

Microbiota. A microbial community; commonly referred to according to the habitat that it occupies (e.g. the gut microbiota).

Mutualism. A relationship between two different organisms, where both members benefit from the interaction.

Parasitism. A relationship between two different organisms where one member benefits and the other is adversely affected.

Phylotype (or operational taxonomic unit, OTU). A group of microbes, commonly defined by the level of sequence similarity (percentage identity) between small subunit (16S) rRNA genes (e.g. ≥ 97% for a 'species'-level phylotype).

Prebiotics. Dietary supplements, frequently indigestible by host enzymes, which are used with the intent of modifying the composition and activity of the human microbiome.

Probiotics. Live bacteria, administered through the diet or topically, which are frequently used with the intent of modifying human health either directly or indirectly by affecting other members of the human microbiota.

Symbiosis. More general term describing a relationship between two or more organisms, that may or may not benefit each member. Comprises commensal, mutualistic and parasitic relationships.

Broadly defined, metagenomics encompasses the culture-independent analysis of the structures and functions of microbial communities and their interactions with the habitats they occupy. Together, these studies incorporate a wide range of techniques, from (i) shotgun sequencing of microbial DNA isolated directly from a given environment (sequence-based metagenomics), (ii) screening of expression libraries, constructed from cloned community DNA, to identify specific functions such as antibiotic resistance (functional metagenomics), (iii) profiling of RNAs and proteins produced by a microbiome (metatranscriptomics and metaproteomics) and (iv) characterization of a community's metabolic network (metabolomics) (Turnbaugh and Gordon, 2008).

Recently, a series of international Human Microbiome Projects (HMPs) have been launched to (i) characterize the organismal and genetic diversity across multiple human body habitats, (ii) generate a set of reference genomes from cultured isolates and (iii) determine what features of the microbiome are shared between individuals and which are associated with disease (Turnbaugh *et al.*, 2007). In addition to these large-scale projects, a wide range of laboratories have begun to couple these metagenomic methods to hypothesis-driven experimental microbiology, revealing many novel findings and emphasizing the role for the human microbiota in health and disease. In this chapter, we will highlight some of the notable recent advancements in our understanding of the human microbiome and discuss some promising new directions that lay ahead in the years to come.

16S rDNA: the barcode of life

The first major breakthrough permitting the culture-independent analysis of complex microbial communities started with the work of Norm Pace and colleagues, when they began to analyse 16S 'rDNA' sequences directly amplified from community DNA isolated from environmental samples (Lane *et al.*, 1985). This method takes advantage of the ubiquitous nature of the rDNA, which encodes for the ribosomal RNA, essential for translation in cells from all three domains of life. Additionally, the 16S rDNA contains highly conserved regions that can be used to amplify the gene with 'universal' primers, and variable regions that can be used to distinguish one group of microbes from another (Pace, 1997; Woese, 2000). One of the initial findings with this work was referred to as 'the great plate count anomaly', wherein only a small fraction of the bacterial species observed in each habitat had been successfully cultured (Staley and Konopka, 1985; Grimes *et al.*, 1986; Torsvik *et al.*, 1990). This finding underscored the biased view of the microbial world found in culture-based studies, and suggested that the full extent of microbial diversity was largely underexplored.

Prior to this, a series of experimental tools were developed to analyse the human microbiota, including anaerobic cultivation methods allowing the culture of extremely oxygen-sensitive bacteria (Hungate *et al.*, 1966), the use of animal models to define relationships between microbes and their animal hosts (Dubos *et al.*, 1966), and gnotobiology, the study of germ-free animals or those colonized with a defined microbiota (Gordon and Pesti, 1971). The application of anaerobic culturing techniques, despite their biases, revealed for the first time that obligate anaerobes outnumbered facultative anaerobes by three orders of magnitude, depriving *Escherichia coli* of its previous position as the most abundant member of the gut microbiota (Bryant, 1972). Research with animal models illustrated that specific bacterial morphotypes were associated with different body habitats (Savage *et al.*, 1968), that the bacterial colonization of the digestive tract of a newborn followed a reproducible succession influenced by the changes in diet and the environment (Schaedler *et al.*, 1965), and that encounters with bacteria early in life could have long-term effects on the animal (Dubos *et al.*, 1966). By the end of the 1980s, there was a growing body of knowledge regarding the microbiota of animals, including a number of key observations (Table 11.1).

These findings were an initial hint that many features of the biochemistry, physiology and immunology of animals could be influenced by their associated microbiotas. However, these studies, and others from diverse environments, raised several discrepancies. For example, cultures still did not properly transcribe the

Table 11.1 Important observations regarding the mammalian gut microbiota

Observations	References
The gastrointestinal tract of vertebrate animals harbours a complex microbial community, dominated by obligate anaerobes	Savage (1977), Ley *et al.* (2008)
The colonization of the gastrointestinal tract follows an ecological succession, where qualitative and quantitative changes following birth occur. The animal's diet can influence this succession	Savage (1977), Koenig *et al.* (2010)
The microbiota can influence the physiology of the gastrointestinal tract. Comparisons made between germ-free and conventionally raised animals colonized with a complex microbial community demonstrate differences in the renewal of gut epithelial cells, small bowel motility, and immunological status	Gordon and Pesti (1971), Pull *et al.* (2005)
The microbiota contributes to host nutrition, and modifies dietary and host-derived substances, including therapeutic drugs	Drasar and Hill (1974), Cummings (1983), Backhed *et al.* (2005), Sonnenburg *et al.* (2005), Flint *et al.* (2008), Sousa *et al.* (2008)
Environmental influences, for example changes in nutrition, can affect the activity and composition of the microbiota	Holdeman *et al.* (1976), Tannock (1983), Turnbaugh *et al.* (2009b), Walker *et al.* (2010)
The microbiota can stimulate immune responses	Tannock (1984), Peterson *et al.* (2007), Mazmanian (2009)
Perturbations in the interactions between the microbiota and the host can result in disease	Drasar and Hill (1974), Finegold (1977), Turnbaugh *et al.* (2007), Peterson *et al.* (2008), Khoruts *et al.* (2010), Peterson and Turnbaugh (2010), Vijay-Kumar *et al.* (2010)

bacterial abundance observed by microscopy (Ward *et al.*, 1990; Stein *et al.*, 1996; Apajalahti *et al.*, 2003), few selective culture media were really selective (Silvi *et al.*, 1996; Tannock, 1999), and even related species grow differently in culture (Baya *et al.*, 1986; Apajalahti *et al.*, 2003). Furthermore, culture-based techniques do not provide insight on the phylogenetic relatedness between the detected bacterial species from the same or different individuals (O'Sullivan, 2000). Collectively, it appeared that other techniques were needed to properly describe the microbial community diversity and functions within natural samples, including the human body. This led to the development of 'culture-independent' techniques, based on molecular detection methods largely targeted against the 16S rDNA.

Culture-independent molecular methods are now frequently based on sequencing the 16S rDNA, after amplification by polymerase chain reaction (PCR), but can also comprise fingerprinting methods, fluorescent hybridizations, and DNA–DNA re-association techniques, among others. These methods have been applied to a wide range of ecosystems, including the human body, and have provided many important results. Sequencing the 16S rDNA confirmed that a significant fraction of the human microbiota had not been described previously (Wilson and Blitchington, 1996; Kroes *et al.*, 1999; Suau *et al.*, 1999). Fluorescence *in situ* hybridization (FISH) and quantitative PCR (qPCR) have allowed the quantification of specific bacterial groups in human samples faeces (Langendijk *et al.*, 1995; Harmsen *et al.*, 2000; Matsuki *et al.*, 2003). A wide range of fingerprinting methods, i.e. denaturing gradient gel electrophoresis (DGGE), temperature gradient gel electrophoresis (TGGE), single strand conformation polymorphism (SSCP), and terminal-restriction fragment length polymorphism (T-RFLP), have enabled the analysis of a large number of samples for relatively low cost, but provide limited information on the phylogenetic composition of the human microbiota (Table 11.2).

Table 11.2 Methods used for studying the human microbiome

Method	Benefits	Limitations
Culturing	• Allows *in vitro* characterization and genetics • Enables whole genome sequencing • Low cost	• Biased relative abundance • Many microbes cannot be recovered • Strains may change during process of culturing • Low throughput unless automated
Fingerprinting (DGGE, T-RFLP)	• Low cost • High throughput • Enables description of community ribotype diversity	• Low resolution • Does not allow direct analysis of phylogenetics or taxonomy
Microarray, Phylochip	• Moderate cost • Moderate throughput • Enables comparative genomic hybridization, gene expression profiling, identification of single nucleotide polymorphism	• Limited to known strains • Does not allow direct analysis of phylogenetics or taxonomy
Quantitative-PCR	• Moderate cost • Moderate throughput • Quantitative • Enables the detection and quantification of one or more specific sequences within more complex samples	• Limited to known strains or to broader taxonomic groups
FISH (fluorescence *in situ* hybridization)	• Moderate cost • Quantitative • Allows analysis of spatial relationships • Can be used with sorting to isolate single cells • Enables description of community diversity	• Low throughput • Limited to known strains or to broader taxonomic groups
16S DNA sequencing	• Low–high cost depending on extent of pooling • High throughput • Enables discovery of new species • Enables phylogenetics • Can assess relative abundance of bacterial taxa • Can assess species-level diversity • Many software packages available for analysis	• Complex communities still cannot be sequenced completely • Sequencing and PCR errors can inflate diversity
Metagenomic shotgun sequencing	• Moderate throughput • Can construct small pools (<20 samples) • Allows *de novo* assembly • Allows gene-centric analysis • Enables discovery of new gene families • Population genomics	• High cost • Assemblies can be challenging • Reference genomes are often required • Annotations are often based on distant homology • No ability to retrieve a given sequence using next-generation techniques (no plasmid library) • Computationally challenging • Only allows measurement of 'metabolic potential' not activity

Table 11.2 continued

Method	Benefits	Limitations
Metatranscriptomics	• Enables measurement of gene expression • Enables improvement of gene/operon predictions • Can construct small pools for low-depth screening (<20 samples)	• High cost • Low throughput • rRNA/tRNA must be removed • Reference genomes are important • Still often limited by sequence-based annotations
Metabolomics	• Moderate cost • High throughput • Most direct measurement of metabolic activity	• Difficult to quantify all metabolites • Limited set of known metabolites • Difficult to link metabolites to the source
Functional metagenomics	• Low–high throughput • Low–moderate cost • Enables direct measurement of gene activity • Enables novel annotations of hypothetical genes • Allows *in vitro* studies of activity in host strain	• Limited by available host strains • Appropriate screens can be difficult to set up • Cannot detect activities requiring genes in distant genomic 'neighbourhoods'

Microbial diversity within and between human body habitats

Culture-independent methods utilizing the 16S rDNA have revolutionized our understanding of microbial diversity. One of the major goals of the human microbiome project was to determine whether there is an identifiable 'core microbiome' of abundant shared organisms, genes, or functional capabilities found in a given body habitat of all or the vast majority of humans (Turnbaugh *et al.*, 2007). When exploring this diversity, two different levels of analysis are possible: (i) α-diversity, within a given body habitat from one individual and (ii) β-diversity, between body habitats or different individuals (Table 11.3).

In one of the most comprehensive studies to date, 16S rDNA amplicons were sequenced from 9 individuals across 27 different body habitats, including the gut (faeces), inner ear, nasal cavity, mouth, and multiple locations on the skin (Costello *et al.*, 2009). These analyses revealed that bacterial community composition is largely determined by body habitat, despite high levels of inter-individual variation for each location. Furthermore, microbes can be transferred between some habitats; bacteria from the tongue could be deposited on a disinfected forearm but not the forehead for up to 8 hours. One of the most surprising results was that phylum-level bacterial diversity in the human microbiota is relatively low. Indeed, of the 70 known bacterial phyla (Backhed *et al.*, 2005; Ley *et al.*, 2006a), generally only four are found at high relative abundance in the human microbiota, regardless of the specific body habitat: Actinobacteria, Bacteroidetes, Firmicutes and Proteobacteria (Costello *et al.*, 2009). This is in stark contrast with other environments, such as microbial mats, where 57 bacterial phyla have been described from a single location (Ley *et al.*, 2006a). Despite these low levels of phylum-level diversity, humans are home to a many different bacterial taxa at finer levels (genus, species, or strain), such that each individual harbours a relatively unique collection of bacterial strains and species (Costello *et al.*, 2009).

To date, most large-scale studies of the human microbiota have focused on the gastrointestinal tract. However, in the following sections, we aim to provide an overview of the microbial diversity across multiple body habitats: the gut, the skin, the oral cavity, and the vagina.

Table 11.3 Bacterial diversity of the human microbiota as depicted by 16S DNA sequencing

Body habitat	Bacterial phyla detected	References
Oral cavity	Firmicutes Actinobacteria Proteobacteria Bacteroidetes Fusobacteria TM7 phylum	Aas *et al.* (2005), Costello *et al.* (2009), Kroes *et al.* (1999), Nasidze *et al.* (2009), Zaura *et al.* (2009)
Stomach	Proteobacteria Firmicutes Actinobacteria Bacteroidetes Fusobacteria TM7 phylum Deferribacteres Deinococcus/Thermus	Bik *et al.* (2006)
Distal gut	Firmicutes Bacteroidetes Actinobacteria Proteobacteria Fusobacteria Verrucomicrobia Cyanobacteria Deferribacteres Lentisphaera TM7 phylum	Costello *et al.* (2009), Frank *et al.* (2007), Ley *et al.* (2006c), Turnbaugh *et al.* (2009a, 2010)
Skin	Actinobacteria Proteobacteria Firmicutes Bacteroidetes Cyanobacteria Acidobacteria	Costello *et al.* (2009), Dekio *et al.* (2005), Fierer *et al.* (2008), Grice *et al.* (2008, 2009)
Vagina	Firmicutes Actinobacteria Bacteroidetes	Zhou *et al.* (2004, 2007, 2009b), Ravel *et al.* (2010)

The gastrointestinal microbiota

The human gut harbours the largest collection of microbes in any of our body habitats, and is commonly subdivided into four regions: the oesophagus, the stomach, the small intestine, and the large intestine, composed of the caecum and the colon. Many of these different habitats have been sampled, using biopsies or luminal contents, and a rough outline of microbial diversity in the human gut is now emerging from these analyses. The first results from these molecular analyses indicated that the same bacterial phyla tend to predominate in the stomach, small intestine, caecum, and large intestine (Frank and Pace, 2008). Thus, more than 98% of all species detected belong to four phyla, and the combined results from three large-scale independent studies report that on average Firmicutes (64%) and Bacteroidetes (23%) lead the way in terms of abundance in front of the Proteobacteria (8%) and Actinobacteria (3%) in the human gastrointestinal tract (Eckburg *et al.*, 2005; Ley *et al.*, 2006c; Frank *et al.*, 2007).

The upper digestive tract, and most notably the oesophagus, has been less frequently sampled, as culture-based studies on samples obtained from washes of the luminal oesophageal contents identified low numbers of bacteria (Pajecki *et al.*, 2002). Pei and colleagues (Pei *et al.*, 2004), used 16S rDNA sequencing to examine oesophageal biopsies from four healthy individuals, and revealed a more complex microbiota than previously thought. Members of Firmicutes, Bacterioides, Actinobacteria, Proteobacteria, Fusobacteria and TM7 (an uncultured bacterial phylum) were predominantly detected in all four patients, and 41 genera were detected within these six phyla. Surprisingly, 82.1% of the identified bacterial species were known and previously cultured (Pei *et al.*, 2004), contrasting with results from other parts of the gastrointestinal tract.

The human stomach may also harbour an unexpected level of bacterial diversity (Monstein *et al.*, 2000; Bik *et al.*, 2006). A study of mucosal gastric biopsies from 23 individuals determined that members of the phyla Proteobacteria (with *Helicobacter pylori* as the most abundant representative), Firmicutes, and Bacteroidetes dominated the stomach microbiota, while representatives of the phyla Actinobacteria, Fusobacteria, TM7, Deferribacteres, and Deinococcus/Thermus were also detected at low relative abundance (Bik *et al.*, 2006). Within these eight phyla, 128 different bacterial phylotypes were detected. Furthermore, 50% of these phylotypes were derived from uncultured bacteria, whereas 10% had not been previously identified (Bik *et al.*, 2006).

In the distal gut (caecum, colon, and faecal samples) the predominant bacterial phyla are typically the Firmicutes and Bacteroidetes, in addition to one Archaeal species, *Methanobrevibacter smithii* (Backhed *et al.*, 2005; Eckburg *et al.*, 2005; Ley *et al.*, 2006c; Frank *et al.*, 2007; Turnbaugh *et al.*, 2009a). The remaining phyla, including Proteobacteria, Actinobacteria, Fusobacteria, TM7 and Verrucomicrobia, are detected at much lower frequencies. Despite the restricted number of dominant bacterial phyla found in the distal gut, each individual harbours a remarkable number of species. A recent deep dive into the faecal microbiotas of an adult monozygotic female twin pair revealed an estimated 800–900 bacterial species in each co-twin, less than half of which were shared by both individuals (Turnbaugh *et al.*, 2010). Expanding this analysis to include a shallower sampling of 281 faecal samples obtained from 54 mono- and dizygotic twin pairs and their mothers revealed > 4000 species-level bacterial phylogenetic types (phylotypes). However, of the 134 bacterial species whose relative abundance was > 0.1% in at least one faecal community, only 37 were detected in more than half of the analysed samples (Turnbaugh *et al.*, 2010).

The skin surface microbiota

The human body is covered by keratinized skin, making it one of the largest human-associated microbial habitats. The skin is composed of a wide variety of body habitats (regions with broad ranges of pH, temperature, moisture and sebum content) and cell structures (hair follicles, sebaceous, eccrine and apocrine glands). Additionally, human skin is frequently submitted to a variety of environmental exposures, including skin shedding, UV exposure, soaps, and detergents.

Several studies have explored the diversity of the skin surface microbiota, by swabbing the skin of healthy adults in multiple locations, such as the forehead (Dekio *et al.*, 2005), the forearm (Gao *et al.*, 2007), the inner elbow (Grice *et al.*, 2008), and the palm of the hands (Fierer *et al.*, 2008). Together, these studies reveal the common presence of the same major bacterial phyla, notably the Actinobacteria, Proteobacteria, Firmicutes and Bacteroidetes, but their relative proportions and their composite species sharply differ according to the location sampled. For example, the *Propionibacteria* genus, member of the Actinobacteria phylum, dominates the bacterial communities of the forehead, hands and forearm, whereas the Proteobacteria phylum dominates the inner elbow with the genera *Pseudomonas* and *Janthinobacterium* (Grice *et al.*, 2008). Furthermore, skin bacterial communities are not uniformly distributed across the body, and it seems that skin surfaces closer in proximity may harbour more similar bacterial communities. For example, the bacterial diversity found on the hand surface appears to be more similar to the one from the forearm than the forehead (Fierer *et al.*, 2008).

Collectively, these studies indicate the existence of a common shared skin microbiome at the phylum level, with substantial differences among individuals and locations at the genus and the species levels (Grice *et al.*, 2009). The fact that only 13% of the bacterial phylotypes on the palm surface are shared between any two individuals, or that women harbour bacterial communities significantly more diverse than men, clearly illustrates this heterogeneity (Fierer *et al.*, 2008). Furthermore, even the same body habitat on different sides of the body can have discrepant microbial communities, as only 17% of the phylotypes were shared between both hands of the same individual. Skin bacterial communities are at least partially resilient to perturbations; the microbes found on an individual's palm are altered by washing, but rapidly recover (Fierer *et al.*, 2008). Finally, interpersonal variation in community composition generally exceeds temporal variation within people (Costello *et al.*, 2009; Grice *et al.*, 2009).

The high level of inter-individual variation in the skin microbiota may suggest that each individual has a unique and specific microbial signature that could be harnessed for a novel type of forensic analysis. Comparisons of bacteria associated with each individual's fingers and their computer keyboard and mouse revealed similarities that might be used to distinguish the owner of each device (Fierer *et al.*, 2010). These initial findings prompt a variety of future directions for research addressed at determining the relative proportions of the different bacterial species, their dynamics and their relationships with physiological and physical states (sex, handedness) and external environmental parameters (washing, use of cosmetics), and establishing a healthy baseline from which to detect and understand microbial community differences associated with a wide variety of human skin diseases.

The oral cavity microbiota

A wide variety of habitats are found in the oral cavity: mucosal sites, solid surfaces, and crevices, all in contact with saliva. Based on 16S rDNA analysis, six major bacterial phyla are found in the oral cavity: the Firmicutes, Actinobacteria, Proteobacteria, Bacteroidetes, Fusobacteria, and TM7, comprising 101 known genera and 64 novel genera (Aas *et al.*, 2005; Nasidze *et al.*, 2009; Zaura *et al.*, 2009). These studies indicate that only a few phylotypes can dominate the oral microbiome and that most sequences recovered are rare (Kroes *et al.*, 1999; Nasidze *et al.*, 2009; Zaura *et al.*, 2009). In addition, more than half of the described bacterial phylotypes have not yet been cultured (Kroes *et al.*, 1999; Aas *et al.*, 2005). Previous studies revealed that oral bacteria preferentially colonize different surfaces of the oral cavity, as a result of the complementary binding between bacterial adhesins and oral surface receptors (Gibbons *et al.*, 1976; Gibbons, 1989). This was confirmed by recent sequencing-based analyses, where significant differences in bacterial profiles according to the sampling location (e.g. tongue dorsum, lateral tongue surface, saliva and tonsils) were reported (Aas *et al.*, 2005). As the oral cavity represents a common entry point of bacteria into the body, there may be important interactions between this microbiota and other communities in the human body, in particular that of the intestinal tract.

The vaginal microbiota

The vaginal microbiota is very dynamic, and is influenced by age, menarche, time in the menstrual cycle, pregnancy, infections, methods of birth control, sexual behaviours, as well as various habits and practices (Schwebke *et al.*, 1999; Eschenbach *et al.*, 2000; Burton and Reid, 2002; Clarke *et al.*, 2002; Ness *et al.*, 2002). The vaginal community is primarily composed of members of the Firmicutes phylum, with a strong predominance of *Lactobacillus* species, along with some members of the Actinobacteria and Bacteroidetes phyla (Verhelst *et al.*, 2004; Zhou *et al.*, 2004). Unlike many other body habitats or locations, the vaginal microbiota is primarily constituted of *Lactobacillus*, namely *L. crispatus*, *L. gasseri*, *L. iners* and *L. jensenii* (Verhelst *et al.*, 2004; Zhou *et al.*, 2004; Ravel *et al.*, 2010). There have also been reports of other types of lactic acid-producing bacteria, mainly reflecting the fact that the production of lactic acid is a key ecological function of the vaginal microbiota (Ravel *et al.*, 2010).

The species diversity within the *Lactobacillus* genus varies significantly among individuals, and indicates that there is no single core microbiome for the human vagina (Zhou *et al.*, 2004; Ravel *et al.*, 2010). Rather, it seems that this microbiota carries out core functions conserved among communities, despite differences or changes in bacterial species composition (Ravel *et al.*, 2010). Interestingly, a large-scale study on North American women equally representing four ethnic groups (i.e. white, black, Asian, and Hispanic) revealed significant differences in species composition of the vaginal microbiota, with *Lactobacillus* dominating this microbiota in more than 80% of Asian and white women, and only reached high abundance in 60% of black and Hispanic women

(Ravel *et al.*, 2010). This was also reported by Zhou and colleagues, who studied the vaginal microbiota of white, black, and Japanese women (Zhou *et al.*, 2007, 2009a). One important factor that correlates with the composition of the vaginal microbiota is variation in vaginal pH: the sampled black and Hispanic women had higher pH values than Asian and white women (Ravel *et al.*, 2010).

While the molecular mechanisms underlying these inter-individual differences are unknown, possibilities include host factors (such as differences in immune systems, the composition of vaginal secretions, and ligands on epithelial surfaces), or human habits and practices (such as personal hygiene, methods of birth control, and sexual behaviour) (Schwebke, 2009). Studies have also associated the vaginal microbiota with the occurrence of infections or disorders, such as candidiasis or vaginosis; *Gardnerella vaginalis*, *Peptostreptococcus anaerobius*, and *Prevotella* spp. have all been linked to disease (Verhelst *et al.*, 2004; Zhou *et al.*, 2009b). Collectively, it appears that more research is necessary to better assess the complex interactions between the members of the vaginal microbiota, and the impact of external factors on its composition and capacity of resilience.

The human microbiome project: generating a catalogue of genes and genomes from multiple human body habitats

The initial results from two large-scale HMPs have begun to describe an extensive catalogue of references genes and genomes from the human microbiome (Fig. 11.1) (Qin *et al.*, 2010; Nelson *et al.*, 2010).

Faecal samples were collected from 124 European individuals, community DNA was extracted, and Illumina-based metagenomic sequencing was used to generate a massive dataset comprising 576.7 Gbp of sequences representing 3.3 million non-redundant genes. Despite inter-individual variation in the relative abundance of each bacterial species, a large set of 'core' genes was shared between all individuals: when considering characterized gene families, hypothetical genes found in reference genomes, or novel genes unique to the sampled microbiomes (Qin *et al.*, 2010). These observations are consistent with previous studies of 18 monozygotic twins and their mothers, based on 100 Mbp of sequence per individual (Turnbaugh *et al.*, 2009a). Together, these results indicate that although there may not be a core

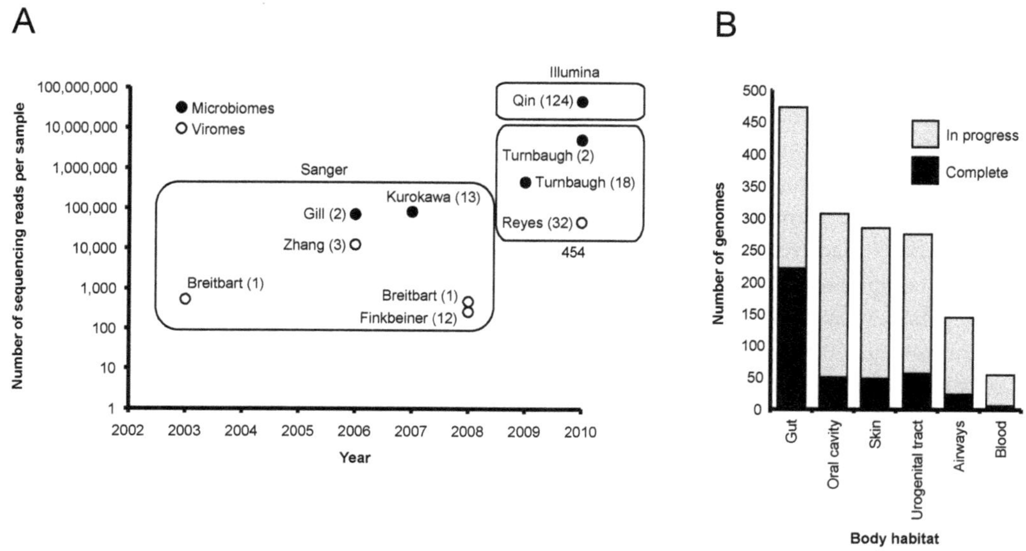

Figure 11.1 A representative sampling of metagenomic studies of the human microbiome performed from 2002 to 2010. (A) Number of sequencing reads generated per sample by a selection of gut microbiomes (black) and viromes (white). The primary author is shown for each manuscript (see references), and the number of samples is shown in parentheses. Note: some studies include multiple samples from the same individual. Studies are separated by sequencing methodology: Sanger, 454, or Illumina. (B) The number of reference genomes complete (black) and in progress (grey) from a variety of body habitats. Source as of 9/1/2010: http://www.hmpdacc-resources.org.

microbiome with respect to a shared set of abundant bacterial species, there is an identifiable set of core genes, representing key metabolic functions, some of which may be essential for life in the distal gut. Similar analyses of the microbiomes found in other body habitats will provide a new perspective on the gene sets common to bacteria in general, relative to the gene sets unique to a given habitat. Additionally, these studies suggest the presence of bacterial 'guilds' or collections of species with at least partially redundant functional capabilities that together fulfil a given niche or profession in the human gut microbiome (Tschöp *et al.*, 2009).

One of the major challenges to metagenomic sequencing of any environment, is the availability of a suitable collection of reference genomes with which to anchor the largely unassembled metagenomic sequencing reads and genome fragments. However, given the thousands of strains that have been isolated from the human gut and other body habitats, it becomes challenging to determine the best method of prioritizing genomes for sequencing. One option is to use phylogenetics to guide the selection of largely uncharacterized bacterial species for sequencing, by first building a phylogenetic tree relating all known cultured isolates and uncultured strains based on a shared marker gene, e.g. the 16S rDNA. This approach was recently used in the first phase of the 'Genomic Encyclopedia of Bacteria and Archaea' (GEBA) project (Wu *et al.*, 2009). A collection of 56 cultured species of Bacteria and Archaea were selected based on maximizing phylogenetic distance, and the resulting analysis emphasized the benefits of such an approach, including aiding in the reconstruction of phylogenetic history, identifying novel protein families, and predicting gene function in other organisms (Wu *et al.*, 2009).

An analogous approach was used to select an initial set of 100 genomes from the human gut (Gordon, 2005), and formed the initial set of isolates that seeded the National Institute of Health HMP Jumpstart Initiative (Fig. 11.1B) (Nelson *et al.*, 2010). In total, 178 genomes were sequenced, containing 547,968 predicted proteins. Novel predicted proteins were identified by comparing this gene set to the NCBI non-redundant protein database, resulting in a set of 30,867 proteins, 29,286 of which were unique at 98% sequence identity (28,469 at 90% identity). This level of novelty was roughly 2-fold higher than that of an equivalent number of randomly selected genomes from the reference database.

These new reference genomes were valuable tools in re-analysing 20 published human gut microbiomes (Gill *et al.*, 2006; Turnbaugh *et al.*, 2009a). Sixty-two different genomes had a significant level of recruitment, comprising 11.3 million of the 16.8 million reads tested. Of these, 41% of the mapped reads showed closest homology to one of the HMP genomes; 20–40% of the reads would not have been recruited without these new genomes (Nelson *et al.*, 2010). Furthermore, these efforts allow a clear link between each bacterial strain and its gene content, a method that can be complicated from metagenomic datasets obtained from complex microbial communities.

The human virome

One relatively unexplored component of the microbiome is the diverse collection of host-associated viruses, the human virome. Work in other habitats has emphasized the importance of bacterial–viral interactions, which often occur in the context of Lotka–Volterra (or 'kill the winner') dynamics. This model describes bacteriophages as key players responsible for maintaining a high level of bacterial diversity within natural environments (Thingstad and Lignell, 1997; Winter *et al.*, 2010). Viral infection is host density dependent (Murray and Jackson, 1992), and is thought to be species specific (Ackerman and Dubow, 1987). Consequently, when the abundance of a given group of bacteria increases, so does the contact rate between them and their phages, leading to more successful infections. These numerically dominant bacteria will then be lysed by their phages, releasing their cell content in the system upon cell lysis. This will allow for other groups of bacteria to increase and occupy this vacant niche, benefiting either from the release of competition, or from the supply of metabolites from the dead cells. This will lead once again to increased rates of infection of the most abundant bacteria, and so on. This mathematical model has been validated both by *in situ* and experimental evidence, but several studies have demonstrated that there might be more complex interactions occurring within this general theoretical framework (Schwalbach *et al.*, 2004; Bouvier and del Giorgio, 2007).

To date, there has been a much more limited set of studies analysing the gut viromes of healthy or diseased humans, than that applied to study the microbial members of the gut microbiome. The initial metagenomic analyses of the human gut virome focused on a small number of individuals, without a concomitant analysis of the microbial community (Breitbart *et al.*, 2003; Zhang *et al.*, 2006; Breitbart *et al.*, 2008). These studies revealed a diverse assemblage of DNA viruses, mostly siphophages, many of which were uncultured, with an estimated total of 1200 viral genotypes in a single faecal sample (Breitbart *et al.*, 2003). Furthermore, a time series analysis of a single infant, revealed a viral community with low diversity, dominated by phages that did not originate from breast milk or formula. Although some viruses were consistently found one and two weeks after birth, an analysis utilizing a custom microarray suggested that there was significant temporal variability between these two time-points (Breitbart *et al.*, 2008).

In addition to DNA viruses, the human gut also contains a suite of RNA viruses, many of which may have originated from the diet (Zhang *et al.*, 2006). Three faecal samples from two healthy individuals were analysed by concentrating the viral fraction by tangential flow filtration, extracting viral RNA, and sequencing viral cDNA (complementary DNA) libraries. Of the 36,769 viral sequences obtained, most showed sequence similarity to plant pathogenic viruses, the most abundant of which was the pepper mild mottle virus (PMMV). This virus was found at high concentration, as high as 10^9 virions per gram of dry weight faecal matter, in 12 of 18 faecal samples collected from healthy individuals across two continents, and was detected in pepper-containing foods (e.g. bottled hot chilli sauce). Furthermore, these faecal viruses were still able to infect host plants; Hungarian wax pepper (*Capsicum anuum*) plants exposed to a small amount of faecal suspension developed typical symptoms of PMMV infection after 1 week. Together, these results suggest that the human gut can serve as a reservoir for plant pathogens, and that the diet may be one source of RNA viruses (Zhang *et al.*, 2006).

Metagenomics can also be used as a tool for viral pathogen discovery, by screening a collection of diseased subjects that test negative for known pathogens. This approach was recently used to analyse RNA viruses isolated from the faecal samples of 12 children suffering from diarrhoea (Finkbeiner *et al.*, 2008). Unlike previous work in healthy adults, this analysis resulted in the identification of multiple known enteric viruses, in addition to novel viruses with limited sequence similarity to those in the NCBI database.

The most comprehensive survey to date of the human virome involved the analysis of faecal samples obtained from healthy adult female monozygotic twins and their mothers at three different time-points over a year (Reyes *et al.*, 2010). DNA viral profiles were also compared to bacterial community structure via 16S rDNA sequencing and bacterial gene content obtained from shotgun sequencing of total community DNA. The identified viruses were predominantly prophages or phages classified as temperate, and 81% of the reads did not match any previously characterized virus. As expected, the predicted hosts of these viruses matched the predominant bacterial phyla in the human gut: Firmicutes and Bacteroidetes. Previous studies have demonstrated that family members share an increased level of similarity in the structure of their gut bacterial communities, and that the bacterial taxa can show large fluctuations in relative abundance over time (Turnbaugh *et al.*, 2009a, 2010). As with the bacterial members of the community, inter-individual variation in the specific viruses found is high. However, the faecal virome deviates from the bacterial community in a number of important respects: viromes are surprisingly stable over time and family members do not share a detectable similarity in their viral profiles. These observations lead the authors to suggest that unlike other ecosystems, the faecal virome does not show signs of Lotka–Volterra dynamics, whereby viral and bacterial populations rapidly shift based on predator–prey interactions, prompting accelerated evolution of key genes, e.g. phage tail fibres (Reyes *et al.*, 2010).

One hypothesis explaining the predominance of temperate phages that are stable over time in the faeces is that they could be induced due to energy limitation. To investigate this hypothesis, gnotobiotic mice were co-colonized for 2 weeks *with Marvinbryantia formatexigens,* a human gut acetogen with three predicted prophages, and *Bacteroides thetaiotaomicron,* a saccharolytic bacterium with two predicted prophages. Interestingly, one of the three prophages in *M. formatexigens* was induced in all three faecal samples and only one of the three caecal samples analysed in mice (Reyes *et al.*, 2010). These findings are consistent with the hypothesis that phages may be induced during transit along the distal gut. Follow-up studies will be critical to investigate the mechanisms that govern virome assembly and diversity along the gastrointestinal tract and in other body habitats. These studies, coupled with an analysis of the bacterial members of these communities, promise to provide a broad perspective of host–viral–bacterial interactions in the human microbiome.

One initial study of the virome of another body habitat, the human oral cavity, led to the detection of phage-encoded platelet-binding factors, suggesting that the oral virome may be an under-explored reservoir of genes involved in pathogenesis (Willner *et al.*, 2010). Viral DNA was analysed from 19 pooled oropharyngeal swabs and sequenced. The identified viruses were primarily phages, and the complete genomes of multiple phages were assembled: *E. coli* phage T3, *Propionibacterium acnes* phage PA6, and *Streptococcus mitis* phage SM1. Interestingly, genes encoding platelet-binding factors that mediate the attachment of *S. mitis* to platelets (pblA and pblB of phage SM1) were observed. Additionally, the consumption of nicotine and soy sauce were shown to induce phage SM1, raising the possibility that diet could influence phage-mediated pathogenesis (Willner *et al.*, 2010).

Moving beyond metagenomics

One of the major challenges today is to move beyond describing which microbes or genes are present (their metabolic potential), to measuring their *in situ* metabolic activities (Fig. 11.2).

Multiple complementary approaches that have only recently been developed involve

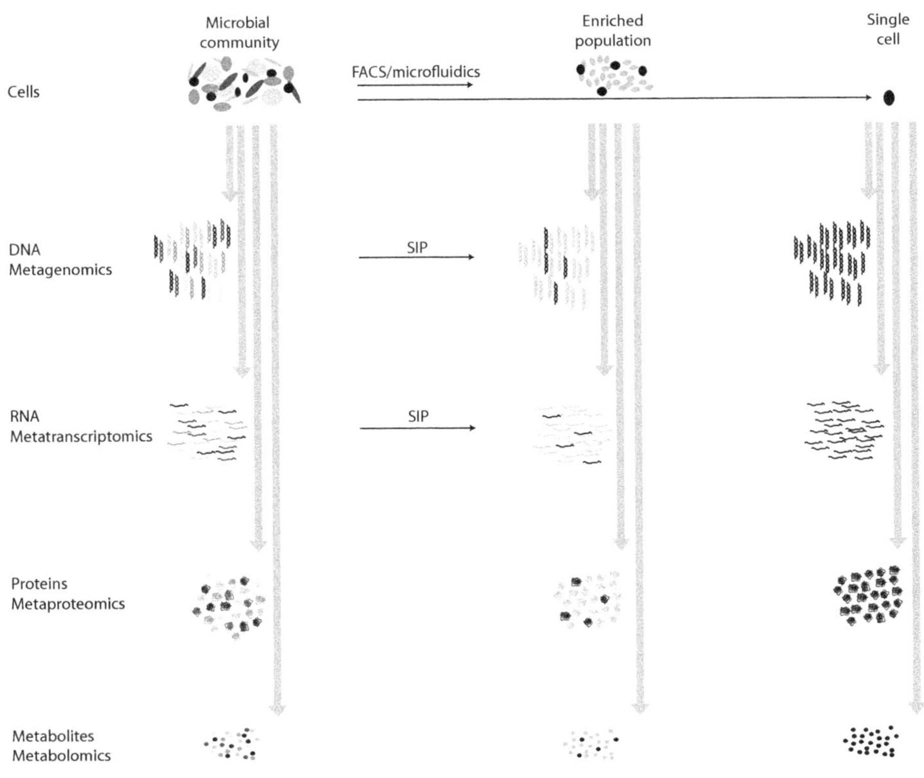

Figure 11.2 A systems biology approach to studying the human microbiome at many different levels.

profiling gene expression (metatranscriptomics and metaproteomics) and metabolic activities (metabolomics) of the human microbiome.

Metatranscriptomics, which has been extensively validated on pure cultures and in oceanic microbial communities (Frias-Lopez *et al.*, 2008; Rey *et al.*, 2010; Stewart *et al.*, 2010), was first applied to a study of germ-free mice that had been colonized with a human gut microbiota, referred to as 'humanized mice' (Turnbaugh *et al.*, 2009b). These humanized mouse models bridge the gap between controlled mouse experiments and studies of human subjects, as they have a reproducible community structure similar to that of the human donor within an experimentally tractable model organism. Multiplex pyrosequencing of 16S rDNA was performed on mice after colonization with fresh and frozen human faecal samples, after transfer to a second generation of germ-free recipients, during the establishment of a gut microbiota early in life, and on samples taken along the entire length of the gastrointestinal tract. The transfer of mice from a standard low-fat/high-polysaccharide (LF/PP) diet to a high-fat/high-sugar 'Western' diet resulted in dramatic changes in community structure after only a single day on this diet, with an increased relative abundance of Firmicutes and a decreased relative abundance of Bacteroidetes. Shotgun sequencing of community DNA revealed that samples taken from mice clustered according to their current diet, with an enrichment for predicted ABC transporter and phosphotransferase system (PTS) genes in the gut microbiome of mice on the Western diet. A novel isolate representative of the bacterial species found at higher abundance on the Western diet (*C. innocuum* strain SB23) was cultured from a humanized mouse gut sample and subjected to whole genome shotgun sequencing, revealing a diverse array of predicted PTS, ABC transporters, and carbohydrate-active enzymes. Meta-transcriptomics was then applied to samples taken from the caecum of mice fed the LF/PP and Western diet. As with gene content, the transcriptional profiles clustered according to diet, when considering community gene expression or the expression of *C. innocuum* SB23. A total of 69 genes in the SB23 genome were significantly up-regulated on the Western diet, including key genes predicted to be involved in carbohydrate transport and metabolism; these trends were subsequently validated by qRT-PCR. When taking into account community gene expression, a total of 520 clusters of genes were differentially expressed between the two diets (Turnbaugh *et al.*, 2009b).

More recently, two monozygotic female twins were deeply surveyed, at the level of community structure, gene content, and transcriptional activity (Turnbaugh *et al.*, 2010). A total of 1.2–1.5 million bacterial 16S rDNA sequencing reads were obtained from each individual, representing an estimated 800 species-level phylotypes. Deep sequencing of community DNA allowed the assembly and binning of genomic fragments into genus-level bins, while sequencing of the metatranscriptome provided insight into the expressed genes in each community. These results revealed marked differences in gene abundance and gene expression, with some genus-level bins having higher relative gene expression compared to their DNA-level gene abundance: e.g. the *Bifidobacterium* had the lowest level of relative expression in both microbiomes. 6961–7893 genes in each microbiome had high or low relative expression (defined as a 10-fold difference between RNA and DNA abundance). These highly expressed genes represent hypothetical genes, genes responsible for essential cellular functions (e.g. translation and energy metabolism), and genes encoding proteins that may be adaptive for life in the human gut (e.g. carbohydrate metabolism, ABC transport systems, methanogenesis, and acetogenesis). These trends were consistent across five additional samples taken from two sets of monozygotic co-twins and one unrelated individual (Turnbaugh *et al.*, 2010).

Although meta-transcriptomics is valuable for determining the expression of genes in the human microbiome, it does not allow the direct inference of the abundance of each protein. The first large-scale analysis of a human-associated metaproteome employed a non-targeted shotgun mass spectrometry-based whole community proteomics method (metaproteomics), allowing the measurement of thousands of proteins from two faecal samples taken from a single monozygotic twin pair (Verberkmoes *et al.*, 2009). The functional distribution of proteins were skewed relative to community gene content, with an increased abundance of proteins for translation, energy production, and carbohydrate metabolism. Additionally, host proteins such as antimicrobial peptides were observed, potentially allowing the simultaneous measurement of host and microbial proteins.

In order to obtain a comprehensive view of the human microbiome, it will be important to develop tools to analyse the metabolic activities of microbial communities and how they vary across space and time (Fig. 11.2). This 'marriage' of metagenomics and metabolomics (Turnbaugh and Gordon, 2008), has already begun to reveal interesting new findings. For example, a recent time series analysis of samples taken from patients following small bowel transplantation, has demonstrated a microbial community dominated by facultative anaerobes (*Lactobacilli* and Enterobacteria), presumably due to the introduction of oxygen through the ileostomy created to monitor the progress of the allograft (Hartman *et al.*, 2009). Metabolomic profiling of this altered community revealed enrichment for metabolites involved in aerobic respiration, along with several other unidentified compounds.

A series of studies from Jeremy Nicholson and colleagues has demonstrated the role that the gut microbiome plays in influencing the range of metabolites found in human plasma, urine, and gastrointestinal contents (Nicholson *et al.*, 2005; Nicholson, 2006). For example, a recent study provided evidence for an impact of the gut microbiome on the metabolism of a widely used analgesic drug, acetaminophen; individuals with a high pre-dose urinary level of *p*-cresol sulfate had a low post-dose urinary ratios of acetaminophen sulfate to acetaminophen glucuronide (Clayton *et al.*, 2009). One explanation for these results may be that individuals with a gut microbiota that generated a high level of *p*-cresol would result in competitive O-sulfonation of *p*-cresol, outcompeting the drug. Thus, these studies are consistent with the hypothesis that the gut microbiota is an important mediator of drug metabolism, even for orally administered therapeutics that can be absorbed rapidly into circulation.

Linking phylogenetic identity to function

Describing the phylogenetic diversity, gene content, and transcriptional activity of the human microbiome, although powerful, still leaves many questions unanswered. One of the next major challenges is to assign ecological roles (also referred to as niches or functional capabilities) to the uncultured organisms revealed by metagenomics. Bacteria carry out key functions in numerous environmental processes, and many of the genes, enzymes, and metabolic pathways involved remain to be identified. Such enzymes also constitute a largely untapped resource for biotechnology applications, making enzyme and natural product discovery an active research area. In this section, we will discuss some of the techniques that have allowed us to gain insight into certain microbial processes, even in the human body, and the upcoming techniques that promise to be valuable for linking phylogenetics to function.

Currently, there are many different approaches for the culture-independent functional analysis of complex microbial communities, including (i) the analysis and functional annotation of nucleotide sequences, (ii) assays that measure enzymatic activity or another functional property (also called functional screening or functional metagenomics), (iii) measurement of Substrate-Induced Gene EXpression (SIGEX), (iv) the analysis of the mobile metagenome, and (v) stable-isotope probing (SIP).

The former, and perhaps most widely used, approach can be used in a targeted fashion, relying on conserved DNA sequences to design hybridization probes or PCR primers to screen metagenomic libraries for clones that contain sequences of interest. This approach has proven valuable when PCR is used for the direct retrieval of target DNA sequences, leading to the retrieval of catechol 2, 3-dioxygenase genes from environmental samples (Okuta *et al.*, 1998). It is also possible to randomly sequence clones from metagenomic libraries, without any prior knowledge of the sequence. This direct sequencing approach can identify homologues of useful genes that would not have been detected otherwise, and complements the results from hybridization or

PCR-based experiments. In one well-known case, sequencing of metagenomic libraries showed that microbial rhodopsins are not restricted to Archaea, with important implications for phototrophy in oceanic surface waters (Beja *et al.*, 2000). However, this type of approach relies to some degree on prior knowledge; for example, the identification of genes with shared sequence characteristics to known genes, or those annotated via sequence-independent approaches, i.e. co-occurrence or neighbourhood methods. This can make it difficult to infer the activity of totally novel genes, many of which may be responsible for the interesting metabolic properties of a given habitat.

In the second approach, functional metagenomics, a shotgun library is constructed and used to screen for clones expressing a desired activity or functional property. This approach can identify clones that have potential applications in medicine, agriculture, or industry by focusing on natural products or enzymes with useful activities. Functional screening has allowed the identification of several new enzymes of industrial importance (Gillespie *et al.*, 2002; Courtois *et al.*, 2003; Brady and Clardy, 2004) as well as genes with novel DNA sequences (Diaz-Torres *et al.*, 2003; Riesenfeld *et al.*, 2004). Indeed, analysis of complex soil bacterial communities revealed the existence of molecules with broad-spectrum antibiotic activity (Gillespie *et al.*, 2002). Furthermore, the identification of new polyketide synthase genes and other compounds, including alcohol isomers and molecules with biological activity, strongly illustrates the potential of functional screening of complex environmental samples, such as soils, for the identification, and eventually industrial production, of compounds of interest.

However, there are some important limitations to functional screens. For example, screening is dependent on the expression of cloned genes in a culturable and genetically manipulable host, generally *Escherichia coli*, and the cloning of all of the genes required for the function (Uchiyama *et al.*, 2005; Yun and Ryu, 2005). However, fortunately some progress has been made to facilitate activity-based screening in different host species, and to expand the range of expression in *E. coli* (Schloss and Handelsman, 2003). To date, libraries have been created in diverse Proteobacteria, including *Agrobacterium tumefaciens*, *Burkholderia graminis*, *Caulobacter vibrioides*, *Pseudomonas putida*, and *Ralstonia metallidurans*, allowing the identification of unique sets of clones containing genes involved in small molecule biosynthesis (Craig *et al.*, 2010).

Functional metagenomics has recently been used to explore the antibiotic resistance reservoir of the human microbiome (Sommer *et al.*, 2009). Metagenomic libraries were constructed in *E. coli* from DNA isolated from saliva and faecal samples of two unrelated healthy humans who had not been treated with antibiotics for 1 year. Clones conferring antibiotic resistance were selected by exposing each library to a panel of 13 antibiotics of multiple drug classes: amino acid derivatives, aminoglycosides, amphenicols, beta-lactams, and tetracyclins. Sequencing of each identified insert revealed 95 unique sequences, with an average similarity of 69.5% at the nucleotide level to sequences in the NCBI database. 22% of the identified inserts had close homology to known antibiotic resistance genes, while some had 100% identity to hypothetical proteins in cultured reference genomes. On the other hand, a similar analysis of the antibiotic resistance genes found in isolates cultured from these samples under aerobic conditions, revealed that half of the genes were nearly identical to resistance genes found in known pathogens (Sommer *et al.*, 2009). These results suggest a powerful approach to identify novel resistance genes, and that the human microbiome contains a large reservoir of resistance genes, including those also found in pathogens.

The third approach, SIGEX, aims to circumvent some of the previously cited technical drawbacks and improve the efficiency of screening by using the fact that the expression of catabolic genes is often induced by specific substrates or metabolites, and that the expression of these genes can be controlled by regulatory elements located in proximity of the gene (Uchiyama *et al.*, 2005). This allows the identification of clones inducible by the presence of specific substrates. To make SIGEX a high-throughput process, an operon-trap *gfp*-expression vector, suitable for subsequent cloning, is used. The expression of catabolic genes in cloned metagenomic DNA is determined by *gfp* expression in the presence of the substrate, and the positive clones are separated by fluorescence activated cell sorting (FACS) (Uchiyama *et al.*, 2005). For example, SIGEX has allowed the isolation of 33 clones induced by benzoate and 2 clones induced by naphthalene from over 150,000 clones, as well as the identification of a novel enzyme Bzo71-8 P450 (Uchiyama *et al.*, 2005). However, SIGEX also presents limitations: it cannot detect genes that are constitutively expressed, it is sensitive to the orientation and structure of the desired gene and the *gfp* reading frame, and it requires substrates that can enter the cytoplasm.

The fourth approach focuses on studying the 'mobile metagenome' (Jones and Marchesi, 2007; Jones *et al.*, 2010). Bacterial plasmids encode a wide variety of functions, including virulence factors, resistance to antibiotics, and production of antimicrobials (Jones and Marchesi, 2007). Furthermore, mobile genetic elements (MGE) such as plasmids, transposons, insertion sequences and integrons impact gene flow between bacterial species, potentially making them key players in the adaptation of bacteria to their ecosystem (Jones and Marchesi, 2007). For example, several studies have highlighted the role of MGE in the spread of antibiotic resistance genes, such as erythromycin resistance genes among *Bacteriodes* spp. or tetracycline resistance genes in the gut microbial community (Smalla *et al.*, 2000; Diaz-Torres *et al.*, 2003).

In addition to standard metagenomic approaches, TRansposon Aided CApture methods (TRACA) can be used to specifically target bacterial plasmids from a complex microbial community (Jones and Marchesi, 2007). Briefly, total community DNA is isolated, genomic DNA is digested using plasmid-safe DNase, transposons containing selectable markers are inserted into each plasmid *in vitro*, and the resulting pool is transformed into a host strain (Jones and Marchesi, 2007). This method was recently used to analyse the human gut microbiome, by acquiring plasmids from a sample collected from a healthy adult male volunteer. Four plasmids were sequenced, two of which were homologous to sequences from previously published gut microbiome datasets, but were less abundant in the murine gut, ocean, or soil microbiomes (Jones *et al.*, 2010). These results provide an initial view of the mobile metagenome, and provide a proof-of-principle for additional studies of MGE from multiple individuals.

The final approach, stable-isotope probing (SIP), allows the identification of microbes involved in the metabolism of labelled substrates, thereby directly linking microbial diversity to function (Radajewski *et al.*, 2003; Dumont and Murrell, 2005). The rational behind this procedure is that labelled substrates will be 'heavier' than their non-labelled counterparts, and bacteria actively metabolizing the labelled substrates will thus be separable from the inactive bacteria (Neufeld *et al.*, 2007a). The most commonly used label is ^{13}C, but several studies have used with success other molecules, such as ^{15}N (Lechene *et al.*, 2007; Neufeld *et al.*, 2007b).

Different biomarker molecules can be used, such as phospholipid-derived fatty acid (PLFA-SIP), which were the first molecules used; nucleic acids (DNA-SIP); and ribonucleic acids (RNA-SIP) (Neufeld *et al.*, 2007a). PLFA-SIP does not require long incubations times, because biosynthesis of lipids does not require cell replication. It is also the most sensitive method, as it does not depend on extensive incorporation of the labelled molecule, unlike DNA- and RNA-SIP (Kreuzer-Martin, 2007; Neufeld *et al.*, 2007a). However, lipids cannot be used to easily determine the phylogenetic identity of the microbes involved and little is known about the fatty acid profiles of uncultured microorganisms (Kreuzer-Martin, 2007).

The use of labelled nucleic acids as biomarkers has the potential to identify a wider range of bacteria with a greater degree of confidence, although they require higher concentrations of labelled substrates, potentially leading to cross-feeding of substrates (Neufeld *et al.*, 2007a). DNA-SIP has enabled the analysis of both 16S rDNA and physiological genes from organisms that grow on specific carbon substrates, but this method requires long incubations times, as cell division is necessary for DNA synthesis (Neufeld *et al.*, 2007a). RNA-SIP on the other hand does not require cell replication and has enabled the subsequent analysis of 16S RNA (Manefield *et al.*, 2002; Neufeld *et al.*, 2007a). Performing SIP incubations before proceeding with metagenomics or high-throughput sequencing has the potential to reduce community complexity, which can improve assembly efforts and may allow the detection of low-abundance species (Chen and Murrell, 2010). For example, a recent study used isotope labelled starch to identify bacteria responsive for the degradation of this substrate in an *in vitro* model of the human gut (Kovatcheva-Datchary *et al.*, 2008). The application of SIP to the exploration of the human microbiome will be one critical method for linking microbial phylogenetics to function.

Succession of the gut microbiome in infants

One of the fundamental questions regarding the human microbiome relates to the colonization of each host shortly after birth, given the observation that humans and other mammals are born germ-free. One model system that can be used to investigate such questions under controlled conditions is the 'humanized' mouse model; germ-free mice can be stably colonized with a human gut microbiome and the re-establishment of this community in their offspring can be analysed over time (Turnbaugh *et al.*, 2009b). Humanized mice were sampled beginning at postnatal day 14 through 85. The overall level of bacterial diversity increased over time from days 14 through 24, approaching levels found in adult animals, and corresponding to a shift in the relative abundance of multiple classes of bacteria. Notably, there was a drop in γ-Proteobacteria (the dominant class at day 14), and Bacilli, with a rise in the Bacteroidetes at the time of weaning (Turnbaugh *et al.*, 2009b).

These culture-independent methods have also been applied to study the development of the human infant gut microbiota (Palmer *et al.*, 2007; Koenig *et al.*, 2010). In a study performed by Palmer *et al.*, a microarray containing probes targeted towards different taxonomic groups of bacteria based on 16S rDNA sequences was used to profile the microbiotas of 26 faecal samples each from 14 healthy infants, including a single pair of dizygotic twins. The resulting time series dataset, starting at the first stool post birth and continuing through the first year of life, was suggestive of a dynamic pattern of assembly with large inter- and intra-individual variations over time. Interestingly, the dizygotic twins had a very similar pattern of colonization, suggesting that either shared environmental exposures or similar host genotype contributed to each pattern (Palmer *et al.*, 2007).

One hypothesis regarding the temporal changes in community structure during the first year or two of life is that this represents an orderly progression based on environmental exposures and host developmental stage. Alternatively, these changes could represent stochastic fluctuations prior to the gut community reaching a 'stable state'. Evidence in support of the former hypothesis was shown in a recent analysis of a single infant across 2.5 years, spanning sixty faecal samples sampled with 16S rDNA sequencing and 12 samples analysed by metagenomic shotgun sequencing (Koenig *et al.*, 2010). The results indicate that the human gut microbiota gradually increases in phylogenetic diversity over time and that changes in community composition follow a smooth temporal gradient. As in the previous study, dramatic shifts in the relative abundance of bacterial taxa were observed, but each alteration to the gut microbiota could be attributed to a change in diet or health status. An analysis of the co-occurrence of bacterial species revealed that temporal assembly of the gut microbiota is non-random, suggesting that the developing infant microbiota contains interacting bacteria consortia. Furthermore, the analysis of community gene content revealed that genes for degrading complex plant polysaccharides were present prior to weaning, potentially priming the gut microbiome to rapidly access solid foods (Koenig *et al.*, 2010). Together, these results are consistent with the hypothesis that the gut microbiome is established in an orderly fashion, largely driven by environmental exposures throughout development. Follow-up studies with defined interventions, either in human infants or humanized animal models, should provide more insight into the key factors that regulate these changes and their temporal dependencies.

One of these important host factors that could impact the developing gut microbiota is likely to be the human milk glycobiome:

the diverse array of complex oligosaccharides found in human milk (HMOs) (Zivkovic *et al.*, 2010). Recent studies have shown that milk is comprised of an array of oligosaccharides that are indigestible by the developing infant, and may instead be provided to select for a specific consortia of colonizing bacteria. For example, multiple strains of *Bifidobacterium longum* subsp. *infantis* will preferentially consume oligosaccharides that are found at high levels early in the lactation cycle. Some of these oligosaccharides can also bind to intestinal receptors for pathogens, thus protecting infants from disease. The specific set of HMOs varies between mothers, the reasons for which are largely unexplored but could be related to diet, lifestyle, ethnicity, or other factors. Furthermore, whole genome sequencing of strains isolated from infants relative to those found in adults, has revealed a relative enrichment for genes targeted at host-derived carbohydrates, suggesting that the nutritional milieu of breast milk could select for infant-adapted strains (Zivkovic *et al.*, 2010).

Gut microbes as one component of host energy balance

A variety of recent studies of humans and mice have implicated the gut microbiome in some of the symptoms associated with metabolic syndrome, defined as a collection of conditions (abdominal obesity, hypertension, high triglycerides, low levels of high density lipoproteins, and glucose intolerance) that increase the risk of vascular disease (e.g. heart disease and stroke). The study of the microbial component of these conditions was initiated by a series of observations in germ-free and colonized mice; germ-free animals have lower total body fat than conventionally raised animals, and this difference can be reversed within 2 weeks after colonization (Backhed *et al.*, 2004). This happens in both male and female animals, and in multiple mouse strains. Additionally, germ-free or antibiotic treated mice are at least partially resistant to diet-induced obesity caused by the consumption of a high-fat, high-sugar 'Western' diet (Backhed *et al.*, 2007; Cani *et al.*, 2008). The mechanism for this microbial contribution to adiposity is thought to impact both sides of the energy balance equation, by contributing additional available calories through the microbial degradation of dietary polysaccharides that are inaccessible by the host, and by regulating the expression and activity of host genes that promote deposition of adipose tissue (Backhed *et al.*, 2004, 2007).

These initial findings prompted a simple question: are the microbes found in the distal gut of obese animals different from those in lean animals? Mouse studies have shown reproducible trends with an increased relative abundance of Firmicutes and a decreased relative abundance of Bacteroidetes in the gut microbiota of mice with genetically induced obesity, caused by a deficiency in the gene encoding leptin (Ley *et al.*, 2005; Turnbaugh *et al.*, 2006) and mice with diet-induced obesity (Turnbaugh *et al.*, 2008). Metagenomic analysis of the gene content of the gut microbiomes of genetically induced obese mice and their lean littermates, coupled to biochemical analyses, were suggestive of an increased capacity for dietary energy harvest by the gut microbiota that might contribute to increased adiposity (Turnbaugh *et al.*, 2006).

Strikingly, the first culture-independent 16S rDNA analysis of the gut microbiomes of obese and lean humans revealed trends in microbial ecology consistent with mouse studies; obese individuals had high levels of Firmicutes and low levels of Bacteroidetes. Furthermore, the low level of Bacteroidetes increased as these individuals lost weight on either a low-fat or low-carbohydrate diet (Ley *et al.*, 2006c). However, these trends should be treated with caution given the limited number of subjects analysed (n=12), and the mixed results obtained in follow-up studies that used a variety of different culture-independent methods (Ley, 2010).

Although these differences in microbial ecology are intriguing, they provide a chicken and egg question: does the microbiome contribute to obesity or does obesity cause shifts in the microbiome, potentially mediated by shifts in dietary habits? Although these two are not necessarily mutually exclusive, the question of causation was initially addressed through 'microbiota transplantation' experiments. In these studies, a panel of lean adult germ-free mouse recipients is colonized with distal gut samples taken from either an obese or lean donor. These studies have repeatedly shown that mice colonized with an obesity-associated gut microbiome have a greater increase in adiposity than those colonized with a sample taken from a lean donor. To date, donors have included mice with genetically induced obesity (Turnbaugh *et al.*, 2006), diet-induced obesity (Turnbaugh *et al.*, 2008), and diet-induced obesity after colonization with a human faecal sample (Turnbaugh *et al.*, 2009b).

More recently, mouse studies have linked the innate immune system, the gut microbiome, and metabolic syndrome. Mice with a mutation in the toll-like receptor 5 (TLR5), a key component of the innate immune system that fights infection and is expressed in the gut mucosa, develop some of the symptoms of metabolic syndrome, including hyperlipidaemia, hypertension, insulin resistance, and increased adiposity. These changes are accompanied by shifts in the composition of the gut microbiota, while microbiota transplantation experiments have shown that these phenotypes are transmissible to germ-free mouse recipients. Furthermore, treatment with broad-spectrum antibiotics upon weaning for 12 weeks prevented these phenotypes in TLR5-deficient mice (Vijay-Kumar *et al.*, 2010).

Together, these studies present an intriguing example of a complex disease (metabolic syndrome) that may at least in part be mediated by the gut microbiome, along with more traditional factors such as host genetics, diet, and energy expenditure. It also provides a challenging paradigm to explore in human populations, given high levels of inter-individual variation in the gut microbiome (see 'The gastrointestinal microbiome' section above) and the ill-defined factors that govern the composition of each microbial community. Prospective human trials attempting to link specific microbial communities to the emergence of disease are needed, coupled to human case-control studies where energy balance is strictly measured along with defined diets. The substantial progress that has been made so far in animal studies suggests that these models may allow the further dissection of the mechanism of these host–microbial interactions, allowing follow-up studies in humans addressing more specific hypotheses.

Applying theories from macro-ecology to the human microbiome

Both macro- and micro-ecology focus on understanding how the abundances and distributions of species vary at different spatial and temporal scales. While it is obvious that the distribution of trees within a forest will span a larger spatial scale than the distribution of bacteria on the human skin surface, it is possible that some of the same ecological principles will apply to both relationships. The same examples could apply for the temporal scale (i.e. centuries, decades, weeks, or hours). The fact that bacterial diversity is extremely high within the human microbiota, and that it is involved in a number of functions, have prompted a series of fundamental ecological questions, similar to those framed by studies of macro-ecology: how are we colonized by this microbiota, and how have we co-evolved with it? What external and internal factors allow its establishment? What is the level of functional redundancy between members of this community? What factors can alter our microbiota? Clearly, there are strong ecological forces in place allowing us to develop our microbiota, and the afore-mentioned questions show a shift in our view and comprehension of the human microbiota. Both basic and medical studies of the human microbiome have much to benefit from such an ecological framework; determining which theories can be applied to the human microbiome, and to what extent this can affect our health.

One important question to start with is how can we describe the interactions between humans and their microbiota, on an evolutionary scale? Shortly after birth, humans are colonized by diverse microbial communities, eventually forming our resident microbiota. This initial microbiota is influenced by the mode of delivery (Bennet and Nord, 1987; Hallstrom *et al.*, 2004; Penders *et al.*, 2006; Dominguez-Bello *et al.*, 2010), and changes drastically during the first years of life (Benno *et al.*, 1984; Penders *et al.*, 2006; Palmer *et al.*, 2007; Koenig *et al.*, 2010) (see the 'Succession of the gut microbiome in infants', above). Yet, the initial results from studies of the human microbiome suggest that humans share a small set of abundant bacterial phyla, albeit with a unique collection at the strain or species level.

Clearly, the interactions between humans and their microbiota are complex, but can be regrouped under the general term of 'symbiosis', that describes intimate interactions between unlike organisms (Walter *et al.*, 2010, citing Anton de Bary). Symbiosis thus includes mutualism, where both partners derive benefit; commensalism, where one partner benefits and the other is unaffected; and parasitism, where the benefit of one partner is costly to the other (Box 11.1). This cost is generally measured with, or linked to changes, in an individual's fitness, an ecological measure of the reproductive success of an individual and of his offspring.

While these distinctions and categorization are relatively easy with macro-organisms, they become much more difficult to apply to the microbial world. Indeed, many of the normal members of the gut microbiota can become pathogenic, if changes in the genotype or environment occur. In addition, many common human pathogens are closely related to non-pathogenic members of our microbiota: examples are found in the genera *Staphylococcus*, *Streptococcus*, *Neisseria* and *Enterococcus*, as well as in the Enterobacteriaceae family (Robinson and Enright, 2004; Wirth *et al.*, 2004; Callaghan *et al.*, 2006; Leavis *et al.*, 2006). Also, the biochemical mechanisms for sensing host environments, interacting with host surfaces and even communicating with the host are often the same in human pathogens and commensal microorganisms (Sperandio *et al.*, 2003; Shiner *et al.*, 2005; Callaghan *et al.*, 2006). The absence of clear frontier between these different host–microbial interactions suggests that our interactions with our microbiome should be viewed as a continuum between mutualistic and pathogenic interactions (Walter *et al.*, 2010).

Another key question is how are we colonized by our microbial partners and how does our body deal with the immense external microbial diversity to only maintain a few dominant bacterial phyla? The specificity of our different body habitats (e.g. pH, anaerobic conditions for the gut, and the dryness of skin) provides one set of possible selective constraints influencing the microbiota's ability to colonize and remain attached to these different locations. Considering the human gut as an example, this habitat-selection includes the need to encode a wide array of enzymes to process our diversified diet, to evade an active immune system, and to deal with the mechanical processes of digestion. There could also be a 'functional' selection, whereby humans have co-evolved with bacteria that increase host fitness; for example, by performing metabolic processes that are not encoded by the host genome our microbiome has reduced some metabolic costs. Several cues argue for this co-evolution between humans and their microbiota: the existence of shared abundant bacterial phyla, albeit with unique proportions of the different members; the site-specificity of those bacteria, and the distinction of our microbiota from that of other vertebrates (Dethlefsen *et al.*, 2007, and references therein).

Humans and other mammals are born germ-free and colonized shortly after birth. This phenomenon suggests that colonization will be dependent on our surrounding environment, and that members of the same family will share a more similar microbiome than unrelated individuals. These shared features between family members have been reported for humans in studies of multiple body habitats, including the ear canal (Frank *et al.*, 2003) and the gut microbiome (Zoetendal *et al.*, 2001; Turnbaugh *et al.*, 2009a). Although these familial similarities could also be explained by shared early environmental exposures, it remains likely that at least some of the microbiota is transmitted between family members, especially given the fact that many of the microbes found in the human gut cannot grow or survive under aerobic conditions, thus relying on close contact for transmission.

Once established, members of the microbiota are subjected to a wide variety of selection pressures through interactions with the host and other microbes. Indeed, each bacterial cell is part of a community providing services for the host (e.g. dietary energy harvest and protection against pathogens), but each cell is also in competition with the others for available nutrients. Thus, the structure and composition of the gut microbiota reflect selections at two levels: at the microbial level, considering the fitness of an individual bacterium within a competitive ensemble; and at the host level, with potential impacts on host fitness. These interactions can be discussed in terms of 'top-down' and 'bottom-up' forces (Ley *et al.*, 2006b; Kirchman, 2008). The host imposes top-down forces on the microbiota, potentially selecting for communities with high functional redundancy. Redundancy within a bacterial community is thought to promote stability, also referred to as the insurance hypothesis (Yachi and Loreau, 1999). In this case, keystone processes or metabolisms can still be performed by other members of the community, providing the host with alternative routes when faced with factors altering microbial diversity.

In contrast, bottom-up forces select for specialization of the members of the human microbiota, due to competition between individual microbes. Although functional redundancy may be the dominant strategy for our microbiota, there are cases where one unique organism appears to be irreversibly adapted to a niche, such as the case for *Methanobrevibacter smithii* in the gut. This Archaea competes directly for H_2 with sulfate-reducing bacteria, but is thought to hold a competitive advantage *in vivo* (Strocchi *et al.*, 1994). Furthermore, there are other interactions occurring between the bacterial members of our microbiota, whereby transient metabolic products can promote cooperation or enhance competition between them. An example of cooperation between members of the microbiota takes place in the colon, where numerous species are required for the complete degradation of polysaccharides to provide the host with butyrate (Flint, 2004; Flint *et al.*, 2007). Some strains are not directly involved in the degradation process, but use the resulting sugars to provide growth factors to the strains responsible for degradation (Flint, 2004). In this case, the members of the community cooperate to provide the host with a specific metabolite, butyrate. Cooperation also takes place between the human host and its microbiota, when some bacteria within the community interfere with the colonization of pathogens, which is also known as pathogen interference (Servin, 2004; Brook, 2005). This case may be one example of an indirect increase in bacterial fitness through an increase in host fitness (Dethlefsen *et al.*, 2007).

Collectively, many ecological theories have been explored in the context of the human microbiome, and have revealed new perspectives on our microbial partners. However, an important set of ecological hypotheses cannot be easily tested on human individuals, and the use of gnotobiotic animals, such as mice, remain an essential tool (Falk *et al.*, 1998; Savage, 2001). These mice can then be subsequently 'humanized' to explore the microbial ecology in the context of the human microbiome (Turnbaugh *et al.*, 2009b) (see the 'Succession of the gut microbiome in infants' and 'Moving beyond metagenomics' sections above).

The use of multiplex barcoded 16S rDNA sequencing to analyse the human microbiome has allowed the study of large sets of spatially and temporally related samples necessary to begin to couple metagenomics with fundamental questions in ecology (Costello *et al.*, 2009). The resulting view of microbial ecology can be framed at two levels: the diversity within a community (α-diversity), and the diversity between different communities (individuals/body habitats) or along an environmental gradient (β-diversity) (Lozupone and Knight, 2008). Studies of the human microbiota have already begun to reveal the diversity between different body habitats, within one individual over time, and within a single sample through deep sequencing (Costello *et al.*, 2009; Turnbaugh *et al.*, 2010). However, numerous theories from macro-ecology remain to be tested, as we are not yet able to directly observe the microorganisms or their precise locations in our body. The ongoing development of new imaging tools, and the combined use of different analytical tools such as Raman microspectroscopy and fluorescence *in situ* hybridization (FISH) (Huang *et al.*, 2007); secondary ion mass spectrometry imaging (SIMS) (Wagner, 2009); or stable-isotope probing (SIP) in combination with metagenomic analyses promise to provide important insights regarding the interactions between humans and their microbiomes.

Taking a systems biology approach to metagenomics

Another major challenge for metagenomics is to develop predictive mathematical models that incorporate these disparate types of data. These types of modelling experiments could be developed and used to predict the operation of microbial systems across both micro- and macro-scales. For example, studies of the marine bacterium *Prochlorococcus* have led to the development of a model incorporating biogeochemical data that is able to predict the global distribution of this organism (Follows *et al.*, 2007). Similar approaches linking the metabolic properties of each individual's body habitats to the distribution of organisms across space and time, could provide a deeper understanding of the human microbiome, at scales ranging from one body habitat in one individual, to many body habitats across globally distributed populations.

A potential application of such a 'systems' approach is that biomarkers spanning multiple levels of the human microbiome could be identified and associated with specific disease states, such as obesity, body odour, gum disease, or inflammatory bowel disease. These efforts will hopefully expand the current toolbox for personalized medicine, allowing both the host genotype and aspects of the microbiome to be taken into account.

Additionally, a broader understanding of the impacts of defined changes in dietary macro- and micro-nutrient composition could allow for customized diets with the goal of promoting a specific community composition and/or metabolic activity. This form of 'personalized nutrition' may even be possible prior to consumption of solid foods through manipulation of the glycans found in milk (Zivkovic *et al.*, 2010), and would provide a non-invasive means of modifying part of the human metagenome.

Synthetic metagenomics

One of the most recent emerging areas of future interest has been referred to as 'synthetic metagenomics' (Bayer *et al.*, 2009). This topic of study incorporates the rapidly growing toolkit used for synthetic biology, an emerging field focused on building and designing biological systems from a collection of simple and modular building blocks. For the most part, these components come from a small number of well-characterized model organisms, such as *S. cerevisiae* or *E. coli*. However, recent efforts based on functional metagenomics are expanding the number of genes with experimentally validated activities that could be harnessed for industrially relevant processes, such as the production of biofuels (Sommer *et al.*, 2010).

To date, most metagenomic analyses are either aimed at randomly sequencing DNA or cDNA from a given body habitat, or constructing a plasmid library that is screened for inserts of a given activity (See 'Linking phylogenetic identity to function' section). However, because much of the DNA sequencing is now done with technology that does not require the construction of a plasmid library, it is not possible to store or retrieve specific genes identified with sequencing approaches. Synthetic metagenomics bridges this gap, by allowing the rapid *de novo* synthesis of genes identified in large scale sequencing projects, their expression in a bacterial host strain, and subsequent analysis of the activity.

In the first demonstration of this approach, members of the Voigt laboratory at the University of California San Francisco focused on the synthesis of methyl halides, molecules used as agricultural fumigants that can be catalytically converted to chemicals or fuels (Bayer *et al.*, 2009). This metabolic pathway was attractive given its industrial relevance, the lack of known enzymes that are able to perform the conversion at yields high enough for large-scale production, and the simplicity of the conversion. Only a single enzyme is required to transfer a methyl group for the ubiquitous S-adenosylmethionine to a halide ion. The gene sequence encoding a known methyl halide transferase was used to retrieve 89 putative genes from public sequence databases, many with distant homology. This panel of genes was then chemically synthesized and cloned into *E. coli* in order to determine the rates of CH_3Cl, CH_3Br, and CH_3I production. Of the 89 expressed genes, 56% were active on chloride, 85% on bromide, and 69% on iodide. The gene with the highest activity in *E. coli* was then transferred into *S. cerevisiae*, resulting in the production of 190 mg/l-h of methyl halides from glucose and sucrose. Finally, co-culturing experiments were performed with this novel genetically engineered yeast and a cellulolytic bacterium *Actinotalea fermentans*, allowing the production of methyl halides from a variety of unprocessed sources: maize stover, sugar cane, poplar, and switchgrass (Bayer *et al.*, 2009).

These experiments provide a template for a discovery pipeline that could be applied to a series of metabolic pathways found in the human microbiome, ranging from carbon fixation (methanogenesis and acetogenesis), to the degradation of cellulose and other complex polysaccharides, to the conversion of xenobiotics and secondary metabolites. In addition to the potential for validating and making experimentally tractable a previously inaccessible suite of enzymes for industrial applications, these studies could provide new insight into the metabolic capabilities of members of the human microbiome, prompting fundamental insights into microbial community structure and function.

Manipulating the human microbiome

Despite the wealth of unanswered questions regarding the human microbiome, there is an active interest in intentionally manipulating our microbial partners, either to support a healthy lifestyle or to intervene in microbe-related diseases. These interventions come in three basic categories: (i) prebiotics, food supplements that stimulate specific strains or species of bacteria, (ii) probiotics, live microbes administered orally or topically and (iii) drug therapies aimed at key host or microbial targets.

Prebiotics

Prebiotics provide a promising approach to manipulating gut microbial ecology without the need for dramatic alterations to an individual's diet. Much of this work has been focused on linking the abundance of specific types of dietary fibre with increases in the abundance of bacterial species that are thought to be beneficial. For example, a recent study assessed temporal changes in the gut microbiota of 14 overweight men fed with a diet high in resistant starch, high in non-starch polysaccharides, or low in carbohydrates (Walker *et al.*, 2010). As expected from previous studies (Ley *et al.*, 2006c), samples clustered by individual when taking into account the overall microbial community structure. However, analyses using quantitative PCR for specific bacterial groups revealed 'blooms' that occurred following some of the dietary shifts that were reversible by subsequent shifts. *Ruminococcus* were found in high abundance in most individuals fed a resistant starch diet, representing on average 17% of the total bacteria as opposed to 3.8% on the non-starch polysaccharide diet. Interestingly, Bifidobacterium had a variable response to the resistant starch diet, increasing in one individual but not in the others (Walker *et al.*, 2010).

These findings, together with recent studies of humanized mouse models on defined diets (Turnbaugh *et al.*, 2009b), emphasize the critical role that diet plays in shaping the gut microbiome, while highlighting the fact that the baseline composition of each individual's microbiome may play an important role in determining the response to a given diet. In the near future, it may be possible to perform an initial diagnostic survey of the structure and/or gene content of a subject's microbiome, prior to assigning a recommended dose and composition of the administered prebiotics.

Probiotics

Probiotics are a widely used attempt at manipulating the human microbiome, and ultimately human health, through the consumption of live bacterial species. Despite the widespread nature of this approach, there remains a limited molecular understanding of the impact that these organisms have on the microbiota, host gene expression, or host physiology. One approach to tackle this complicated problem is to study gnotobiotic mice colonized with a limited set of bacterial strains under controlled conditions. In a recent study, germ-free mice were colonized with *Bacteroides thetaiotaomicron*, a saccharolytic bacterium commonly found in the human gut, and *Bifidobacterium longum*, a widely used probiotic strain (Sonnenburg *et al.*, 2006). The addition of *B. longum* to the distal gut resulted in an expansion in the polysaccharide-degrading capability of *B. thetaiotaomicron*, a phenomenon that was shown to be independent of host genotype, and also occurred upon colonization with a strain commonly found in dairy products, *Lactobacillus casei*. The addition of *B. longum* also induced the expression of host genes for innate immunity, demonstrating that probiotic strains have the potential to influence both microbial and host gene expression, at least in the context of a two-component model community (Sonnenburg *et al.*, 2006).

More detailed analyses of the impact of probiotics on the human microbiome, at multiple levels (Fig. 11.2), coupled to studies in mouse models, will be necessary to determine the mechanisms underlying probiotics and to develop a more refined protocol for their administration. These studies also provide an intriguing problem from the standpoint of microbial ecology, as they often involve exposing a complex microbial community comprised of trillions of microbes to a relatively small inoculating dose of live probiotics. Do these probiotic strains represent keystone species that have a major impact on microbial ecology and function despite their low abundance? What are the parameters that govern how each individual responds to a given probiotic? In the end, it may also be necessary to develop a diagnostic test aimed at predicting which probiotic and which dose will be most appropriate.

One extreme example of probiotics is in the recent use of 'bacteriotherapy' for the treatment of recurrent *Clostridium difficile*-associated diarrhoea (Khoruts *et al.*, 2010). The composition of the distal gut microbiota was monitored before and after treatment with a faecal transplantation from a healthy donor. Before therapy, the patient's microbiota had low levels of Firmicutes and Bacteroidetes, two of the commonly abundant bacterial phyla in the gut, while after therapy the composition of the community was dominated by Bacteroidetes, similar to that of the donor's community. These changes, evident after only 14 days, were accompanied by a resolution of the patient's symptoms (Khoruts *et al.*, 2010). Although it is unknown how general these phenomena will be, whether the therapy replaced the recipient's microbial community with that of the donor, or merely encouraged a recovery of the recipient's resident microbiota, or if these changes are permanent, these findings provide a fascinating alternative to a more defined cocktail of probiotic strains.

Drug therapies targeted at host or microbial gene products

The final option, direct manipulation of host or microbial gene products through the administration of drugs targeting key steps in the links between the human microbiome and disease, is perhaps the most promising but also least understood of the three possible interventions. Although there are multiple potential problems with any antibiotic-based treatment strategy, especially for long-term administration in chronic diseases such as obesity or diabetes, more specific drug therapies could either block the host response to microbial signals or even disrupt components of the microbial community without encouraging resistance (Bollenbach *et al.*, 2009). These approaches will require a thorough molecular analysis of the relevant microbial metabolic pathways, and will at least partially capitalize on new innovations in the field of metagenomics and our expanding view of the human gut microbiome.

Conclusions

In the past decade our knowledge of the microbial diversity found across the planet has increased exponentially, largely due to the continued decrease in sequencing cost, increase in high-throughput methods, and the development of a suite of culture-independent metagenomic tools for analysing microbial community structure, gene content, and activity. Laboratories around the world now have access to hundreds, and soon to be thousands, of reference genomes, millions of genes, and the ability to rapidly sample the microbial communities found across multiple body habitats over space and time.

These datasets have already begun to reveal the high level of variation in community structure between different individuals, the shared genetic and metabolic features of our microbiome, some of the factors governing the assembly and succession of these communities in infants, the potential role of gut microbes in obesity and other diseases, and the importance of the virome. Functional metagenomic analyses have for their part revealed new genes for antibiotic resistance and polysaccharide degradation, expanding the toolkit of biological parts available for synthetic biology initiatives.

A major challenge moving ahead will be to integrate the data coming from disparate methodological approaches across a range of studies of diverse populations of humans and animals. By combining these datasets into predictive mathematical models and through developing testable hypotheses related to ecological principles concerning community structure and function, we will hopefully move towards the ultimate goal of understanding, predicting, and manipulating the composition and activity of our microbial partners.

Web resources

The Kyoto Encyclopedia of Genes and Genomes: http://www.genome.jp/kegg/
The STRING database of known and predicted protein–protein interactions: http://string.embl.de/
QIIME: quantitative insights into microbial ecology toolkit: http://qiime.sourceforge.net
Ribosomal database project: http://rdp.cme.msu.edu/
Greengenes 16S rDNA tools: http://greengenes.lbl.gov
Carbohydrate-Active enZYmes database: http://www.cazy.org/
RAST: Rapid Annotation using Subsystem Technology: http://rast.nmpdr.org/
MG-RAST: Metagenome Rapid Annotation using Subsystem Technology: http://metagenomics.nmpdr.org/
MetaHIT: EU-funded Human Microbiome Project: http://www.metahit.eu/
HMP: NIH-funded Human Microbiome Project: http://nihroadmap.nih.gov/hmp/

References

Aas, J.A., Paster, B.J., Stokes, L.N., Olsen, I., and Dewhirst, F.E. (2005). Defining the normal bacterial flora of the oral cavity. J. Clin. Microbiol. *43*, 5721–5732.

Ackerman, H.W., and DuBow, M.S. (1987). Viruses of prokaryotes: general properties of bacteriophages. (Boca Raton, Fla, CRC Press, Inc.).

Apajalahti, J.H.A., Kettunen, A., Nurminen, P.H., Jatila, H., and Holben, W.E. (2003). Selective plating underestimates abundance and shows differential recovery of Bifidobacterial species from human feces. Appl. Environ. Microbiol. *69*, 5731–5735.

Backhed, F., Ding, H., Wang, T., Hooper, L.V., Koh, G.Y., Nagy, A., Semenkovich, C.F., and Gordon, J.I. (2004). The gut microbiota as an environmental factor that regulates fat storage. Proc. Natl. Acad. Sci. U.S.A. *101*, 15718–15723.

Backhed, F., Ley, R.E., Sonnenburg, J.L., Peterson, D.A., and Gordon, J.I. (2005). Host-bacterial mutualism in the human intestine. Science *307*, 1915–1920.

Backhed, F., Manchester, J.K., Semenkovich, C.F., and Gordon, J.I. (2007). Mechanisms underlying the resistance to diet-induced obesity in germ-free mice. Proc. Natl. Acad. Sci. U.S.A. *104*, 979–984.

Bayer, T.S., Widmaier, D.M., Temme, K., Mirsky, E.A., Santi, D.V., and Voigt, C.A. (2009). Synthesis of methyl halides from biomass using engineered microbes. In. J. Am. Chem. Soc., pp. 6508–6515.

Beja, O., Aravind, L., Koonin, E.V., Suzuki, M.T., Hadd, A., Nguyen, L.P., Jovanovich, S.B., Gates, C.M., Feldman, R.A., Spudich, J.L. *et al.* (2000). Bacterial rhodopsin: evidence for a new type of phototrophy in the sea. Science *289*, 1902–1906.

Bennet, R., and Nord, C.E. (1987). Development of the faecal anaerobic microflora after caesarean section and treatment with antibiotics in newborn infants. Infection *15*, 332–336.

Benno, Y., Sawada, K., and Mitsuoka, T. (1984). The intestinal microflora of infants: composition of fecal flora in breast-fed and bottle-fed infants. Microbiol. Immunol. *28*, 975–986.

Bik, E.M., Eckburg, P.B., Gill, S.R., Nelson, K.E., Purdom, E.A., Francois, F., Perez-Perez, G., Blaser, M.J., and Relman, D.A. (2006). Molecular analysis of the bacterial microbiota in the human stomach. Proc. Natl. Acad. Sci. U.S.A. *103*, 732–737.

Bollenbach, T., Quan, S., Chait, R., and Kishony, R. (2009). Nonoptimal microbial response to antibiotics underlies suppressive drug interactions. Cell *139*, 707–718.

Bouvier, T., and del Giorgio, P.A. (2007). Key role of selective viral-induced mortality in determining marine bacterial community composition. Environ. Microbiol. *9*, 287–297.

Brady, S.F., and Clardy, J. (2004). Palmitoylputrescine, an antibiotic isolated from the heterologous expression of DNA extracted from bromeliad tank water. J. Nat. Prod. *67*, 1283–1286.

Breitbart, M., Haynes, M., Kelley, S., Angly, F., Edwards, R.A., Felts, B., Mahaffy, J.M., Mueller, J., Nulton, J., Rayhawk, S. *et al.* (2008). Viral diversity and dynamics in an infant gut. Res. Microbiol. *159*, 367–373.

Breitbart, M., Hewson, I., Felts, B., Mahaffy, J.M., Nulton, J., Salamon, P., and Rohwer, F. (2003). Metagenomic analyses of an uncultured viral community from human feces. J. Bacterio.l *185*, 6220–6223.

Brook, I. (2005). The role of bacterial intererence in otitis, sinusitis and tonsilitis. Otolaryng. Head Neck *133*, 139–146.

Bryant, M.P. (1972). Commentary on the Hungate technique for culture of anaerobic bacteria. Am. J. Clin. Nutri. *25*, 1324–1328.

Burton, J.P., and Reid, G. (2002). Evaluation of the bacteria vaginal flora of 20 postmenopausal women by direct (nugent score) and molecular (polymerase chain reaction and denaturing gradient gel electrophoresis) techniques. J. Infect. Dis. *186*, 1770–1780.

Callaghan, M.J., Jolley, K.A., and Maiden, M.C. (2006). Opacity-associated adhesin repertoire in hyperinvasive *Neisseria* meningitidis. Infect. Immun. *74*, 5085–5094.

Cani, P.D., Bibiloni, R., Knauf, C., Waget, A., Neyrinck, A.M., Delzenne, N.M., and Burcelin, R. (2008). Changes in gut microbiota control metabolic endotoxemia-induced inflammation in high-fat diet-induced obesity and diabetes in mice. Diabetes *57*, 1470–1481.

Chen, Y., and Murrell, J.C. (2010). When metagenomics meets stable-isotope probing: progress and perspectives. Trends Microbiol. *18*, 157–163.

Clarke, J.G., Peipert, J.F., Hillier, S.L., Heber, W., Boardman, L., Moench, T.R., and Mayer, K. (2002). Microflora changes with the use of a vaginal microbicide. Sex Transmitted Dis. *29*, 288–293.

Clayton, T.A., Baker, D., Lindon, J.C., Everett, J.R., and Nicholson, J.K. (2009). Pharmacometabonomic identification of a significant host-microbiome metabolic interaction affecting human drug metabolism. Proc. Natl. Acad. Sc.i U S A *106*, 14728–14733.

Costello, E.K., Lauber, C.L., Hamady, M., Fierer, N., Gordon, J.I., and Knight, R. (2009). Bacterial community variation in human body habitats across space and time. Science *326*, 1694–1697.

Courtois, S., Cappellano, C.M., Ball, M., Francou, F.X., Normand, P., Helynck, G., Martinez, A., Kolvek, S.J., Hopke, J., Osburne, M.S. *et al.* (2003). Recombinant environmental libraries provide access to microbial diversity for drug discovery from natural products. Appl. Environ. Microbiol. *69*, 49–55.

Craig, J.W., Chang, F.Y., Kim, J.H., Obiajulu, S.C., and Brady, S.F. (2010). Expanding small-molecule functional metagenomics through parallel screening of broad-host-range cosmid environmental DNA libraries in diverse proteobacteria. Appl. Environ. Microbiol. *76*, 1633–1641.

Cummings, J.H. (1983). Fermentation in the human large intestine: evidence and implications for health. The Lancet *1*, 1206–1209.

Dekio, I., Hayashi, H., Sakamoto, M., Kitahara, M., Nishikawa, T., Suematsu, M., and Benno, Y. (2005). Detection of potentially novel bacterial components of the human skin microbiota using culture-independent molecular profiling. J. Med. Microbiol. *54*, 1231–1238.

Dethlefsen, L., McFall-Ngai, M., and Relman, D.A. (2007). An ecological and evolutionary perspective on human-microbe mutualism and disease. Nature *449*, 811–818.

Diaz-Torres, M.L., McNab, R., Spratt, D.A., Villedieu, A., Hunt, N., Wilson, M., and Mullany, P. (2003). Novel tetracycline resistance determinant from the oral metagenome. Antimicrob. Agents Chem. *47*, 1430–1432.

Dominguez-Bello, M.G., Costello, E.K., Contreras, M., Magris, M., Hidalgo, G., Fierer, N., and Knight, R. (2010). Delivery mode shapes the acquisition and structure of the initial microbiota across multiple body habitats in newborns. Proc. Natl. Acad. Sci. U.S.A.*107*, 11971–11975.

Drasar, B.S., and Hill, M.J., eds. (1974). Human intestinal flora (London, Academic Press).

Dubos, R., Savage, D., and Schaedler, R. (1966). Biological freudianism- lasting effects of early environmental influences. Pediatrics *38*, 789–800.

Dumont, M.G., and Murrell, J.C. (2005). Stable isotope probing – linking microbial identity to function. Nature Rev. Microbiol. *3*, 499–504.

Eckburg, P.B., Bik, E.M., Bernstein, C.N., Purdom, E., Dethlefsen, L., Sargent, M., Gill, S.R., Nelson, K.E., and Relman, D.A. (2005). Diversity of the human intestinal microbial flora. Science *308*, 1635–1638.

Eschenbach, D.A., Thwin, S.S., Patton, D.L., Hooton, T.M., Stapleton, A.E., Agnew, K., Winter, C., Meier, A., and Stamm, W.E. (2000). Influence of the normal menstrual cycle on vaginal tissue, discharge, and microflora. Clin. Infect. Dis. *30*, 901–307.

Falk, P.G., Hooper, L.V., Midtvedt, T., and Gordon, J.I. (1998). Creating and maintaining the gastrointestinal ecosystem: What we know and need to know from gnotobiology. Microbiol. Mol. Biol. Rev. *62*, 1157–1170.

Fierer, N., Hamady, M., Lauber, C.L., and Knight, R. (2008). The influence of sex, handedness, and washing on the diversity of hand surface bacteria. Proc. Natl. Acad. Sci. U.S.A. *105*, 17994–17999.

Fierer, N., Lauber, C.L., Zhou, N., Mcdonald, D., Costello, E.K., and Knight, R. (2010). Forensic identification using skin bacterial communities. Proc. Natl. Acad. Sci., *107*, 6477–6481.

Finegold, S.M. (1977). Anaerobic bacteria in human disease (New York, Academic Press).

Finkbeiner, S.R., Allred, A.F., Tarr, P.I., Klein, E.J., Kirkwood, C.D., and Wang, D. (2008). Metagenomic analysis of human diarrhea: viral detection and discovery. PLoS Pathog *4*, e1000011.

Flint, H.J. (2004). Polysaccharide breakdown by anaerobic microorganisms inhabiting the mammalian gut. Adv. Appl. Microbiol. *56*, 89–120.

Flint, H.J., Bayer, E.A., Rincon, M.T., Lamed, R., and White, B.A. (2008). Polysaccharide utilization by gut bacteria: potential for new insights from genomic analysis. Nat. Rev. Microbiol. *6*, 121–131.

Flint, H.J., Duncan, S.H., Scott, K.P., and Louis, P. (2007). Interactions and competition within the microbial community of the human colon: links between diet and health. Environ. Microbiol. *9*, 1101–1111.

Follows, M.J., Dutkiewicz, S., Grant, S., and Chisholm, S.W. (2007). Emergent biogeography of microbial communities in a model ocean. Science *315*, 1843–1846.

Frank, D.N., and Pace, N.R. (2008). Gastrointestinal microbiology enters the metagenomics era. Curr. Opin. Gastroenterol. *24*, 4–10.

Frank, D.N., Spiegelman, G.B., Davis, W., Wagner, E., Lyons, E., and Pace, N.R. (2003). Culture-independent molecular analysis of microbial constituents of the healthy human outer ear. J. Clin. Microbiol. *41*, 295–303.

Frank, D.N., St Amand, A.L., Feldman, R.A., Boedeker, E.C., Harpaz, N., and Pace, N.R. (2007). Molecular-phylogenetic characterization of microbial community imbalances in human inflammatory bowel diseases. Proc. Natl. Acad. Sci. U.S.A. *104*, 13780–13785.

Frias-Lopez, J., Shi, Y., Tyson, G.W., Coleman, M.L., Schuster, S.C., Chisholm, S.W., and Delong, E.F. (2008). Microbial community gene expression in ocean surface waters. Proc. Natl. Acad. Sci. U.S.A. *105*, 3805–3810.

Gao, Z., Tseng, C.H., Pei, Z., and Blaser, M.J. (2007). Molecular analysis of human forearm superficial skin bacterial biota. Proc. Natl. Acad. Sci. U.S.A.*104*, 2927–2932.

Gibbons, R.J. (1989). Bacterial adhesion to oral tissues: a model for infectious diseases. J. Dental Res. *68*, 238–246.

Gibbons, R.J., Spinell, D.M., and Skobe, Z. (1976). Selective adherence as a determinant of the host tropisms of certain indigenous and pathogenic bacteria. Infection and Immunity *13*, 238–246.

Gill, S.R., Pop, M., Deboy, R.T., Eckburg, P.B., Turnbaugh, P.J., Samuel, B.S., Gordon, J.I., Relman, D.A., Fraser-Liggett, C.M., and Nelson, K.E. (2006). Metagenomic analysis of the human distal gut microbiome. Science *312*, 1355–1359.

Gillespie, D.E., Brady, S.F., Bettermann, A.D., Cianciotto, N.P., Liles, M.R., Rondon, M.R., Clardy, J., Goodman, R.M., and Handelsman, J. (2002). Isolation of antibiotics turbomycin A and B from a metagenomic library of soil microbial DNA. App. Environ. Microbiol. *68*, 4301–4306.

Gordon, H.A., and Pesti, L. (1971). The gnotobiotic animal as a tool in the study of host microbial relationships. Bacteriol. Rev. *35*, 390–429.

Gordon, J.I. *et al.* (2005). Extending Our View of Self: the Human Gut Microbiome Initiative (HGMI). http://www.genome.gov/Pages/Research/Sequencing/SeqProposals/HGMISeq.pdf.

Grice, E.A., Kong, H.H., Conlan, S., Deming, C.B., Davis, J., Young, A.C., Bouffard, G.G., Blakesley, R.W., Murray, P.R., Green, E.D. *et al.* (2009). Topographical and temporal diversity of the human skin microbiome. Science *324*, 1190–1192.

Grice, E.A., Kong, H.H., Renaud, G., Young, A.C., Bouffard, G.G., Blakesley, R.W., Wolfsberg, T.G., Turner, M.L., and Segre, J.A. (2008). A diversity profile of the human skin microbiota. Genome Res. *18*, 1043–1050.

Grimes, D.J., Atwell, R.T., Brayton, P.R., Palmer, L.M., Rollins, D.M., Roszak, D.B., Singleton, F.L., Tamplin, M.L., and Colwell, R.R. (1986). The fate of enteric pathogenic bacteria in estuarine and marine environments. Microbiological Sciences 3, 324–329.

Hallstrom, M., Eerola, E., Vuento, R., Janas, M., and Tammela, O. (2004). Effects of mode of delivery and necrotising enterocolitis on the intestinal microflora in preterm infants. Eur. J. Clin. Microbiol. Infect. Dis. *23*, 463–470.

Harmsen, H.J., Wildeboer-Veloo, A.C., Raangs, G.C., Wagendorp, A.A., Klijn, N., Bindels, J.G., and Welling, G.W. (2000). Analysis of intestinal flora development in breast-fed and formula-fed infants by using molecular identification and detection methods. J. Pediatr. Gastr. Nutr. *30*, 61–67.

Hartman, A.L., Lough, D.M., Barupal, D.K., Fiehn, O., Fishbein, T., Zasloff, M., and Eisen, J.A. (2009). Human gut microbiome adopts an alternative state following small bowel transplantation. Proc. Natl. Acad. Sci. U.S.A. *106*, 17187–17192.

Holdeman, L.V., Good, I.J., and Moore, W.E. (1976). Human fecal flora: variation in bacterial composition within individuals and a possible effect of emotional stress. Appl. Environ. Microbiol. *31*, 359–375.

Huang, W.E., Stoecker, K., Griffiths, R., Newbold, L., Daims, H., Whitely, A.S., and Wagner, M. (2007). Raman-FISH: combining stable-isotope Raman spectroscopy and fluorescence in situ hybridization for the single cell analysis of identity and function. Environ. Microbiol. *9*, 1878–1889.

Hungate, R.E., Smith, W., and Clarke, R.T. (1966). Suitability of butyl rubber stoppers for closing anaerobic roll culture tubes. J. Bacteriol. *91*, 908–909.

Jones, B.V., and Marchesi, J.R. (2007). Accessing the mobile metagenome of the human gut microbiota. Mol. Biosyst. *3*, 749–758.

Jones, B.V., Sun, F., and Marchesi, J.R. (2010). Comparative metagenomic analysis of plasmid encoded functions in the human gut microbiome. BMC Genomics *11*, 46.

Kendall, A. (1909). Some observations on the study of the intestinal bacteria. J. Biol. Chem. *6*, 499–507.

Khoruts, A., Dicksved, J., Jansson, J.K., and Sadowsky, M.J. (2010). Changes in the composition of the human fecal microbiome after bacteriotherapy for recurrent Clostridium difficile-associated diarrhea. J. Clin. Gastroenterol. *44*, 354–360.

Kirchman, D.L. (2008). Microbial ecology of the oceans, 2nd edition (New Jersey, John Wiley & Sons).

Koenig, J.E., Spor, A., Scalfone, N., Fricker, A.D., Stombaugh, J., Knight, R., Angenent, L.T., and Ley, R.E. (2010). Succession of microbial consortia in the developing infant gut microbiome. Proc. Natl. Acad. Sci. U.S.A. *108*, Suppl 1, 4578–4585.

Kovatcheva-Datchary, P., Egert, M., Maathuis, A., Rajilic-Stojanovic, M., de Graaf, A.A., Smidt, H., de Vos, W.M., and Venema, K. (2008). Linking phylogenetic identities of bacteria to starch fermentation in an *in vitro* model of the large intestine by RNA-based stable isotope probing. Environ. Microbiol. *11*, 914–926.

Kreuzer-Martin, H.W. (2007). Stable isotope probing: linking functional activity to specific members of microbial communities. Soil Sci. Soc. Am. J. *71*, 611–619.

Kroes, I., Lepp, P.W., and Relman, D.A. (1999). Bacterial diversity within the human subgingival crevice. Proc. Natl. Acad. Sci. U.S.A. *96*, 14547–14552.

Lane, D.J., Pace, B., Olsen, G.J., Stahl, D.A., Sogin, M.L., and Pace, N.R. (1985). Rapid determination of 16S ribosomal RNA sequences for phylogenetic analyses. Proc. Natl. Acad. Sci. U.S.A. *82*, 6955–6959.

Langendijk, P.S., Schut, F., Jansen, G.J., Raangs, G.C., Kamphuis, G., Wilkinson, M.H.F., and Welling, G.W.

(1995). Quantitative fluorescence *in situ* hybridization of Bifidobacterium spp. with genus-specific 16S rRNA targeted probe and its application in fecal samples. Appl Environ. Microbiol. *61*, 3069–3075.
Leavis, H.L., Bonten, M.J., and Willems, R.J. (2006). Identification of high-risk enterococcal clonal complexes: global dispersion and antibiotic resistance. Curr. Opin. Microbiol. 9, 454–460.
Lechene, C.P., Luyten, Y., McMahon, G., and Distel, D.L. (2007). Quantitative imaging of nitrogen fixation by individual bacteria within animal cells. Science *317*, 1563–1566.
Ley, R.E. (2010). Obesity and the human microbiome. Curr. Opin. Gastroenterol. *26*, 5–11.
Ley, R.E., Backhed, F., Turnbaugh, P., Lozupone, C.A., Knight, R.D., and Gordon, J.I. (2005). Obesity alters gut microbial ecology. Proc. Natl. Acad. Sci. U.S.A. *102*, 11070–11075.
Ley, R.E., Harris, J.K., Wilcox, J., Spear, J.R., Miller, S.R., Bebout, B.M., Maresca, J.A., Bryant, D.A., Sogin, M.L., and Pace, N.R. (2006a). Unexpected diversity and complexity of the Guerrero Negro hypersaline microbial mat. Appl. Environ. Microbiol. *72*, 3685–3695.
Ley, R.E., Lozupone, C.A., Hamady, M., Knight, R., and Gordon, J.I. (2008). Worlds within worlds: evolution of the vertebrate gut microbiota. Nat. Rev. Microbio.l *6*, 776–788.
Ley, R.E., Peterson, D.A., and Gordon, J.I. (2006b). Ecological and evolutionary forces shaping microbial diversity in the human intestine. Cell *124*, 837–848.
Ley, R.E., Turnbaugh, P.J., Klein, S., and Gordon, J.I. (2006c). Microbial ecology: human gut microbes associated with obesity. Nature *444*, 1022–1023.
Lozupone, C.A., and Knight, R. (2008). Species divergence and the measurement of microbial diversity. FEMS Microbiol. Rev. *32*, 557–578.
Manefield, M., Whiteley, A.S., Griffiths, R.I., and Bailey, M.J. (2002). RNA stable isotope probing, a novel means of linking microbial community function to Phylogeny. Appl. Environ. Microbiol. *68*, 5367–5373.
Matsuki, T., Watanabe, K., and Tanaka, R. (2003). Genus- and species-specific PCR primers for the detection and identification of bifidobacteria. Curr. Issues Intestinal Microbiol. *4*, 61–69.
Mazmanian, S.K. (2009). Gut immune balance is as easy as S-F-B. Immunity *31*, 536–538.
Monstein, H.J., Tiveljung, A., Kraft, C.H., Borch, K., and Jonasson, J. (2000). Profiling of bacterial flora in gastric biopsies from patients with Helicobacter pylori-associated gastritis and histologically normal control individuals by temperature gradient gel electrophoresis and 16S rDNA sequence analysis. J. Med. Microbiol. *49*, 817–822.
Murray, A.G., and Jackson, G.A. (1992). Viral dynamics – a model of the effects of size, shape, motion and abundance of single-celled planktonic organisms and other particles. Mar. Ecol. Prog. Ser. *89*, 103–116.
Nasidze, I., Li, J., Quinque, D., Tang, K., and Stoneking, M. (2009). Global diversity in the human salivary microbiome. Genome Res. *19*, 636–643.
Nelson, K.E., Weinstock, G.M., Highlander, S.K., Worley, K.C., Creasy, H.H., Wortman, J.R., Rusch, D.B., Mitreva, M., Sodergren, E., Chinwalla, A.T. *et al.* (2010). A Catalog of Reference Genomes from the Human Microbiome. Science, *328*, 994–999.
Ness, R.B., Hillier, S.L., Richter, H.E., Soper, D.E., Stamm, C., McGregor, J., Bass, D.C., Sweet, R.L., and Rice, P. (2002). Douching in relation to bacterial vaginosis, lactobacilli, and facultative bacteria in the vagine. Obstet. Gynecol. *100*, 765–772.
Neufeld, J.D., Dumont, M.G., Vohra, J., and Murrell, J.C. (2007a). Methodological considerations for the use of stable isotope probing in microbial ecology. Microb. Ecol. 53, 435–442.
Neufeld, J.D., Vohra, J., Dumont, M.G., Lueders, T., Manefield, M., Friedrich, M.W., and Murrell, J.C. (2007b). DNA stable-isotope probing. Nature Protocols 2, 860–866.
Nicholson, J.K. (2006). Global systems biology, personalized medicine and molecular epidemiology. Mol. Syst. Biol. 2, 52.
Nicholson, J.K., Holmes, E., and Wilson, I.D. (2005). Gut microorganisms, mammalian metabolism and personalized health care. Nat. Rev. Microbiol. 3, 431–438.
O'Sullivan, D.J. (2000). Methods for analysis of the intestinal microflora. Current Issues in Intestinal Microbiology *1*, 39–50.
Okuta, A., Ohnishi, K., and Harayama, S. (1998). PCR isolation of catechol 2,3-dioxygenase gene fragments from environmental samples and their assembly into functional genes. Gene *212*, 221–228.
Pace, N.R. (1997). A molecular view of microbial diversity and the biosphere. Science *276*, 734–740.
Pajecki, D., Zilberstein, B., dos Santos, M.A., Ubriaco, J.A., Quintanilha, A.G., Cecconello, I., and Gama-Rodrigues, J. (2002). Megaesophagus microbiota: a qualitative and quantitative analysis. J. Gastrointest. Surg. *6*, 723–729.
Palmer, C., Bik, E.M., Digiulio, D.B., Relman, D.A., and Brown, P.O. (2007). Development of the Human Infant Intestinal Microbiota. PLoS Biol. *5*, e177.
Pei, Z., Bini, E.J., Yang, L., Zhou, M., Francois, F., and Blaser, M.J. (2004). Bacterial biota in the human distal esophagus. Proc. Natl. Acad. Sci. U.S.A. *101*, 4250–4255.
Penders, J., Thijs, C., Vink, C., Stelma, F.F., Snijders, B., Kummeling, I., van den Brandt, P.A., and Stobberingh, E.E. (2006). Factors influencing the composition of the intestinal microbiota in early infancy. Pediatrics *118*, 511–521.
Peterson, D.A., Frank, D.N., Pace, N.R., and Gordon, J.I. (2008). Metagenomic approaches for defining the pathogenesis of inflammatory bowel diseases. Cell Host Microbe 3, 417–427.
Peterson, D.A., McNulty, N.P., Guruge, J.L., and Gordon, J.I. (2007). IgA response to symbiotic bacteria as a mediator of gut homeostasis. Cell Host Microbe *2*, 328–339.

Peterson, D.A., and Turnbaugh, P.J. (2010). A microbe-dependent viral key to Crohn's box. Sci. Transl. Med. *2*, 43ps39.

Pull, S.L., Doherty, J.M., Mills, J.C., Gordon, J.I., Stappenbeck, T.S. (2005). Activated macrophages are an adaptive element of the colonic epithelial progenitor niche necessary for regenerative responses to injury. Proc. Natl Acad. Sci. U.S.A. *102*, 99–104.

Qin, J., Li, R., Raes, J., Arumugam, M., Burgdorf, K.S., Manichanh, C., Nielsen, T., Pons, N., Levenez, F., Yamada, T., *et al.* (2010) A human gut microbial gene catalogue established by metagenomic sequencing. Nature *464*, 59–65.

Radajewski, S., McDonald, I.R., and Murrell, J.C. (2003). Stable-isotope probing of nucleic acids: a window to the function of uncultured microorganisms. Curr. Opin. Biotechnol. *14*, 296–302.

Ravel, J., Gajer, P., Abdo, Z., Schneider, G.M., Koenig, S.S., McCulle, S.L., Karlebach, S., Gorle, R., Russell, J., Tacket, C.O., *et al.* (2010). Vaginal microbiome of reproductive-age women. Proc. Natl. Acad. Sci. U.S.A. *108* Suppl 1, 4680–4687.

Rey, F.E., Faith, J.J., Bain, J., Muehlbauer, M.J., Stevens, R.D., Newgard, C.B., and Gordon, J.I. (2010). Dissecting the *in vivo* metabolic potential of two human gut acetogens. J. Biol. Chem. *285, 22082–22090.*

Reyes, A., Haynes, M., Hanson, N., Angly, F.E., Heath, A.C., Rohwer, F., and Gordon, J.I. (2010). Viruses in the faecal microbiota of monozygotic twins and their mothers. Nature *466*, 334–338.

Riesenfeld, C.S., Goodman, R.M., and Handelsman, J. (2004). Uncultured soil bacteria are a reservoir of new antibiotics resistance genes. Environ. Microbiol. *6*, 981–989.

Robinson, D.A., and Enright, M.C. (2004). Multilocus sequence typing and the evolution of methicillin-resistant *Staphylococcus aureus*. Clin. Microbiol. Infect. *10*, 92–97.

Savage, D.C. (1977). Microbial ecology of the gastrointestinal tract. Annu. Rev. Microbiol. *31*, 107–133.

Savage, D.C. (2001). Microbial biota of the human intestine: a tribute to some pioneering scientists. Curr. Issues Intestinal Microbiol. *2*, 1–15.

Savage, D.C., Dubos, R., and Schaedler, R.W. (1968). The gastrointestinal epithelium and its autochthonous bacterial flora J. Exp. Med. *127*, 67–76.

Schaedler, R.W., Dubos, R., and Costello, R. (1965). The development of the bacterial flora in the gastrointestinal tract of mice. J. Exp. Med. *122*, 59–66.

Schloss, P.D., and Handelsman, J. (2003). Biotechnological prospects from metagenomics. Curr. Opin. Biotechnol. *14*, 303–310.

Schwalbach, M.S., Hewson, I., and Fuhrman, J.A. (2004). Viral effects on bacterial community composition in marine plankton microcosms. Aquat. Microb. Ecol. *34*, 117–127.

Schwebke, J.R. (2009). New concepts in the etiology of bacterial vaginosis. Curr. Infect. Dis. Rep. *11*, 143–147.

Schwebke, J.R., Richey, C.M., and Weiss, H.L. (1999). Correlation of behaviours with microbiological changes in vaginal flora. J. Infect. Dis. *180*, 1632–1636.

Servin, A.L. (2004). Antagonistic activities of lactobacilli and bifidobacteria against microbial pathogens. FEMS Microbiol. Rev. *28*, 405–440.

Shiner, E.K., Rumbaugh, K.P., and Williams, S.C. (2005). Inter-kingdom signaling: deciphering the language of acyl homoserine lactones. FEMS Microbiol. Rev. *29*, 935–947.

Silvi, S., Rumney, C.J., and Rowland, I.R. (1996). An assessment of three selective media for bifidobacteria in faeces. J. Appl. Bacteriol. *81*, 561–564.

Smalla, K., Osburne, A.M., and Wellington, E.M.H. (2000). Isolation and characterization of plasmids from bacteria. In The horizontal gene pool, bacterial plasmids and gene spread, Thomas, C.M., ed. (Amsterdam, Harwood Academic Publishers).

Sommer, M.O., Church, G.M., and Dantas, G. (2010). A functional metagenomic approach for expanding the synthetic biology toolbox for biomass conversion. Mol. Syst. Biol. *6*, 360.

Sommer, M.O., Dantas, G., and Church, G.M. (2009). Functional characterization of the antibiotic resistance reservoir in the human microflora. Science *325*, 1128–1131.

Sonnenburg, J.L., Chen, C.T., and Gordon, J.I. (2006). Genomic and metabolic studies of the impact of probiotics on a model gut symbiont and host. PLoS Biol. *4*, e413.

Sonnenburg, J.L., Xu, J., Leip, D.D., Chen, C.H., Westover, B.P., Weatherford, J., Buhler, J.D., and Gordon, J.I. (2005). Glycan foraging *in vivo* by an intestine-adapted bacterial symbiont. Science *307*, 1955–1959.

Sousa, T., Paterson, R., Moore, V., Carlsson, A., Abrahamsson, B., and Basit, A.W. (2008). The gastrointestinal microbiota as a site for the biotransformation of drugs. Int. J. Pharm. *363*, 1–25.

Sperandio, V., Torres, A.G., Jarvis, B., Nataro, J.P., and Kaper, J.B. (2003). Bacteria-host communication: the language of hormones. Proc. Natl. Acad. Sci. U.S.A. *100*, 8951–8956.

Staley, J.T., and Konopka, A. (1985). Measurement of *in situ* activities of nonphotosynthetic microorganisms in aquatic and terrestrial habitats. Annu. Rev. Microbiol. *39*, 321–346.

Stein, J.L., Marsh, T.L., Wu, K.Y., Shizuya, H., and DeLong, E.F. (1996). Characterization of uncultivated prokaryotes: isolation and analysis of a 40-kilobase-pair genome fragment from a planktonic marine archaeon. J. Bacteriol. *178*, 591–599.

Stewart, F.J., Ottesen, E.A., and DeLong, E.F. (2010). Development and quantitative analyses of a universal rRNA-subtraction protocol for microbial metatranscriptomics. Isme J. *4*, 896–907.

Strocchi, A., Furne, J., Ellis, C., and Levitt, M. (1994). Methanogens outcompete sulphate reducing bacteria for H2 in the human colon. Gut 35, 1098–1101.

Suau, A., Bonnet, R., Sutren, M., Godon, J.J., Gibson, G.R., Collins, M.D., and Dore, J. (1999). Direct analysis of genes encoding 16S rRNA from complex communities reveals many novel molecular species within the human gut. Appl. Environ. Microbiol. *65*, 4799–4807.

Tannock, G.W. (1983). Effect of dietary and environmental stress on the gastrointestinal microbiota. In Human intestinal microflora in health and disease, Hentges, D.J., ed. (New York, Academic Press), pp. 517–539.

Tannock, G.W. (1984). Control of gastrointestinal pathogens by normal flora. In Current perspectives in microbial ecology, Klug, M.J., and Reddy, C.A., eds (Washington DC, American Society for Microbiology), pp. 374–382.

Tannock, G.W. (1999). Identification of lactobacilli and bifidobacteria. Curr. Issues Mol. Biol. *1*, 53–56.

Thingstad, T.F., and Lignell, R. (1997). Theoretical models for the control of bacterial growth rate, adundance, diversity and carbon demand. Aquat. Microb. Ecol. *13*, 19–27.

Torsvik, V., Goksoyr, J., and Daae, F.L. (1990). High diversity in DNA of soil bacteria. Appl. Environ. Microbiol. *56*, 782–787.

Tschöp, M., Hugenholtz, P., and Karp, C. (2009). Getting to the core of the gut microbiome. Nat. Biotechnol. *27*, 344–346.

Turnbaugh, P.J., Backhed, F., Fulton, L., and Gordon, J.I. (2008). Diet-induced obesity is linked to marked but reversible alterations in the mouse distal gut microbiome. Cell Host Microbe *3*, 213–223.

Turnbaugh, P.J., and Gordon, J.I. (2008). An invitation to the marriage of metagenomics and metabolomics. Cell *134*, 708–713.

Turnbaugh, P.J., Hamady, M., Yatsunenko, T., Cantarel, B.L., Duncan, A., Ley, R.E., Sogin, M.L., Jones, W.J., Roe, B.A., Affourtit, J.P. *et al.* (2009a). A core gut microbiome in obese and lean twins. Nature *457*, 480–484.

Turnbaugh, P.J., Ley, R.E., Hamady, M., Fraser-Liggett, C.M., Knight, R., and Gordon, J.I. (2007). The human microbiome project. Nature *449*, 804–810.

Turnbaugh, P.J., Ley, R.E., Mahowald, M.A., Magrini, V., Mardis, E.R., and Gordon, J.I. (2006). An obesity-associated gut microbiome with increased capacity for energy harvest. Nature *444*, 1027–1031.

Turnbaugh, P.J., Quince, C., Faith, J.J., McHardy, A.C., Yatsunenko, T., Niazi, F., Affourtit, J., Egholm, M., Henrissat, B., Knight, R. *et al.* (2010). Organismal, genetic, and transcriptional variation in the deeply sequenced gut microbiomes of identical twins. Proc. Natl. Acad. Sci. U.S.A. *107*, 7503–7508.

Turnbaugh, P.J., Ridaura, V.K., Faith, J.J., Rey, F.E., Knight, R., and Gordon, J.I. (2009b). The effect of diet on the human gut microbiome: a metagenomic analysis in humanized gnotobiotic mice. Sci. Transl. Med. *1*, 6ra14.

Uchiyama, T., Abe, T., Ikemura, T., and Watanabe, K. (2005). Substrate-induced gene-expression screening of environmental metagenome libraries for isolation of catabolic genes. Nat Biotechnol. *23*, 88–93.

Verberkmoes, N.C., Russell, A.L., Shah, M., Godzik, A., Rosenquist, M., Halfvarson, J., Lefsrud, M.G., Apajalahti, J., Tysk, C., Hettich, R.L. *et al.* (2009). Shotgun metaproteomics of the human distal gut microbiota. Isme J. *3*, 179–189.

Verhelst, R., Verstraelen, H., Claeys, G., Verschraegen, G., Delanghe, J., Van Simaey, L., De Ganck, C., Temmerman, M., and Vaneechoutte, M. (2004). Cloning of 16S rRNA genes amplified from normal and disturbed vaginal microflora suggests a strong association between Atopobium vaginae, Gardnerella vaginalis and bacterial vaginosis. BMC Microbiol. *4*, 16.

Vijay-Kumar, M., Aitken, J.D., Carvalho, F.A., Cullender, T.C., Mwangi, S., Srinivasan, S., Sitaraman, S.V., Knight, R., Ley, R.E., and Gewirtz, A.T. (2010). Metabolic syndrome and altered gut microbiota in mice lacking Toll-like receptor 5. Science *328*, 228–231.

Wagner, M. (2009). Single-cell ecophysiology of microbes as revealed by Raman microspectroscopy or secondary ion mass spectrometry imaging. Annu. Rev. Microbiol. *63*, 411–429.

Walker, A.W., Ince, J., Duncan, S.H., Webster, L.M., Holtrop, G., Ze, X., Brown, D., Stares, M.D., Scott, P., Bergerat, A. *et al.* (2010). Dominant and diet-responsive groups of bacteria within the human colonic microbiota. ISME J. *5*, 220–230.

Walter, J., Britton, R.A., and Roos, S. (2010). Host–microbial symbiosis in the vertebrate gastrointestinal tract and the Lactobacillus reuteri paradigm. Proc. Natl. Acad. Sci. U.S.A. *108* Suppl 1, 4645–4652.

Ward, D.M., Weller, R., and Bateson, M.M. (1990). 16S rRNA sequences reveal numerous uncultured organisms. Nature *345*, 63–65.

Willner, D., Furlan, M., Schmieder, R., Grasis, J.A., Pride, D.T., Relman, D.A., Angly, F.E., McDole, T., Mariella, R.P., Jr., Rohwer, F. *et al.* (2010). Metagenomic detection of phage-encoded platelet-binding factors in the human oral cavity. Proc. Natl. Acad. Sci. U.S.A. *108* Suppl 1, 4547–4553.

Wilson, K.H., and Blitchington, R.B. (1996). Human colonic biota studied by ribosomal DNA sequence analysis. Appl. Environ. Microbiol. *62*, 2273–2278.

Winter, C., Bouvier, T., Weinbauer, M.G., Thingstad, T.F. (2010). Trade-offs between competition and defense specialists among unicellular planktonic organisms: the 'killing the winner' hypothesis revisited. Microbiol. Mol. Biol. Rev. *74*, 42–57.

Wirth, T., Wang, X., Linz, B., Novick, R.P., Lum, J.K., Blaser, M., Morelli, G., Falush, D., and Achtman, M. (2004). Distinguishing human ethnic groups by means of sequences from Helicobacter pylori: Lessons from Ladakh. Proc. Natl. Acad. Sci. U.S.A.. *101*, 4746–4751.

Woese, C.R. (2000). Interpreting the universal phylogenetic tree. Proc. Natl. Acad. Sci. U.S.A. *97*, 8392–8396.

Wu, D., Hugenholtz, P., Mavromatis, K., Pukall, R., Dalin, E., Ivanova, N.N., Kunin, V., Goodwin, L., Wu, M., Tindall, B.J. *et al.* (2009). A phylogeny-driven genomic encyclopaedia of Bacteria and Archaea. Nature, *462*, 1056–1060.

Yachi, S., and Loreau, M. (1999). Biodiversity and ecosystem productivity in a fluctuating environment: The insurance hypothesis. Proc. Natl. Acad. Sci. U.S.A. *96*, 1463–1468.

Yun, J., and Ryu, S. (2005). Screening for novel enzymes from metagenome and SIGEX, as a way to improve it. Microbial Cell Factories 4.

Zaura, E., Keijser, B.J., Huse, S.M., and Crielaard, W. (2009). Defining the healthy 'core microbiome' of oral microbial communities. BMC Microbiol. *9*, 259.

Zhang, T., Breitbart, M., Lee, W.H., Run, J.Q., Wei, C.L., Soh, S.W., Hibberd, M.L., Liu, E.T., Rohwer, F., and Ruan, Y. (2006). RNA viral community in human feces: prevalence of plant pathogenic viruses. PLoS Biol. *4*, e3.

Zhou, X., Bent, S.J., Schneider, M.G., Davis, C.C., Islam, M.R., and Forney, L.J. (2004). Characterization of vaginal microbial communities in adult healthy women using cultivation-independent methods. Microbiology *150*, 2565–2573.

Zhou, X., Brown, C.J., Abdo, Z., Davis, C.C., Hansmann, M.A., Joyce, P., Foster, J.A., and Forney, L.J. (2007). Differences in the composition of vaginal microbial communities found in healthy Caucasian and black women. ISME J. *1*, 121–133.

Zhou, X., Hansmann, M.A., Davis, C.C., Siuzuki, H., Brown, C.J., Schutte, U., Pierson, J.D., and Forney, L.J. (2009a). The vaginal bacterial communities of Japanese women resemble those of women in other racial groups. FEMS Immunol. Med. Microbiol. *58*, 169–181.

Zhou, X., Westman, R., Hickey, R., Hansmann, M.A., Kennedy, C., Osborn, T.W., and Forney, L.J. (2009b). Vaginal microbiota of women with frequent vulvovaginal candidiasis. Infect. Immun. *77*, 4130–4135.

Zivkovic, A.M., German, J.B., Lebrilla, C.B., and Mills, D.A. (2010). Human milk glycobiome and its impact on the infant gastrointestinal microbiota. Proc. Natl. Acad. Sci. U.S.A. *108* Suppl 1, 4653–4658.

Zoetendal, E.G., Akkermans, A.D.L., Akkermans-van Vliet, W.M., de Visser, J.A.G.M., and de Vos, W.M. (2001). The host genotype affects the bacterial community in the human gastrointestinal tract. Microb. Ecol. Health Dis. *13*, 129–134.

Metagenomics and Integrative '-omics' Technologies in Microbial Bioremediation: Current Trends and Potential Applications

12

Varun Shah, Kunal Jain, Chirayu Desai and Datta Madamwar

Abstract

Implementation of efficacious bioremediation strategies relies heavily on intrinsic microbial community dynamics, structure and function. Any one particular microorganism is incapable of processing all the metabolic reactions to degrade environmental pollutants, however a group of diverse organisms form a community and collectively process all the metabolic reactions for bioremediation. Therefore, metagenomics-based analyses of entire microbial community genomes becomes imperative to delineate the metabolic pathways responsible for biodegradation. The essential genes for bioremediation may be present, however to ascertain how many of them are involved in bioremediation we need high throughput metatranscriptomics and metaproteomics, transcriptome and proteome analyses of entire community, respectively. Metametabolomics, analyses of the entire repertoire of microbial community metabolites and fluxomics, real time flux analysis of molecules/metabolites over a time period provide the missing links about regulation of metabolism of anthropogenic/xenobiotic compounds. Interactive studies between metagenomics, metatranscriptomics, metaproteomics and metametabolomics have become a trend in microbial bioremediation. In this chapter, we discuss the potential of recent innovative breakthroughs in molecular and 'omics' technologies such as molecular profiling, ultrafast pyro-sequencing, microarrays, mass spectrometry and other novel techniques and applications along with bioinformatics tools to gain insights of indigenous microbial communities and their mechanism in bioremediation of environmental pollutants.

Introduction

The 20th century industrialization has now compelled us to think about developing environmental remediation strategies on the priority basis to save the basic essential components of life. The initially developed physical and chemical environmental clean-up technologies are expensive, energy consuming and most importantly are not completely environment-friendly. Consequently, the focus shifted on developing self-sustainable and eco-friendly microbial bioremediation technologies. Prior to designing bioremediation strategies, it is necessary to inspect and assess the contaminated site. Factors like climate, soil characteristics, geological aspects, water levels, waste and disposal facilities and such other factors play an important role in success of bioremediation treatments. Implementation of efficacious bioremediation strategies relies heavily on indigenous microbial community dynamics – structure and function. Microbial communities are fundamental components of ecosystems playing critical role in the metabolism and detoxification of anthropogenic/xenobiotic compounds. Microbial bioremediation strategies can be applied either *ex situ* or *in situ* in order to restore a contaminated environment. *Ex situ* treatment involves an off-site separate treatment facility usually a bioreactor or an effluent treatment plant for degradation of contaminants. *In situ* treatment involves on-site bioremediation of the contaminants by monitoring the efficiencies of natural attenuation (progressive removal of contaminant) and then facilitating the degradation by intentional biostimulation of indigenous microbial communities (by providing nutrients and electron acceptors/donors) or bioaugmentation (by adding microbes and nutrients) (Desai *et al.*, 2010). Bioremediation has been successfully applied for oil spill removal, hydrocarbon degradation, wastewater treatments, metal removal, xenobiotic/recalcitrant compound degradation and many other such contaminant removal applications (Ham and Bonner, 1997).

A multilevel molecular biology approach, combining molecular- (including DNA, mRNA, protein and metabolites), cellular-, individual- and community-level data represents a powerful new multi disciplinary approach to decipher complex biological systems. 'Omics' platform that have been used in microbial systems biology include genomics – which determines the sequence of either whole genome or group of genes; transcriptomics – which measures mRNA transcript levels; proteomics – which quantifies protein abundance; metabolomics – which determines abundance of small cellular metabolites; interactomics – which resolves the whole set of molecular interactions in cells; and fluxomics – which establishes dynamic changes of molecules within a cell over a time. However, no single 'omics' analysis can fully unravel the complexities of fundamental microbial ecology. Therefore, integration of multiple layers of information, the multi-'omics' approach, is required to acquire a precise picture of living micro-organisms and their metabolisms (Zhang *et al.*, 2010).

Genomics and metagenomics

Genomics is the analysis of genome of an organism and metagenomics is to access the total genomic potential of an environmental sample either directly or after enrichment of indigenous microbial communities. The term 'metagenomics' (also known as community genomics, ecogenomics, or environmental genomics) has had the greatest impact within the last few years as it has enabled to explore the uncultivable microorganisms and understand their role in various metabolic pathways. Cultivation-independent analyses of the microbial community structures at contaminated sites have augmented to our understanding of the community dynamics, relative abundance and distribution of micro-organisms actively involved in bioremediation (Desai *et al.*, 2010). The genes responsible for biodegradation pathways are usually arranged in clusters that comprise (i) catabolic genes encoding the enzymatic steps of the catabolic pathway; (ii) transport genes responsible for active uptake of the compound; and (iii) regulatory genes that adjust expression of the catabolic and transport genes to the presence of the compound to be degraded (Diaz and Prieto, 2000; Diaz, 2004).

Of all the 'omic' studies, genomics and metagenomics have had the greatest share of focus in the last two decades, mainly because of rapid advances in several molecular technologies. These technologies have completely revolutionized the approaches of studies and have provided enormous information about the microbial genome(s). However, two technologies – amplification (PCR) and sequencing need a special mention as they have conferred the major thrust to genomics and metagenomics. As a corollary of revolutionary developments in high-throughput DNA sequencing technologies, more than thousand microbial genomes from almost all known major phylogenetic lineages have been fully sequenced, many are nearing completion and many more are regularly initiated all over the world. The computational-based annotation and comparative genomic analyses of DNA sequences have provided biologists with information regarding gene function, genome structures, biological pathways, metabolic and regulatory networks, and evolution of microbial genomes, which has greatly enhanced our understanding of microbial metabolism (Schoolnik, 2001; Ward and Fraser, 2005; Sharan and Ideker, 2006; Cardenas and Tiedje, 2008; Rocha, 2008; Zhang *et al.*, 2010).

A significant achievement in the field of microbial ecology was the finding of highly conserved as well as variable gene sequences that are present in all microorganisms, most notable of them is the 16S rRNA gene, which has been considered as a 'gold standard' for characterizing phylogenetic affiliations of micro-organisms that comprise microbial communities (Lane *et al.*, 1985; Amann *et al.*, 1995; Lovley, 2003). Therefore, by analysing 16S rRNA sequences in contaminated environments it was possible to determine definitively the phylogenetic placement of the microorganisms that are associated with bioremediation processes (Watanabe *et al.*, 2000; Lovley, 2003; Rogers *et al.*, 2003). Bittner *et al.* (2010) have provided a hypothesis that integrative metagenomic studies – aiming at the simultaneous examination (or even better at the integration) of observations about the elements, functions and evolutionary processes captured by the massive sequencing of multiple markers – should be preferred over DNA barcoding projects and over metagenomic projects based on a single marker. Moreover, such metagenomic data should be studied with novel inclusive network-based approaches, designed to draw inferences both on the many units and on the many processes present in the environments. Improving knowledge of biodiversity will benefit conservation biology, enhance bioremediation studies, and will lead to new treatments (Bittner *et al.*, 2010).

The development of microbial ecological DNA microarrays has enabled researchers to simultaneously analyse thousands of phylogenetic links and functional genes in order to characterize microbial communities involved in bioremediation (He *et al.*, 2007). Moreover, the functional screening of metagenomic libraries has led to the discovery of novel genes encoding polyphenol oxidase (Beloqui *et al.*, 2006), ester and glycosyl hydrolase (Ferrer *et al.*, 2005) and also given indications on the diversity of extradiol dioxygenases in coke plant wastewater (Suenaga *et al.*, 2007) and diversity of extradiol dioxygenases at a contaminated site under bioremediation treatment was analysed (Brennerova *et al.*, 2009).

Metagenomic analyses in bioremediation entail (i) identification of genes involved in bioremediation; (ii) characterization of genes and complete genomes; and (iii) phylogenetic affiliation of key catabolic genes.

Conversely, sheer detection of a gene or a cluster of genes does not bequeath with a complete picture as the information about the mRNA expression level, the amount of protein produced and its location, biological activity or functional relationship with metabolomes are required for understanding any metabolic reaction. Moreover, in cells many levels of regulation occur after genes have been transcribed, such as post-transcriptional, translational and post-translational regulation and all forms of biochemical control such as allosteric or feedback regulation. Hence, to fully elucidate microbial metabolism and its responses to environmental factors, it is necessary to include functional characterization and accurate quantification of all levels of gene products, mRNA, proteins and metabolites, as well as their interaction (Zhang *et al.*, 2010).

Transcriptomics and metatranscriptomics

The genes for bioremediation may be present but not expressed. Therefore there has been an increased emphasis on quantifying the levels of mRNA for key bioremediation genes. Transcriptomics is analysis of transcripts of a single organism and metatranscriptomics is analyses of transcripts of the entire interacting community. Transcriptomics, also called as gene expression profiling, provides the insights into the up- or down- regulation of genes under stress conditions in environmental microbial communities. The first regulatory point for successful synthesis of a protein from a gene is regulation of gene expression, one of the key processes for adapting to sustain in affected environmental conditions. Often increased mRNA concentration can be at least qualitatively associated with higher rates of contaminant degradation (Schneegurt *et al.*, 1998; Lovley, 2003). Although gene expression is affected by many environmental factors, a subset of genes with altered expression can inform on stress responses. The potential utility here is to improve biomarker identification and to identify patterns of gene expression associated with different types of pollutants. Analysis of the mRNA concentration for genes other than those directly involved in bioremediation might yield additional insights into the factors that control the rate and extent of bioremediation (Lovley, 2003).

Transcriptome analyses have mainly been dependent on technologies such as PCR, microarray and sequencing. Highly sensitive methods that can detect mRNA for key bioremediation genes in single cells are now available (Bakermans and Madsen, 2002). Conversely, the only major precaution required is that RNA have a very short life (some transcripts have life of less than a minute). Consequently, RNA storage after sample collection and preparation is very important step and strategies for transcriptome analysis need to be designed accordingly.

Metatranscriptome analyses in bioremediation entail (i) identification of differentially expressed mRNA and their respective genes; and (ii) assessing changes in the expression of previously characterized key genes (biomarkers).

Proteomics and metaproteomics

Proteomics is the analyses of proteins of a single organism and metaproteomics is to analyse the protein expression profile of the whole community. Proteins, the major components of cell, serve as catalytic enzymes in metabolic pathways and in signal transduction of regulatory pathways of cells (Graham *et al.*, 2007; Zhang *et al.*, 2010). The cellular expression of proteins in an organism varies with environmental conditions. The changes in physiological response may occur due to the organism's adaptive responses to different external stimuli such as presence of toxic chemicals in the environment. Developments in functional proteomics approach for environmental remediation will further pave the way towards cell free bioremediation (Singh and Nagaraj, 2006). However, today's technologies are capable of resolving only a minute fraction of metaproteome of complex environments, consequently substantial improvements in the technologies for protein extraction, separation and identification are necessary to encompass a complete picture (Wilmes and Bond, 2006).

Proteome analyses have revolved around two techniques – Two dimensional gel electrophoresis (2 DE) and mass spectrometry (MS). MS has been integrated not just with 2 DE but also with different chromatographs and various other techniques.

Metaproteome analyses in bioremediation entail: (i) identification of differentially expressed proteins and their respective genes; (ii) assessing changes in the abundance of previously characterized key proteins (biomarkers); and (iii) protein structure and function characterization.

Metabolomics, metametabolomics and fluxomics

Metabolomics is analysis of entire repertoire of cellular metabolites and metametabolomics is analyses of metabolites of entire interacting community. Real time flux analysis of cellular molecules/metabolites within a cell/community over a time period is known as fluxomics (Wiechert *et al.*, 2007).

A microbial cell releases a number of low molecular weight primary and secondary metabolites in response to an environment challenge or stress. The influence of the local environment on the metabolome of an organism/community can be exploited in bioremediation to characterize the effects of xenobiotic compounds. Metabolites can be characterized by mass spectrometry and various spectroscopic techniques. Villas-Boas and Bruheim (2007) have discussed the scenario of metabolome analysis in bioremediation. They have focused on potential of various experimental and conceptual approaches developed for metabolomics to be applied in bioremediation research, such as strategies for elucidation of biodegradation pathways using isotope distribution analysis and molecular connectivity analysis, the assessment of mineralization process using metabolic footprinting analysis, and the improvement of the biodegradation process via metabolic engineering.

'Meta'metabolome analyses in bioremediation entail: (i) identification of differential metabolic profile; and (ii) quantifying functional roles of metabolites. Advances in molecular technologies over last few decades have removed the boundaries between genomics, transcriptomics, proteomics and metabolomics. Functional analyses of microbial communities at different molecular levels of 'transcriptomes, proteomes and metabolomes' in coordination with genome fingerprint have been used to predict microbial biodegradation pathways. In this chapter we discuss some basic concepts of various experimental 'omics' approaches and recent application of the integrated 'omics' for exploring metabolic and regulatory mechanisms in microbial bioremediation. Online bioinformatics tools and databases that is available for application of 'omics' technologies in bioremediation are also briefly discussed.

Milestones in 'omic' studies and technologies

There have been numerous discoveries and innovations in technologies which have changed the course of scientific research. Besides these, even some novel ideas and/or approaches have given new insights and dimensions to the scientific thoughts. These technologies and approaches revolutionize not just one specific scientific field of area or study but all the scientific fields gain benefit. Table 12.1 describes some of the important milestones that have revolutionized the approach of molecular biology.

Integrating molecular and '-omics' technologies to gain insights into microbial bioremediation

Integration of various approaches and technologies is essential to obtain a more comprehensive view of the scientific questions. Vilchez-Vargas *et al.* (2010) have presented the current status of the capabilities to assess and predict catabolic potential of environmental sites by applying gene fingerprinting, catabolome arrays, metagenomics and complementary 'omics' technologies in their review analysis. Fig. 12.1 presents an overall framework for integration of molecular and 'omic' approaches to gain insight into microbial bioremediation. In the following sections various technologies and their applications in bioremediation are described with special emphasis on applications wherein two technologies have been used to complement each other. This will be of assistance in developments of improved bioremediation strategies.

Nucleic acid-based molecular techniques for fingerprinting

Nucleic acid-based molecular techniques such as denaturing gradient gel electrophoresis (DGGE), temperature gradient gel electrophoresis (TGGE), terminal restriction fragment length polymorphism (T-RFLP), amplified ribosomal DNA restriction analysis (ARDRA), random amplified polymorphic DNA (RAPD), amplified fragment length polymorphism (AFLP), automated rRNA intergenic spacer analysis (ARISA) and single strand conformation polymorphism (SSCP) enable us to monitor the changes in bacterial community in detail. Such advanced molecular microbiological techniques provide new insights into bioremediation in terms of process optimization, validation, and the impact on the ecosystem, which are indispensable data to make the technology reliable and safe.

Denaturing gradient gel electrophoresis (DGGE) and temperature gradient gel electrophoresis (TGGE) are forms of electrophoresis and are used to separate polymerase chain reaction (PCR)-generated DNA products. The polymerase chain reaction of environmental DNA can generate templates of differing DNA sequence that represent many of the dominant microbial organisms. During DGGE, PCR products encounter increasingly higher concentrations of chemical denaturant as they migrate through a polyacrylamide gel. Upon reaching a threshold denaturant concentration, the weaker melting domains of the double-stranded PCR product will begin to denature at which time migration slows dramatically. Differing sequences of DNA (from different bacteria) will denature at different denaturant concentrations resulting in a pattern of bands. Once generated, fingerprints can be uploaded into databases in which fingerprint similarity can be assessed to determine microbial structural differences between environments or among treatments. Furthermore, with the range of PCR primers available, DGGE can also be used to investigate broad phylogenies or specific xenobiotic degraders. Whereas, in TGGE the only change is that PCR products encounter increasingly higher temperature as they migrate through a polyacrylamide gel.

Table 12.1 Milestones in 'omic' studies and technologies

(a) Molecular and 'omic' studies

Genomics	Fred Sanger and his group established techniques of sequencing, genome mapping, data storage and bioinformatics analyses. They sequenced complete genome of virus and mitochondrion. The first bacterial genome to be completed was that of *Haemophilis influenzae* in 1995.	1970s-1980s
Metabolomics	The concept that individuals might have a metabolic profile was introduced by Roger Williams in late 1940s. However, the term 'metabolic profile' was introduced by Horning in 1971.	1971
Bioinformatics	The term bioinformatics was coined by Paulien Hogeweg.	1979
16S rRNA gene	The pioneering work of Carl Woese, using 16SrRNA as 'evolutionary chronometers' in deducing the bacterial phylogeny changed the way we visualize the microbial world.	1977
Ribosomal Database Project	The Ribosomal Database Project (RDP) provides ribosome related data and services to the scientific community, including online data analysis and aligned and annotated Bacterial and Archaeal small-subunit 16S rRNA sequences. Moreover, the database is regularly augmented and upgraded for new sequences and better analysis.	1992
Proteomics	The term proteomics was coined to make analogy with genomics.	1997
Metagenomics	Norman Pace and co-workers were the first to clone genes directly from environmental samples in 1991. The term was coined by Jo Handelsman in 1998.	2000s
Metaproteomics	The term metaproteomics was proposed by Francisco Rodríguez-Valera to describe the genes and/or proteins most abundantly expressed in environmental samples.	2004
Multi'omics'	Researchers are developing platforms to integrate all the studies and technologies.	2005 onwards

(b) Technologies

Mass spectrometry	The first application of mass spectrometry for analysis of amino acids and peptides was reported in 1958. Mass spectrometry is a combination of many technologies. Many Nobel prizes in physics and chemistry have been awarded to studies linked to mass spectrometry and this in itself highlights the importance of this technology.	1958
Two-dimensional gel electrophoresis	O'Farrell developed this technique and revolutionized protein purification and characterization.	1975
Sequencing	Maxam–Gilbert and chain termination methods, developed between 1975 and 1980, were the first true and feasible sequencing technologies.	1975–1980
PCR	Kary Mullis developed PCR technology and this invention revolutionized the genomic era and provided a large thrust to many of the scientific discoveries.	1983
Microarray	Microarray evolved from Southern blotting and the use of collection of distinct DNAs was first time described in 1987.	1987
Pyrosequencing	The technique was developed by Mostafa Ronaghi in 1996. However, the major applications started only after 2005. Robert Edwards and his colleagues published the first sequences of environmental samples generated with so-called next-generation sequencing, chip-based pyrosequencing developed by 454 Life Sequences.	1996

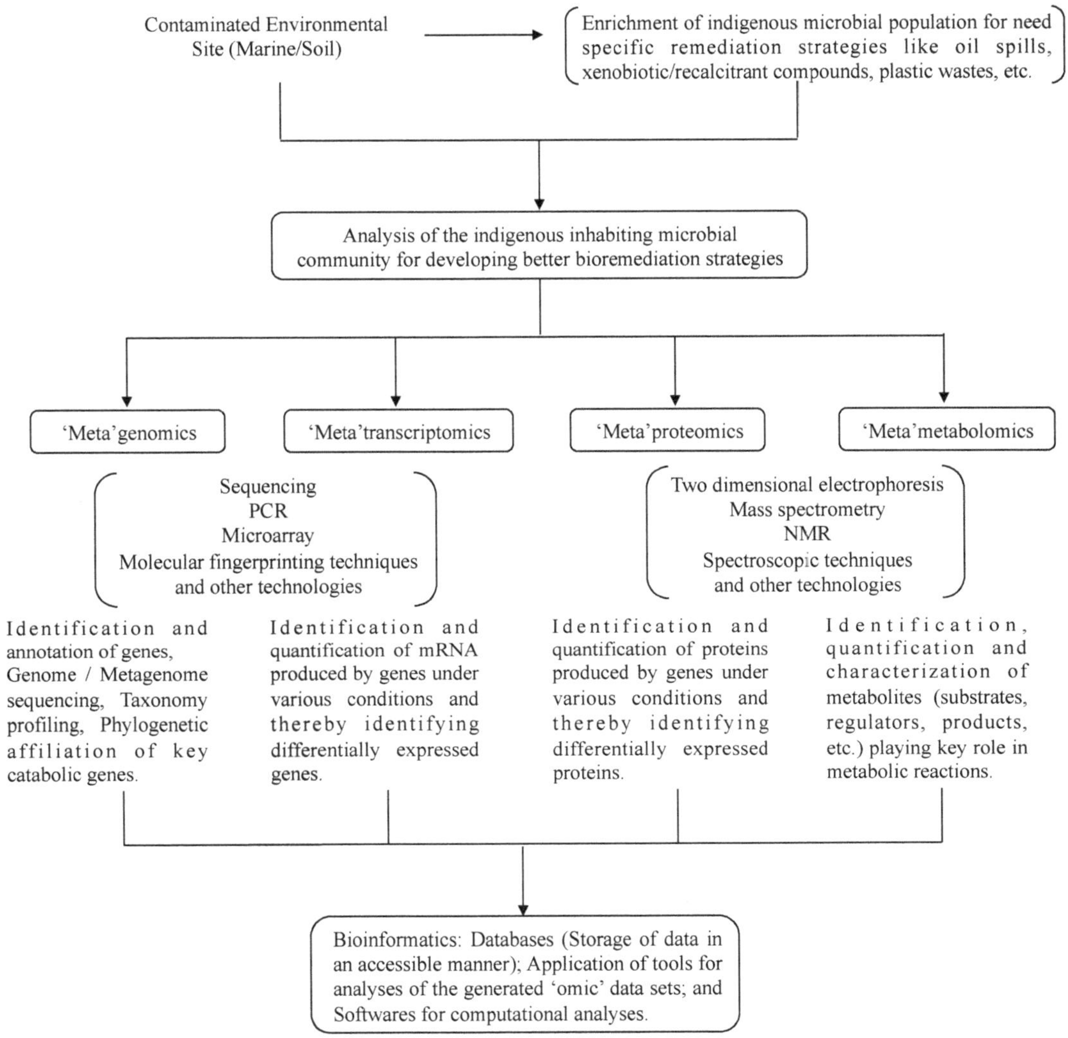

Figure 12.1 Integrating molecular and 'omic' approaches to gain insight into microbial bioremediation.

Wakase *et al.* (2008) used polymerase chain reaction-denaturing gradient gel electrophoresis (PCR-DGGE) for detection of all the predominant species in the microbial succession during composting process. Li *et al.* (2006) applied DGGE fingerprinting for detecting shifts in microbial community diversity in Cu-, Zn- and Cd- contaminated paddy soils. Geets *et al.* (2006) applied dsrB gene-based DGGE for community and diversity surveys of sulfate-reducing bacteria. When real-time polymerase chain reaction (RT-PCR) was applied in conjunction with denaturing gradient gel electrophoresis (DGGE) to study a petroleum-hydrocarbon contaminated site, the genes of phenol hydroxylase (PHE), ring-hydroxylating toluene monooxygenase (RMO), naphthalene dioxygenase (NAH), toluene monooxygenase (TOL), toluene dioxygenase (TOD), and biphenyl dioxygenase (BPH4) were quantified by real-time PCR and variations in microbial community were analysed by DGGE. Knowledge and comprehension obtained from this study was helpful in evaluating the occurrence and effectiveness of intrinsic bioremediation of petroleum-hydrocarbon contaminated groundwater (Kao *et al.*, 2010). Denaturing gradient gel electrophoresis (DGGE) along with microsensor techniques have been widely used in microbial ecology. In one such study carried out by Abed and Grotzschel (2005), DGGE provides information on microbial diversity of polluted ecosystems and microsensors detect the microenvironment of contaminated sediments and monitor changes in various microbial metabolic processes such as respiration, photosynthesis and sulfate reduction.

Terminal-restriction fragment length polymorphism (T-RFLP) is a technique for profiling microbial communities based on the position of restriction sites. This method is based on amplification of marker genes by primers tagged with fluorescent labels. The resultant mixture of amplicon products are digested by one or more restriction enzymes to generate fragments that are separated by capillary electrophoresis. Every single fragment represents a particular phylogenetic lineage or operational taxonomic unit (OTU). The size and relative abundance can be detected using an automated DNA sequencing instrument. The result of a T-RFLP profiling is a graph called electropherogram which is an intensity plot representation of peaks corresponding to bands on gel. The peak patterns can be compared to databases by several web-based tools such as PAT, TRUFFLER, APLAUS. Many online tools such as TAP-TRFLP (http://rdp.cme.msu.edu), torast (http://www.torast.de) and MiCA (http://mica.ibest.uidaho.edu) have been developed to perform in-silico analysis for choosing appropriate combinations of primers and restriction enzymes. Kim *et al.* (2008a) monitored microbial community composition by bacteria-specific and archaea-specific PCR-TRFLP to find out the impact of nutrients in contaminated marine harbour sediments. Scully *et al.* (2005) monitored anaerobic wastewater treatment failure using TRFLP analysis.

ARDRA, RAPD and AFLP are other PCR-based molecular techniques useful for microbial community analyses in bioremediation applications. Automated ribosomal intergenic spacer analysis (ARISA) is based on the variation of species-specific amplicon lengths of the intergenic transcribed spacers (ITS) located between 16S and 23S rRNA genes. The technique involves amplification of ITS from microbial community DNA using fluorescent labelled primers followed by separation and detection of fluorescent ITS amplicons in an automated sequencing system (Nocker *et al.*, 2007, Desai *et al.*, 2010). Single strand conformation polymorphism (SSCP) is the electrophoretic separation of single-stranded nucleic acids based on subtle differences in sequence (often a single base pair) which results in a different secondary structure and a measurable difference in mobility through a gel. Zhao *et al.* (2006) investigated the relationship between microbial community structure and status of processes in various wastewater treatment processes using SSCP. Kiesel *et al.* (2008) used SSCP to study microbial community shifts as a response to efficient degradation of chlorobenzene under hypoxic conditions.

Thus, nucleic acid-based molecular techniques for fingerprinting have been regularly applied for detection of predominant species, analysis of microbial community diversity and shifts. These techniques when used in conjunction with PCR technologies, microsensors and other methods, provide more information besides microbial community analysis at polluted sites, such as up/down regulations of genes; microenvironment of the sites; and impact of nutrients on bioremediation.

Fluorescence *in situ* hybridization (FISH)

The fluorescence *in situ* hybridization technique (FISH) is based on hybridization of fluorescent dye-labelled oligonucleotide probes to specific DNA sequences attached to membrane filters or glass slides and visualized or quantified using epifluorescence microscopy, confocal laser scanning microscopy or flow cytometry techniques. There have been numerous modifications in the use of FISH and many of them have been a great aid in microbial ecological studies. Dijk *et al.* (2008) developed catalysed reporter deposition fluorescence *in situ* hybridization (CARD-FISH), an *in situ* signal amplification technique based on the deposition of fluorochrome-labelled reporter molecules (such as tyramine) due to catalytic activity of horseradish peroxidase (HRP) enzyme coupled with oligonucleotide probes which leads to enhanced fluorescence signal intensity. Lee *et al.* (1999) used fluorescence *in situ* hybridization in combination with microautoradiography (FISH-MAR) for structure–function analyses in microbial ecology. They demonstrated the potential of this method by visualizing the uptake of organic and inorganic radiolabelled substrates ([^{14}C]acetate, [^{14}C]butyrate, [^{14}C]bicarbonate, and $^{33}P_i$) in probe-defined populations from complex activated sludge microbial communities by using aerobic as well as anaerobic incubation conditions (with and without nitrate). The FISH-MAR technique involves short incubation of the environment sample with radioactively labelled substrate, followed by identification of microbial populations using FISH and in-parallel processing of identified microbial cells using radiation-sensitive photographic silver emulsions. Consequently, the silver particles deposited around the actively growing cells are visualized under transmission electron microscopy (TEM) to determine whether the microbe identified using rRNA-targeted probes was functionally or metabolically active in consuming the radio-labelled substrate offered at the time of incubation. Therefore, coupling of FISH with microautoradiography facilitates both phylogenetic as well as functional identification of substrate-active cells (Rogers *et al.*, 2007).

Fluorescence *in situ* hybridization when used in conjunction with many other tools has brought about many novel applications. FISH is used in the field of bioremediation for species identification and defining spatial–temporal patterns of gene expression.

Nucleic acid-based stable isotope probing

Nucleic acids-based stable isotope probing (SIP) is a method which attempts to link an organism's identity with its biological function under conditions approaching those *in situ*. For example, addition of ^{13}C labelled substrate as a carbon source to an environmental sample results in ^{13}C labelling of actively dividing bacteria and then nucleic acids can be separated by density gradient ultra-centrifugation. Consequently, functional microbial communities that utilized heavy isotope labelled substrates can be identified.

Singleton *et al.* (2005) supplied [^{13}C6] salicylate, [U-^{13}C]naphthalene and [U-^{13}C] phenanthrene into PAH-contaminated soils of a bioreactor to identify enriched bacterial degraders of PAHs within the bioreactor community DNA. In this DNA-SIP experiment, analysis of 16S rDNA-based taxa using DGGE fingerprinting technique identified *Pseudomonas* spp., *Ralstonia* spp. and *Acidovorax* spp. as degraders of naphthalene, salicylate and phenanthrene, respectively. Bombach *et al.* (2010) developed and characterized a toluene degrading microbial consortium by combining *in situ* microcosms and stable isotope probing techniques. The application of metagenomics, the culture-independent approach and subsequent analysis of genomic DNA from the environment, has greatly expanded our knowledge of the diversity of microbes and microbial protein families; however, the metabolic functions of many microorganisms remain largely unknown. DNA stable isotope probing (DNA-SIP) is a recently developed method in which the incorporation of stable isotope from a labelled substrate is used to identify the function of microorganisms in the environment. The technique has now been used in conjunction with metagenomics to establish links between microbial identity and particular metabolic functions. The combination of DNA-SIP and metagenomics not only permits the detection of rare low-abundance species but also facilitates the detection of novel enzymes and bioactive compounds (Chen and Murrell, 2010).

Nucleic acids based stable isotope probing technique has been applied using wide variety of xenobiotic compounds to delineate the active microbial populations that utilize these compounds as substrates in cellular metabolism (Desai *et al.*, 2010). SIP has been used in conjunction with DGGE and microcosms, as a result we can identify the microorganisms which play role in bioremediation.

Metagenomic libraries

More than 99% of the microbes that exist in the environment cannot be cultivated easily (Schloss and Handelsman, 2005). Thus, most of the microbes in the environment have not been described and accessed for biotechnology or basic research. Abundance and diversity of unculturable bacteria in almost all environmental niches have led to the understanding that the so-called 'unculturable' bacteria actually multiply in their natural environment and if suitable culture conditions were provided it should be possible to cultivate them in the laboratory. In the earliest cultivation attempts, media with very low nutrient are used considering the high nutrient contents of common laboratory media as compared to those present in the natural environments (Sharma *et al.*, 2005).

One way around this problem is metagenomics, which aims to access the genomic potential of an environmental sample either directly or after enrichment for specific purpose. Soil/marine metagenome can be enriched in various ways keeping in mind specific gene targets to be screened such as encoding promising biocatalysts, novel antibiotics and other such targets that have potential applications in bioremediation, industry, medicine or agriculture. Bacteria capable of xenobiotic degradation are widely distributed in the environment. These bacteria have evolved to utilize a variety of compounds that are present in the environment. Hence environmental samples are considered to be a reservoir of useful enzymes for bioremediation as well as industrial catalysis. The metagenomic approach has also demonstrated that novel metabolic genes play an important role in the biodegradation of the compounds.

The first and foremost step in metagenomic analyses consists of isolating high molecular weight DNA from environmental samples in an unbiased manner. Besides this, many reports have highlighted the challenges in obtaining high quality (free from interference of co-contaminants like humic acids in case of soil) DNA from polluted sites (Fortin *et al.*, 2004; Desai and Madamwar, 2007). Metagenomic libraries are prepared by cloning of DNA fragments isolated from an environmental sample in a suitable vector [e.g. plasmid, phage, fosmid, cosmid, or bacterial artificial chromosome (BAC)], which is then transformed into a suitable host strain. Two approaches, the function-driven analysis and the sequence-driven analysis, have emerged to extract biological information from metagenomic libraries. The function driven analysis is based on identification of clones that express the desired trait, followed by characterization of the active clones by sequence and/or biochemical analysis. The limitations of this approach are that it requires expression of the function of interest in the host cell and clustering of all of the genes required for the function. It also depends on the availability of an high-throughput assay for the function of interest that can be performed efficiently on vast libraries, because the frequency of active clone is quite low. Many approaches such as improved systems for heterologous gene expression with shuttle vectors that facilitate screening of the metagenomic DNA in diverse host species and modifications of *Escherichia coli* to expand the range of gene expression are being developed to alleviate these limitations (Schloss and Handelsman, 2003). Conversely, the sequence-driven analysis relies on the use of conserved DNA sequences to design hybridization probes or PCR primers to screen metagenomic libraries for clones that contain sequence of interest. Significant discoveries have also resulted from random sequencing of metagenomic clones. However, there has been disagreement about the utility of random sequencing of metagenomic clones as according to some scientists this approach is too undirected to yield biological understanding, while others point out that there is very little information about some divisions of Bacteria that any genomic sequence is helpful in designing of experiments to discover their characteristics (Schloss and Handelsman, 2003).

Tyson *et al.* (2004) used the random shotgun sequencing approach to recover the genomes of uncultivated species. They obtained two near-complete genomes and three partial genomes from an environmental sample. Brennerova *et al.* (2009) revealed the diversity and abundance of meta-cleavage pathways in microbial communities from soil highly contaminated with jet fuel under air-sparging bioremediation using metagenomics. The extradiol dioxygenase diversity of a site highly contaminated with aliphatic and aromatic hydrocarbons under air-sparging treatment was assessed by functional screening of a fosmid library in *Escherichia coli* with catechol as substrate. Kim *et al.* (2010) prepared a metagenomic library in fosmid vector from a completely fermented compost and functionally screened and characterized a novel family VIII alkaline esterase.

Metagenomics is a burgeoning area that is generating enormous amounts of biological information. In addition, the development of new bioinformatics approaches and tools is allowing innovative mining of both existing and new data (Ward, 2006). However metagenomics is still a developing technology with limitations to be overcome, such as gene enrichment, DNA extraction, host-vector design, library construction, screening, and the phylogenetic affiliation of isolated genes.

Nucleic acid sequencing technologies

Sequencing of DNA and/or RNA of humans and many other animals, plants and microbes have revolutionized the biological research. Sanger developed the chain termination method in 1975 and two years after that Maxam and Gilbert developed a method based on chemical modifications. The major revolution started only after the improvements in chain-termination method. Conversely, automated analysis played the biggest role as it made the sequencing easier and faster. Capillary electrophoresis replaced slab gel electrophoresis and became an integral part of automated sequencing analysis. Even matrix-assisted laser desorption ionization time-of-flight mass spectrometry (MALDI-TOF MS) has been explored for DNA sequencing. MALDI-TOF MS produces mass spectra of these DNA sequencing fragments with nearly digital resolution, allowing accurate determination of the mixed bases and completely eliminates compressions associated with gel-based systems. Mass spectrometry-based sequencing has focused on the detection of frame shift mutations and single nucleotide polymorphisms (SNPs), making it an ideal choice for resequencing projects (Edwards *et al.*, 2005).

However, the era's demand of high throughput sequencing at low costs led to constant upgradation of technologies. Next-generation sequencing has propelled biological research (Schuster, 2008). Today a number of commercial next-generation DNA sequencing systems are available, namely: Roche's 454 GS FLX Genome Analyzer (Pyrosequencing) (Margulies *et al.*, 2005), Illumina's Solexa sequencer (Bentley *et al.*, 2008), Applied Biosystem's SOLiD system (McKernan *et al.*, 2009), Helicos HeliScope (Shendure and Ji, 2008), Complete Genomics (Drmanac *et al.*, 2010) and many more are coming in the near future, such as Ion-Torrent Semiconductor sequencing (www.iontorrent.com), Pacific Biosciences SMRT (www.pacificbiosciences.com) and Oxford Nanopore sequencing (www.nanoporetech.com).

Pyrosequencing, based on 'sequencing by synthesis' principle, involves sequencing the single stranded DNA by synthesizing the complementary strand and detection of pyrophosphate released on addition of each nucleotide. The templates for pyrosequencing are made either by solid phase template preparation (streptavidin-coated magnetic beads) or enzymatic template preparation (apyrase and exonuclease). The templates are hybridized to a sequencing primer and incubated with the enzymes DNA polymerase, ATP sulfurlyase, luciferase and apyrase, and with the substrates adenosine 5′ phosphosulfate (APS) and luciferin. Detection of pyrophosphate, liberated during DNA synthesis, is possible by cascade of enzymatic reactions followed by detection with photodiode, photomultiplier tubes, or a charge-coupled device (CCD) camera. Currently, a limitation of the method is that the lengths of individual reads of DNA sequence are in the neighbourhood of 300–500 nucleotides, shorter than the 800–1000 obtained with chain termination methods.

The sequence information generated can be used to prepare functional microarrays to characterize active microbial community population. Microarrays based on sequence analysis have been used to monitor genes involved in 2,4-dichlorophenoxyacetic acid (2,4-D) degradation in environmental samples (Dennis *et al.*, 2003). Desai *et al.* (2009a) tracked the influence of long-term chromium pollution on soil microbial communities by sequencing 16S rRNA gene clone libraries and observed community shifts from Proteobacteria to Firmicutes at chromium polluted sites. There are numerous reports on sequencing of complete genome or functionally essential genes involved in metabolism of biodegradation (Nelson *et al.*, 2002; Kim *et al.*, 2008b; Mattes *et al.*, 2008; Desai *et al.*, 2010). The genome sequence of *Pseudomonas* sp. has revealed details about oxygenases, oxidoreductases, ferredoxins and cytochromes, dehydrogenases, sulfur metabolism proteins, etc. Moreover it also possesses many operons coding for the metabolism of a large number of aromatic compounds and gene clusters encoding for enzymes which are predicted to be involved in the metabolism of non-natural substrates (Nelson *et al.*, 2002). Ozsolak *et al.* (2009) reported direct single molecule RNA sequencing without prior conversion of RNA to cDNA. They applied this technology to sequence femtomole quantities of poly(A) *Saccharomyces cerevisae* RNA using a surface coated with poly(dT) oligonucleotides to capture the RNAs at their natural poly(A) tails and initiate sequencing by synthesis. They observed transcript 3′ end heterogeneity and polyadenylated small nucleolar RNAs. This study provides a path to high-throughput and low-cost direct RNA sequencing and achieving the ultimate goal of a comprehensive and bias-free understanding of transcriptomes.

Metagenomics and Metatranscriptomics, which till now have dealt with apprehension about the analysis of generated data, however, as a result of feasibility of sequencing in large numbers using these latest sequencing technologies, these techniques are gaining more importance in the field of microbial ecology. These advanced sequencing technologies have made it possible to sequence data directly from environments, bypassing lab cultivation step and avoiding the use of cloning step. Moreover it has provided a better understanding of microbial ecology.

Polymerase chain reaction (PCR) technologies

The polymerase chain reaction (PCR) is a technique whereby a single or few copies of nucleic acid can be amplified to a large number of copies (increase by thousands/millions times). This technique has revolutionized the approaches to study nucleic acids and over the years several modifications such as hot start, multiplex, real time, reverse transcription, nested and many others have been developed. Besides novel approaches, the basic method of PCR has also seen many upgrades. The major aspects that have led to modifications are unbiased detection, specificity and quantification.

Real-time quantitative PCR (qPCR)

Real time PCR is a technique wherein we can quantify the amount of targeted DNA molecule at every cycle of amplification. Quantification is based on real-time detection of a reporter molecule whose fluorescence increases as PCR products accumulate at each amplification cycle. Two common methods for detection of products are (1) non-specific dsDNA intercalating dyes [SYBR Green along with carboxy-X-rhodamine (ROX) as a passive reference dye] or (2) sequence specific hybridization probes consisting of oligonucleotides that are labelled with a fluorescent reporter which permits detection only after hybridization of the probe with its complementary DNA target (TaqMan-molecular beacons with fluorescence resonance energy transfer).

Real-time PCR can be used to quantify nucleic acids by two strategies – relative quantification and absolute quantification. Relative quantification measures the fold-difference (2×, 3×, etc.) in the target amount, whereas absolute quantification measures the exact number of target molecules present by comparing with known standards. Quantification of relative concentrations of DNA present during the exponential phase of the reaction is determined by plotting fluorescence against cycle number on a logarithmic scale (so an exponentially increasing quantity will give a straight line). A threshold for detection of fluorescence above background is determined and the cycle at which the fluorescence from a sample crosses the threshold is called the cycle threshold, C_t. The quantity of DNA theoretically doubles every cycle during the exponential phase and relative amounts of DNA can be calculated, e.g. a sample whose C_t is three cycles earlier than another's has $2^3 = 8$ times more template. Since all sets of primers don't work equally well, one has to calculate the reaction efficiency first. Thus, by using this as the base and the cycle difference $C(_t)$ as the exponent, the precise difference in starting template can be calculated. Amounts of RNA or DNA are then determined by comparing the results to a standard curve produced by real-time PCR of serial dilutions of a known amount of RNA or DNA. To accurately quantify gene expression it is necessary to normalize for possible variation in the amount and quality of RNA between different samples. Normalization is achieved by dividing the measured amount of RNA from the gene of interest by the amount of RNA from a housekeeping gene measured in the same sample. Thus normalization permits accurate comparison of expression of the gene of interest between different samples, provided that the expression of the reference (housekeeping) gene used in the normalization is very similar across all the samples. Selecting a reference gene fulfilling this criterion is therefore of high importance, and often challenging, because only very few genes show equal levels of expression across a range of different conditions. The quality of standard is important for accurate quantification.

Mesarch *et al.* (2000) analysed petroleum-amended soils and detected an increase in catechol 2,3-dioxygenase genes by using the dioxygenase-specific primers. Benzene, toluene, xylenes, phenol, naphthalene and biphenyl are among a group of compounds that have at least one reported pathway for biodegradation involving catechol 2,3-dioxygenase enzymes and thus detection of catechol 2,3-dioxygenase genes can serve as a basis for identifying and quantifying bacteria that have these catabolic activities. Jorgensen (2008) in his review has shown that real-time PCR in actual bioremediation studies have mainly been concerned with naphthalene (Debruyn *et al.*, 2007), nitro aromatics (Indest *et al.*, 2007), arsenic (Sun *et al.*, 2004) and for the anaerobic degradation of hydrocarbons (Beller *et al.*, 2002). Baek *et al.* (2009) evaluated the activity and abundance of the crude oil degrading bacterium *Nocardia* sp. H17-1 during bioremediation of oil-contaminated soil, by quantifying the genes encoding 16S ribosomal RNA (16S rRNA), alkane monooxygenase (alkB4), and catechol 2,3-dioxygenase (23CAT) using real-time PCR. Nyyssonen *et al.* (2006) validated a sensitive real-time PCR assay for quantification of naphthalene-hydroxylating dioxygenase (*nahAc*) genes of naphthalene-degrading Proteobacteria within soil samples and also reported that qPCR assay is more sensitive than hybridization-based analysis in monitoring bioremediation processes. Desai *et al.* (2009b) determined abundance of active bacterial populations of an enriched bacterial consortium during the steady-state condition within continuous bioreactors treating Cr(VI) and azo dye mixtures by calculating 16S rRNA gene copy numbers using qPCR assays. Cebron *et al.* (2008) showed that PAH-bacterial degraders could represent up to 1% of the total bacterial community in the PAH-contaminated sites by the ratio of PAH genes relative to the 16S rRNA genes quantified by real-time PCR.

Nanolitre volume PCR on OpenArray platform

Development of quantitative PCR (QPCR) assays typically requires extensive screening within and across a given species to ensure specific detection and lucid identification. To minimize screening requirements, Stedtfeld *et al.* (2008) targeted multiple virulence and marker genes (VMGs) simultaneously to enhance reliability. A predictive threshold cycle (C_T) equation was developed to calculate the number of starting copies based on an experimental C_T. The empirical equation was developed with Sybr green detection in nanolitre-volume QPCR chambers (OpenArray) and tested with 220 previously unvalidated primer pairs targeting 200 VMGs from 30 pathogens. A high correlation (R^2=0.816) was observed between the predicted and experimental C_Ts based on the organism's genome size, guanine and cytosine (GC) content, amplicon length, and stability of the primer's 3′-end. The performance of the predictive C_T equation was tested using 36 validation samples consisting of pathogenic organisms spiked into genomic DNA extracted from three environmental waters. In addition, the primer success rate was dependent on the GC content of the target organisms and primer sequences. Targeting multiple assays per organism and using the predictive C_T equation are expected to reduce the extent of the validation necessary when developing QPCR arrays for a large number of pathogens or other targets (Stedtfeld *et al.*, 2008).

Real-time reverse-transcription PCR (qRT-PCR)

Real-time reverse-transcription PCR is a combination of reverse transcription PCR and real-time quantitative PCR, useful in bioremediation studies for analysis of mRNA expression of key genes involved in metabolism. The first step is reverse transcription, in which RNA is reverse transcribed to cDNA using reverse transcriptase. This is followed by real time polymerase chain reaction based on detection and quantification of specific targets. Increased concentrations of mRNA for *nahA* – a gene involved in aerobic degradation of naphthalene were positively correlated with rates of naphthalene degradation in hydrocarbon-contaminated soils (Fleming *et al.*, 1993).

Differential display reverse transcription PCR (DDRT-PCR)

Differential display, also referred to as DDRT-PCR, is the technique by which one can identify and analyse altered gene expression at the mRNA level. The environmental samples having morphological, genetic or other experimental differences can be studied for the changes in gene expression patterns. The method involves reverse transcription using a set of anchored primers, followed by cDNA amplification with additional upstream primers and then electrophoretic separation of the fragments. For characterization, the differentially expressed genes are reamplified, cloned and sequenced. Fleming *et al.* (1998) optimized the DDRT-PCR for prokaryotic mRNA for its application to pure cultures and soil microcosms. Bordenave *et al.* (2009) analysed differential display of cDNA involved in microbial mat response after heavy fuel contamination.

The PCR technologies have been used in bioremediation studies in conjunction with techniques such as DGGE, sequencing technologies, microarrays and many others. The PCR technologies have helped in detection and quantification of many genes or transcripts involved in biodegradation pathways (Nyyssonen *et al.*, 2006; Cebron *et al.*, 2008; Desai *et al.*, 2009b).

Microarray

Microarray is a multiplex technology that evolved from Southern Blotting. It consists of arrayed series of features where large number of probes (oligonucleotides, cDNAs, PCR products, genes or other known genomic fragments) are attached either on a solid surface (glass, plastic or a silicon chip) by a covalent bond to a chemical matrix (epoxy-silane, amino-silane, lysine, polyacrylamide or others) or on coded beads (polystyrene beads). As microarray contains thousand of different probes, therefore a single experiment can accomplish many tests in parallel. The principle behind microarrays is hybridization of target DNA molecules to probes. The labelled target sequences that bind to a probe sequence generate a signal that depends on the strength of the hybridization determined by the number of paired bases, the hybridization conditions (such as temperature), and washing (which removes non-specific bonding sequences) after hybridization. Probe–target hybridization is usually detected and quantified by detection of fluorophore-, silver-, or chemiluminescence-labelled targets to determine relative abundance of nucleic acid sequences in the target DNA. Based on the probes utilized in the fabrication of an array or depending on their applications in microbial bioremediation, microarrays can be classified into several types such as phylogenetic oligonucleotide arrays (POA), functional gene arrays (FGA) and whole genome arrays (WGA) (Bae and Park, 2006; Desai *et al.*, 2010).

Phylogenetic oligonucleotide arrays (POA) or PhyloChips are constructed using short stretches of oligonucleotide sequences based on rRNA genes from different microbial phyla and are used to identify microbial populations in environmental samples. Loy *et al.* (2005) used a 16S rRNA gene-targeted oligonucleotide microarray (RHC-PhyloChip) for cultivation-independent diversity analysis of betaproteobacterial order 'Rhodocyclales' in activated sludge from an industrial wastewater treatment plant. The microarray results were confirmed and extended by 16S rRNA gene sequencing and quantitative fluorescence *in situ* hybridization analyses. Adamczyk *et al.* (2003) developed an Isotope array approach where PhyloChip was used to monitor labelled-substrate utilization and thus linked the microbial community structure with function.

Functional gene arrays (FGA) are constructed using PCR-amplified products or oligonucleotides derived from functional genes and are used to identify genes encoding key enzymes involved in various ecological and environmental processes such as carbon fixation, nitrification, denitrification, sulphate reduction and contaminant degradation (Rhee *et al.*, 2004). These types of arrays are useful in studying the physiological status and functional activities of microbial communities in environments. Designing such arrays is possible by analysis and alignment studies of gene and protein sequences available in ever augmenting public databases. He *et al.* (2007) developed GeoChip, a comprehensive microarray containing 24,243 oligonucleotide (50 mer) probes and covering more than 10,000 genes in more than 150 functional groups involved in nitrogen, carbon, sulphur and phosphorus cycling, metal reduction and resistance, and organic contaminant degradation. The developed GeoChip was successfully used for tracking the dynamics of metal-reducing bacteria and associated communities for an *in situ* bioremediation study. He *et al.* (2010) developed a new generation gene array, GeoChip 3.0 with around 28,000 probes covering approximately 57,000 gene variants from 292 functional gene families. GeoChip 3.0 is a high-throughput powerful tool for studying microbial community functional structure, and linking microbial communities to ecosystem processes and functioning.

Whole genome arrays (WGA) or also known as Community genome arrays (CGA) are constructed using complete genome of a particular micro-organism and used to correlate or contrast genomes of samples. West *et al.* (2008) developed a whole genome microarray based on genome of '*Dehalococcoides ethenogenes*' 195 and used it to compare and contrast genes involved in dehalogenation of chlorinated solvents, such as tetrachloroethene and trichloroethene from an enrichment culture. Gao *et al.* (2007) carried out a microarray-based analysis of whole community RNA extracted by using whole community RNA amplification (WCRA) method from a uranium-contaminated groundwater sample from a site biostimulated by ethanol feeding. Yergeau *et al.* (2009) compared two distinct large-scale field bioremediation experiments, located at the Canadian high-Arctic stations of Alert (*ex situ* approach) and Eureka (*in situ* approach) and assessed the bacterial community structure and function using microarrays targeting the 16S rRNA genes of bacteria found in cold environments and hydrocarbon degradation genes as well as quantitative reverse transcriptase PCR targeting key functional genes.

Microarrays have been used in conjunction with PCR technologies (Yergeau *et al.*, 2009), sequencing technologies and FISH (Loy *et al.*, 2005). Consequently these techniques can be used in bioremediation studies to identify microbial populations in ecosystems, assess shifts in microbial populations due to xenobiotic compounds, measure changes in expression levels (gene expression profiling), co-relate and compare genomes, identify genes encoding enzymes, detect mutations and single nucleotide polymorphisms (SNPs).

Serial analysis of gene expression

Serial analysis of gene expression (SAGE) is a technique involving multiple enzymatic, PCR amplification, purification and cloning steps. SAGE is based on two basic principles. The first principle is that a short oligonucleotide sequence, defined by a specific restriction endonuclease (anchoring enzyme, AE) at a fixed distance from the poly(A) tail, can uniquely identify mRNA transcripts. [Theoretically, a 10-bp sequence tag can give 4^{10} (10, 48, 576) different sequence combinations, which is sufficient to discriminate all the transcripts derived from the genome.] The second principle is that the end-to-end concatenation of these short oligonucleotides allows multiple transcript detection per sequencing reaction. The SAGE protocol starts with the purification of mRNA bound to solid phase oligo(dT) magnetic beads. The cDNA is synthesized directly on the oligo(dT) bead and then digested with the anchoring enzyme (AE) to reveal the 3′-most restriction site anchored to the oligo(dT) bead. Most SAGE experiments use the 4-bp recognition site anchoring enzyme, predicted to occur after every 256 bp and thus present on most mRNA species. Next, the sample is equally divided into two separate tubes and ligated to two different linkers, A or B. Both linkers contain the recognition site for an enzyme that cuts 10-bp 3′ from the anchoring enzyme recognition site and generates a unique oligonucleotide known as the SAGE tag, hence called the tagging enzyme (TE). The SAGE tags released from the oligo(dT) beads are then separated, blunted, and ligated to each other to give rise to ditags. The ditags are PCR amplified, released from the linkers, gel purified, serially ligated, cloned, and sequenced using an automated sequencer (Patino *et al.*, 2002).

Several variants have been developed, most notably a more robust version is LongSAGE wherein the tagging enzyme cuts 17-bp 3′ from the anchoring site (Saha *et al.*, 2002), RL-SAGE (Gowda *et al.*, 2004) wherein the tag length is 21 bp and the most recent SuperSAGE (Matsumura *et al.*, 2005) wherein the tag length is 26 bp. In general many of these have improved the technique with the capture of longer tags, enabling more confident identification of a source gene. Serial analysis of gene expression was used to profile transcript levels in *Arabidopsis* roots and assess their responses to 2,4,6-trinitrotoluene (TNT) exposure. Identifying transcriptome-level responses to TNT exposure will define the metabolic pathways plants use to detoxify this xenobiotic compound, which should help in improving the phytoremediation strategies directed at TNT and other nitroaromatic compounds (Ekman *et al.*, 2003).

Serial analysis of gene expression technique is a combination of PCR technologies, cloning strategies and sequencing methods. SAGE can be used in bioremediation strategies to identify the gene and measure the gene expression, thus leading to a better understanding of the metabolic pathways.

ChIP-Chip technology

ChIP-on-chip is a technique that combines chromatin immunoprecipitation ('ChIP') with microarray technology ('chip') and is used to investigate interactions between proteins and DNA *in vivo*. The technique involves at first binding of protein of interest with DNA, followed by fragmentation and then filtering of protein–DNA complexes by specific antibodies. The complex is debound and DNA is amplified and this oligonucleotide stretch is used as a probe in DNA microarray. Grainger *et al.* (2004) showed that *Escherichia coli* MelR protein is a transcription activator that is essential for melibiose-dependent expression of the melAB genes. Further analysis of immunoprecipitated DNA, by using an Affymetrix GeneChip array, showed that the melAB promoter is the major, if not the sole, target in *E. coli* for MelR. This was confirmed by a transcriptomics experiment to analyse RNA in cells either with or without melR. Bruscella *et al.* (2008) used this technique to characterize the PpsR regulon and to ascertain the significance of degeneracy within the PpsR binding sequence *in vivo* in *R. sphaeroides*. They demonstrated that *in vivo* there was direct and significant binding by PpsR to newly identified genes involved in microaerobic respiration and periplasmic stress resistance, as well as to photosynthesis genes.

ChIP-Chip technology involves immunoprecipitation technique and microarray technology and can be used in bioremediation studies to investigate protein–DNA interactions which might be involved in some regulatory pathways.

Two-dimensional gel electrophoresis (2-DE)

Two-dimensional gel electrophoresis also known as 2-DE is a technique wherein proteins are separated based on two properties in two dimensions. In the first dimension proteins are separated based on isoelectric point. After which the native charge is nullified by addition of SDS and the gel is rotated 90 degree and run on regular polyacrylamide gel so as to carry out separation based on protein mass. This way of separating proteins by 2D-PAGE is highly useful as proteins differing by either of the properties – charge or mass also get separated. Kim *et al.* (2009) in his review has given a summary on proteomic approaches used to study bacterial degradation of aromatic hydrocarbons. Wilmes and Bond (2004) successfully carried out extraction and purification of the entire proteome from a laboratory-scale activated sludge system optimized for enhanced biological phosphorus removal, separation by two-dimensional polyacrylamide gel electrophoresis and the mapping of this metaproteome. Highly expressed protein spots were excised and identified using quadrupole time-of-flight mass spectrometry with *de novo* peptide sequencing. Roma-Rodrigues *et al.* (2010) applied two dimensional gel electrophoresis to quantitatively compare the membrane proteins of *Pseudomonas putida* KT2440, before and after phenol exposure. This will help in understanding the *Pseudomonas* adaptation to solvents, and possible impact in biodegradation, bioremediation and biocatalysis. Singh *et al.* (2010) have discussed optimized ideal directions for 2DE, followed by easy-to-follow directions for a protein identification strategy using MALDI-TOF and targeting novel proteins/enzymes to identify the biomolecules and their molecular mechanisms in bioremediation. They applied two-dimensional gel electrophoresis (2DE) in conjunction with mass spectrometry (MS) and identified a set of catabolic proteins, enzymes, and heat shock molecular chaperones associated with the regulatory network that was found to be overexpressed under phenol-stressed conditions.

Two-dimensional gel electrophoresis has been used in conjunction with mass spectrometry, peptide sequencing and other proteomic applications. It can be used in bioremediation to identify the up- or down-regulated proteins, purify and characterize the key enzymes involved in biodegradation pathways.

Mass spectrometry (MS)

Mass spectrometry (MS) is an analytical technique for the determination of the elemental composition of a sample or molecule. It is also used for elucidating the chemical structures of molecules such as peptides and other chemical compounds and masses of particles. The principle consists of ionizing chemical compounds to generate charged molecules or molecule fragments and measurement of their mass-to-charge ratios. The procedure involves conversion of gas phase molecules into ions followed by sorting of ions on the basis of masses by application of electromagnetic fields and finally detection resulting in a mass spectrum. This technique can be applied for both qualitative as well as quantitative purposes and the main applications are identification of unknown compounds, determining isotopic components of elements in a molecule, structure determination and quantification of a compound in a sample.

Mass spectrometry is not any one particular instrument and nor is it based on just one principle. Many different techniques are available for ionization, mass analysis and detection. Consequently, depending on the application the combinations of these three main components are decided. Generally for ionization of liquid and solid biological samples, electrospray ionization (ESI) and matrix-assisted laser desorption ionization (MALDI) are the two techniques preferred out of the numerous options available. The ions are then transported by magnetic or electric fields to the mass analyser. Mass analysers separate the ions according to their mass-to-charge ratio, using either static or dynamic fields and magnetic or electric fields. There are many kinds of mass analysers available, with unique strengths and weaknesses and hence unique applications. The time-of-flight (TOF) analyser uses an electric field to accelerate the ions by the same potential, and then measures the time they take to reach the detector. Particles having same charge have identical kinetic energies, and hence their velocities will depend only on their masses. Quadrupole mass analyser uses oscillating electrical fields to selectively stabilize or destabilize the paths of ions passing through a radio frequency quadrupole field created between four parallel rods. Ions in a certain range of mass/charge ratio are allowed to pass through the system at a particular time. However, changing the potentials on the rods allow a wide range of m/z values to be swept rapidly, either continuously or in a succession of discrete hops. The quadrupole ion trap works on the same physical principles as the quadrupole mass analyser, but the only difference here is that ions are trapped and sequentially ejected. Many mass spectrometers use two or more mass analysers for tandem mass spectrometry (MS/MS) wherein, the first mass analyser can separate one peptide from a bunch entering a mass spectrometer and the second mass analyser then stabilizes the peptide ions while they are fragmented by collision with a gas [collision-induced dissociation (CID)] or other fragmentation methods. The third mass analyser then sorts the fragments produced from the peptides. Lastly, the detector measures the charge induced or the current produced when an ion passes by or hits a surface and produces a mass spectrum of mass/charge (m/Q) ratio. The ion detectors are generally microchannel plate or a fast secondary emission multiplier (SEM). The electrical signal from the detector is recorded by means of a time-to-digital converter (TDC) or a fast analogue-to-digital converter (ADC). However, tandem MS can also be done in a single mass analyser over a time period using quadrupole ion trap as a mass analyser as describer earlier. Tandem mass spectrometry has an important application in protein identification.

Mass spectrometry is an important method for the characterization of proteins. The intact proteins are ionized either directly or the proteins are enzymatically digested into smaller peptides using proteases such as trypsin or pepsin, either in solution or in gel after electrophoretic separation and the collection of peptide products are then introduced to the mass analyser. When the characteristic pattern of peptides is used for the identification of the protein the method is called peptide mass fingerprinting (PMF).

Chromatographic techniques combined with mass spectrometry

Over the years, for enhanced separation and better resolution, chromatographic techniques are used in combination with mass spectrometry (Garcia *et al.*, 2008). Gas chromatography–mass spectrometry (GC/MS) and liquid chromatography–mass spectrometry (LC/MS) are the two most widely preferred techniques. In GC/MS first gas chromatography (mobile phase is gas) separates the compounds which are then passed into ionization chamber of mass spectrometry. Similarly, in LC/MS initially the liquid mobile phase separates the compounds which are then passed into electrospray ionization (most preferred for LC/MS) of mass spectrometry.

Ram *et al.* (2005) carried out an extensive metaproteomic study using genomics and mass spectrometry-based methods. They evaluated gene expression, identified key activities, and examined partitioning of metabolic functions in a natural acid mine drainage (AMD) microbial biofilm community. Schulze *et al.* (2005) applied mass spectrometry to analyse proteins isolated from dissolved organic matter of lake, deciduous forest, spruce forest at various seasons and from various depths. Their aim was to determine the number of proteins and also to elucidate they belong to which phylogenetic groups. The comparison of proteome fingerprints helped in understanding the presence and activity of organisms in an ecosystem. Wilmes *et al.* (2008) applied 2-DE along with MALDI-TOF MS and Q-TOF MS/MS to identify highly expressed proteins in a mixed culture activated sludge system and this study provided key functional proteins necessary for enhanced biological phosphorous removal (EPBR). Benndorf *et al.* (2007) applied 2-DE along with LC-ESI/MS for functional analysis of metaproteomes of soil enriched with 2,4-dichlorophenoxy acetic acid (2,4-D) and groundwater from the aquifer of a chlorobenzene-contaminated site. Keum *et al.* (2008) evaluated metabolic profiles including polar metabolites, fatty acids, and polyhydroxyalkanoates through untargeted metabolome analyses. Intracellular metabolomes during the degradation of phenanthrene were compared with those from natural carbon sources. Among 207 gas chromatography-mass spectrometry peaks from the polar metabolite fraction, 60% of the peaks were identified and compared. Several intermediates of tricarboxylic acid cycles and glycolysis, such as trehalose, and amino acids, such as glycine, homoserine and valine, were increased while more than 70% of identified metabolites were decreased during the phenanthrene metabolism. Accumulation of sulfur amino acids and nicotinic acid suggested the possible oxidative stress conditions during phenanthrene metabolism. Tang *et al.* (2009) performed a fluxomics analysis on *Shewanella* spp. known to have co-metabolic pathways for bioremediation of toxic metals, radionuclides, and halogenated organic compounds. In the metabolic flux analysis of *Shewanella* spp. using GC-MS and statistical, biochemical and genetic algorithms it was deduced that *Shewanella* spp. display a relatively flexible metabolism fluxes when adapting to different carbon sources.

Mass spectrometry has been used in conjunction with two-dimensional gel electrophoresis, liquid chromatography, gas chromatography, protein/peptide sequencing and many other techniques. They can be used in bioremediation studies for quantification of gene expression, identification, purification and characterization of key enzymes involved in biodegradation pathways. Mass spectrometry has been a boon to 'omic' studies such as proteome profiling, metabolome profiling and real time metabolite profiling known as fluxomics, as it can be used in conjunction with almost all other techniques and thus bestows with complete facts and scenario.

Labelling-based or label-free techniques for quantitative proteomics

Two-dimensional difference gel electrophoresis (2D-DiGE)

Two-dimensional difference gel electrophoresis (2D-DiGE) is another modified version of 2-DE wherein different protein samples are labelled with different fluorescent dyes, mixed together and resolved in a single gel. After the gel electrophoresis, the gel is scanned with the excitation wavelength of each dye one after the other. It overcomes the traditional gel-to-gel variations, the most severe problem in the gel-based proteomics. Trautwein *et al.* (2008) applied 2D-DiGE to measure the proteomic response of the denitrifying bacterium '*Aromatoleum aromaticum*' strain EbN1 to solvent stress. They determined up-regulation of PHB granule-associated phasins, cytochrome cd1 nitrite reductase of denitrification, and several proteins involved in oxidative (e.g. SodB) and general (e.g. ClpB) stress responses.

Stable isotope labelling-based isotope-coded affinity tags (ICAT)

Isotope-coded affinity tags (ICATs) is a method that relies on chemical labelling reagents. These chemical probes consist of three general elements: a reactive group capable of labelling a defined amino acid side chain (e.g. iodoacetamide modifies cysteine residues), an isotopically coded linker, and a tag (e.g. biotin) for the affinity isolation of labelled proteins/peptides. For the quantitative comparison of two proteomes, one sample is labelled with the isotopically light eight hydrogen atoms (d0) probe and the other with the isotopically heavy eight deuterium atoms (d8). Both samples are then mixed, digested with a protease (i.e. trypsin), and subjected to avidin affinity chromatography to isolate peptides labelled with isotope-coded tagging reagents. These peptides are then analysed by liquid chromatography–mass spectrometry (LC-MS). The ratios of signal intensities of differentially mass-tagged peptide pairs are quantified to determine the relative levels of proteins in the two samples. The present form of the ICAT reagent differs slightly from the original. It contains the biotin affinity tag which is attached to an acid cleavable linker, making it easier to remove; the light and heavy isotopically labelled region contains either nine ^{12}C or nine ^{13}C atoms which overcomes slight differential elution problems observed when using hydrogen and deuterium, however the thiol-specific labelling group remains the same (Goshe and Smith, 2003; Tao and Aebersold, 2003; Graham *et al.*, 2007). Hagglund *et al.* (2008) applied iodoacetamide-based isotope-coded affinity tag (ICAT) reagents for quantification of Trx-mediated target disulfide reduction based on thiol-specific differential labelling. Protein extract of embryos from germinated barley seeds was treated with or without Trx, and thiols released from target protein disulfides were irreversibly blocked with iodoacetamide. The remaining cysteine residues in the Trx-treated and the control (–Trx) samples were then chemically reduced and labelled with the 'light' (^{12}C) and 'heavy' (^{13}C) ICAT reagent, respectively. The extent of Trx-mediated reduction was thus quantified for individual cysteine residues based on ratios of tryptic peptides labelled with the two ICAT reagents as measured by liquid chromatography coupled with mass spectrometry (LC-MS). Wu *et al.* (2010) performed systematic analyses at the transcriptomic and proteomic [SDS-PAGE and isotope-coded affinity Tag (ICAT)] levels to investigate the global expression changes that take place in a high Mn environment, which results in a Mn-dependent oxidative stress resistance phenotype. These studies revealed that under conditions of increased Mn concentration, there was a change in regulation of proteins involved in virulence (e.g. pilin, a key adhesin), oxidative stress defence (e.g. superoxide dismutase), cellular metabolism, protein synthesis, RNA processing and cell division. Mn regulation of inorganic pyrophosphatase (Ppa) indicated the potential involvement of phosphate metabolism in the Mn-dependent oxidative stress defence. A detailed analysis of the role of Ppa and polyphosphate kinase (Ppk) in the gonococcal oxidative stress response revealed that ppk and ppa mutant strains showed increased resistance to oxidative stress. Investigation of these mutants grown with high Mn suggests that phosphate and pyrophosphate are involved in Mn-dependent oxidative stress resistance.

Kim *et al.* (2006) carried out proteomic analysis of aromatic pathways in *Pseudomonas putida* KT 2440 using 2-DE/MS and cleavable isotope-coded affinity tag (ICAT) to determine whether proteins involved in aromatic compound degradation pathways were altered as predicted by genomic analysis carried out by Jiménez *et al.* (2002). Many unique proteins were identified by 2-DE/MS as well as cleavable isotope-coded affinity tag (ICAT) and thus proving that the techniques complement each other and it is beneficial to use them in conjunction. Moreover, these results suggest that proteome analysis complements and supports predictive information obtained by genomic sequence analysis.

Isobaric tag for relative and absolute quantification (iTRAQ)

Isobaric tag for relative and absolute quantitation (iTRAQ) is a technique used to identify and quantify proteins from different sources in one single experiment. The method is based on the covalent labelling of the N-terminus and side chain amines of peptides from protein digestions with tags of varying mass. There are currently two mainly used reagents, 4-plex and 8-plex, which can be used to label all peptides from different samples/treatments. The iTRAQ reagent has several advantages over ICAT: four samples/states or, in the most recent version, eight samples/states rather than two can be measured; and free amine groups rather than reduced cysteines (present in only 95% of proteins) are labelled. The 4-plex reagent contains an amine specific reactive group, a balance group and a reporter group that can have a mass of 114, 115, 116 or 117. Samples of proteins from up to four states are first digested with trypsin, resulting in a peptide mixture with each cleaved peptide having a free amine group. Each sample then is labelled with one of the specific reagents by attachment of the label via the amine specific reactive group. All four samples are then mixed, separated by liquid chromatography and introduced into the mass spectrometer. During tandem mass spectrometry of the labelled peptides the reporter group is released, and measurement of the peak areas of these resultant ions gives an assessment of the abundance of that particular peptide under each condition (Washburn *et al.*, 2002; Graham *et al.*, 2007). As iTRAQ reagents can distinguish eight samples at a time, consequently this technique has bright potential applications in bioremediation studies, wherein contaminated sites can be analysed and compared for their differential expressions.

Quantitative protein profiling by mass spectrometry using label-free proteomics

Currently, the MS-based methods utilize mainly stable isotope labels (e.g. ICAT, iTRAQ) that enable easy identification of differentially expressed proteins in two or more samples. 'Label-free' MS-based methods would alleviate several limitations of the labelling methods, provided that relative quantitative profiling of proteins among multiple MS runs is achievable. Haqqani *et al.* (2008) describe the use of custom-developed MatchRx software in quantitative comparison of multiple label-free MS runs along with details of sample preparation, fractionation, statistical analysis, and protein database searching for label-free comparative quantitative proteomics, as well as the application of a 'targeted' MS approach, which includes quantification of the samples using MS followed by selective identification of only the differentially expressed peptides using tandem MS (MS/MS).

Exponentially modified protein abundance index (emPAI)

Protein abundance index (PAI) values obtained at different concentrations of serum albumin show a linear relationship with the logarithm of protein concentration in LC-MS/MS experiments (Rappsilber *et al.*, 2002). However, they later on developed exponentially modified protein abundance index (emPAI) for quantification of absolute protein contents in complex mixture. emPAI was applied to comprehensive protein expression analysis and for a comparison study between gene and protein expression in a human cancer cell line, HCT116. The values of emPAI are easily calculated and add important quantitative information to proteomic experiments; therefore their application in large scale proteomic identification projects will be of great assistance (Ishihama *et al.*, 2005).

These techniques of quantitative proteomics are used in conjunction with LC-MS, 2-DE and other proteome profiling techniques. These labelling or label-free methods are of importance in bioremediation studies as they can allow the quantification of proteins besides identification and characterization achieved by other techniques used in combination.

Protein sequencing

Protein sequencing, in real sense amino acid sequencing, can be carried out by various N-terminal amino acid analysis methods, C-terminal amino acid analysis methods and many other approaches. Moreover, it is also possible to infer the amino acid sequence from the DNA or RNA sequence encoding the protein. However, they are susceptible to the discrepancies of codon usage. The unordered amino acid composition analysis by hydrolysis followed by separation is also possible.

However, two major techniques – Edman degradation and mass spectrometry have been used extensively because of their advantages over other approaches. Edman degradation remains the most important reaction for protein sequencing. In this method, the N-terminal residue is reacted with phenylisothiocyanate under mildly alkaline condition to form a cyclical phenylthiocarbamoyl derivative. Then, under acidic conditions, this derivative of the terminal amino acid is cleaved as a thiazolinone derivative without disrupting the peptide bonds between other amino acid residues. The thiazolinone amino acid is then selectively extracted into an organic solvent and treated with acid to form the more stable phenylthiohydantoin amino acid derivative that can be identified by using either chromatography or electrophoresis. D'Souza-Ticlo *et al.* (2009) purified and characterized a thermostable metal-tolerant laccase having bioremediation potential. They used automated protein sequencing system for N-terminal sequencing and for internal peptide sequencing the protein Lac IId was digested *in-gel* (2DPAGE) by trypsin, the peptides were extracted by sonication and spotted on the α-cyano-4-hydroxycinnamic acid (HCCA) matrix for a MALDI TOF analysis.

The protein sequencing is applied in conjunction with 2-DE, MS or other blotting techniques. The protein sequencing can help in bioremediation studies as sequences of key enzymes can be used to study protein folding and structure analysis. This will provide more information about catalytic and active sites.

Spectroscopy

Spectroscopy is the study of the interaction between radiation and matter as a function of wavelength. Spectroscopy can be classified based on many aspects. Moreover, there are many kinds of spectroscopic techniques. However, here we will be discussing those spectroscopic techniques that can be applied in studies useful for bioremediation and environmental restoration.

Nuclear magnetic resonance (NMR)

Nuclear magnetic resonance is a property that nuclei possess in a magnetic field or applied electromagnetic pulses. The nuclei absorb energy from the EM pulse and radiate this energy back at specific resonance frequency depending on the strength of magnetic field and spin of the nuclei. Many techniques apply the NMR phenomena to study the molecular aspects and properties of atomic nucleus. Boersma *et al.* (2001) illustrated the potential of 19F NMR technique for metabolic characterization of newly isolated microorganisms and also for identification of relatively unstable pathway intermediates like fluoromuconolactones and fluoromaleylacetates. Durand *et al.* (2010) applied *ex situ* nuclear magnetic resonance (NMR) and liquid chromatography-NMR (LC-NMR) as complementary tools to LC-MS to define the biodegradation pathway of mesotrione, a herbicide, by the bacterial strain *Bacillus* sp. 3B6. The complementarities of *ex situ* and LC-NMR identified six metabolites whereas the structures of only four metabolites were suggested by LC-MS. The presence of a new metabolic pathway was evidenced by NMR. The results demonstrate that NMR and LC-NMR spectroscopy provide unambiguous structural information for xenobiotic metabolic profiling. Viggi *et al.* (2010) analysed the metabolic necessities for the removal of formic acid by unacclimated activated sludge using ^{13}C-NMR. These studies suggest that NMR can be applied for gaining insights into metabolic processes of biodegradation pathways.

Vibrational spectroscopy

Vibrational spectroscopy is used for analysis of vibrational states of a molecule. Vibrations of atoms in a molecule affect the electromagnetic radiation. Infrared spectroscopy and Raman spectroscopy are the two techniques which can measure the vibrations. In IR spectroscopy light of all frequencies is passed through a sample and the intensity of transmitted light is measured at each frequency and thus determining the frequency at which light is absorbed. Whereas in Raman spectroscopy vibrating group in the molecule interacts with the light and changes the frequency of the light. Difference in the frequency of a particular fraction of light is due to the corresponding vibrational energies. Raman spectroscopy has been used to monitor the appearance of surface biopolymers (including DNA/RNA, lipids, proteins, and carbohydrates) of microorganisms.

Fourier transform spectroscopy

Fourier transform spectroscopy is a measurement technique whereby spectra are collected based on time-domain or space-domain measurements of coherence of the electromagnetic radiation or other types of radiation. It can be applied to a variety of spectroscopy including optical spectroscopy, infrared spectroscopy (FTIR, FT-NIRS), nuclear magnetic resonance (NMR) and magnetic resonance spectroscopic imaging (MRSI), mass spectrometry and electron spin resonance spectroscopy. Fourier transform spectroscopy rather than allowing only one wavelength at a time to pass through to the detector, it allows a beam containing many different wavelengths of light at once, and measures the total beam intensity. Between the light source and the detector, there is a certain configuration of mirrors that allows some wavelengths to pass through and blocks others (because of wave interference). The beam is modified for each new data point to contain a different combination of wavelengths by moving one of the mirrors, which changes the set of wavelengths that can pass through. This process is repeated many times. Afterwards, a computer takes all this data and works backwards to infer how much light is there at each wavelength. The processing required turns out to be a common algorithm called the Fourier transform (hence the name, 'Fourier transform spectroscopy').

Wharfe *et al.* (2010) used FT-IR as a metabolite fingerprinting tool for monitoring the phenotypic and biochemical changes in complex bacterial communities capable of degrading phenol. In this investigation the activated sludge from an industrial bioreactor capable of degrading phenol was incubated in laboratory and monitored for its ability to degrade phenol over a 48-h period. Multiple samples were taken across the time-course and analysed by Fourier transform infrared (FT-IR) spectroscopy. This study demonstrates that FT-IR spectroscopy when combined with chemometric analysis is a very powerful high throughput screening approach for assessing the metabolic capability of complex microbial communities. Reddy *et al.* (2010) studied biodegradation of phenanthrene in conjunction with biosurfactant production. The bacterial strain *Brevibacillus* sp. PDM-3 degraded 93% of phenanthrene in 6 days, as analysed by HPLC. The ability to produce biosurfactant during phenanthrene degradation was detected by the surface tension measurements of the culture supernatant and the emulsification index. The biosurfactant was identified by its functional groups through FT-IR spectroscopy. Wang *et al.* (2010) applied Fourier transform infrared spectroscopy (FT-IR) and GC-MS for analysis of degradation of lube oil and TDOC (total dissolved organic carbon) by a mixed bacterial consortium. In addition, they analysed that mixed bacterial consortium can degrade benzene and its derivatives and aromatic ring organic matters more than 97%.

The spectroscopic techniques are used in conjunction with liquid chromatography, gas chromatography, mass spectrometry and other techniques having applications in proteome and metabolome profiling. Spectroscopic techniques can be applied in bioremediation studies for identification of biodegradation pathways, analysis and characterization of intermediates and final products, monitoring phenotypic and biochemical changes in bacterial communities and other such applications.

Bioinformatics – databases, software, tools and approaches

Development of bioinformatics techniques has propelled the biological research by a large magnitude as it has led to automated, precise and faster analyses of the generated data. Many novel approaches of analysis have been developed. Though analysis of the generated data may be the last step, yet it is the most important, as the wet-lab-data is meaningless, if not completely evaluated. Many bioinformatic infrastructures are available for analysis of genomics, transcriptomics, proteomics or metabolomics. The major revolutions have been automated genome sequencing, genome comparisons to identify the genome function and microbial evolution, derivation of metabolic and regulatory pathways, gene expression analysis, statistical tools, data mining techniques to derive protein–protein and protein–DNA interactions, modelling of 3D structure of proteins and 3D docking between proteins and biochemicals for rational drug design. Despite the fast paced global effort, the current analysis is limited by the lack of available gene-functionality from the wet-lab data, the lack of computer algorithms to explore vast amount of data with unknown functionality, limited availability of protein–protein and protein–DNA interactions, and the lack of knowledge of temporal and transient behaviour of genes and pathways (Bansal, 2005). However, now the focus in '*in silico*' developments is to modify the existing tools or develop new tools to facilitate integrated analyses of 'omic' studies. The integrated analyses of data sets will assist in understanding the changes taking place at various levels, beginning from DNA which contains the basic genetic information, followed by RNA and finally the products protein or metabolites. Today scientific community is on the verge of using all this knowledge to understand cellular mechanisms at the systemic level. Bioinformatics requires the study of microbial genomics, transcriptomics, proteomics, systems biology, computational biology, phylogenetic trees and data mining as a base for application of major bioinformatic tools for determining the structures and biodegradative pathways of xenobiotic compounds (Fulekar and Sharma, 2008).

McKenna *et al.* (2010) have developed Genome Analysis Tool Kit (GATK), a structured programming framework designed to ease the development of efficient and robust analysis tools for next-generation DNA sequencers using the functional programming philosophy of MapReduce. Prokaryotic genomes can be annotated based on their structural, operational and functional properties and Qiu *et al.* (2010) have described a approach to simultaneously determine the structural and operational annotation of the *Geobacter sulfurreducens* genome. Several *in silico* models and methods such as Poisson regression model to address the issue of undetected proteins, multivariate statistical methods to study gene expression and protein abundance data, mechanistic genomewide model for translation that provides mapping between changes in mRNA and protein levels have been used for integrating transcriptomics and proteomics (Mehra *et al.*, 2003; Nie *et al.*, 2006; Fagan *et al.*, 2007; Zhang *et al.*, 2010). The multivariate analysis methods can also be applied to integrate transcriptome profile with metabolic variables. Some general statistical tools such as unsupervised multivariate exploratory data mining, correlation network topology analysis, pattern recognition, principal components analysis and sparse partial least squares regression have been used for integrative proteomic and metabolomic analysis (Weckwerth *et al.*, 2004; Weckwerth and Morgenthal, 2005; Morgenthal *et al.*, 2006, 2007; Wienkoop *et al.*, 2008; Le^ Cao *et al.*, 2008, 2009; Zhang *et al.*, 2010). In another study, Mogilevskaya *et al.* (2009) presented an approach to developing mathematical description of metabolic pathways, and then to integrate reaction kinetics and different types of experimental data including transcriptomic, proteomic and metabolic data.

Bioinformatic resources exclusively for bioremediation studies and analyses have been and are being developed. The University of Minnesota Biocatalysis/Biodegradation Database (UM-BBD, http://umbbd.msi.umn.edu/) was developed in 1995 and is regularly updated. It contains information on 1240 compounds, 864 enzymes, 1337 reactions, 274 biotransformation rules and 510 microorganism entries (as on August 2010). Besides these data, it includes a Biochemical Periodic Table (UM-BPT) and a rule-based Pathway Prediction System (UM-PPS) (http://umbbd.msi.umn.edu/predict/) that predicts plausible pathways for microbial degradation of organic compounds. Public access to UM-BBD data is increasing and UM-BBD compound data are now contributed to PubChem and ChemSpider, the public chemical databases. A new mirror website of the UM-BBD, UM-BPT and UM-PPS is being developed at ETH Zürich to improve speed and reliability of online access from anywhere in the world (Gao *et al.*, 2010). Urbance *et al.* (2003) have developed a freely accessible Biodegradative Strain Database (BSD) (http://bsd.cme.msu.edu) within the phylogenetic framework of the Ribosomal Database Project II (RDPII: http://rdp.cme.msu.edu/html) to provide detailed information on degradative bacteria and the hazardous substances they degrade, including corresponding literature citations, relevant patents and links to additional web-based biological and chemical data. Pazos *et al.* (2003) studied the characteristics of the global biodegradation network by analysing a large set of chemical reactions that are implicated in biodegradation to obtain quantitative insights in its organization and possible mechanism of evolution. Pazos *et al.* (2005) developed a system, MetaRouter for maintaining heterogeneous information related to bioremediation in a framework that allows its query, administration and mining. The system can be accessed and administrated through a web interface and is available at http://pdg.cnb.uam.es/MetaRouter and additional material is available at http://www.pdg.cnb.uam.es/biodeg_net/MetaRouter. Arora *et al.* (2009) compiled a database of biodegradative oxygenases (OxDBase) (http://www.imtech.res.in/raghava/oxdbase/), which provides a compilation of the oxygenase data as sourced from primary literature in the form of web accessible database. There are two separate search engines for searching into the database – mono and di oxygenases database, respectively. Each enzyme entry contains its common name and synonym, reaction in which enzyme is involved, family and subfamily, structure and gene link and literature citation. The entries are also linked to several external database including BRENDA, KEGG, ENZYME and UM-BBD providing wide background information. At present (August 2010) the database contains information of over 235 oxygenases including both dioxygenases and monooxygenases. OxDBase is the first database that is dedicated to oxygenases and provides comprehensive information about them. Due to the importance of the oxygenases in chemical synthesis of drug intermediates and oxidation of xenobiotic compounds, OxDBase database would be very useful tool in the field of synthetic chemistry as well as bioremediation. Moriya *et al.* (2010) described PathPred (http://www.genome.jp/tools/pathpred/), a web-based server to predict plausible pathways of multi-step reactions starting from a query compound, based on the local RDM pattern match and the global chemical structure alignment against the reactant pair library. This server focuses on predicting pathways for microbial biodegradation of environmental compounds and biosynthesis of plant secondary metabolites, which correspond to characteristic RDM patterns in 947 and 1397 reactant pairs, respectively. The server provides transformed compounds and reference transformation patterns in each predicted reaction, and displays all predicted multi-step reaction pathways in a tree-shaped graph.

The transformation of xenobiotic compounds by microorganisms is essential for the bioremediation of contaminated environments. There is no single resource that provides information about environmental contaminants as well as microorganisms with biodegradative capabilities. A website that consolidates the detailed information about chemical compound and reference data related to biocatalysis, biotransformation, biodegradation and bioremediation would be an invaluable tool for academic and industrial researchers and environmental engineers (Urbance *et al.*, 2003).

Scenario and future perspectives

Bioremediation strategies are multidisciplinary involving ecology, geology, physiology, genetics, biochemistry, engineering and microbiology. The molecular biology and 'omics' technologies have led to tremendous progress in exploration of polluted sites and understanding the catabolic reactions occurring in bioremediation processes. The data generated by present 'omics' technologies are huge and the foremost need that has arisen is how to store this huge information and above all how to correlate and integrate them for comprehensive analyses. The use of massively parallel sequencing technologies is expanding the generated data exponentially. Therefore, the development of novel and sophisticated bioinformatic tools dedicated to bioremediation studies is necessary for analysis and integration of omic data sets involving identification of genes, phylogenetic affiliation of sequence reads and assignment of functions to expressed proteins. This integrated 'omics' technology-based functional analysis from genes via transcripts to proteins and metabolites at 'meta' level will answer all the relevant queries and mysteries that have hampered bioremediation strategies till now. At present there is a dire need for high-throughput methods for functional screening and characterization of key enzymes and metabolites involved in stress response metabolic networks. Developments in existing approaches and innovations in technologies will lead to novel and extensive approaches. Conversely, the technologies will have to be economical and applicable at high throughput level for increased access and applications. In future, analyses by 'meta' approaches will be successfully used in devising site/contamination specific bioremediation strategies.

Concluding annotations

The increasing amount of information available about the microorganisms and their genome, transcript, proteome and metabolome profiles along with reactions implicated in microbial biodegradation of xenobiotic compounds provides us with the building blocks for formulating a 'biodegradation network'. Review of literature has revealed that more and more 'omic' technologies in an integrative manner have been applied in bioremediation studies and applications in last few years. However, still efforts are needed to integrate all the levels of 'omic' studies-, which will enable us to get a complete picture of the plausible metabolic pathways.

Acknowledgements

The authors acknowledge the financial support provided by Department of Biotechnology, Government of India (India), and thank Ms Asha Parmar, BRD School of Biosciences, Sardar Patel University, Gujarat, India, for critical suggestions.

References

Abed, R.M., and Grotzschel, S. (2005). Applications of denaturing gradient gel electrophoresis (DGGE) and microsensor techniques in oil biodegradation studies. In Oil Pollution and its Environmental Impact in the Arabian Gulf Region (Developments in Earth and Environmental Sciences Series), Al-Azab, M., El-Shorbagy, W., and Al-Ghais, S. eds. (Oxford: Elsevier Ltd), pp. 75–91.

Adamczyk, J., Hesselsoe, M., Iversen, N., Horn, M., Lehner, A., Nielsen, P.H., Schloter, M., Roslev, P., and Wagner, M. (2003). The isotope array, a new tool that employs substrate-mediated labeling of rRNA for determination of microbial community structure and function. Appl. Environ. Microbiol. *69*, 6875–6887.

Amann, R.I., Ludwig, W., and Schleifer, K.H. (1995). Phylogenetic identification and *in situ* detection of individual microbial cells without cultivation. Microbiol. Rev. *59*, 143–169.

Arora, P.K., Kumar, M., Chauhan, A., Raghava, G.P., and Jain, R.K. (2009). OxDBase: a database of oxygenases involved in biodegradation. BMC Res. Notes *2*, 67.

Bae, J.-W., and Park, Y.-H. (2006). Homogeneous versus heterogeneous probes for microbial ecological microarrays. Trends Biotechnol. *24*, 318–323.

Baek, K.H., Yoon, B.D., Cho, D.H., Kim, B.H., Oh, H.M., and Kim, H.S. (2009). Monitoring bacterial population dynamics using real-time PCR during the bioremediation of crude-oil-contaminated soil. J. Microbiol. Biotechnol. *19*, 339–345.

Bakermans, C., and Madsen, E.L. (2002). Detection in coal tar waste-contaminated groundwater of mRNA transcripts related to naphthalene dioxygenase by fluorescent *in situ* hybridization with tyramide signal amplification. J. Microbiol. Methods *50*, 75–84.

Bansal, A.K. (2005). Bioinformatics in microbial biotechnology – a mini review. Microb. Cell. Fact. *4*, 19.

Beller, H.R., Kane, S.R., Legler, T.C., and Alvarez, P.J.J. (2002). A real-time polymerase chain reaction method for monitoring anaerobic, hydrocarbon-degrading bacteria based on a catabolic gene. Environ. Sci. Technol. *36*, 3977–3984.

Beloqui, A., Pita, M., Polaina, J., Martinez-Arias, A., Golyshina, O.V., Zumarraga, M., Yakimov, M.M., García-Arellano, H., Alcalde, M., Fernández, V.M. *et al.* (2006). Novel polyphenol oxidase mined from a metagenome expression library of bovine rumen: biochemical properties, structural analysis, and phylogenetic relationships. J. Biol. Chem. *281*, 22933–22942.

Benndorf, D., Balcke, G.U., Harms, H., and von Bergen, M. (2007). Functional metaproteome analysis of protein extracts from contaminated soil and groundwater. ISME J. *1*, 224–234.

Bentley, D.R., Balasubramanian, S., Swerdlow, H.P., Smith, G.P., Milton, J., Brown, C.G., Hall, K.P., Evers, D.J., Barnes, C.L., Bignell, H.R. *et al.* (2008). Accurate whole human genome sequencing using reversible terminator chemistry. Nature *456*, 53–59.

Bittner, L., Halary, S., Payri, C., Cruaud, C., de Reviers, B., Lopez, P., and Bapteste, E. (2010). Some considerations for analyzing biodiversity using integrative metagenomics and gene networks. Biol. Direct 5, 47.

Boersma, M.G., Solyanikova, I.P., Van Berkel, W.J., Vervoort, J., Golovleva, L.A., and Rietjens, I.M. (2001). 19F NMR metabolomics for the elucidation of microbial degradation pathways of fluorophenols. J. Ind. Microbiol. Biotechnol. *26*, 22–34.

Bombach, P., Chatzinotas, A., Neu, T.R., Kästner, M., Lueders, T., and Vogt, C. (2010). Enrichment and characterization of a sulfate-reducing toluene-degrading microbial consortium by combining *in situ* microcosms and stable isotope probing techniques. FEMS Microbiol. Ecol. *71*, 237–246.

Bordenava, S., Goni-Urizza, M., Caumette, P., and Duran, R. (2009). Differential display analysis of cDNA involved in microbial mats response after heavy fuel oil contamination. J. Microbial Biochem. Technol. *1*, 1–4.

Brennerova, M.V., Josefiova, J., Brenner, V., Pieper, D.H., and Junca, H. (2009). Metagenomics reveals diversity and abundance of meta-cleavage pathways in microbial communities from soil highly contaminated with jet fuel under air-sparging bioremediation. Environ. Microbiol. *11*, 2216–2227.

Bruscella, P., Eraso, J.M., Roh, J.H., and Kaplan, S. (2008). The use of chromatin immunoprecipitation to define PpsR binding activity in Rhodobacter sphaeroides 2.4.1. J. Bacteriol. *190*, 6817–6828.

Cardenas, E., and Tiedje, J.M. (2008). New tools for discovering and characterizing microbial diversity. Curr. Opin. Biotechnol. *19*, 544–549.

Cébron, A., Norini, M.-P., Beguiristain, T., and Leyval, C. (2008). Real-time PCR quantification of PAH-ring hydroxylating dioxygenase (PAH-RHDa) genes from Gram positive and Gram negative bacteria in soil and sediment samples. J. Microbiol. Methods 73, 148–159.

Chen, Y., and Murrell, J.C. (2010). When metagenomics meets stable-isotope probing: progress and perspectives. Trends Microbiol. *18*, 157–163.

Debruyn, J.M., Chewning, C.S., and Sayler, G.S. (2007). Comparative quantitative prevalence of Mycobacteria and functionally abundant nidA, nahAc, and nagAc dioxygenase genes in coal tar contaminated sediments. Environ. Sci. Technol. *41*, 5426–5432.

Dennis, P., Edwards, E.A., Liss, S.N., and Fulthorpe, R. (2003). Monitoring gene expression in mixed microbial communities by using DNA microarrays. Appl. Environ. Microbiol. *69*, 769–778.

Desai, C., and Madamwar, D. (2007). Extraction of inhibitor-free metagenomic DNA from polluted sediments, compatible with molecular diversity analysis using adsorption and ion-exchange treatments. Bioresour. Technol. *98*, 761–768.

Desai, C., Jain, K., Patel, B., and Madamwar, D. (2009b). Efficacy of bacterial consortium-AIE2 for contemporaneous Cr(VI) and azo dye bioremediation in batch and continuous bioreactor systems, monitoring steady-state bacterial dynamics using qPCR assays. Biodegradation *20*, 813–826.

Desai, C., Parikh, R.Y., Vaishnav, T., Shouche, Y.S., and Madamwar, D. (2009a). Tracking the influence of long-term chromium pollution on soil bacterial community structures by comparative analyses of 16S rRNA gene phylotypes. Res. Microbiol. *160*, 1–9.

Desai, C., Pathak, H., and Madamwar, D. (2010). Advances in molecular and '-omics' technologies to gauge microbial communities and bioremediation at xenobiotic/anthropogen contaminated sites. Bioresour. Technol. *101*, 1558–1569.

Díaz, E. (2004). Bacterial degradation of aromatic pollutants: a paradigm of metabolic versatility. Int. Microbiol. *7*, 173–180.

Díaz, E., and Prieto, M.A. (2000). Bacterial promoters triggering biodegradation of aromatic pollutants. Curr. Opin. Biotechnol. *11*, 467–475.

Dijk, J.A., Breugelmans, P., Philips, J., Haest, P.J., Smolders, E., and Springael, D. (2008). Catalyzed reporter deposition-fluorescent *in situ* hybridization (CARD-FISH) detection of Dehalococcoides. J. Microbiol. Methods 73, 142–147.

Drmanac, R., Sparks, A.B., Callow, M.J., Halpern, A.L., Burns, N.L., Kermani, B.G., Carnevali, P., Nazarenko, I., Nilsen, G.B., Yeung, G. *et al.* (2010). Human genome sequencing using unchained base reads on self-assembling DNA nanoarrays. Science *327*, 78–81.

D'Souza-Ticlo, D., Sharma, D., and Raghukumar, C. (2009). A thermostable metal-tolerant laccase with bioremediation potential from a marine-derived fungus. Mar. Biotechnol. (NY). *11*, 725–737.

Durand, S., Sancelme, M., Besse-Hoggan, P., and Combourieu, B. (2010). Biodegradation pathway of mesotrione: complementarities of NMR, LC-NMR and LC-MS for qualitative and quantitative metabolic profiling. Chemosphere *81*, 372–380.

Edwards, J.R., Ruparel, H., and Ju, J. (2005). Mass-spectrometry DNA sequencing. Mutat. Res. *573*, 3–12.

Ekman, D.R., Lorenz, W.W., Przybyla, A.E., Wolfe, N.L., and Dean, J.F. (2003). SAGE analysis of transcriptome responses in Arabidopsis roots exposed to 2,4,6-trinitrotoluene. Plant Physiol. *133*, 1397–1406.

Fagan, A., Culhane, A.C., and Higgins, D.G. (2007). A multivariate analysis approach to the integration of proteomic and gene expression data. Proteomics 7, 2162–2171.

Ferrer, M., Golyshina, O.V., Chernikova, T.N., Khachane, A.N., Reyes-Duarte, D., Santos, V.A., Strompl, C., Elborough, K., Jarvis, G., Neef, A. *et al.* (2005). Novel hydrolase diversity retrieved from a metagenome library of bovine rumen microflora. Environ. Microbiol. 7, 1996–2010.

Fleming, J.T., Sanseverino, J., and Sayler, G.S. (1993). Quantitative relationship between naphthalene catabolic gene frequency and expression in predicting PAH degradation in soils at town gas manufacturing sites. Environ. Sci. Technol. *27*, 1068–1074.

Fleming, J.T., Yao, W.H., and Sayler, G.S. (1998). Optimization of differential display of prokaryotic mRNA: application to pure culture and soil microcosms. Appl. Environ. Microbiol. *64*, 3698–3706.

Fortin, N., Beaumier, D., Lee, K., and Greer, C.W. (2004). Soil washing improves the recovery of total community DNA from polluted and high organic content sediments. J. Microbiol. Methods 56, 181–191.

Fulekar, M.H., and Sharma, J. (2008). Bioinformatics applied in bioremediation. Innov. Romanian Food Biotechnol. 2, 28–36.

Gao, H., Yang, Z.K., Gentry, T.J., Wu, L., Schadt, C.W., and Zhou, J. (2007). Microarray-based analysis of microbial community RNAs by whole-community RNA amplification, Appl. Environ. Microbiol. 73, 563–571.

Gao, J., Ellis, L.B., and Wackett, L.P. (2010). The University of Minnesota Biocatalysis/Biodegradation Database: improving public access. Nucleic Acids Res. *38*, D488-D491.

Garcia, D.E., Baidoo, E.E., Benke, P.I., Pingitore, F., Tang, Y.J., Villa, S., and Keasling, J.D. (2008). Separation and mass spectrometry in microbial metabolomics. Curr. Opin. Microbiol. *11*, 233–239.

Geets, J., Borremans, B., Diels, L., Springael, D., Vangronsveld, J., Van der lelie, D., and Vanbroekhoven, K. (2006). DsrB gene based DGGE for community and diversity surveys of sulfate-reducing bacteria. J. Microbiol. Methods. *66*, 194–205.

Goshe, M.B., and Smith, R.D. (2003). Stable isotope-coded proteomic mass spectrometry. Curr. Opin. Biotechnol. *14*, 101–109.

Gowda, M., Jantasuriyarat, C., Dean, R.A., and Wang, G.L. (2004). Robust-LongSAGE (RL-SAGE): a substantially improved LongSAGE method for gene discovery and transcriptome analysis. Plant Physiol. *134*, 890–897.

Graham, R.L.J., Graham, C., and McMullan, G. (2007). Microbial proteomics: a mass spectrometry primer for biologists. Microb. Cell Fact. *6*, 26.

Grainger, D.C., Overton, T.W., Reppas, N., Wade, J.T., Tamai, E., Hobman, J.L., Constantinidou, C., Struhl, K., Church, G., and Busby, S.J. (2004). Genomic studies with *Escherichia coli* MelR protein: applications of chromatin immunoprecipitation and microarrays. J. Bacteriol. *186*, 6938–6943.

Hägglund, P., Bunkenborg, J., Maeda, K., and Svensson, B. (2008). Identification of thioredoxin disulfide targets using a quantitative proteomics approach based on isotope-coded affinity tags. J. Proteome Res. 7, 5270–5276.

Ham, J.G., and Bonner, J.S. (1997). Environmental site assessments and bioremediation. In Perspectives in Bioremediation Technologies for Environmental Improvement, Wild, J.R., Varfolomeyev, S.D., and Scozzafava, A., eds (Netherlands: Kluwer Academic Publishers), pp. 1–12.

Haqqani, A.S., Kelly, J.F., and Stanimirovic, D.B. (2008). Quantitative protein profiling by mass spectrometry using label-free proteomics. Methods Mol. Biol. *439*, 241–256.

He, Z., Deng, Y., Van Nostrand, J.D., Tu, Q., Xu, M., Hemme, C.L., Li, X., Wu, L., Gentry, T.J., Yin, Y. *et al.* (2010). GeoChip 3.0 as a high-throughput tool for analyzing microbial community composition, structure and functional activity. ISME J. *4*, 1167–1179.

He, Z., Gentry, T.J., Schadt, C.W., Wu, L., Liebich, J., Chong, S.C., Huang, Z., Wu, W., Gu, B., Jardine, P. *et al.* (2007). GeoChip: a comprehensive microarray for investigating biogeochemical, ecological and environmental processes. ISME J. *1*, 67–77.

Indest, K.J., Crocker, F.H., and Athow, R. (2007). A TaqMan polymerase chain reaction method for monitoring RDX degrading bacteria based on the xplA functional gene. J. Microbiol. Meth. *68*, 267–274.

Ishihama, Y., Oda, Y., Tabata, T., Sato, T., Nagasu, T., Rappsilber, J., and Mann, M. (2005). Exponentially modified protein abundance index (emPAI) for estimation of absolute protein amount in proteomics by the number of sequenced peptides per protein. Mol. Cell Proteomics. *4*, 1265–1272.

Jiménez, J.I., Miñambres, B., García, J.L., and Díaz, E. (2002). Genomic analysis of the aromatic catabolic pathways from *Pseudomonas* putida KT2440. Environ. Microbiol. *4*, 824–841.

Jorgensen, K.S. (2008). Advances in monitoring of catabolic genes during bioremediation. Indian J. Microbiol. *48*, 152–155.

Kao, C.M., Chen, C.S., Tsa, F.Y., Yang, K.H., Chien, C.C., Liang, S.H., Yang, C.A., and Chen, S.C. (2010). Application of real-time PCR, DGGE fingerprinting, and culture-based method to evaluate the effectiveness of intrinsic bioremediation on the control of petroleum-hydrocarbon plume. J. Hazard. Mater. *178*, 409–416.

Keum, Y.S., Seo, J.S., Qing, X.L., and Kim, J.H. (2008). Comparative metabolomic analysis of Sinorhizobium sp. C4 during the degradation of phenanthrene. Appl. Microbiol. Biotechnol. *80*, 863–872.

Kiesel, B., Balcke, G.U., Dietrich, J., Vogt, C., and Geyer, R., (2008). Microbial community shifts as a response to efficient degradation of chlorobenzene under hypoxic conditions. Biodegradation *19*, 435–446.

Kim, M., Bae, S.S., Seol, M., Lee, J.H., and Oh, Y.S. (2008a). Monitoring nutrient impact on bacterial community composition during bioremediation of anoxic PAH-contaminated sediment. J. Microbiol. *46*, 615–623.

Kim, S.J., Kweon, O., and Cerniglia, C.E. (2009). Proteomic applications to elucidate bacterial aromatic hydrocarbon metabolic pathways. Curr. Opin. Microbiol. *12*, 301–309.

Kim, S.J., Kweon, O., Jones, R.C., Ricky, D.E., and Cerniglia, C.E. (2008b). Genomic analysis of polycyclic aromatic hydrocarbon degradation in Mycobacterium vanbaalenii PYR-1. Biodegradation *19*, 859–881.

Kim, Y.H., Cho, K., Yun, S.H., Kim, J.Y., Kwon, K.H., Yoo, J.S., and Kim, S.I. (2006). Analysis of aromatic catabolic pathways in *Pseudomonas* putida KT 2440 using a combined proteomic approach: 2DE/MS and cleavable isotope-coded affinity tag analysis. Proteomics *6*, 1301–1318.

Kim, Y.H., Kwon, E.J., Kim, S.K., Jeong, Y.S., Kim, J., Yun, H.D., and Kim, H. (2010). Molecular cloning and characterization of a novel family VIII alkaline esterase from a compost metagenomic library. Biochem. Biophys. Res. Commun. *393*, 45–49.

Lane, D.J., Pace, B., Olsen, G.J., Stahl, D.A., Sogin, M.L., and Pace, N.R. (1985). Rapid determination of 16S ribosomal RNA sequences for phylogenetic analyses. Proc. Natl. Acad. Sci. U.S.A. *82*, 6955–6959.

Le˄ Cao, K.A., Martin, P.B., Robert-Granie´, C., and Besse, P. (2009). Sparse canonical methods for biological data integration: application to a cross-platform study. BMC Bioinformatics *10*, 34.

Le˄ Cao, K.A., Rossouw, D., Robert-Granie´, C., and Besse, P. (2008). A sparse PLS for variable selection when integrating omics data. Stat. Appl. Genet. Mol. Biol. *7*, Article 35.

Lee, N., Nielsen, P.H., Andreasen, K.H., Juretschko, S., Nielsen, J.L., Schleifer, K.H., and Wagner, M. (1999). Combination of fluorescent *in situ* hybridization and microautoradiography-a new tool for structure-function analyses in microbial ecology. Appl. Environ. Microbiol. *65*, 1289–1297.

Li, Z., Xu, J., Tang, C., Wu, J., Muhammad, A., and Wang, H. (2006). Application of 16S rDNA-PCR amplification and DGGE fingerprinting for detection of shift in microbial community diversity in Cu-, Zn-, and Cd-contaminated paddy soils. Chemosphere *62*, 1374–1380.

Lovley, D.R. (2003). Cleaning up with genomics: applying molecular biology to bioremediation. Nat. Rev. Microbiol. *1*, 35–44.

Loy, A., Schulz, C., Lucker, S., Schopfer-Wendels, A., Stoecker, K., Baranyi, C., Lehner, A., and Wagner, M. (2005). 16S rRNA gene-based oligonucleotide microarray for environmental monitoring of the betaproteobacterial order 'Rhodocyclales'. Appl. Environ. Microbiol. *71*, 1373–1386.

Margulies, M., Egholm, M., Altman, W.E., Attiya, S., Bader, J.S., Bemben, L.A., Berka, J., Braverman, M.S., Chen, Y.-J., Chen, Z. *et al.* (2005). Genome sequencing in microfabricated high-density picolitre reactors. Nature *437*, 376–380.

Matsumura, H., Ito, A., Saitoh, H., Winter, P., Kahl, G., Reuter, M., Krüger, D.H., and Terauchi, R. (2005). SuperSAGE. Cell Microbiol. *7*, 11–18.

Mattes, T.E., Alexander, A.K., Richardson, P.M., Munk, C., Han, C.S., Stothard, P., and Colemans, N.V. (2008). The genome of Polaromonas sp. strain JS666: insights into the evolution of a hydrocarbon- and xenobiotic-degrading bacterium, and features of relevance to biotechnology. Appl. Environ. Microbiol. *74*, 6405–6416.

McKenna, A., Hanna, M., Banks, E., Sivachenko, A., Cibulskis, K., Kernytsky, A., Garimella, K., Altshuler, D., Gabriel, S., Daly, M. *et al.* (2010). The Genome Analysis Toolkit: a MapReduce framework for analyzing next-generation DNA sequencing data. Genome Res. *20*, 1297–1303.

McKernan, K.J., Peckham, H.E., Costa, G.L., McLaughlin, S.F., Fu, Y., Tsung, E.F., Clouser, C.R., Duncan, C., Ichikawa, J.K., Lee, C.C. *et al.* (2009). Sequence and structural variation in a human genome uncovered by short-read, massively parallel ligation sequencing using two-base encoding. Genome Res. *19*, 1527–1541.

Mehra, A., Lee, K.H., and Hatzimanikatis, V. (2003). Insights into the relation between mRNA and protein expression patterns: I. Theoretical considerations. Biotechnol. Bioeng. *84*, 822–833.

Mesarch, M.B., Nakatsu, C.H., and Nies, L. (2000). Development of catechol 2,3-dioxygenase-specific primers for monitoring bioremediation by competitive quantitative PCR. Appl. Environ. Microbiol. *66*, 678–683.

Mogilevskaya, E., Bagrova, N., Plyusnina, T., Gizzatkulov, N., Metelkin, E., Goryacheva, E., Smirnov, S., Kosinsky, Y., Dorodnov, A., Peskov, K. *et al.* (2009). Kinetic modeling as a tool to integrate multilevel dynamic experimental data. Methods Mol. Biol. *563*, 197–218.

Morgenthal, K., Weckwerth, W., and Steuer, R. (2006). Metabolomic networks in plants: Transitions from pattern recognition to biological interpretation. Biosytems *82*, 108–117.

Morgenthal, K., Wienkoop, S., Wolschin, F., and Weckwerth, W. (2007). Integrative profiling of metabolites and proteins: improving pattern recognition and biomarker selection for systems level approaches. Methods Mol. Biol. *358*, 57–75.

Moriya, Y., Shigemizu, D., Hattori, M., Tokimatsu, T., Kotera, M., Goto, S., and Kanehisa, M. (2010). PathPred: an enzyme-catalyzed metabolic pathway prediction server. Nucleic Acids Res. *38*, 1–6.

Nelson, K.E., Weinel, C., Paulsen, I.T., Dodson, R.J., Hilbert, H., Martins dos Santos, P., Fouts, D.E., Gill, S.R., Pop, M., Holmes, M. *et al.* (2002). Complete genome sequence and comparative analysis of metabolically versatile *Pseudomonas* putida KT2440. Environ. Microbiol. *4*, 799–808.

Nie, L., Wu, G., and Zhang, W. (2006). Correlation of mRNA expression and protein abundance affected by multiple sequence features related to translational efficiency in Desulfovibrio vulgaris: a quantitative analysis. Genetics *174*, 2229–2243.

Nocker, A., Burr, M., and Camper, A., (2007). Genotypic microbial community profiling: a critical technical review. Microb. Ecol. *54*, 276–289.

Nyyssonen, M., Piskonen, R., and Itavaara, M., (2006). A targeted real-time PCR assay for studying naphthalene degradation in the environment. Microb. Ecol. *52*, 533–543.

Ozsolak, F., Platt, A.R., Jones, D.R., Reifenberger, J.G., Sass, L.E., McInerney, P., Thompson, J.F., Bowers, J., Jarosz, M., and Milos, P.M. (2009). Direct RNA sequencing. Nature *461*, 814–818.

Patino, W.D., Mian, O.Y., and Hwang, P.M. (2002). Serial analysis of gene expression: technical considerations and applications to cardiovascular biology. Circ. Res. *91*, 565–569.

Pazos, F., Guijas, D., Valencia, A., and De Lorenzo, V. (2005). MetaRouter: bioinformatics for bioremediation. Nucleic Acids Res. 33, D588–D592.

Pazos, F., Valencia, A., and De Lorenzo, V. (2003). The organization of the microbial biodegradation network from a systems-biology perspective. EMBO Rep. *4*, 994–999.

Qiu, Y., Cho, B.-K., Park, Y.S., Lovley, D., Palsson, B.O., and Zengler, K. (2010). Structural and operational complexity of the Geobacter sulfurreducens genome. Genome Res. *20*, 1304–1311.

Ram, R.J., Verberkmoes, N.C., Thelen, M.P., Tyson, G.W., Baker, B.J., Blake, R.C. 2nd., Shah, M., Hettich, R.L., and Banfield, J.F. (2005). Community proteomics of a natural microbial biofilm. Science *308*, 1915–1920.

Rappsilber, J., Ryder, U., Lamond, A.I., and Mann, M. (2002). Large-scale proteomic analysis of the human spliceosome. Genome Res. *12*, 1231–1245.

Reddy, M.S., Naresh, B., Leela, T., Prashanthi, M., Madhusudhan, N.C., Dhanasri, G., and Devi, P. (2010). Biodegradation of phenanthrene with biosurfactant production by a new strain of Brevibacillus sp. Bioresour. Technol. [Epub ahead of print PMID 20627713]

Rhee, S.K., Liu, X., Wu, L., Chong, S.C., Wan, X., and Zhou, J. (2004). Detection of genes involved in biodegradation and biotransformation in microbial communities by using 50-mer oligonucleotide microarrays. Appl. Environ. Microbiol. *70*, 4303–4317.

Rocha, E.P. (2008). The organization of the bacterial genome. Annu. Rev. Genet. *42*, 211–223.

Rogers, S.L., and McClure, N. (2003). The role of microbiological studies in bioremediation process optimisation. In Bioremediation: A critical review, Head, I.M., Singleton, I., and Milner, M., eds. (Norfolk, UK: Horizon Scientific Press), pp. 27–59.

Rogers, S.W., Mooreman, T.B., and Onge, S.K. (2007). Fluorescent *in situ* hybridization and microautoradiography applied to ecophysiology in soil. Soil Sci. Soc. Am. J. *71*, 620–631.

Roma-Rodrigues, C., Santos, P.M., Benndorf, D., Rapp, E., and Sá-Correia, I. (2010). Response of *Pseudomonas* putida KT2440 to phenol at the level of membrane proteome. J. Proteomics 73, 1461–1478.

Saha, S., Sparks, A.B., Rago, C., Akmaev, V., Wang, C.J., Vogelstein, B., Kinzler, K.W., and Velculescu, V.E. (2002). Using the transcriptome to annotate the genome. Nat. Biotechnol. *20*, 508–512.

Schloss, P.D., and Handelsman, J. (2003). Biotechnological prospects from metagenomics. Curr. Opin. Biotechnol. *14*, 303–310

Schloss, P.D., and Handelsman, J. (2005). Metagenomics for studying unculturable microorganisms: cutting the Gordian knot. Genome Biol. *6*, 229.

Schoolnik, G.K. (2001). The accelerating convergence of genomics and microbiology. Genome Biol. 2 REPORTS 4009.

Schulze, W.X., Gleixner, G., Kaiser, K., Guggenberger, G., Mann, M., and Schulze, E.D. (2005). A proteomic fingerprint of dissolved organic carbon and of soil particles. Oecologia. *142*, 335–343.

Schuster, S.C. (2008). Next-generation sequencing transforms today's biology. Nat. Methods *5*, 16–18.

Scneegurt, M.A., and Kulpa, C.F. (1998). The application of molecular techniques in environmental biotechnology for monitoring microbial systems. Biotechnol. Appl. Biochem. *27*, 73–79.

Scully, C., Collins, G., and O'Flaherty, V. (2005). Assessment of anaerobic wastewater treatment failure using terminal restriction fragment length polymorphism analysis. J. Appl. Microbiol. *99*, 1463–1471.

Sharan, R., and Ideker, T. (2006). Modeling cellular machinery through biological network comparison. Nat. Biotechnol. *24*, 427–433.

Sharma,R., Ranjan, R., Kapardar, R.K., and Grover, A. (2005). 'Unculturable' bacterial diversity: An untapped resource. Curr. Sci. *89*, 72–77.

Shendure, J., and Ji, H. (2008). Next-generation DNA sequencing. Nat. Biotechnol. *26*, 1135–1145.

Singh, O.V., and Chandel, A.K. (2009). Two-dimensional gel electrophoresis: discovering biomolecules for environmental bioremediation. In Bioremediation: Methods and Protocols (Methods in Molecular Biology Series), Cummings, S.P., eds. (New Jersey, USA: Humana Press), pp. 141–156.

Singh, O.V., and Nagaraj, N.S. (2006). Transcriptomics, proteomics and interactomics: unique approaches to track the insights of bioremediation. Brief Funct Genomic Proteomic *4*, 355–362.

Singleton, D.R., Powell, S.N., Sangaiah, R., Gold, A., Ball, L.M., and Aitken, M.D. (2005). Stable-isotope probing of bacteria capable of degrading salicylate, naphthalene, or phenanthrene in a bioreactor treating contaminated soil. Appl. Environ. Microbiol. *71*, 1202–1209.

Stedtfeld, R.D., Baushke, S.W., Tourlousse, D.M., Miller, S.M., Stedtfeld, T.M., Gulari, E., Tiedje, J.M., and Hashsham, S.A. (2008). Development and experimental validation of a predictive threshold cycle equation for quantification of virulence and marker genes by high-throughput nanoliter-volume PCR on the OpenArray platform. Appl. Environ. Microbiol. *74*, 3831–3838.

Suenaga, H., Ohnuki, T., and Miyazaki, K. (2007). Functional screening of a metagenomic library for genes involved in microbial degradation of aromatic compounds. Environ. Microbiol. *9*, 2289–2297.

Sun, Y., Polishchuk, E.A., Radoja, U., and Cullen, W.R. (2004). Identification and quantification of arsC genes in environmental samples by using real-time PCR. J. Microbiol. Meth. *58*, 335–349.

Tang, Y.J., Martin, H.G., Dehal, P.S., Deutschbauer, A., Llora, X., Meadows, A., Arkin, A., and Keasling, J.D. (2009). Metabolic flux analysis of *Shewanella* spp. reveals evolutionary robustness in central carbon metabolism. Biotechnol. Bioeng. *102*, 1161–1169.

Tao, W.A., and Aebersold, R. (2003). Advances in quantitative proteomics via stable isotope tagging and mass spectrometry. Curr. Opin. Biotechnol. *14*, 110–118.

Trautwein, K., Kühner, S., Wöhlbrand, L., Halder, T., Kuchta, K., Steinbüchel, A., and Rabus, R. (2008). Solvent stress response of the denitrifying bacterium 'Aromatoleum aromaticum' strain EbN1. Appl. Environ. Microbiol. *74*, 2267–2274.

Tyson, G.W., Chapman, J., Hugenholtz, P., Allen, E.E., Ram, R.J., Richardson, P.M., Solovyev, V.V., Rubin, E.M., Rokhsar, D.S., and Banfield, J.F. (2004). Community structure and metabolism through reconstruction of microbial genomes from the environment. Nature *428*, 37–43.

Urbance, J.W., Cole, J., Saxman, P., and Tiedje, J.M. (2003). BSD: the Biodegradative Strain Database. Nucleic Acids Res. *31*, 152–155.

Viggi, C.C., Dionisi, D., Miccheli, A., Valerio, M., and Majone, M. (2010). Metabolic analysis of the removal of formic acid by unacclimated activated sludge. Water Res. *44*, 3393–3400.

Vilchez-Vargas, R., Junca, H., and Pieper, D.H. (2010). Metabolic networks, microbial ecology and 'omics' technologies: towards understanding *in situ* biodegradation processes. Environ. Microbiol. *12*, 3089–3104.

Villas-Boas, S.G., and Bruheim, P. (2007). The potential of metabolic tools in bioremediation studies. OMICS: A Journal of Integrative Biol. *11*, 305–313.

Wakase, S., Sasaki, H., Itoh, K., Otawa, K., Kitazume., Nonaka, J., Satoh, M., Sasaki, T., and Nakai, Y. (2008). Investigation of the microbial community in a microbiological additive used in a manure composting process.akai, Y. Bioresour. Technol. *99*, 2687–2693.

Wang, H., Xu, R., Li, F., Qiao, J., and Zhang, B. (2010). Efficient degradation of lube oil by a mixed bacterial consortium. J. Environ. Sci. (China). *22*, 381–388.

Ward, N. (2006). New directions and interactions in metagenomics research. FEMS Microbiol. Ecol. *55*, 331–338.

Ward, N., and Fraser, C.M. (2005). How genomics has affected the concept of microbiology. Curr. Opin. Microbiol. *8*, 564–571.

Washburn, M.P., Ulaszek, R., Deciu, C., Shcieltz, D.M., and Yates, J.R. III (2002). Analysis of quantitative proteomic data generated via multidimensional protein identification technology. Anal. Chem. *74*, 1650–1657.

Watanabe, K., and Baker, P.W. (2000). Environmentally relevant microorganisms. J. Biosci. Bioeng. *89*, 1–11.

Weckwerth, W., and Morgenthal, K. (2005). Metabolomics: from pattern recognition to biological interpretation. Drug Discov. Today *10*, 1551–1558.

Weckwerth, W., Wenzel, K., and Fiehn, O. (2004). Process for the integrated extraction, identification and quantification of metabolites, proteins and RNA to reveal their co-regulation in biochemical networks. Proteomics *4*, 78–83.

West, K.A., Johnson, D.R., Hu, P., DeSantis, T.Z., Brodie, E.L., Lee, P.K.H., Feil, H., Andersen, G.L., Zinder, S.H., and Alvarez-Cohen, L. (2008). Comparative genomics of "Dehalococcoides ethenogenes' 195 and an enrichment culture containing unsequenced "Dehalococcoides' strains. Appl. Environ. Microbiol. *74*, 3533–3540.

Wharfe, E.S., Jarvis, R.M., Winder, C.L., Whiteley, A.S., and Goodacre, R. (2010). Fourier transform infrared spectroscopy as a metabolite fingerprinting tool for monitoring the phenotypic changes in complex bacterial communities capable of degrading phenol. Environ Microbiol. *12*, 3253–3263.

Wiechert, W., Schweissgut, O., Takanaga, H., and Frommer, W.B. (2007). Fluxomics: mass spectrometry versus quantitative imaging. Curr. Opin. Plant Biol. *10*, 323–330.

Wienkoop, S., Morgenthal, K., Wolschin, F., Scholz, M., Selbig, J., and Weckwerth, W. (2008). Integration of metabolomic and proteomic phenotypes. Mol. Cell Proteomics *7*, 1725–1736.

Wilmes, P., and Bond, P.L. (2004). The application of two-dimensional polyacrylamide gel electrophoresis and downstream analyses to a mixed community of prokaryotic microorganisms. Environ. Microbiol. *6*, 911–920.

Wilmes, P., and Bond, P.L. (2006). Metaproteomics: studying functional gene expression in microbial ecosystems. Trends Microbiol. *14*, 92–97.

Wilmes, P., Wexler, M., and Bond, P.L. (2008). Metaproteomics provides functional insight into activated sludge wastewater treatment. PLoS One 12 (e1778), 1–11.

Wu, H.J., Seib, K.L., Srikhanta, Y.N., Edwards, J., Kidd, S.P., Maguire, T.L., Hamilton, A., Pan, K.T., Hsiao, H.H., Yao, C.W. *et al.* (2010). Manganese regulation of virulence factors and oxidative stress resistance in *Neisseria* gonorrhoeae. J. Proteomics. *73*, 899–916.

Yergeau, E., Arbour, M., Brousseau, R., Juck, D., Lawrence, J.R., Masson, L., Whyte, L.G., and Greer, C.W. (2009). Microarray and real-time PCR analyses of the responses of high-arctic soil bacteria to hydrocarbon pollution and bioremediation treatments. Appl. Environ. Microbiol. *75*, 6258–6267.

Zhang, W., Li, F., and Nie, L. (2010). Integrating multiple 'omics' analysis for microbial biology: application and methodologies. Microbiology *156*, 287–301.

Zhao, Y.G., Wang, A.J., Ren, N.Q., and Zhao, Y. (2006). Microbial community structure in different wastewater treatment processes characterized by single-strand-conformation polymorphism (SSCP) technique. Huan Jing Ke Xue *27*, 1429–1433.

Escherichia coli Host Engineering for Efficient Metagenomic Enzyme Discovery

13

Reia Hosokawa-Okamoto and Kentaro Miyazaki

Abstract

Enzymes are environmentally friendly biocatalysts that are widely used in modern life, e.g. in food processing, laundry detergent, and production of medicinal compounds. An increasing demand to shift focus from petrochemicals to biotechnology-based industries has expanded the use of enzymes. To date, most industrially relevant enzymes are of microbial origin. Therefore, mining for microbial enzymes is key to the development of the biotechnology industry; however, less than 1% of environmental bacteria can be cultured in the laboratory at present. To accelerate the discovery of industrially relevant enzymes, it would therefore be advantageous to employ a metagenomic approach to extend the available microbial sources to presently 'unculturable' taxa. However, such an approach also risks a low hit rate; typically, only a few positives are obtained from the hundreds of thousands of library clones screened. This is largely because of the discrepancy between the host's transcriptional/translational machineries and the genetic signals present in the metagenomes. *Escherichia coli* has long been used as a generic host for cloning and production purposes and is, in most instances, suitable for this purpose. However, several modifications are necessary to overcome the problems of heterologous gene expression. In this chapter, we discuss several approaches that may be useful for developing the utility of *E. coli* as a host for efficient functional screening of metagenomic libraries.

Introduction

Industrial enzymes

Enzymes have been used since ancient times for the production of fermented food products such as cheese, sourdough, beer, wine, and vinegar. Humans have enjoyed the benefits of this biotechnology in terms of transformation of raw materials into value-added products with superior flavour, texture, and longevity. Beginning in the second half of the twentieth century, developments in microbial screening, breeding, and fermentation technologies allowed large-scale manufacture of enzymes as purified, well-characterized preparations. Concurrent development of related technologies (i.e. genetic engineering, protein purification, protein immobilization, etc.) has established enzymes as an indispensable workhorse in various sectors, including the chemical industry. The total market for industrial enzymes reached \$2 billion in 2000 and has risen to \$2.5 billion today (Demain and Vaishnav, 2009), of which the largest sectors are the food and feed industries. The majority of industrial enzymes are the hydrolases that are used for the degradation of various natural substances. Proteases are the dominant industrial enzymes, accounting for 57% of the market, and are used extensively in detergents and the dairy industry. Carbohydrolases, primarily amylases and cellulases, are the second largest group, used primarily in starch and the textile, detergent, and baking industries. Others include glucoamylase, xylose isomerase, lactase, lipase, pullulanase, and xylanase. Microorganisms (i.e. bacteria, yeast, and moulds) are the major sources of these enzymes, whereas enzymes with animal and plant origins constitute 8 and 4% of the total, respectively. More details on the current status of industrial enzymes and their applications can be found in Kirk *et al.* (2002).

One roadblock to more widespread use of enzymes in industry is their lack of stability under non-native conditions, e.g. extremes of pH, temperature, and ionic strength that are common to industrial bioprocesses. This problem can be partly solved by genetically modifying the protein via rational design or directed evolution (Arnold, 2001). Such protein engineering may also improve other enzymatic properties (e.g. enhanced substrate specificity, expression level, and reduced product inhibition, etc.) that are crucial to industrial bioprocesses.

No matter how useful protein engineering may be, however, the key to development of enzyme-based industrial bioprocesses is discovery of novel enzymes. To this end, we must develop a technology for accelerated discovery of novel enzymes from the natural environment, and especially from microorganisms, the dominant supplier of industrial enzymes.

Enzyme discovery by metagenomics

Microorganisms can be found in almost every natural environment on our planet (Curtis *et al.*, 2002; Ward, 2002; Schloss and Handelsman, 2004). The traditional method for obtaining enzymes from the natural environment is to search for microorganisms that exhibit the desired activity. Once the strain has been identified, it is cultured in the laboratory, and expression of the target enzyme is optimized. The original strain is often bred for higher expression. To further enhance production of the enzyme, the gene is cloned and overexpressed as a recombinant protein in an appropriate host. However, the fact that the majority of microorganisms cannot be cultivated in the laboratory (Amann *et al.*, 1995) limits discovery of novel enzymes (or strains) by means of the above-mentioned traditional approach.

Metagenomics is as an alternative to cultivation-based approaches and allows investigation of collective microbial genomes derived from the environment. Overcoming the problems associated with cultivation allows, in theory, exploration of the entire natural microbiome, that is dominated (≥99%) by unculturable taxa. Metagenomics is now driven by increasing biotechnological demands for novel enzymes and biomolecules and contributes to a better understanding of global microbial ecology.

Metagenomes are at present screened based on either their sequence or function. Sequence-based screening is readily available and is used by many researchers (Francis *et al.*, 2005; Marzorati *et al.*, 2007; Xu *et al.*, 2008). A set of degenerate primers are designed based on the consensus amino acid sequences of target enzymes and used to clone homologous genes by PCR. The resulting amplicons are recombinantly expressed in a surrogate host, and gene products are characterized. However, cloned genes do not always display the expected activities.

Compared with the sequence-driven approach, function-based screening is straightforward but more labour- and time-intensive. This method involves construction of genomic expression libraries in vectors possessing strong expression signals, followed by functional screening, often automated and in a high-throughput fashion. Because this method does not rely on sequence homologies, it often leads to identification of novel classes of enzymes. Functional screening of metagenomic libraries has to date led to discovery of novel antibiotic resistance genes, hydrolases (e.g. protease, esterase, amylase, lipase, and xylanase), and enzymes involved in degradation of aromatic compounds.

The largest problem with function-based screening of metagenomes is the low hit rate. For instance, in a search for lipases in the metagenome of German soil, only a single positive was recovered from 730,000 clones screened (Henne *et al.*, 2000). Similarly, a North American soil metagenome library produced 29 clones expressing haemolytic activity out of ~25,000 screened (Rondon *et al.*, 2000). Such low hit rates are a common problem in metagenomic library screening, irrespective of target or source environment (Lorenz and Eck, 2005; Uchiyama and Miyazaki, 2009).

High-throughput screening is a possible solution to the low hit-rate problem. Agar-plate screening is the most commonly used technique. It is extremely high throughput, cost-effective, and convenient. Clones are cultivated on agar containing the appropriate substrate, and positive colonies are identified by the presence of a clear or coloured zone (halo) around colonies. However, this approach often results in identification of no positives due to a lack of sensitivity. To improve sensitivity, we used cell lysate-based screening in 96-well plates instead of agar plates (Suenaga *et al.*, 2007). A total of 91 catechol 2,3-dioxygenase genes were obtained; none of these showed yellow colour on catechol-containing agar plates. Another strategy is to use a reporter assay. Green fluorescent protein (Uchiyama *et al.*, 2005; Guan *et al.*, 2007) and β-galactosidase (Mohn *et al.*, 2006; Schipper *et al.*, 2009) are the most commonly used reporters whose biological activities are linked to substrate-induced gene expression (Uchiyama *et al.*, 2005) or product formation (Uchiyama and Miyazaki, 2010) to aid detection of positive clones. All of these approaches are effective in retrieving enzymes from metagenomes; however, the problem of expression of heterologous genes remains.

Improved expression of metagenomic genes by host engineering

The 'low hit-rate problem' often encountered during metagenomic library screening is assumed to be due in large part to insufficient expression in *E. coli* (Daniel, 2004; Handelsman, 2004; Gabor *et al.*, 2007; Uchiyama and Miyazaki, 2009). Gabor *et al.* (2004) analysed 32 complete microbial genomes for the presence of expression signals (transcriptional promoters and ribosome binding sites) that could be recognized by *E. coli*. The results suggested that due to the lack of suitable expression signals, only 40% of enzymatic activities would be functional. It is believed that culturable microorganisms are phylogenetically distinct from unculturable ones; thus, unculturable microorganisms are likely to contain expression signals that are not recognized by *E. coli*. This, together with the data of Gabor and co-workers, suggests that successful expression of foreign genes in *E. coli* is problematic.

Historically, *E. coli* has been the workhorse of recombinant DNA technology. To date, many *E. coli*–based genetic systems have been developed, including genetic components (antibiotic resistance genes, replication origins, and inducible promoters), hosts (e.g. BL21[DE3], JM109), and vectors (e.g. pET, pUC) that are useful for enhanced insert expression. Moreover, recent progress in gene synthesis technology enables use of genes whose nucleotide sequences have been optimized for maximal expression (Fuglsang, 2003; Welch *et al.*, 2009). These technologies are useful for expression of genes whose sequences are known. However, in functional metagenomics genomic fragments of unknown sequence are frequently the targets. Therefore, it is necessary to develop a system to express genes in the absence of sequence information. A suitable host for metagenomic library screening would be capable of recognizing a broad range of expression signals from diverse taxa.

One obvious solution to this problem would be to use a host that is phylogenetically distant from *E. coli*. *Bacillus subtilis*, *Pseudomonas putida*, *Streptomyces lividans*, and *Rhizobium leguminosarum* have been proposed to provide superior gene expression characteristics in case that *E. coli* expression system does not work efficiently (Wang *et al.*, 2000; Daniel, 2004; Schmeisser *et al.*, 2007). Wexler *et al.* (2005) reported that alcohol/aldehyde dehydrogenases were not functionally expressed in *E. coli*; this was overcome by using *R. leguminosarum*. Courtois *et al.* (2003) demonstrated that *S. lividans* is useful for functional screening of soil metagenomic libraries for novel polyketide synthase genes. Martinez *et al.* (2004) used three different hosts to screen a metagenomic library; the genes retrieved depended on the host microbe (*E. coli*, *P. putida* and *S. lividans*). All of these studies clearly suggest the presence of host bias in the expression of foreign genes. Thus, parallel construction of libraries using phylogenetically diverse hosts can overcome individual deficiencies. However, the transformation efficiencies of these organisms are usually low, and so they are unsuitable for construction of large-scale genomic libraries. Thus, the problem of heterologous gene expression seems insurmountable.

To our knowledge, no studies that aimed to develop hosts for non-biased expression of foreign genes have been published yet. In this chapter, we collected studies on *E. coli* host engineering aimed at improving heterologous gene expression through modification of transcriptional and translational machineries as well as by using molecular chaperones. Papers of particular interest are summarized in Table 13.1.

These approaches should open up new possibilities for engineering *E. coli* strains suitable for creation of metagenomic libraries and better expression of genes from various source organisms. The concept of *E. coli* engineering for screening metagenomic libraries is described in Fig. 13.1.

Engineering *E. coli* transcriptional systems

RNA polymerase (RNAP) is a multisubunit enzyme whose subunit composition is $\alpha_2\beta\beta'\omega$. One of seven sigma factors (the determinant of promoter specificity) binds to this complex, recognizes appropriate promoters, and initiates transcription (Ishihama, 2000). Promoter sequences and sigma factors vary widely across organisms. However, one sigma factor, known as RpoD or σ^{70} in *E. coli*, is well conserved among species. The discrepancy between *E. coli* RNAP and promoter sequences in the metagenomes can lead to inefficient gene expression. The current approach to overcoming this problem is to use vector-derived promoters to transcribe insert gene fragments.

Experimental and bioinformatics studies

To evaluate the cross-reactivity of transcriptional machineries between species (not restricted to microorganisms), Warren *et al.* (2008) compared transcription levels of *H. influenzae*, *P. aeruginosa*, and human genes in *E. coli* using programmable microarrays and RT-PCR. Highly transcribed genes frequently had promoter sequences similar to the *E. coli* consensus promoter sequence, which is recognized by the major sigma factor, RpoD. Compared with *H. influenzae*, a smaller portion of *P. aeruginosa* genes was transcribed. This may be due to weaker conservation of regulatory elements in *P. aeruginosa* genes, and because *H. influenzae* genes are AT-rich (61.9%), which may facilitate binding to the AT-rich *E. coli* core promoter sequence. Punctuated transcription of human DNA was observed. This report demonstrated that expression of genes decreases with taxonomic distance from the host microorganism.

Table 13.1 Various approaches to host engineering for improvement of heterologous gene expression

Step	Target	Method	Host	Reference
Transcription	RNAP (sigma factor 70)	Mutagenesis	*E. coli*	Alper and Stephanopoulos (2007)
Transcription	RNAP (α-subunit)	Mutagenesis	*E. coli*	Klein-Marcuschamer *et al.* (2009)
Transcription	RNAP (β-subunit)	Mutagenesis	*S. lividans*	Hu *et al.* (2002)
Translation	Ribosomal protein S1	Mutagenesis	*E. coli*	Berstein *et al.* (2007)
Translation	Ribosomal protein S12	Mutagenesis	*B. subtilis*	Kurosawa *et al.* (2006)
Translation	IF3 (N-terminal domain)	Mutagenesis	*E. coli*	Maar *et al.* (2008)
Translation	IF3 (C-terminal domain)	Mutagenesis	*E. coli*	O'Connor *et al.* (2001)
Translation	tRNA (rare codons)	Co-expression of tRNAs	*E. coli*	Fu *et al.* (2008) Xiao *et al.* (2008)
Protein folding	Chaperone (DnaK-DnaJ-GrpE, GroES-GroEL, TF)	Co-expression of chaperone	*E. coli*	Nishihara *et al.* (1998) Nishihara *et al.* (2000)
Protein folding	Chaperone	Co-expression of chaperone	*E. coli*	Ferrer *et al.* (2004)

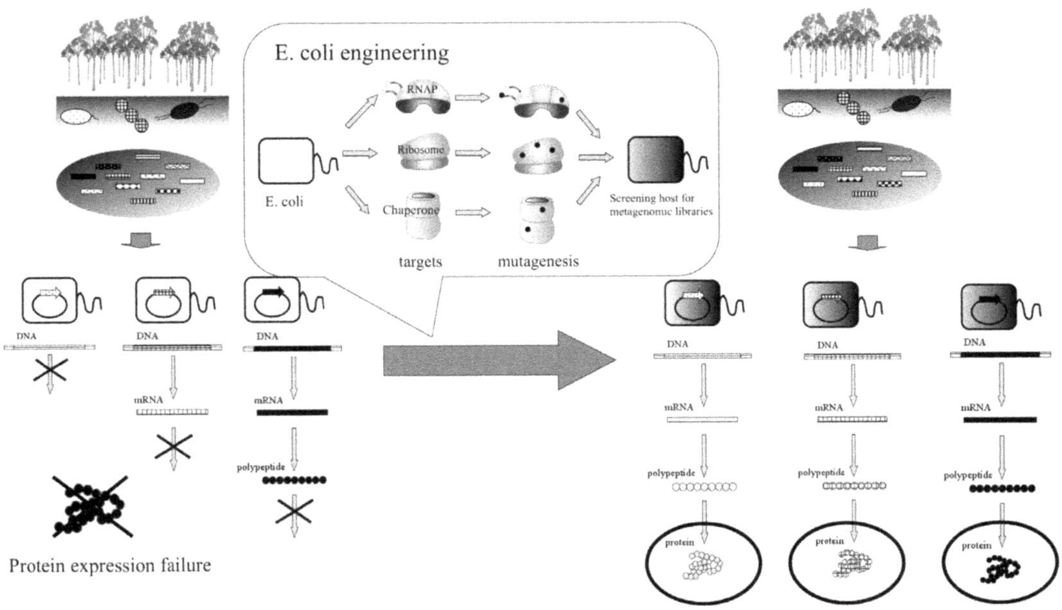

Figure 13.1 Schematic overview of *E. coli* host engineering for enhanced expression of metagenomic library genes. The 'low hit-rate problem' of metagenomics is thought to be due primarily to a lack of expression signals recognizable by *E. coli*. Thus, engineering the transcriptional/translational machinery may facilitate metagenome expression.

Park *et al.* (2007) compared the promoter regions of metagenome libraries collected from the GenBank database with 4243 ORFs from *E. coli* MG1655. The TATA box sequence at the −10 region was moderately conserved, whereas no distinct conservative patterns were detected in the −35 region. Thus, native *E. coli* RNAP might bind weakly or inappropriately to the promoter regions residing in metagenomic gene fragments. Engineering RNAP to recognize a wider range of promoter sequences may facilitate more efficient transcription.

Mutations in sigma factors can alter the phenotypic profile

Sigma factors are interesting targets for mutagenesis due to their role in promoter recognition and binding. Alper and Stephanopoulos (2007) produced a mutant library of *E. coli* rpoD, which encodes the main sigma factor σ^{70}. They screened this library for phenotypes such as tolerance to ethanol and SDS and overproduction of lycopene. Mutation of rpoD resulted in global perturbations in the transcriptome, which led to varied and complex phenotypes. Moreover, σ^{70}-targeted engineering outperformed the traditional whole-organism-based breeding approach; mutants were quickly and more effectively identified in σ^{70}-targeted libraries.

Mutagenesis of subunits other than σ may also be effective. Klein-Marcuschamer *et al.* (2009) screened RpoA (RNAP α-subunit) mutant libraries generated by error-prone PCR for three industrially relevant phenotypes: improved butanol tolerance and increased production of L-tyrosine and hyaluronic acid. The RpoA library delivered improvements in all three phenotypes. The L33 mutant grew faster and exhibited a higher accumulated cell mass than the control in the presence of butanol, and the rpoA14 strain exhibited a 91% increase in titre when cultured in 790 mg/l L-tyrosine. A ~50% increase in hyaluronic acid production was observed in an rpoA-HA mutant. It is expected that this technique will be useful for improving a broad spectrum of other traits. This capacity is likely related to the function of the carboxyl-terminal domain of the α-subunit, as all mutations found in this study mapped to this region.

In *S. lividans*, mutations in the RNAP β-subunit enhanced expression of genes that were expressed at low levels in the native strain. Ochi and colleagues (Hu *et al.*, 2002) showed that certain mutations in the RNAP β-subunit could activate *S. lividans* antibiotic genes. Data suggested that these mutations activated transcription of key regulatory genes of actinorhodin and undecylprodigiosin. This research group also investigated the mechanism of RNAP mutants activating cryptic genes (Hosaka *et al.*, 2009). First, they selected two Streptomyces strains producing novel antibacterial compounds, both of which carried single mutations in RNAP β-subunit. To investigate the mechanism of activation of antibacterial biosynthetic pathways, they determined the affinities of the mutant RNAPs for promoters of genes known to be highly expressed during the exponential-growth phase. Mutant RNAPs isolated from stationary-phase cells showed higher binding affinities than those from exponentially growing bacteria, whereas the opposite was observed for the wild-type protein. An RNAP mutation that did not activate antibacterial production showed affinity similar to that of wild-type RNAP. These findings suggest that the increased affinity of mutant RNAPs for promoters can enhance transcription of certain genes that would otherwise be 'silent'.

Engineering *E. coli* translational systems

Ribosome binding

Translational initiation in bacteria is triggered by formation of the ribosome–mRNA complex. The ribosome-binding site, also known as the Shine–Dalgarno (SD) sequence, is generally located 8 bp upstream of the initiation codon and contains purine nucleotides (typically AGGAGG) that are complementary to the 3′-terminal of the 16S rRNA (anti-SD, UCCUCC) (Shine and Dalgarno, 1974). The SD and anti-SD sequence pair, the translation initiation factors, as well as the initiator tRNA (fMet-tRNA) are then recruited to the ribosome.

Because the SD sequence plays a critical role in translation, less attention has been paid to the existence of SD-independent initiation pathways. However, increasing numbers of genes that do not possess the SD sequence are being documented. Chang *et al.* (2006) analysed 162 complete genomes and reported that the proportion of genes that contained the SD sequence was 11.6–90.8%, suggesting that translation of the majority of genes is not SD dependent. Similarly, Nakagawa *et al.* (2010) examined the ubiquity of the SD sequence among prokaryotes among 277 species (249 eubacteria and 28 archaebacteria). They found that the proportion of SD-containing genes varied among prokaryotic taxa, but were similar within phyla. Ma *et al.* (2002) reported that fast-growing bacteria, Gram-negative thermophiles, spirochetes, and methanogens possessed a relatively high proportion of SD-led genes, whereas that of obligate intracellular parasites, surface parasites, pathogens, and cyanobacteria was lower. In particular, 89.4% of *B. subtilis*, but only 26% of cyanobacterial (*Synechosystis* spp.), genes possessed the SD sequence. These data suggest that a significant proportion of metagenomic genes do not possess SD, expression of which in *E. coli* are problematic. Thus, investigation of the mechanisms of SD-independent translation and development of *E. coli* strains capable of translating genes via SD-independent routes is vital. This applies to metagenomes as well as culturable taxa. Park *et al.* (2007) compared the ribosome binding sites (RBS) of 4,243 ORFs of *E. coli* and metagenomic libraries obtained from GenBank. A number of *E. coli* genes possessed a purine-rich region from −12 bp to −7 bp of the initiation codon (the SD sequence), whereas no such pattern was detected in metagenomic libraries. This suggests that most metagenomic genes do not possess the SD sequence.

The major difference between the translational apparatus of *Synechosystis* spp. and that of *B. subtilis* is the presence and absence, respectively, of the ribosomal protein S1. S1 is an integral component of the translation machinery of all proteobacteria, cyanobacteria, and many other prokaryotic taxa, according to the NCBI protein sequence database, but is absent from low-GC Gram-positives (Roberts and Rabinowitz, 1989). It is essential for translation of most transcripts in *E. coli* (Sorensen *et al.*, 1998) and raises the nonspecific affinity of the ribosome for single-stranded RNA (Draper and Reynaldo, 1999) via six repeats of the oligonucleotide-binding (OB) fold. Farwell *et al.* (1992) have designed a set of nine plasmids containing the reporter cat (chloramphenicol acetyltransferase) gene with one of three SD sequences (weak, strong, or stronger) and one of three initiation codons (AUG, GUG, or UUG). *E. coli* ribosomes were able to translate all nine transcripts with variable efficiencies. *B. subtilis* and S1-depleted *E. coli* ribosomes were similar, and they differed from non-depleted *E. coli* ribosomes in that they required strong or stronger SD sequences and were unable to translate any of the weak transcripts. The addition of either *E. coli* or *Micrococcus luteus*-derived S1 enabled S1-depleted *E. coli* ribosomes to translate mRNAs with weak SD sequences. Therefore, construction of an S1 protein that can bind a range of mRNA may enhance translational initiation of foreign genes.

Bernstein *et al.* (2007) demonstrated the potential of engineered S1 protein. They evolved S1 of *E. coli* and succeeded in obtaining variants able to recognize genes of *Rhodopseudomonas palustris*, which has a high GC content, with 12 times greater efficiency. Eight of 14 amino acid mutations mapped to the third or fourth repeats of the OB fold. It is possible that these two domains represent a hot spot of mutation for enhanced expression. Simultaneous mutations are believed to be essential for tight RNA binding. They concluded that the variant S1 protein became able to recognize the 5′-untranslated region of GC-rich genes, including the ribosome-binding site.

S12 may also be a target for mutation intended to enhance heterologous gene expression. Ochi's research group reported greater production of particular enzymes in an S12 mutant of *B. subtilis* 168 (Kurosawa *et al.*, 2006). Previous studies by the same group (Shima *et al.*, 1996; Hesketh and Ochi, 1997) suggested that some S12 mutations induced dramatic activation of antibiotic production, mainly due to maintenance of high levels of protein synthesis in late growth phase. Based on this, they hypothesized that bacterial gene expression can be dramatically altered by modifying ribosomal proteins. As the proteins used in this study, α-amylase and protease, are produced during late growth phase, it seemed likely that synthesis of these enzymes would be enhanced by mutation of the S12 protein. Indeed, a 1.3- to 1.4-fold increase in α-amylase and protease production was observed.

Initiation codon

The translational initiation codon is heavily biased towards AUG in *E. coli* (91%); however, other organisms favour GUG and UUG. The initiation codon is an important determinant of translational efficiency. O'Donnell and Janssen (2001) compared the translational efficiencies of AUG, GUG, UUG, and CUG for the naturally leaderless cI mRNA from bacteriophage λ. In a cI-lacZ translational fusion, only AUG supported a high level of expression; GUG supported only a low level of expression, and UUG and CUG expression was barely above background levels. Addition of an untranslated lac leader and SD sequence to cI increased expression, but the dependence on AUG for maximum expression remained.

Villegas and Kropinski (2008) examined initiation codon usage using data from 620 bacterial chromosomes. They found no correlation between initiation codon usage and GC content. Most bacteria that used mainly GUG as a start codon were members of the Actinobacteria. For instance, more than half of the initiation codons of *Kineococcus radiotolerans* were GUG. In contrast, no such correlation was found between UUG frequency and phylogenetic properties. The most UUG-favouring microbe, *Leptospira interrogans*, belonged to the Spirochetes, while another, *Tropheryma whipplei*, was an Actinobacterium; the highest UUG usage was ~30%. This study demonstrated that initiation codon usage varies among taxa; the initiation codons of considerable numbers of metagenomic genes are other than AUG.

Bacterial translation initiation factor-3 (IF3) functions as an 'anti-association' factor and antagonizes the association between the 30S and the 50S ribosomal subunits. It is involved in the fidelity of translational initiation at several levels, including start-codon discrimination, mRNA translation, and initiator-tRNA selection. Thus, mutagenesis of IF3 may enhance translational efficiency of genes with a non-AUG initiation codon. Genetic analysis has shown that mutations in either the N-terminal (NTD) or C-terminal (CTD) domains of IF3 can perturb start-codon discrimination *in vivo*, suggesting that both domains are necessary for sensing interactions between the initiator tRNA and the start codon. Maar *et al.* (2008) reported that IF3 with the Y75N mutation (NTD) was defective in start-codon discrimination, inhibition of initiation on leaderless mRNA, and initiator-tRNA selection. This mutation also increased the rate of translational initiation from a non-canonical initiation codon (AUU) by about fourfold. O'Connor *et al.* (2001) showed that CTD mutations also perturb IF3's role in initiator-tRNA selection *in vivo*, consistent with earlier *in vitro* observations by Hartz *et al.* (1989).

Mutational studies on IF3s are mostly focused on investigating the mechanisms to recognize non-canonical start codons such as AUU and ACG. However, considering the fact that many bacteria favour GUG and UUG (Villegas and Kropinski, 2008) and inefficient recognition of such minor start codons in *E. coli*, it would be advantageous for metagenomic gene expression to engineer IF3 (or some other factors related to translational initiation) to recognize the non-AUG initiation codons.

Codon bias and mRNA secondary structure

Genes highly expressed in *E. coli* have been found to use almost exclusively codons translated by abundant tRNAs, whereas genes that are not highly expressed appear to be less biased in their codon usage (Sharp and Li, 1987). It is possible to supplement with these rare tRNAs to complement the deficiency of 'rare codons' in *E. coli*. Such modifications can improve gene expression significantly (Del Tito *et al.*, 1995; Kane, 1995; Burgess-Brown *et al.*, 2008). Rosetta is a commercial strain that is frequently used to express genes that are produced at low levels in *E. coli* by providing tRNA for seven rare codons (AGA, AGG, AUA, CUA, GGA, CCC, and CGG) (Fu *et al.*, 2008; Xiao *et al.*, 2008). By using Rosetta, expression of native genes can be enhanced in the absence of sequence information, irrespective of their natural codon usage. This may also facilitate enhanced expression of metagenomic genes.

In an analysis of 318 bacterial genomes, Willenbrock *et al.* (2006) showed that codon-usage preference provides an environmental signature by which it is possible to group bacteria according to their lifestyle, for instance, soil bacteria and soil symbionts, spore formers, enteric bacteria, aquatic bacteria, and intracellular and extracellular pathogens. They also confirmed that fast-replicating bacteria are more likely to have a dominant translational codon usage bias. Thus, efficient expression of (i) low-abundance genes, (ii) those originating from slow-replicating bacteria, and (iii) phylogenetically distant genes in *E. coli* may be challenging.

Kudla *et al.* (2009) demonstrated, using 154 synonymous variants of GFP, that a high codon-adaptation index (CAI, Carbone *et al.*, 2003; 2005) increased the elongation rate. However, given that translational initiation is the rate-limiting step, this did not significantly affect expression. On the other hand, rapid elongation sequestered fewer ribosomes on the mRNA, thereby increasing the rate of protein synthesis and cell growth.

The authors also demonstrated that mRNAs with stronger secondary structure (long hairpin loop) produced lower levels of protein expression. In particular, RNA stem loops, formed around the ribosome-binding site, restricted the decoding of mRNAs. The strong correlation between mRNA folding and fluorescence suggested that tightly folded messages obstruct translation initiation and thereby reduce protein synthesis. RNA stem-loop reduced protein expression when formed in the 5′-untranslated region, but not when formed downstream of the initiation codon. Paulus *et al.* (2004) introduced a synthetic RNA stem loop in front of various gene sequences. In each case, the stem loop was inserted 15 nt downstream from the initiation codon, and enhanced expression of all genes was investigated.

Protein folding

Molecular chaperones play a central role in conformational quality control by supporting proper folding, translocation, and stabilization of a wide range of polypeptides. Although constitutively expressed under balanced growth conditions, many chaperones are upregulated upon heat shock or other insults that increase cellular protein misfolding, and they are therefore classified as heat-shock proteins (Hsps) (Baneyx and Mujacic, 2004). Overproduced recombinant proteins commonly fail to fold correctly, resulting in formation of insoluble aggregates (Makrides, 1996). In some cases, co-expression of large amounts of chaperones can overcome this limitation (Roman *et al.*, 1995; Georgiou and Valax, 1996). Yura's group constructed plasmids harbouring selected chaperones, DnaK-DnaJ-GrpE, GroEL-GroES, and trigger factor (TF) (Nishihara *et al.*, 1998, 2000), commercially available from TaKaRa Bio (Shiga, Japan), as a 'chaperone plasmid set.' Nascent chains probably interact with TF, and most small proteins (~70%) fold rapidly upon synthesis without further assistance (Niwa *et al.*, 2009). Longer chain proteins (10–20%) subsequently interact with DnaK and DnaJ and fold correctly after one or more cycles of ATP-dependent binding and release. About 10–15% of polypeptide chains transit the GroEL/GroES chaperonin system for folding. GroEL does not bind to nascent chains and thus likely receives its substrates after their interaction with DnaK (Hartl and Hayer-Hartl, 2002). Overexpression of these plasmids in *E. coli* had marked effects on the production of many aggregation-prone proteins, such as mouse endostatin, human oxygen-regulated protein ORP150, human lysozyme, and Japanese cedar pollen allergen.

Ferrer *et al.* (2004) introduced a foreign chaperone (cpn) cloned from psychrophilic bacteria to improve expression of cold-active esterase (EstRB8) in *E. coli* at temperatures <15°C. This strain exhibited high growth rates at 4–15°C. Purified EstRB8 was up to 180-fold more active than that from a wild-type *E. coli* strain grown at 37°C. Additionally, target protein expression levels and biochemical properties were not affected by co-expression of cpn. Although many proteins are more soluble when *E. coli* is cultured at low temperatures (16–18°C), the *E. coli* cpn$^+$ strain was also more efficient (in terms of rate of production and protein expression) than a wild-type strain grown at suboptimal temperatures. This approach represents an alternative to existing methodologies for expression of thermosensitive proteins in mesophilic hosts with improved solubility and activity.

Conclusions

Metagenomics is anticipated to accelerate enzyme discovery by expanding the search space to unculturable microorganisms. However, many technological barriers remain; the most serious is the discrepancy between the host's transcriptional/translational machineries and the expression signals (i.e. promoter, RBS, initiation codon, codon usage, etc.) present in genes of interest. For functional exploration of potentially useful genes, it is thus necessary to systematically revise the machineries so that they are able to recognize signals from diverse taxa. The engineering direction is completely different from that applied to overexpression of cloned genes (or open-reading frames). As summarized in Table 13.1, various approaches have been taken to alter the *E. coli* transcription–translation–export machineries. Step-by-step improvements and subsequent integration of individually optimized protein production components should convert *E. coli* into a more capable host.

Acknowledgement

This work was supported by Grant-in-Aid for JSPS Fellows.

References

Alper, H., and Stephanopoulos, G. (2007). Global transcription machinery engineering: a new approach for improving cellular phenotype. Metab. Eng. 9, 258–267.

Amann, R.I., Ludwig, W., and Schleifer, K.H. (1995). Phylogenetic identification and *in situ* detection of individual microbial cells without cultivation. Microbiol. Rev. 59, 143–169.

Arnold, F.H. (2001). Combinatorial and computational challenges for biocatalyst design. Nature *409*, 253–257.

Baneyx, F., and Mujacic, M. (2004). Recombinant protein folding and misfolding in *Escherichia coli*. Nat. Biotechnol. 22, 1399–1408.

Bernstein, J.R., Butler, T., Shen, C.R., and Liao, J.C. (2007). Directed evolution of ribosomal protein S1 for enhanced translational efficiency of high GC *Rhodopseudomonas palustris* DNA in *Escherichia coli*. J. Biol. Chem. 282,18929–18936.

Burgess-Brown, N.A., Sharma, S., Sobott, F., Loenarz, C., Oppermann, U. (2008). Codon optimization can improve expression of human genes in *Escherichia coli*: A multi-gene study. Protein Expr. Purif. 59, 94–102

Carbone, A., Zinovyev, A., and Kepes, F. (2003). Codon adaptation index as a measure of dominating codon bias. Bioinformatics *19*, 2005–2015

Carbone, A., Kepes, F., Zinovyeva, A. (2005) Codon bias signatures, organization of microorganisms in codon space, and lifestyle. Mol. Biol. Evol. *22*, 547–561.

Chang, B., Halgamuge, S., and Tang, S.L. (2006). Analysis of SD sequences in completed microbial genomes: Non SD-led genes are as common as SD-led genes. Gene *373*, 90–99.

Courtois, S., Cappellano, C.M., Ball, M., Francou, F.X., Normand, P., Helynck, G., Martinez, A., Kolvek, S.J., Hopke, J., Osburne, M.S. *et al.* (2003). Recombinant environmental libraries provide access to microbial diversity for drug discovery from natural products. Appl. Environ. Microbiol. *69*, 49–55.

Curtis, T.P., Sloan, W.T., and Scannell., J.W. (2002). Estimating prokaryotic diversity and its limit. Proc. Natl. Acad. Sci. U.S.A. *99*, 10494–10499.

Daniel, R. (2004). The soil metagenome – a rich resource for the discovery of novel natural products. Curr. Opin. Biotechnol. *15*, 199–204.

Del Tito, B.J., Ward, J.M., Hodgson, J., Gershater, C.J., and Edwards, H. (1995). Effects of a minor isoleucyl tRNA on heterologous protein translation in *Escherichia coli*. J. Bacteriol. 177, 7086–7091.

Demain, A.L., and Vaishnav, P. (2009). Production of recombinant proteins by microbes and higher organisms. Biotechnol. Adv. *27*, 297–306.

Draper, D.E., and Reynaldo, L.P. (1999). RNA binding strategies of ribosomal proteins. Nucleic Acids Res. *27*, 381–388.

Farwell, M.A., Roberts, M.W., and Rabinowitz, J.C. (1992). The effect of ribosomal protein S1 from *Escherichia coli* and *Micrococcus luteus* on protein synthesis *in vitro* by *E. coli* and *Bacillus subtilis*. Mol. Microbiol. *6*, 3375–3383.

Ferrer, M., Chernikova, T.N., Timmis, K.N., and Golyshin P.N. (2004). Expression of a temperature-sensitive esterase in a novel chaperone-based *Escherichia coli* strain. Appl. Environ. Microbiol. *70*, 4499–4504.

Francis, C.A., Roberts, K.J., Beman, J.M., Santoro, A.E., and Oakley, B.B. (2005). Ubiquity and diversity of ammonia-oxidizing archaea in water columns and sediments of the ocean. Proc. Natl. Acad. Sci. U.S.A. *102*, 14683–14688.

Fu, W., Lin, J., and Cen, P. (2008). Expression of a hemA gene from *Agrobacterium* radiobacter in a rare codon optimizing *Escherichia coli* for improving 5-aminolevulinate production. Appl. Biochem. Biotechnol. *160*, 456–466.

Fuglsang, A. (2003). Codon optimizer: a freeware tool for codon optimization. Protein Expr. Purif. *31*, 247–249.

Gabor, E., Liebeton, K., Niehaus, F., Eck, J., and Lorenz, P. (2007). Updating the metagenomics toolbox. Biotechnol. J. *2*, 201–206.

Gabor, E.M., Alkema, W.B.L and Janssen, D.B. (2004). Quantifying the accessibility of the metagenome by random expression cloning techniques. Environ. Microbiol. *6*, 879–886.

Georgiou, G., and Valax, P. (1996). Expression of correctly folded proteins in *Escherichia coli*. Curr. Opin. Biotechnol. *7*, 190–197.

Guan, C., Ju, J., Borlee, B.R., Williamson, L.L., Shen, B., Raffa, K.F., and Handelsman, J. (2007). Signal mimics derived from a metagenomic analysis of the gypsy moth gut microbiota. Appl. Environ. Microbiol. *73*, 3669–3676.

Handelsman, J. (2004). Metagenomics: Application of genomics to uncultured microorganisms. Microbiol. Mol. Biol. Rev. *68*, 669–685.

Hartl, F.U., and Hayer-Hartl, M. (2002). Molecular chaperones in the cytosol: from nascent chain to folded protein. Science *295*, 1852–1858.

Hartz, D., McPheeters, D.S., and Gold, L. (1989). Selection of the initiator tRNA by *Escherichia coli* initiation factors. Genes Dev. *3*, 1899–1912.

Henne, A., Schmitz, R.A., Bomeke, M., Gottschalk, G., and Daniel, R. (2000). Screening of environmental DNA libraries for the presence of genes conferring lipolytic activity of *Escherichia coli*. Appl. Environ. Microbiol. *66*, 3113–3116.

Hesketh, A., and Ochi, K. (1997). A novel method for improving *Streptomyces coelicolor* A3(2) for production of actinorhodin by introduction of rpsL (encoding ribosomal protein S12) mutations conferring resistance to streptomycin. J. Antibiot. *50*, 532–535.

Hosaka, T., Ohnishi-Kameyama, M., Muramatsu, H., Murakami, K., Tsurumi, Y., Kodani, S., Yoshida, M., Fujie, A., and Ochi, K. (2009). Antibacterial discovery in Actinomycetes strains with mutations in RNA polymerase or ribosomal protein S12. Nat. Biotechnol. *27*,462–464.

Hu, H., Zhang, Q., and Ochi, K. (2002). Activation of antibiotic biosynthesis by specified mutations in the rpoB gene (encoding the RNA polymerase b subunit) of *Streptomyces lividans*. J. Bacteriol. *184*, 3984–3991.

Ishihama, A. (2000). Functional modulation of *Escherichia coli* RNA polymerase. Annu. Rev. Microbiol. *54*, 499–518.

Kane, J.F. (1995) Effects of rare codon clusters on high-level expression of heterologous proteins in *Escherichia coli*. Curr. Opin. Biotechnol. *6*, 494–500.

Kirk, O., Borchert, T.V., and Fuglsang, C.C. (2002). Industrial enzyme applications. Curr. Opin. Biotechnol. *13*, 345–351.

Klein-Marcuschamer, D., Santos, C.N.S., Yu, H., and Stephanopoulos, G. (2009). Mutagenesis of the bacterial RNA polymerase alpha subunit for improvement of complex phenotypes. Appl. Environ. Microbiol. 75, 2705–2711.

Kudla, G., Murray, A.W., Tollervey, D., and Plotkin, J.B. (2009). Coding-sequence determinants of gene expression in *Escherichia coli*. Science *324*, 255–258.

Kurosawa, K., Hosaka, T., Tamehiro, N., Inaoka, T., and Ochi, K. (2006). Improvement of α-amylase production by modulation of ribosomal component protein S12 in *Bacillus subtilis* 168. Appl. Environ, Microbiol. *72*, 71–77.

Lorenz, P., and Eck, J. (2005). Metagenomics and industrial applications. Nat. Rev. Microbiol. 3, 510–516.

Ma, J., Campbell, A., and Karlin, S. (2002). Correlations between Shine-Dalgarno sequences and gene features such as predicted expression levels and operon structures. J. Bacteriol. *184*, 5733–5745.

Maar, D., Liveris, D., Sussman, J.K., Ringquist, S., Moll, I., Heredia, N., Kil, A., Blasi, U., Schwartz, I., and Simons, W. (2008). A single mutation in the IF3 N-terminal domain perturbs the fidelity of translation initiation at three levels. J. Mol. Biol. *383*, 937–944.

Makrides, S.C. (1996). Strategies for achieving high-level expression of genes in *Escherichia coli*. Microbiol. Rev. *60*, 512 –538.

Martinez, A., Kolvek, S.J., Yip, C.L.T., Hopke, J., Brown, K.A., MacNeil, I.A., and Osburne, M.S. (2004). Genetically modified bacterial strains and novel bacterial artificial chromosome shuttle vectors for constructing environmental libraries and detecting heterologous natural products in multiple expression hosts. Appl. Environ. Microbiol. *70*, 2452–2463.

Marzorati, M., de Ferra, F., Raemdonck, H.V., Borin, S., Allifranchini, E., Carpani, G., Serbolisca, L., Verstraete, W., Boon, N., and Daffonchio, D. (2007). A novel reductive dehalogenase, identified in a contaminated groundwater enrichment culture and in *Desulfitobacterium dichloroeliminans* strain DCA*1*, is linked to dehalogenation of *1*,2-dichloroethane. Appl. Environ. Microbiol. *73*, 2990–2999.

Mohn, W.W., Garmendia, J., Galvao, T.C., and de Lorenzo, V. (2006). Surveying biotransformations with *à la carte* genetic traps: translating dehydrochlorination of lindane (gamma-hexachlorocyclohexane) into lacZ-based phenotypes. Environ. Microbiol. *8*, 546–555.

Nakagawa, S., Niimura, Y., Miura, K., and Gojobori, T. (2010). Dynamic evolution of translation initiation mechanisms in prokaryotes. Proc. Natl. Acad. Sci. U.S.A. *107*, 6382–6387.

Nishihara, K., Kanemori, M., Kitagawa, M., Yanagi, H., and Yura, T. (1998). Chaperone coexpression plasmids: differential and synergistic roles of DnaK-DnaJ-GrpE and GroEL-GroES in assisting folding of an allergen of Japanese cedar pollen, Cryj2, in *Escherichia coli*. Appl. Environ. Microbiol. *64*, 1694–1699.

Nishihara, K., Kanemori, M., Yanagi, H., and Yura, T. (2000). Overexpression of trigger factor prevents aggregation of recombinant proteins in *Escherichia coli*. Appl. Environ. Microbiol. *66*, 884–889.

Niwa, T., Ying, B., Saito, K., Jin, W.Z., Takada, S., Ueda, T., and Taguchi, H. (2009). Bimodal protein solubility distribution revealed by an aggregation analysis of the entire ensemble of *Escherichia coli* proteins. Proc. Natl. Acad. Sci. U.S.A. *106*, 4201–4206.

O'Connor, M., Gregory, S.T., Rajbhandary, U.L., and Dahlberg, A.E. (2001). Altered discrimination of start codons and initiator tRNAs by mutant initiation factor 3. RNA *7*, 969–978.

O'Donnell, S.M., and Janssen, G.R. (2001). The initiation codon affects ribosome binding and translational efficiency in *Escherichia coli* of cI mRNA with or without the 5′ untranslated leader. J. Bacteriol. *183*, 1277–1283.

Park, H.S., Cheong, D.E., Lee, J.Y., Han, S.S., Lee, J.H., and Kim, G.J. (2007). Analyses of the structural organization of unidentified open reading frames from metagenome. Biochem. Biophys. Res. Commun. *356*, 961–967.

Paulus, M., Haslbeck, M., and Watzele, M. (2004). RNA stem-loop enhanced expression of previously non-expressible genes. Nucleic Acids Res. *32*, e78.

Roberts, M.W., and Rabinowitz, J.C. (1989). The effect of *Escherichia coli* ribosomal protein S1 on the translational specificity of bacterial ribosomes. J. Biol. Chem. *264*, 2228–2235.

Roman, L.J., Sheta, E.A., Martasek, P., Gross, S.S., Liu, Q., and Masters, B.S.S. (1995). High-level expression of functional rat neuronal nitric oxide synthase in *Escherichia coli*. Proc. Natl. Acad. Sci. U.S.A. *92*, 8428–8432.

Rondon, M. R., August, P.R., Bettermann, A.D., Brady, S.F., Grossman, T.H., Liles, M.R., Loiacono, K.A., Lynch, B.A., MacNeil, I.A., Minor, C. *et al.* (2000). Cloning the soil metagenome: a strategy for accessing the genetic and functional diversity of uncultured microorganisms. Appl. Environ. Microbiol. *66*, 2541–2547.

Schipper, C., Hornung, C., Bijtenhoom, P., Quitschau, M., Grond, S., and Streit, W.R. (2009). Metagenome-derived clones encoding two novel lactonase family proteins involved in biofilm inhibition in *Pseudomonas aeruginosa*. Appl. Environ. Microbiol. *75*, 224–233.

Schloss, P.D., and Handelsman, D. (2004). Status of the microbial census. Microbiol. Mol. Biol. Rev. *68*, 686–691.

Schmeisser, C., Steele, H., and Streit, W. (2007). Metagenomics, biotechnology with non-culturable microbes. Appl. Microbiol. Biotechnol. *75*, 955–962.

Sharp, P.M., and Li, W.H. (1987). The codon adaptation index-a measure of directional synonymous codon usage bias, and its potential applications. Nucleic Acids Res. *15*,1281–1295.

Shima, J., Hesketh, A., Okamoto, S., Kawamoto, S., and Ochi, K. (1996). Induction of actinorhodin production by rpsL (encoding ribosomal protein S12) mutations that confer streptomycin resistance in *Streptomyces lividans* and *Streptomyces coelicolor* A3(2). J. Bacteriol. *178*, 7276–84.

Shine, J., and Dalgarno, L. (1974). The 3′-terminal sequence of *Escherichia coli* 16S ribosomal RNA: complementarity to nonsense triplets and ribosome binding site. Proc. Natl. Acad. Sci. U.S.A. *71*, 1342–1346.

Sorensen, M.A., Fricke, J., and Pedersen S. (1998). Ribosomal protein S1 is required for translation of most, if not all, natural mRNAs in *Escherichia coli in vivo*. J. Mol. Biol. *280*, 561–569.

Suenaga, H., Ohnuki, T., and Miyazaki, K. (2007). Functional screening of a metagenomic library for genes involved in microbial degradation of aromatic compounds. Environ. Microbiol. *9*, 2289–2297.

Uchiyama, T., Abe, T., Ikemura, T., and Watanabe, K. (2005). Substrate-induced gene expression screening of environmental metagenome libraries for isolation of catabolic genes. Nat. Biotechnol. *23*, 88–93.

Uchiyama, T., and Miyazaki, K. (2009). Functional metagenomics for enzyme discovery: challenges

to efficient screening. Curr. Opin. Biotechnol. *20*, 616–622.
Uchiyama, T., and Miyazaki, K. (2010). Product-induced gene expression, a product-responsive reporter assay used to screen metagenomic libraries for enzyme-encoding genes. Appl. Envion. Microbiol. *76*, 7029–7035.
Villegas, A., and Kropinski, A.M. (2008). An analysis of initiation codon utilization in the domain bacteria-concerns about the quality of bacterial genome annotation. Microbiology *154*, 2559–2561.
Wang, G.Y.S., Graziani, E., Waters, B., Pan, W., Li, X., McDermott, J., Meurer, G., Saxena, G., Anderson, R.J., and Davies, J. (2000). Novel natural products from soil DNA libraries in a Streptomycete host. Org. Lett. *2*, 2401–2404.
Ward, B.B. (2002). How many species of prokaryotes are there? Proc. Natl. Acad. Sci. U.S.A. *99*, 10234–10236.
Warren, R.L., Freeman, J.D., Levesque, R.C., Smailus, D.E., Flibotte, S., and Holt, R.A. (2008). Transcription of foreign DNA in *Escherichia coli*. Genome Res. *18*, 1798–1805.
Welch, M., Govindarajan, S., Ness, J.E., Villalobos, A., Gurney, A., Minshull, J., and Gustafsson, C. (2009). Design parameters to control synthetic gene expression in *Escherichia coli*. PloS One *4*, e7002.
Wexler, M., Bond, P.L., Richardson, D.J., and Johnston, A.W.B. (2005). A wide range-host metagenomic library from a wastewater treatment plant yields a novel alcohol/aldehyde dehydrogenase. Environ. Microbiol. *7*, 1917–1926.
Willenbrock, H., Friis, C., Juncker, A.S., and Ussery, D. (2006). An environmental signature for 323 microbial genomes based on codon adaption indices. Genome Biol. *7*, R114.
Xiao, Z., Boyd, J., Grosse, S., Beauchemin, M., Coupe, E., and Lau, P.C.K. (2008). Mining *Xanthomonas* and *Streptomyces* genomes for new pectinase-encoding sequences and their heterologous expression in *Escherichia coli*. Appl. Microbiol. Biotechnol. *78*, 973–981.
Xu, M., Xiao, X., and Wang, F. (2008). Isolation and characterization of alkane hydroxylases from a metagenomic library of Pacific deep-sea sediment. Extremophiles *12*, 255–262.

Recent Contributions of Metagenomics to Studies on Quorum-sensing and Plant–Pathogen Interactions

14

Denis Faure, Mélanie Tannières, Samuel Mondy and Yves Dessaux

Abstract

Though metagenomics is a novel tool in the field of plant–microbe interaction, the technique has already led to remarkable advances. Among these, the identification of yet-uncultivable phytopathogens or the description of the plant and rhizosphere microflora, are two features that may lead to a better description of the quality of agricultural lands, for instance in the case of disease suppressive soils. At a more molecular level, the identification of novel density-dependent regulatory functions (quorum sensing) and antagonizing elements (quorum quenching) that may be used to develop sustainable, biological control strategies directed at plant pathogen, are examples of valuable outcomes of metagenomics in the plant–microbe interaction field.

Introduction

In this chapter, the term 'metagenomics' is used according to the definition proposed by Riesenfeld and colleagues (Riesenfeld *et al.*, 2004), i.e. a 'functional and sequenced-based analysis of the collective microbial genomes contained in an environmental sample'. In this definition, we included direct sequencing of DNA or RNA extracted from environmental samples, and construction, sequencing and functional screening of libraries, as well as PCR-based analysis of gene markers (e.g. *rrs* gene encoding 16S RNA).

With respect to pathogenic microbe–plant interactions, metagenomics has emerged as a novel approach within the last 5 years or so, and major accomplishments are therefore still limited hence the reduced size of this chapter. Recent metagenomic studies in the field however reported: (i) the discovery and characterization of novel pathogens, as well as in-depth analysis of pathogen functions; (ii) novel strategies to possibly control the virulence of some pathogens, (iii) novel tools to evaluate the environmental impact of plant-protection strategies. With respect to quorum sensing, metagenomics also proved useful to identify novel components involved in this bacterial cell–cell communication process. These components were elements that either permit this regulatory process to occur or elements that antagonize it. Examples of these various achievements are described below.

Metagenomics studies on quorum sensing and quorum quenching

A brief view at quorum sensing (QS)

Bacteria have evolved sophisticated mechanisms to coordinate gene expression at population and community levels upon different environmental constraints. Numerous bacteria, including plant pathogens can synthesize and sense diffusible molecules which act as signals to trigger gene expression when they reach a threshold concentration. In confined environments, accumulation of the signals reflects bacterial cell density (i.e. the quorum) hence this regulatory process was termed quorum-sensing (QS) by Fuqua *et al.* (1994). In open environments, concentration of the same signal also reflects diffusion coefficient;

the term of diffusion sensing was proposed (Redfield, 2002). Quorum and diffusion sensing contribute to optimize expression of key functions at population level when appropriate conditions are completed (Hense *et al.*, 2007).

Plant pathogens and quorum-sensing

The QS signals produced by bacterial plant pathogens are diverse (Von Bodman *et al.*, 2003; Uroz *et al.*, 2009). They encompass the N-acylhomoserine lactones (AHSLs) which had been found in more than 70 bacterial species in alpha-, beta- and gamma-proteobacteria such as *Agrobacterium tumefaciens, Burkholderia glumae, Pantoea stewartii, Pectobacterium atrosepticum* (a non-limitative list); the 3-hydroxypalmitic acid methyl ester (PAME) in *Ralstonia solanacearum;* the diffusible signal factors (DSF) identified as cis-11-methyl-2-dodecanoic acid in *Xanthomonas campestris;* and the *Pseudomonas* quinolone signal (PQS) in *Pseudomonas aeruginosa* (Dessaux *et al.*, 2010). This list is non limitative and novel signal molecules are regularly reported (e.g. Schaefer *et al.*, 2008). Some pathogens produce a combination of different QS signals such as several AHSLs in *Pectobacterium atrosepticum,* and AHSL and PAME in *Ralstonia solanacearum.* The QS-regulated functions are diverse, including the production of extracellular enzymes and proteins (pectynolytic, cellulolytic, proteolytic enzymes), that of antibiotic or antifungal molecules, the modification of surface properties (expolysaccharides), or horizontal transfer of the plasmids harbouring the virulence genes (Uroz *et al.*, 2009). Noticeably, expression of the QS-regulated genes may be modulated in some instances by: (i) QS-independent pathways which reflect the physiological state of bacterial cells, e.g. RpoS, RsmA, in different proteobacteria (Cui *et al.*, 1995; Whiteley *et al.*, 2000), and, (ii) plant host molecules, such as GABA and free proline in *Agrobacterium tumefaciens* (Chevrot *et al.*, 2006; Haudecoeur *et al.*, 2009).

Modulation of QS regulation by quorum quenching

QS is also modulated by the bacterial community that surrounds the pathogen population and the plant host. Some non-pathogenic bacteria can produce QS-signal which may act synergistically with those of the pathogens while others can produce compounds and enzymes such as amido-hydrolases or lactonases that can disrupt QS signals emitted and sensed by the pathogen (Uroz *et al.*, 2009). Plant host may also respond to the presence of QS signals (Mathesius *et al.*, 2003), produce anti-QS compounds (Manefield *et al.*, 1999) as well as enzymes degrading QS signals (Delalande *et al.*, 2005). Such natural anti-QS activities, which are also termed as quorum-quenching (QQ), may be adapted to propose novel anti-virulence strategies (Dong *et al.*, 2001; Uroz *et al.*, 2003; review in Zhang, 2003). It should be important to consider that, in some plant-growth promoting rhizobacteria (PGPR), such as *Serratia plymuthica, Pseudomonas aureofaciens* and *P. chlororaphis,* the antagonistic activities directed at pathogenic micro-organisms are positively regulated by QS (Wood *et al.*, 1997; Müller *et al.*, 2008). From a practical point of view, the implementation of QQ-strategies that target a given pathogen should take in account these data, not to be detrimental to naturally occurring PGPR populations antagonizing other pathogens (Molina *et al.*, 2003).

Metagenomics, quorum-sensing and quorum quenching

Until the year 2005, studies on QS and QQ only relied upon cultivation-based approaches, therefore ignoring a large part of the microbial diversity of complex environments such as bare soil or plant rhizosphere (Nichols, 2007). Metagenomic therefore appeared as a pertinent strategy to address cell–cell communication at the global microbiome level. Several libraries were screened for: (i) identification of functions and genes which are implicated in the biosynthesis of QS-signals as well as (ii) QQ-enzymes that inactivate QS-signals. Several genes involved in the biosynthesis of QS-signals AHSLs were identified (Williamson *et al.*, 2005). Some of them exhibit a weak homology with those previously identified in cultured bacteria and encode the synthesis of unsaturated and substituted long chain AHSLs that have not been described earlier (Hao *et al.*, 2010). Furthermore, functional metagenomics allowed the identification of QS-mimics. A screen of 800,000 metagenomic clones obtained from DNA produced from a insect gut microbiota revealed that one clone encodes the synthesis of an indole compound that induce the expression of AHSL-regulated genes in bioindicator strains. This compound is an oxidative derivative of indoxyl (Guan *et al.*, 2007). This type of AHSL mimics has not been reported earlier. Interestingly, because indole itself is a signal involved in the resistance of bacterial population to antibiotics (Lee *et al.*, 2010), the above results revealed a possible cross-talk between different signalling pathways during host–microbe interactions. These examples emphasize the value of metagenomic approaches to identify previously unknown genes and functions involved in QS.

Screening of soil metagenomics libraries also allowed the extension of the diversity of the QQ-enzymes which are able to cleave a large spectrum of AHSL signals. In a library of 10,121 fosmids with 30–50 kb inserts, a single gene expressing a QQ activity was identified. It encodes a lactonase which is only distantly related to previous known latonase inactivating AHSLs. Interestingly, analysis of genes in the vicinity of the lactonase gene permitted to propose that the cloned DNA region originate from an Acidobacteria while the lactonase gene possibly resulted from a horizontal gene transfer from an alpha-proteobacteria (Riaz *et al.*, 2008). In a second analysed library of 7392 plasmids with 2.5–6 kb inserts, three novel lactonase genes were indentified following a similar functional screening (Schipper *et al.*, 2009). In both reports, the lactonases were able to block the expression of virulence in plant and animal pathogens such as *Pectobacterium* and *Pseudomonas*, a feature that may have quite interesting biotechnological outcomes. The apparent different screening efficiency of small insert – versus large insert – libraries may be explained by the necessity to drive the expression of soil DNA genes from a promoter located at the cloning site of the vector in an *E. coli* host. To circumvent this problem, heterologous expression of library in different hosts was proposed (Craig *et al.*, 2010). Interestingly, a simple estimation of the occurrence frequency of QS and QQ functions in the analysed soil metagenome suggests that the diversity and density of organisms which are able to produce or degrade QS-signals are most likely underestimated when assessed by cultivation-based approaches (D'Angelo-Picard *et al.*, 2005).

Additionally, metagenomics may help to understand variability of QS-linked functions in natural microbial populations. In soil biofilms collected from a mine drainage (Simmons *et al.*, 2008), the authors were able to reconstitute a 'population genome' (20x) which is closely related to *Leptosirillum ferriphilum*. Remarkably, the core QS genes, homologous to the prototypic determinants luxI (AHSL signal synthase) and luxR (regulator of the QS-regulated genes) are located within highly variable regions, suggesting that variants of the populations may be able – or not – to synthesize QS-signals. Such variability of QS functions has also been observed in the pathogen *P. aeruginosa* (Wilder *et al.*, 2009). In another study, metagenomics coupled with competition experiments showed that host-interacting bacteria, such as *Escherichia coli* may sense and use QS-signals produced by host-colonizing bacterial community to optimize its adaptation into specific niches (Hughes *et al.*, 2010).

Metagenomics to study plant–pathogen interactions

For years, investigations on plant–pathogen interactions have relied upon the isolation and cultivation of the pathogen, and been conducted with limited consideration – if any – to the global plant microbial environment. The emergence of the 'omics' techniques, and especially that of metagenomics has changed the situation and allowed important progress in this field. The examples presented below illustrate this assertion.

Genomic analysis of uncultured plant pathogens: the case of the citrus huanglongbing bacteria

Citrus Huanglongbing (HLB, from the words 'huáng lóng bìng', literally the yellow dragon disease in Chinese), also known as citrus greening, is a devastating disease which is associated with the presence in the contaminated plants of three uncultured members of the genus '*Candidatus Liberibacter*': '*Ca. L. asiaticus*', '*Ca. L. americanus*', and '*Ca. L. africanus*'. HLB disease agents are transmitted by phloem-feeding insect vectors such as *Diaphorina citri*. Epidemiological studies on this disease are hampered by the occurrence of a large number or plant reservoirs in which '*Ca. L. asiaticus*', for instance, does not necessarily induce disease. Two consecutive publications (Duan *et al.*, 2009; Tyler *et al.*, 2009) described metagenomic approaches to solve the complete genome of '*Ca. L. asiaticus*'. In the first paper, the authors extracted DNA from 15 insects of the genus *D. citri* which contains up to 10^8 bacterial cells per insect that consist in a mixture of '*Ca. L. asiaticus*' DNA to *D. citri* DNA (approximated 1:1 copy ratio). The sequence genome (one circular chromosome of 1.23 Mb; 36.5 GC%) was obtained by coupling multiple displacement amplification (MDA) and 454-pyrosequencing, and finishing with the Sanger technique. In the second paper, DNA was isolated from the collected phloem (150 mg) of infected citrus plants, and submitted to 454, SOLiD, and Solexa sequencing. The phloem metagenomic data provided 26-fold coverage of the genome that was previously obtained by the insect set of data. These two papers provided estimation of the bacterial cell density per plant cell and complete genome of the pathogen '*Ca. L. asiaticus*', which is phylogenetically related to *Agrobacterium* and *Rhizobium* alpha-proteobacteria.

Towards analysis of plant microbiota

Constructing metagenomic libraries to investigate the plant microbiota is technically difficult, for several reasons. First, metagenomics of rhizospheric community is challenging because of the complexity of root–soil matrix and the huge diversity of soil microbes (Tringe *et al.*, 2005; Von Mering *et al.*, 2007) and the natural occurrence in soil of compounds such as phenolics, humic and fulvic acids (a non limitative list) that impede biochemical reactions widely used in sequencing or cloning techniques, such as PCR (Wilson, 1997).

Construction of metagenomic libraries from plant tissues is also difficult because DNA preparations obtained from this environment generally contain a large (if not extremely large) proportion of undesirable plant DNA that may dilute 'target' microbial DNA. The above quoted papers (Duan *et al.*, 2009; Tyler *et al.*, 2009) illustrated the difficulties associated with the analysis of metagenomes of plant pathogens living inside plant tissues, and/or transmitted by insect vectors. Some technical improvements have been proposed, however, to enrich DNA fractions in microbial DNA during extraction. Wang and colleagues (2008) reported a SDS/NaCl-based treatment, which allows constructing a metagenomic library with 88% bacterial inserts from tissues of the tropical tree *Mallotus nudiflorus*. The library contains DNA from a large variety of bacterial phyla, including *Actinobacteria* (38–56%), Proteobacteria (27 to 39%), *Firmicutes* (2–9%), *Deinococcus–Thermus* (1%), as well as *Chloroflexi, Planctomycetes, Cyanobacteria, Acidobacteria, Gemmatimonadates* and *Bacteroidetes* as minor phyla (<1%). Jiao *et al.* (2006) proposed to lyse plant cell wall enzymatically to release plant nuclear material, and to rely upon a differential centrifugation approach to enrich the preparation in bacterial DNA. Recently, the use of a Nicodenz gradient (Courtois *et al.*, 2001) to recover the bacterial community tightly associated with plant tissues has been successfully implemented (E. Chapelle, unpublished).

Epiphytic community is generally more accessible and has been described in the case of *Arabidopsis thaliana* leaves by combining metagenomics (454 shotgun sequencing and *rrs*-analysis) and metaproteomics (Delmotte *et al.*, 2009) approaches. The recovered microorganisms, of which 99.7% were bacteria, dominantly belonged to (i) the alpha-proteobacteria class (43%) and more precisely to *Methylobacterium* and *Sphingomonas* genera; (ii) *Bacteroidetes* (12%); (iii) *beta-proteoabacteria* (10%); (iv) *Actinobacteria* (10%); (v) *gamma-proteobacteria* (5%); and (vi) *Endoplasmatales* (2%). In this study, metaproteomics revealed a high occurrence of carbohydrate transporters in Sphingomonas suggesting that this feature is an important component of phyllosphere adaptation in this genus.

Metagenomics has also been used to investigate the microbiome of specific plant organs, such as that associated with pitchers of carnivorous plants. A possibility exists that the microbiota of these organs contribute to the ability of the plant to digest trapped insects (Plummer and Jackson,

1963). Catalogues of bacterial diversity were first obtained by cultivation techniques. Recently, the microbial diversity of the pitcher fluid of *Sarracenia alatai* was investigated by a metagenomic approach that revealed the occurrence of over 650 phylogroups at the species level as determined by 454 sequencing of 16S RNA gene fragments. This value must be compared to the 288 taxa identified by automated ribosomal spacer analysis (ARISA) in the same study, and to the 20 species previously identified by cultivation (Koopman *et al.*, 2010), evidencing the critical advancement to the knowledge of microbial diversity that metagenomics allows.

Population genomics of plant pathogens as a support to metagenomics

In silico meta-analysis of pathogen genomes isolated from different samples is also named 'population genomics' (Terauchi and Yoshida, 2010). Even though population genomics is not a metagenomic approach *per se*, sequencing and comparison of a large number of genomes allow a consolidation of genomic databases which are used for metagenomic analysis, as well as the development of tools for genomic comparison at a large scale.

Population genomics investigate evolution of the pathogen populations, including contribution of horizontal transfer, their taxonomic classification as well as identification of key-functions involved in the host–pathogen interactions. Recent original and review papers deciphering this growing field of research in plant–pathogen interaction had been published, providing new insights on various topics such as: (i) the lifestyle of *Pseudomonas syringae* and its metabolic adaptation to the constraint environment that the leaf surface is (Mithani *et al.*, 2010); (ii) the involvement of effector proteins translocated by type III secretion system on the host range of *Xanthomonas citri* subspecies (Moreira *et al.*, 2010); (iii) the identification of the main determinants of pathogenicity of *X. campestris* pv. *campestris* (He *et al.*, 2007); (iv) the involvement of horizontal transfer in genome evolution of *Ralstonia solanacearum* (Guidot *et al.*, 2009); and, (v) the identification of novel effector–effector target genes for instance in the plant/*Magnaporthe oryzae* interaction (Terauchi and Yoshida, 2010).

Metagenomics of disease-suppressive soils

Soils that suppress plant disease, usually named disease-suppressive soils, contain plant growth-promoting rhizobacteria (PGPR) which antagonize plant pathogens (reviews: Weller *et al.*, 2002; Mazzola, 2004). In some of these PGPR, the expression of antagonist functions such as production of the antifungal compounds phenazine may be regulated by QS (Chin-A-Woeng *et al.*, 2001; Haas and Défago, 2005).

A recent study (Sanguin *et al.*, 2009) investigated the succession and diversity of microbial population in a take-all disease suppressive soil. Suppressiveness was established after 10 years of wheat monoculture, and was not only associated with the presence of PGPR pseudomonads as generally reported but also correlated with the prevalence of a broader range of taxa, 'mainly *Acidobacteria, Planctomycetes, Nitrospira, Chloroflexi, Alphaproteobacteria* (notably *Azospirillum*) and *Firmicutes* (notably *Thermoanaerobacter*)'.

Other studies investigated the presence of anti-pathogen functions and genes in disease-suppressive soils via a metagenomic approach. Three libraries of 5000, 15,000, and 100,000 clones were screened for, and allowed the identification of, type I polyketide synthases which are involved in the biosynthesis of various antibiotics and therapeutic compounds (Courtois *et al.*, 2003; Ginolhac *et al.*, 2004; van Elsas *et al.*, 2008). Another work has led to the discovery of antibacterial compounds in a metagenomic library generated from DNA extracted from a rhizospheric soil (Gillespie *et al.*, 2002). These authors isolated three clones (out of 25,000 screened) that produced coloured pigments which were identified upon fractionation as turbomycin A and turbomycin B. Both molecules exhibited broad-spectrum antibiotic activity against Gram-negative and Gram-positive organisms. In another study, Hjort *et al.* (2010) compared the diversity of the amplified chitinase encoding genes in total DNA extracted from soil, in a metagenomic library constructed from soil DNA, and in a collection of bacterial isolates exhibiting chitinase and antifungal activities. These investigations pointed out *Streptomyces* as a dominant soil genus involved in anti-fungal chitinase activity in soil. Another antifungal activity was identified using *Saccharomyces cerevisiae* as a biosensor, upon screening of a ca. 10,000 clone library from a forest soil. Though the active compound(s) was/were not identified, they most likely resulted from the expression of type II polyketide synthase genes (Chung *et al.*, 2008).

The above results demonstrate – should this be necessary – the interest of metagenomic approaches to understand plant-microbe and microbe–microbe interaction in suppressive soils. In agreement, within the last years, an increasing number of reports dealt with functional metagenomics based on the expression of DNA libraries obtained from various soil environments, in single or multiple hosts, to identify potential anti-pathogen, therapeutic or biotechnological functions. Stereoselective amidase, cellulase, alpha-amylase, pectate–pectate lyases (Voget *et al.*, 2003), amidase (Gabor *et al*,. 2004), and unidentified antibacterial molecules (Craig *et al.*, 2010) were just a few examples of the wealth of novel active molecules 'harvested' from soil metagenomic exploration.

Metagenomics and quality of agricultural soils

Because of the multiplicity and complexity of physicochemical and biological factors that typify a soil, the criteria that determine the agricultural quality of a soil remains debated (Batisda *et al.*, 2008). However, soil quality depends in part upon its microbial composition, and the biological function associated with the soil macroflora. Several studies have exemplified the use of metagenomics to investigate soil microbial diversity including that of agricultural soil, the contribution of cultivation-recalcitrant genera to this diversity (Kielak *et al.*, 2009, 2010), and microbial activities favourable to plants such as those described in the above section. Fixed nitrogen (nitrate or ammonium) is often a limiting element within agricultural soils. Denitrification ability is therefore an important aspect of the quality of an agricultural soil. A recent study based on a metagenomic analysis of a 77,000 clone library shed new light on bacterial denitrification, and the diversity of genetic determinants involved in process (Demanèche *et al.*, 2009).

Soil agricultural quality also depends upon physicochemical traits, such as the presence of pollutants deleterious to plant growth. A combination of denaturing gradient gel electrophoresis targeting the 16S rRNA gene and a metagenomic analysis of 16sRNA genes revealed that members of the *Actinobacteria* and *Bacillus* phyla were affected by the presence of monocyclic aromatic compounds in contaminated forest soils (Ji *et al.*, 2007). In addition, a metagenomic study aimed at identifying microbial community associated with pioneer plants growing in heavy metal-contaminated fallow soils (Navarro-Noya *et al.*, 2010). It revealed that the microbial communities associated with two unrelated plant species growing in the contaminated soil, although quite different from the microbial community of the same bare soil (a phenomenon known as the rhizosphere effect; Hartman *et al.*, 2008), exhibited related species richness, relative abundance, and phylogenetic traits. Similar metagenomic-based studies have led to the identification of novel nickel resistance genes in the rhizosphere microbial community of plants adapted to acid mine drainage (Mirete *et al.*, 2007).

Agriculture worldwide heavily relies upon the use of phytochemical compounds, the unwelcome effects of which being possibly deleterious to non target organisms such as microbes. In this respect, a metagenomic approach has been used to evaluate the consequence of the use of chlorinated pesticides on the plant-associated and soil microflora (Manickam *et al.*, 2010). Still in this field, an original work described the biostimulation of anti-pathogen function of the rhizosphere microbiome of *Solanum tuberosum*, upon amendment with gamma-caprolactone. This molecule induces a modification of the structure of the bacterial community that had been evaluated by an *rrs*-based diversity approach that led to an increase density of members of the species *Rhodococcus erythropolis*, the metabolic activity of which interfered with the expression of the pathogenic functions of *Pectobacterium carotovorum* (Cirou et al., 2007).

Further orientation of metagenomic techniques for studying plant–microbe interaction

As stated in the introduction, metagenomics has only emerged recently as a tool to study plant–microbe interactions. However, publications dealing with other environments indicate how much this methodology could bring to this field of investigation. A look at soil (Kielak *et al.*, 2009; 2010), marine waters or sediments (Cuvelier *et al.*, 2010; Quaiser *et al.*, 2010), extreme environments (Wang *et al.*, 2009; Brazelton *et al.*, 2010), or digestive tractus (Ventura *et al.*, 2009) indeed revealed that metagenomics can help us to improve our description of the microbial diversity associated with plants, and especially that of detrimental or beneficial organisms, even yet-uncultured ones (Da Rocha *et al.*, 2009; Duan *et al.*, 2009, Tyler *et al*,. 2009).

Metagenomic studies also provided valuable data on the identification of key functions of the microbiome, again irrespectively of the environment, such as, metal resistance (González-Pastor and Mirete, 2010), production of secondary metabolites (Banik and Brady, 2010), nitrification and denitrification (Wang *et al.*, 2009), etc. More generally, metagenomic also helped us to gain a comprehensive view at the way ecosystems work, respond to environmental changes (Turnbaugh *et al.*, 2009), or dysfunction (Morowitz *et al.*, 2010). Preliminary data obtained in natural or mesocosm ecosystems revealed remarkable shifts affecting the rhizosphere microflora in response to elevated carbon dioxide concentrations (He *et al.*, 2010) or the nitrogen-fixing community upon disturbance and restoration of an alpine grassland ecosystem (Gros *et al.*, 2006).

In agronomy, metagenomics may contribute to a better evaluation of the consequence of agricultural practices on the soil and plant-interacting community, including prokaryotes and eukaryotes. These practices includes the use of phytochemicals (Zaprasis *et al.*, 2009), the likely development of winter soil cover (to reduce nitrate concentration in soil and surface water) and the use or phosphate fertilizers (Carvalho *et al.*, 2007), or the release of genetically modified organisms (S. Mondy and Y. Dessaux, unpublished).

Within the QS and QQ field, metagenomics has already proved to be useful to identify novel genes, signal molecules, signal mimics, or signal degradative activities (see the relevant above section). However, only one class of signals (the AHSLs) has been preferentially investigated, and only a very limited number of environments have been searched. Clearly, exploration of other environments, including the more 'exotic' ones will permit identification of both inter-kingdom communication process and novel molecules and activities of therapeutic and agricultural importance.

Finally, at least, two challenges remain. The fist deals with the various biases associated with metagenomic techniques which forced the investigator to question the representativeness of the investigated extracted, amplified or sequenced DNA (Hong *et al.*, 2009; Gomez-Alvarez *et al.*, 2009; Temperton *et al.*, 2009; Morgan *et al.*, 2010) and the reproducibility of results (Pan *et al.*, 2010). This problem may be solved in part by using combination of protocols, various cloning vectors, and various hosts to investigate heterologous gene expression. The second challenge lies in the integration of metagenomics with other 'omics' techniques (transcriptomics, proteomics, metabolomics) (see Chapters 1 and 12). This expectation faces two difficulties in addition to the above described problems that apply to both RNA and DNA extraction in the case of transcriptomic studies (McIlroy *et al.*, 2009). The first one is to be able to keep the samples under satisfactory physiological conditions, in such a way that the data obtained from proteomics or metabolomics are biological significant (Valenzuela *et al.*, 2006). The second difficulty is to be able to 'extract the information from the soup' (Rosen *et al.*, 2009), a feature that will require a strong ability to 'surf on massive data streams' (Martin and Martin, 2010). It is our opinion that for the time being, the weak joint in further development of 'meta-omics' approaches lies there.

References

Banik, J.J., and Brady, S.F. (2010). Recent application of metagenomic approaches toward the discovery of antimicrobials and other bioactive small molecules. Curr. Opin. Microbiol. *13*, 603–609.

Bastida, F., Zsolnay, A., Hernández, T., and García, C. (2008). Past, present and future of soil quality indices: a biological perspective. Geoderma *147*, 159–171.

Brazelton, W.J., and Baross, J.A. (2010). Metagenomic comparison of two *Thiomicrospira* lineages inhabiting contrasting deep-sea hydrothermal environments. PLoS One 5, e13530.

Carvalho, G., Lemos, P.C., Oehmen, A., and Reis, M.A. (2007). Denitrifying phosphorus removal: linking the process performance with the microbial community structure. Water Res. *41*, 4383–4396.

Chevrot, R., Rosen, R., Haudecoeur, E., Cirou, A., Shelp, B.J., Ron, E., and Faure, D. (2006). GABA controls the level of quorum-sensing signal in *Agrobacterium tumefaciens*. Proc. Natl. Acad. Sci. U.S.A.*103*, 7460–7464.

Chin-A.-Woeng, T.F., van den Broek, D., de Voer, G., van der Drift, K.M., Tuinman, S., Thomas-Oates, J.E., Lugtenberg, B.J., and Bloemberg, G.V. (2001). Phenazine-1-carboxamide production in the biocontrol strain *Pseudomonas chlororaphis* PCL1391 is regulated by multiple factors secreted into the growth medium. Mol. Plant Microbe Interact. *14*, 969–979.

Chung, E.J., Lim, H.K., Kim, J.C., Choi, G.J., Park, E.J., Lee, M.H., Chung, Y.R., and Lee, S.W. (2008). Forest soil metagenome gene cluster involved in antifungal activity expression in *Escherichia coli*. Appl. Environ. Microbiol. *74*, 723–730

Cirou, A., Diallo, S., Kurt, C., Latour, X., and Faure, D. (2007) Growth promotion of quorum-quenching bacteria in the rhizosphere of *Solanum tuberosum*. Environ. Microbiol. *9*, 1511–1522.

Courtois, S., Frostegård, A., Göransson, P., Depret, G., Jeannin, P., and Simonet, P. (2001). Quantification of bacterial subgroups in soil: comparison of DNA extracted directly from soil or from cells reviously released by density gradient centrifugation. Environ. Microbiol. *3*, 431–439.

Courtois, S., Cappellano, C.M., Ball, M., Francou, F.X., Normand, P., Helynck, G., Martinez, A., Kolvek, S.J., Hopke, J., Osburne, M.S. *et al.* (2003). Recombinant environmental libraries provide access to microbial diversity for drug discovery from natural products. Appl. Environ Microbiol. *69*, 49–55.

Craig, J.W., Chang, F.Y., Kim, J.H., Obiajulu, S.C., and Brady, S.F. (2010). Expanding small-molecule functional metagenomics through parallel screening of broad-host-range cosmid environmental DNA libraries in diverse proteobacteria. Appl. Environ. Microbiol. *76*, 1633–1641.

Cui,Y., Chatterjee, A., Liu, Y., Dumenyo, C.K., and Chatterjee, A.K. (1995). Identification of a global repressor gene, rsmA, of *Erwinia carotovora* subsp. carotovora that controls extracellular enzymes, N-(3-oxohexanoyl)-L-homoserine lactone, and pathogenicity in soft-rotting Erwinia spp. J. Bacteriol. *177*, 5108–5115.

Cuvelier, M.L., Allen, A.E., Monier, A., McCrow, J.P., Messié, M., Tringe, S.G., Woyke, T., Welsh, R.M., Ishoey, T., Lee, J.H. *et al.* (2010). Targeted metagenomics and ecology of globally important uncultured eukaryotic phytoplankton. Proc. Natl. Acad. Sci. U.S.A.*107*, 14679–14684.

D'Angelo-Picard, C., Faure, D., Penot, I., and Dessaux, Y. (2005). Diversity of N-acyl homoserine lactone-producing and -degrading bacteria in soil and tobacco rhizosphere. Environ. Microbiol. *7*, 1796–1808.

Da Rocha, U.N., van Overbeek, L., and Van Elsas, J.D. (2009). Exploration of hitherto-uncultured bacteria from the rhizosphere. FEMS Microbiol. Ecol. *69*, 313–328.

Delalande, L., Faure, D., Raffoux, A., Uroz, S., D'Angelo-Picard, C., Elasri, M., Carlier, A., Berruyer, R., Petit, A., Williams, P. *et al.* (2005). N-hexanoyl-L-homoserine lactone, a mediator of bacterial quorum-sensing regulation, exhibits plant-dependent stability and may be inactivated by germinating *Lotus corniculatus* seedlings. FEMS Microbiol. Ecol. *52*, 13–20.

Delmotte, N., Knief; C., Chaffron, S., Innerebner, G., Roschitzki, B., Schlapbach, R., von Mering, C., and Vorholt, J.A. (2009). Community proteogenomics reveals insights into the physiology of phyllosphere bacteria. Proc. Natl. Acad. Sci. U.S.A.*106*, 16428–16433.

Demanèche, S., Philippot, L., David, M.M., Navarro, E., Vogel, T.M., and Simonet, P. (2009). Characterization of denitrification gene clusters of soil bacteria via a metagenomic approach. Appl. Environ. Microbiol. *75*, 534–537.

Dessaux, Y., Chapelle, E., and Faure, D. (2010). Quorum sensing and quorum quenching in soil ecosystems. In Biocommunication in Soil Microorganisms, Witzany, G., ed. (Heidelberg, Dordrecht, London, New York: Springer), pp. 339–367.

Dong, Y.H., Wang, L.H., Xu, J.L., Zhang, H.B., Zhang, X.F., and Zhang, L.H. (2001) Quenching quorum-sensing-dependent bacterial infection by an N-acyl homoserine lactonase. Nature *411*, 813–817.

Duan, Y., Zhou, L., Hall, D.G., Li, W., Doddapaneni, H;, Lin, H., Liu, L;, Vahling, C.M., Gabriel, D.W., Williams, K.P. *et al.* (2009). Complete genome sequence of citrus huanglongbing bacterium, '*Candidatus* Liberibacter asiaticus' obtained through metagenomics. Mol. Plant Microbe Interact. *22*, 1011–1020.

Fuqua, W.C., Winans, S.C., and Greenberg, E.P. (1994) Quorum sensing in bacteria: the LuxR-LuxI family of cell density-responsive transcriptional regulators. J. Bacteriol. *176*, 269–275.

Gabor, E.M., de Vries, E.J., and Janssen, D.B. (2004). Construction, characterization, and use of small-insert gene banks of DNA isolated from soil and enrichment cultures for the recovery of novel amidases. Environ. Microbiol. *6*, 948–958.

Gillespie, D.E., Brady, S.F., Bettermann, A.D., Cianciotto, N.P., Liles, M.R., Rondon, M.R., Clardy, J., Goodman, R.M., and Handelsman, J. (2002). Isolation of antibiotics turbomycin a and B from a metagenomic library of soil microbial DNA. Appl. Environ. Microbiol. *68*, 4301–4306.

Ginolhac, A., Jarrin, C., Gillet, B., Robe, P., Pujic, P., Tuphile, K., Bertrand, H., Vogel, T.M., Perrière, G., Simonet, P. *et al.* (2004). Phylogenetic analysis of polyketide synthase I domains from soil metagenomic libraries allows selection of promising clones. Appl. Environ. Microbiol. *70*, 5522–5527.

Gomez-Alvarez, V., Teal, T.K., and Schmidt, T.M. (2009) Systematic artifacts in metagenomes from complex microbial communities. ISME J. 3, 1314–1317.

González-Pastor, J.E., and Mirete, S. (2010). Novel metal resistance genes from microorganisms: a functional metagenomic approach. Methods Mol. Biol. *668*, 273–285.

Gros, R., Jocteur-Monrozier, L., and Faivre, P. (2006). Does disturbance and restoration of alpine grassland soils affect the genetic structure and diversity of bacterial and N_2-fixing populations? Environ. Microbiol. *8*, 1889–1890.

Guan, C., Ju, J., Borlee, B.R., Williamson, L.L., Shen, B., Raffa, K.F., and Handelsman, J. (2007). Signal mimics derived from a metagenomic analysis of the gypsy moth gut microbiota. Appl. Environ. Microbiol. *73*, 3669–3676.

Guidot, A., Coupat, B., Fall, S., Prior, P., and Bertolla, F. (2009). Horizontal gene transfer between *Ralstonia solanacearum* strains detected by comparative genomic hybridization on microarrays. ISME J. 3, 549–562.

Haas, D., and Défago, G. (2005). Biological control of soil-borne pathogens by fluorescent pseudomonads. Nat. Rev. Microbiol. 3, 307–319

Hao, Y., Winans, S.C., Glick, B.R., and Charles, T.C. (2010). Identification and characterization of new LuxR/LuxI-type quorum sensing systems from metagenomic libraries. Environ. Microbiol. *12*, 105–117.

Hartmann, A., Rothballer, M., and Schmid, M. (2008). Lorenz Hiltner, a pioneer in rhizosphere microbial ecology and soil bacteriology research. Plant Soil *312*, 7–14.

Haudecoeur, E., Planamente, S., Cirou, A., Tannières, M., Shelp, B.J., Moréra, S., and Faure, D. (2009). Proline antagonizes GABA-induced quenching of quorum-sensing in *Agrobacterium tumefaciens*. Proc. Natl. Acad. Sci. U.S.A.*106*, 14587–14592.

He, Y.Q., Zhang, L., Jiang, B.L., Zhang, Z.C., Xu, R.Q., Tang, D.J., Qin, J., Jiang, W., Zhang, X., Liao, J. *et al.* (2007). Comparative and functional genomics reveals genetic diversity and determinants of host specificity among reference strains and a large collection of Chinese isolates of the phytopathogen *Xanthomonas campestris* pv. campestris Genome Biol. *8*, R218.

He, Z., Xu, M., Deng, Y., Kang, S., Kellogg, L., Wu, L., Van Nostrand, J.D., Hobbie, S.E., Reich, P.B., and Zhou, J. (2010). Metagenomic analysis reveals a marked divergence in the structure of below-ground microbial communities at elevated CO_2. Ecol. Lett. *13*, 564–575.

Hense, B.A., Kuttler, C., Müller, J., Rothballer, M., Hartmann, A., and Kreft, J.U. (2007). Does efficiency sensing unify diffusion and quorum sensing? Nat. Rev. Microbiol. 5, 230–239.

Hong, S., Bunge, J., Leslin, C., Jeon, S., Epstein, S.S. (2009). Polymerase chain reaction primers miss half of rRNA microbial diversity. ISME J. 3, 1365–1373.

Hjort, K., Bergström, M., Adesina, M.F., Jansson, J.K., Smalla, K., and Sjöling, S. (2010). Chitinase genes revealed and compared in bacterial isolates, DNA extracts and a metagenomic library from a phytopathogen-suppressive soil. FEMS Microbiol. Ecol. *71*, 197–207.

Hughes, D.T., Terekhova, D.A., Liou, L., Hovde, C.J., Sahl, J.W., Patankar, A.V., Gonzalez, J.E., Edrington, T.S., Rasko, D.A., and Sperandio, V. (2010) Chemical sensing in mammalian host-bacterial commensal associations. Proc. Natl. Acad. Sci. U.S.A.*107*, 9831–9836.

Ji, S.C., Kim,.D., Yoon, J.H., and Lee, C.H. (2007). Metagenomic analysis of BTEX-contaminated forest soil microcosm. J. Microbiol. Biotechnol. *17*, 668–672.

Jiao, J.Y., Wang, H.X., Zeng, Y., and Shen, Y/M. (2006). Enrichment for microbes living in association with plant tissues. J. Appl. Microbiol. *100*, 830–837.

Kielak, A., Pijl, A.S., Van Veen, J.A., and Kowalchuk, G.A. (2009). Phylogenetic diversity of *Acidobacteria* in a former agricultural soil ISME J. 3, 378–382.

Kielak, A., Rodrigues, J.L., Kuramae, E.E., Chain, P.S., van Veen J.A., and Kowalchuk, G.A. (2010). Phylogenetic and metagenomic analysis of *Verrucomicrobia* in former agricultural grassland soil. FEMS Microbiol. Ecol. *71*, 23–33.

Koopman, M.M., Fuselier, D.M., Hird, S., and Carstens, B.C. (2010). The carnivorous pale pitcher plant harbors diverse, distinct, and time-dependent bacterial communities. Appl. Environ. Microbiol. *76*, 1851–1860.

Lee, H.H., Molla, M.N., Cantor, C.R., and Collins, J.J. (2010) Bacterial charity work leads to population-wide resistance. Nature *467*, 82–85.

Manefield, M., de Nys, R., Kumar, N., Read, R., Givskov, M., Steinberg, P., and Kjelleberg, S. (1999). Evidence that halogenated furanones from *Delisea pulchra* inhibit acylated homoserine lactone (AHL)-mediated gene expression by displacing the AHL signal from its receptor protein. Microbiology *145*, 283–291.

Manickam, N., Pathak*1*, A., Saini, H.S., Mayilraj, S., and Shanker, R. (2010). Metabolic profiles and phylogenetic diversity of microbial communities from chlorinated pesticides contaminated sites of different geographical habitats of India. J. Appl. Microbiol. *109*, 1458–1468.

Martin, N.F., and Martin, F. (2010) From galactic archeology to soil metagenomics – surfing on massive data streams. New Phytol. *185*, 343–347.

Mathesius, U., Mulders, S., Gao, M., Teplitski, M., Caetano-Anolles, G., Rolfe, B.G., and Bauer, W.D. (2003). Extensive and specific responses of a eukaryote to bacterial quorum-sensing signals. Proc. Natl. Acad. Sci. U.S.A.*100*, 1444–1449.

Mazzola, M. (2004). Assessment and management of soil microbial community structure for disease suppression. Annu. Rev. Phytopathol. *42*, 35–59.

McIlroy, S.J., Porter, K., Seviour, R.J., and Tillett, D. (2009). Extracting nucleic acids from activated sludge which reflect community population diversity. Antonie Van Leeuwenhoek *96*, 593–605.

Mirete, S., de Figueras, C.G., and González-Pastor, J.E. (2007). Novel nickel resistance genes from the rhizosphere metagenome of plants adapted to acid mine drainage. Appl. Environ. Microbiol. *73*, 6001–6011.

Mithani, A., Hein, J., and Preston, G.M. (2011). Comparative analysis of metabolic networks provides insight into the evolution of plant pathogenic and non-pathogenic lifestyles in Pseudomonas. Mol. Biol. Evol. *28*, 483–499.

Mocali, S., and Benedetti, A. (2010). Exploring research frontiers in microbiology: the challenge of metagenomics in soil microbiology. Res. Microbiol. *161*, 497–505.

Molina, L., Constantinescu, F., Michel, L., Reimmann, C., Duffy, B., and Défago, G. (2003) Degradation of pathogen quorum-sensing molecules by soil bacteria: a preventive and curative biological control mechanism. FEMS Microbiol. Ecol. *45*, 71–81.

Moreira, L.M;, Almeida Jr., N.F., Potnis, N., Digiampietri, L.A., Adi, S.S., Bortolossi, J.C., da Silva, A.C., da Silva, A.M., de Moraes, F.E., de Oliveira, J.C. *et al.* (2010). Novel insights into the genomic basis of citrus canker based on the genome sequences of two strains of *Xanthomonas fuscans* subsp. aurantifolii. BMC Genomics *11*, 238.

Morgan, J.L., Darling, A.E., and Eisen, J.A. (2010). Metagenomic sequencing of an *in vitro*-simulated microbial community. PLoS One. *5*, e10209.

Morowitz, M.J., Poroyko, V., Caplan, M., Alverdy, J., and Liu, D.C. (2010). Redefining the role of intestinal microbes in the pathogenesis of necrotizing enterocolitis. Pediatrics *125*, 777–785.

Müller, H., Westendorf, C., Leitner E, Chernin, L., Riedel, K., Schmidt, S., Eberl, L., Berg, G. (2009). Quorum-sensing effects in the antagonistic rhizosphere bacterium *Serratia plymuthica* HRO-C48. FEMS Microbiol. Ecol. *67*, 468–478.

Navarro-Noya, Y.E., Jan-Roblero, J., González-Chávez Mdel, C., Hernández-Gama, R., and Hernández-Rodríguez, C. (2010). Bacterial communities associated with the rhizosphere of pioneer plants (*Bahia xylopoda* and *Viguiera linearis*) growing on heavy metals-contaminated soils. Antonie Van Leeuwenhoek *97*, 335–349.

Nichols, D. (2007). Cultivation gives context to the microbial ecologist. FEMS Microbiol. Ecol. *60*, 351–357.

Pan, Y., Bodrossy, L., Frenzel, P., Hestnes, A.G., Krause, S., Lüke, C., Meima-Franke, M., Siljanen, H., Svenning, M.M., Bodelier, P.L. (2010). Impacts of inter- and intralaboratory variations on the reproducibility of microbial community analyses. Appl. Environ. Microbiol. *76*, 7451–7458.

Plummer, G.L., and Jackson, T.H. (1963). Bacterial activities within the sarcophagus of the insectivorous plant, *Sarracenia flava*. Am. Midl. Nat. *69*, 462–469.

Quaiser, A., Zivanovic, Y., Moreira, D., and López-García, P. javascript:AL_get(this, 'jour', 'ISME J.');(2011). Comparative metagenomics of bathypelagic plankton and bottom sediment from the Sea of Marmara. ISME J. *5*, 285–304.

Redfield, R.J. (2002). Is quorum sensing a side effect of diffusion sensing? Trends Microbiol. *10*, 365–370.

Remenant, B., Coupat-Goutaland, B., Guidot, A., Cellier, G., Wicker, E., Allen, C., Fegan, M., Pruvost, O., Elbaz, M., Calteau, A. *et al.* (2010). Genomes of three tomato pathogens within the *Ralstonia solanacearum* species complex reveal significant evolutionary divergence. BMC Genomics *11*, 379.

Riaz, K., Elmerich, C., Moreira, D., Raffoux, A., Dessaux, Y., and Faure, D. (2008). A metagenomic analysis of soil bacteria extends the diversity of quorum-quenching lactonases. Environ. Microbiol. *10*, 560–570.

Riesenfeld, C.S., Schloss, P.D., and Handelsman, J. (2004). Metagenomics: genomic analysis of microbial communities. Annu. Rev. Genet. *38*, 525–552.

Rosen, G.L., Sokhansanj, B.A., Polikar, R., Bruns, M.A., Russell, J., Garbarine, E., Essinger, S., and Yok, N. (2009). Signal processing for metagenomics: extracting information from the soup. Curr Genomics *10*, 493–510.

Sanguin, H., Sarniguet, A., Gazengel, K., Moënne-Loccoz, Y., and Grundmann, GL. (2009) Rhizosphere bacterial communities associated with disease suppressiveness stages of take-all decline in wheat monoculture. New Phytol. *184*, 694–707.

Schaefer, A.L., Greenberg, E.P., Oliver, C.M., Oda, Y., Huang, J.J., Bittan-Banin, G., Peres, C.M., Schmidt, S., Juhaszova, K., Sufrin, J.R. *et al.* (2008). A new class of homoserine lactone quorum-sensing signals. Nature *454*, 595–599

Schipper, C., Hornung, C., Bijtenhoorn, P., Quitschau, M., Grond, S., Streit, W.R. (2009). Metagenome-derived clones encoding two novel lactonase family proteins involved in biofilm inhibition in *Pseudomonas aeruginosa*. Appl. Environ. Microbiol. *75*, 224–233.

Simmons, S.L., Dibartolo, G., Denef, V.J., Goltsman, D.S., Thelen, M.P., and Banfield, J.F. (2008). Population genomic analysis of strain variation in *Leptospirillum* group II bacteria involved in acid mine drainage formation. PLoS Biol. 6:e177.

Straight, P.D., and Kolter, R. (2009) Interspecies chemical communication in bacterial development. Annu Rev Microbiol. *63*, 99–118.

Temperton, B., Field, D., Oliver, A., Tiwari, B., Mühling, M., Joint, I., and Gilbert, J.A. (2009) Bias in assessments of marine microbial biodiversity in fosmid libraries as evaluated by pyrosequencing. ISME J. *3*, 792–796.

Terauchi, R., and Yoshida, K. (2010). Towards population genomics of effector–effector target interactions. New Phytol. *187*, 929–939.

Tringe, S.G., von Mering, C., Kobayashi, A., Salamov, A.A., Chen, K., Chang, H.W., Podar, M., Short, J.M., Mathur, E.J., Detter, J.C. *et al.* (2005) Comparative metagenomics of microbial communities. Science. *308*, 554–557.

Turnbaugh, P.J., Ridaura, V.K., Faith, J.J., Rey, F.E., Knight, R., and Gordon, J.I. (2009). The effect of diet on the human gut microbiome: a metagenomic analysis in humanized gnotobiotic mice. Sci. Transl. Med. 1 (6ra14), 1–19.

Tyler, H.L., Roesch, L.F., Gowda, S., Dawson, W.O., and Triplett, E.W. 2009). Confirmation of the sequence of '*Candidatus* Liberibacter asiaticus' and assessment of microbial diversity in Huanglongbing-infected citrus phloem using a metagenomic approach. Mol. Plant Microbe Interact. *22*, 1624–1634.

Uroz, S., D'Angelo-Picard, C., Carlier, A., Elasri, M., Sicot, C., Petit, A., Oger, P., Faure, D., and Dessaux, Y. (2003). Novel bacteria degrading N-acylhomoserine lactones and their use as quenchers of quorum-sensing-regulated functions of plant-pathogenic bacteria. Microbiology *149*, 1981–1989.

Uroz, S., Dessaux, Y., and Oger, P. (2009). Quorum sensing and quorum quenching: the yin and yang of bacterial communication. Chembiochem. *10*, 205–216.

Valenzuela, L., Chi, A., Beard, S., Orell, A., Guiliani, N., Shabanowitz, J., Hunt, D.F., and Jerez, C.A. (2006). Genomics, metagenomics and proteomics in biomining microorganisms. Biotechnol. Adv. *24*, 197–211.

Van Elsas, J.D., Speksnijder, A.J., and Van Overbeek L.S. (2008). A procedure for the metagenomics exploration of disease suppressive soils. J. Microbiological Methods *75*, 515–522.

Voget, S., Leggewie, C., Uesbeck, A., Raasch, C., Jaeger, K.E., and Streit, W.R. (2003). Prospecting for novel biocatalysts in a soil metagenome. Appl. Environ. Microbiol. *69*, 6235–6242.

Von Bodman, S.B., Bauer, W.D., and Coplin, D.L. (2003). Quorum sensing in plant-pathogenic bacteria. Annu. Rev. Phytopathol. *41*, 455–482.

Von Mering, C., Hugenholtz, P., Raes, J., Tringe, S.G., Doerks, T., Jensen, L.J., Ward, N., and Bork, P. (2007). Quantitative phylogenetic assessment of microbial communities in diverse environments. Science *315*, 1126–1130.

Ventura, M., Turroni, F., Canchaya, C., Vaughan, E.E., O'Toole, P.W., and van Sinderen, D. (2009). Microbial diversity in the human intestine and novel insights from metagenomics. Front. Biosci. *14*, 3214–3221.

Wang, F., Zhou, H., Meng, J., Peng, X., Jiang, L., Sun, P., Zhang, C., Van Nostrand, J.D., Deng, Y., He, Z. *et al.* (2009). GeoChip-based analysis of metabolic diversity of microbial communities at the Juan de Fuca Ridge hydrothermal vent. Proc. Natl. Acad. Sci. U.S.A.2009 *106*, 4840–4805.

Wang, H.X., Geng, Z.L., Zeng, Y., and Shen, Y.M. (2008). Enriching plant microbiota for a metagenomic library construction. Environ. Microbiol. *10*, 2684–2691.

Weller, D.M., Raaijmakers, J.M., Gardener, B.B., and Thomashow, L.S. (2002). Microbial populations responsible for specific soil suppressiveness to plant pathogens. Annu. Rev. Phytopathol. *40*, 309–48.

Whiteley, M., Parsek, M.R., and Greenberg, E.P. (2000). Regulation of quorum sensing by RpoS in *Pseudomonas aeruginosa*. J. Bacteriol. *182*, 4356–4360.

Wilder, C.N., Allada, G., and Schuster, M. (2009). Instantaneous within-patient diversity of *Pseudomonas aeruginosa* quorum-sensing populations from cystic fibrosis lung infections. Infect Immun. 77, 5631–5639.

Williamson, L.L., Borlee, B.R., Schloss, P.D., Guan, C., Allen, H.K., and Handelsman, J. (2005). Intracellular screen to identify metagenomic clones that induce or inhibit a quorum-sensing biosensor. Appl. Environ. Microbiol. *71*, 6335–6344.

Wilson, I.G. (1997). Inhibition and facilitation of nucleic acid amplification. Appl. Environ. Microbiol. *63*, 3741–3751.

Wood, D.W., Gong, F., Daykin, M.M., Williams, P., and Pierson 3rd, L.S. (1997). N-acyl-homoserine lactone-mediated regulation of phenazine gene expression by *Pseudomonas aureofaciens* 30–84 in the wheat rhizosphere. J. Bacteriol. *179*, 7663–7670.

Zaprasis, A., Liu, Y.J., Liu, S.J., Drake, H.L., and Horn, M.A. (2009). Abundance of novel and diverse *tfdA*-like genes, encoding putative phenoxyalkanoic acid herbicide-degrading dioxygenases, in soil. Appl. Environ. Microbiol. *76*, 119–128.

Zhang, L.H. (2003). Quorum quenching and proactive host defense. Trends Plant Sci. *8*, 238–244.

Metagenomics Analysis of Below-ground Microbial Communities using Microarrays

15

Joy D. Van Nostrand, Zhili He and Jizhong Zhou

Abstract

The use of microarrays has revolutionized the field of microbiology. While many types of microarrays are available, the functional gene arrays (FGAs) afford a way to link environmental processes with microbial communities. FGAs probe for a wide range of genes involved in functional activities of interest to microbial ecology (e.g. carbon degradation, N_2-fixation, metal resistance) from many different microorganisms, cultured and uncultured. The most comprehensive FGA reported to date are the GeoChip arrays. GeoChip 3.0 targets tens of thousands of genes involved in the geochemical cycling of carbon, nitrogen, phosphorus, and sulphur, metal resistance and reduction, energy processing, antibiotic resistance and contaminant degradation as well as phylogenetic information (gyrB). This technology has been used successfully to examine the effects of global climate change on microbial communities. This chapter provides an overview of FGA development with a focus on the GeoChip. Several GeoChip studies, which have established the GeoChip's worth as a rapid, sensitive and specific tool for the examination of microbial communities, will be highlighted.

Introduction

Microorganisms are the foundation of the Earth's biosphere and play critical and unique roles in ecosystem functioning including biogeochemical cycling of carbon, nitrogen, sulfur, phosphorus, and metals, as well as degradation and/or stabilization of contaminated environments. Microorganisms are among the most phylogenetically and functionally diverse groups of organisms known and have been found in almost every location examined, including the most extreme environments. As such, understanding how microorganisms affect the environment and how the microbial community function, structure, interactions, and populations change temporally and spatially are important goals in microbial ecology. However, there are several challenges related to studying microbial communities. The first is the diversity of microbial communities as 1 g of soil has been estimated to contain 2000 to 50,000 microbial species (Torsvik *et al.*, 1990; Schloss and Handelsman, 2006; Hong *et al.*, 2006; Roesch *et al.*, 2007). This vast diversity makes it difficult to characterize and understand the environmental factors shaping the microbial communities. Over 99% of microorganisms are unculturable (Amann *et al.*, 1995; Furhman and Campbell, 1998; Whitman *et al.*, 1998), so that functional abilities cannot be evaluated directly. While culture-dependent studies are still important, they provide limited information regarding overall microbial community diversity and function. As such, culture-independent approaches are needed to study a majority of environmental microorganisms.

Several 16S rRNA gene-based, culture-independent methods are available for microbial community analysis including cloning, denaturing gradient gel electrophoresis (DGGE), terminal-restriction fragment length polymorphism (T-RFLP), quantitative PCR, and *in situ* hybridization. These methods can also be used with other genes. However, while these are popular methods for microbial community analysis, they all suffer from several important limitations. First, while phylogenetic markers, such as 16S rRNA or DNA gyrase (GyrB) genes are commonly used to examine microbial community structure (Wilson *et al.*, 1990; Amann *et al.*, 1995; Yamamoto and Harayama, 1995; Hugenholtz *et al.*, 1998; Brodie *et al.*, 2006), they provide minimal information on the community's functional abilities and activity. Second, most of these methods require a PCR amplification step, which introduces well-known biases (Suzuki and Giovannoni, 1996; Warnecke *et al.*, 1997; Lueders and Friedrich, 2003) and limits quantitative information. There is also limited resolution and coverage with many of these methods. For example, commonly used sizes for clone libraries may be 10 times (or greater) too small to capture the true diversity of microbial communities (Dunbar *et al.*, 2002). A library of >2000 clones derived from Chesapeake Bay microbial communities was insufficient to cover the nifH diversity of this site (Jenkins *et al.*, 2004). Lastly, the time and expense of these assays can be great when many samples or genes are examined or if multiple runs are necessary to increase coverage.

Many of these limitations can be overcome using microarrays, which can be used to examine thousands of genes at one time, provide specific and sensitive detection at a high resolution, can provide quantitative data if PCR amplification is avoided, and arrays are often more cost-effective than other molecular techniques because of the amount of data provided. The first reported microarray was a gene expression array for the plant *Arabidopsis thaliana* (Schena *et al.*, 1995). Since that publication organism-specific arrays have become a standard tool to examine gene expression for a variety of microorganisms.

Potential applications for microarrays in microbiology were greatly expanded when a 16S rRNA gene array was used to demonstrate that hybridization patterns of DNA and RNA from several different bacterial genera allowed them to distinguish the different bacteria, showing the utility of microarrays for environmental studies (Gushin *et al.*, 1997). Microarrays are ideal for comparing environmental samples from different sites, conditions, or times because arrays have a defined set of genes or microorganisms which all samples are interrogated against. Comparisons between microbial communities can be difficult with other molecular techniques like clone libraries since the sampling process is random, the chance that the same community population will be sampled over multiple sampling events is theoretically quite small (Dunbar *et al.*, 2002; Zhou *et al.*, 2008). These benefits make microarrays excellent tools for examining microbial community structure, function, activities and dynamics in natural settings.

While the use of high throughput sequencing technology has advanced greatly over the past decade, allowing for fairly inexpensive sequencing of thousands to millions of sequences in a single run, there are advantages in using microarrays rather than sequencing for microbial community studies. These include:

1. Fewer random sampling errors. Even though a large amount of sequencing data is provided, only a small portion of the microbial community is actually sampled in most metagenomic studies. For example, ~1400 rRNA gene sequences were obtained per sample from Rhesus macaque gut microbiomes when using 454 sequencing, an underestimate of the total community based on Chao1 estimates (McKenna *et al.*, 2008). Further, based on these finding, 454 pyrosequencing would be expected to greatly underestimate the intestinal microbial community since the human gut microbiome is estimated to have a density of 10^{11} to 10^{12} bacterial cells per ml (Whitman *et al.*, 1998) or 10^9 to 10^{10}/g of stool (Palmer *et al.*, 2007).
2. Ability to accurately compare microbial communities. Since microarrays interrogate communities using the same set of probes across all samples, comparisons between samples and environments can easily be made. In contrast, randomly sequencing a small portion of the microbial communities from different environments or conditions may not be informative or meaningful for comparative purposes due to random sampling errors (Zhou *et al.*, 2008), as

discussed above, unless sufficient sequencing coverage is achieved. However, while there are advantages to using microarrays over high throughput sequencing, there is one major disadvantage in using microrrays: the inability to detect novel sequences. Sequenced-based technologies can find new or novel sequences since there are no limitations on what sequences are sequenced. However, because microarrays are limited to detecting only sequences covered by the probe sequences on an array, detecting new sequences is impossible; although several new developments in microarray technology have discovered ways to find new sequences using microarrays.

Microarrays for microbial community analysis

Phylogenetic oligonucleotide arrays

Several types of arrays are currently available for environmental microbial community analysis (Zhou and Thompson, 2002; Zhou, 2003; Gentry *et al.*, 2006). Phylogenetic oligonucleotide arrays (POA) are designed to determine community composition or phylogenetic relatedness using 16S rRNA or other phylogenetically informative genes (Small *et al.*, 2001; Loy *et al.*, 2002; Wilson *et al.*, 2002; Brodie *et al.*, 2006). The most comprehensive POA is the PhyloChip, which uses the Affymetrix format (Santa Clara, CA) and contains ~300,000 perfect-match and mismatch 16S rRNA gene probes for 842 subfamilies (8741 taxa), covering 121 bacterial and archaeal orders and provides identification at the family to subfamily levels (Brodie *et al.*, 2006; DeSantis *et al.*, 2007; Brodie *et al.*, 2007). While the 16S rRNA gene is most commonly used, the 23S rRNA gene has also been used for POA construction (Lee *et al.*, 2006). The PhyloChip has been used to examine microbial communities from soluble uranium contaminated soils (Brodie *et al.*, 2006), microbial communities associated with aerosols (Brodie *et al.*, 2007), soil and water (DeSantis *et al.*, 2007) and microbial communities from intubated patients colonized with *Pseudomonas aeruginosa* (Flanagan *et al.*, 2007).

Community genome arrays

Community genome arrays (CGA) are used to determine strain identity, community composition, or compare microbial communities or to examine the relatedness of microbial strains using whole genomic DNA of individual species or strains as probes (Wu *et al.*, 2004, 2008; Zhang *et al.*, 2004). CGAs have been used to examine environmental samples from soil, river sediments, and marine sediments (Wu *et al.*, 2004), acid mine drainage (AMD) or bioleaching applications (Chen *et al.*, 2009). CGAs have also been used to determine species relatedness of *Escherichia coli* strains within a reference collection (Zhang *et al.*, 2004) and Wu and colleagues (2008b) used a CGA to examine species relatedness of 55 strains of the genera *Shewanella* (16), *Pseudomonas* (30) and *Azoarcus* (9).

Metagenomic arrays

Metagenomic arrays (MGAs) are made using clone libraries created from environmental DNA as probes and can be used as a high-throughput screening method (Sebat *et al.*, 2003). An MGA with probes made from an enrichment culture-based cosmid library was screened against isolates from the same enrichment culture in order to find uncultured clone sequences (Sebat *et al.*, 2003). Another 'genome-proxy' array was developed using BAC or fosmid clone libraries from Monterrey Bay, Hawaii and the Antarctic shoreline microbial communities to screen other coastal water communities (Rich *et al.*, 2008).

Whole-genome open reading frame arrays

Whole-genome open reading frame (ORF) arrays (WGA) are comprised of probes for all ORFs in one or more genomes (Wilson *et al.*, 1999). This type of array has been used to examine the diversity and relatedness of several metal-reducing *Shewanella* strains (Murray *et al.*, 2001), to examine genes in *Klebsiella pneumoniae* strain 342 using an *E. coli* K-12 WGA (Dong *et al.*, 2001), to evaluate 91 strains of *Pseudomonas syringae* using a WGA of 353 virulence factors to determine which genes were associated with host specificity (Sarker *et al.*, 2006), and to examine chromosome rearrangements in *P. furiosus* and seven

Pyrococcus isolates using a *Pyrococcus furiosusi* DSM 3638 WGA (White *et al.*, 2008).

Functional gene arrays

Functional gene arrays (FGA) are composed of probes for key genes involved in microbial functional processes of interest, such as nitrification, carbon degradation, or metal resistance (Wu *et al.*, 2001; Gentry *et al.*, 2006; He *et al.*, 2007, 2010a). FGAs are especially useful for environmental samples since we know the exact functions of so few microorganisms due to the lack of cultured microorganisms and the high degree of diversity and metabolic flexibility that exists in microbial communities (Fitter *et al.*, 2005). The most comprehensive FGA reported to date is the GeoChip, which contains probes for thousands of functional genes and can link microbial community structure to community function (He *et al.*, 2007, 2010a). Unlike PCR-based techniques that limit the number of genes that can be examined at once, FGAs allow one to simultaneously examine many functional gene groups (Wu *et al.*, 2001; Zhou and Thompson, 2002; Gentry *et al.*, 2006; Wu *et al.*, 2006; He *et al.*, 2007; Wagner *et al.*, 2007; Zhou *et al.*, 2008; Wang *et al.*, 2009). In addition, FGAs have several advantages to other 16S rRNA gene-based microarrays or sequencing technologies:

1 Ability to detect functions. While the 16S rRNA gene can be used to detect the presence of a strain, it does not provide information on a bacterium's functional abilities. In addition, the need for PCR amplification limits the number of genes that can be detected since primers are only available for a limited number of functional genes and the sheer number of reactions to cover many functional genes would be prohibitive. FGAs, however, use functional gene markers and can thus provide information on the potential metabolic functions within a community (Wu *et al.*, 2001; Rhee *et al.*, 2004; Tiquia *et al.*, 2004; He *et al.*, 2007, 2010a).

2 Higher resolution than other 16S rRNA-based methods. The GeoChip FGA provides a resolution at the species-strain level (Tiquia *et al.*, 2004); the PhyloChip array provides a resolution at the family–subfamily level (DeSantis *et al.*, 2007); and 16S rRNA gene-based sequencing generally provides a resolution at the genus-species level (He *et al.*, unpublished).

3 Can provide quantitative information. The need for an initial PCR amplification step for 16S rRNA-based or other similar gene-based methods introduces a well-known amplification bias which limits quantitative information (Suzuki and Giovannoni, 1996; Warnecke *et al.*, 1997; Lueders and Friedrich, 2003). Previous studies with FGAs have shown that hybridization of template DNA with or without random amplification are quite quantitative (Wu *et al.*, 2001, 2006; Tiquia *et al.*, 2004; Rhee *et al.*, 2004; Gao *et al.*, 2007). In addition, whole community genome amplification (WCGA), which can be used to increase the amount of DNA available for hybridization has been shown to produce minimal bias (Wu *et al.*, 2006).

Development of FGAs

The first reported FGA contained PCR-amplicon probes targeting 89 N-cycling genes (nirS, nirK, amoA, and pmoA) derived from pure culture isolates and marine sediment clone libraries (Wu *et al.*, 2001). Additional control probes were included on this array: 10 16S rRNA genes as positive controls and five yeast genes as negative controls. Different hybridization temperatures produced varying levels of specificities. For example, at 45°C all pure culture and environmental samples hybridized to the appropriate gene probes (i.e. target sequences for nirS hybridized to nirS probes) while at 65°C target DNA hybridized to its strain-specific probe. This array provided a sensitivity of approximately 1 ng for pure culture DNA and 10 to 25 ng for community DNA. Once appropriate hybridization conditions were determined, the array was used to examine marine sediment and soil communities. Genes for nirS, nirK, and amoA or pmoA were detected in both samples with higher abundance observed with soil samples than marine sediment. Since this first report, several new FGAs have been reported. These FGAs have generally targeted specific functional groups, genes, or environments.

Nitrogen cycling

Two oligonucleotide (70-mer) arrays containing probes for amoA, nifH, nirS, and nirK was designed to examine denitrification, N-fixation, ammonia oxidation, and nitrite reduction (Taroncher-Oldenburg *et al.*, 2003). This array had the highest specificity when hybridizations were carried out in a commercial hybridization buffer (GlassHyb; Clontech Laboratories) at 65°C. Under these conditions, sequences with at least 84–90% similarity were detected with a detection limit of ~10^7 DNA copies. These arrays were used to examine N-cycling communities from a river system and differences in the denitrifying community were observed based on location and were consistent with differences observed in the overall communities.

A macroarray, containing nifH PCR amplicons as probes on a nylon membrane, was developed to study nitrogenase gene diversity (Steward *et al.*, 2004). At 60°C and 40% formamide sequences hybridized that had >78% sequence similarity and resulted in a detection limit at least an order of magnitude less than 0.13 ng/ml. This macroarray was subsequently expanded and used in a study to examine diazotroph communities in the Chesapeake Bay (Jenkins *et al.*, 2004). Results indicated a high diversity of N_2-fixing microorganisms and differences in specific phylotypes were detected in the surface water communities at different sampling stations across the Chesapeake Bay.

A nitrogen fixation specific array comprised of 56 oligonucleotide probes [15- to 25-mers; melting temperature (T_m) of 65–69°C] for nifH gene variants with 8–12 replicate probes per gene was designed (Zhang *et al.*, 2006). The array was used to examine DNA and RNA extracted from the roots of *Oryza longistaminata*. Array results indicated that nifH-containing strains from the β-Proteobacteria, the Omega subcluster, the anf subcluster, and Gram-positive bacteria were present in the roots and that the Gram-positive bacteria were active within this system, which was similar to results obtained from clone library sequencing.

Another FGA was recently reported that used cDNA clones from extracted microbial community mRNA as probes (McGrath *et al.*, 2010). In this array, 13,056 randomly selected clones obtained from a variety of microbial communities were used to examine *in situ* gene expression in sites with high and low N_2O-producing communities. Results indicated that 153 sequences showed significantly different levels of expression in high versus low N_2O-producing communities. The positive probes were then sequenced to determine clone identity.

Methane production

An array comprising oligonucleotide probes for 59 *pmoA*/*amoA* genes was developed to study methanotrophs (Bodrossy *et al.*, 2003). The array was used to examine the activity of microbial communities from a landfill in Seiberdorf, Austria, using RNA amplified with primers containing the T7 promoter. At 55°C, 50 of the probes were positive for perfect-match or one mismatched sequences. The detection limit for environmental samples for this array under these conditions was approximately 5% of the total population. This array was then expanded to include 68 oligonucleotide probes and was used to study the effects of plant cover on methanotrophic communities from simulated landfills (Stralis-Pavese *et al.*, 2004). Results from the array using DNA indicated that *Methylocystis* spp. and *Methylocaldum* spp. were the dominant members of the methanotrophic community and that supplemental biogas (CH_4 and CO_2 mix, 3:2) increased the overall abundance of methanotrophs. Hybridizations using mRNA indicated that while *Methylocystis* spp. and *Methyocaldum* spp. were the dominant methanotrophic genera in the simulated landfills, *Methyocaldum* spp. activity was quite low and no activity was detected for *Methylocystis* spp. (Bodrossy *et al.*, 2006).

Pathogen detection or testing

An important prospective application of FGAs is for medical diagnosis or water and food safety testing. Several pathogen detection arrays have been tested and shown promise for future applications. A preliminary diagnostic array was developed with oligonucleotide probes (70-mers) targeting *Salmonella*-specific virulence factors (invA and sopB) with an Enterobacteriaceae 16S rRNA probe as a positive control (Kostić *et al.*, 2005). The array was tested using whole genome amplified DNA from *Escherichia coli* DH5α and *Salmonella enterica* serovar Senftenberg. Hybridizations carried out at 42°C and 40% formamide were able to detect the target sequence when it represented as little at 1% of the original community.

Another array [virulence and marker gene (VMG) array] designed to detect waterborne pathogens contained two sets of probes, one ($n=791$) targeting 35 VMGs from 12 pathogens, the second ($n=2034$) targeting 67 VMGs from 17 pathogens (Miller *et al.*, 2008). This VMG array was able to detect pathogens in tap water, wastewater treatment plant effluent, and river water at a relative abundance of as little at 0.01%. Another prototype array, comprising 120 PCR amplicon probes (200–800 bp) for virulence genes, antibiotic resistance genes, and metabolic or structural genes specific to *Staphylococcus aureus, E. coli,* and *Pseudomonas aeruginosa,* was designed to detect bloodstream pathogens (Cleven *et al.*, 2006). The array was tested against 45 clinical or reference strains of the target organisms and hybridizations were specific for each strain tested. The presence of antibiotic resistance genes detected on the array was confirmed for *S. aureus* strains using traditional culture-based methods and agreed with microarray findings.

A larger pathogen detection array containing 930 genes from 10 different pathogens (*Staphylococcus* spp., *Streptococcus* spp., *Enterococcus* spp., *Proteus* spp., *Klebsiella* spp., *Stenotrophomonas* spp., *Enterobacter* spp., *Acinetobacter* spp., *E. coli, Pseudomonas aeruginosa,* and *Candida albicans*) including antimicrobial resistance genes and bacterial 16S rRNA gene probes (Palka-Santini *et al.*, 2009). For this array samples were amplified using a multiplex PCR amplification with primers specific for all genes carried on the array (800 primer pairs). With the multiplex PCR amplification and hybridizations carried out at 42°C and 25% formamide, the detection limit was 1 ng of DNA. Infected wound swabs were used to test the amplification method and the array and positive hybridizations to multiple *Enterococcus faecium* and *Staphylococcus epidermidis* specific probes were observed and positive hybridizations were confirmed by culturing. Singleton positive probes for some pathogens and negative culture results indicated non-infectious agents.

In addition to pathogen detection, arrays could also possibly be used to determine antibiotic resistance. An antibiotic resistance array was developed and was comprised of PCR amplicons ~550 bp in length derived from tetracycline resistance (*tet*) genes (Call *et al.*, 2003). Of 48 isolates tested, 38 showed positive hybridizations to one tet gene, one isolate hybridized to two probes, while the remainder did not have positive hybridizations, agreeing with previous findings for these isolates.

Specific communities or locations

Several FGAs have also been developed to study communities from specific locations or environments. For example, a nodC FGA comprising probes ($n=130$, 41- and 50-mers) designed from two conserved regions of the *nodC* gene was developed to detect known and new nodC sequences (Bontemps *et al.*, 2005). The array was used to characterize rhizobial isolates and subsequent sequencing of positive strains indicated that novel nodC sequences had been detected.

Another array was designed to study AMD communities and contained 50mer oligonucleotide probes for 1071 genes (16S rRNA and functional genes) relevant to AMD and bioleaching systems (Yin *et al.*, 2007). When this array was used to examine samples from an AMD site (pH 2.0), a bioleaching system (pH 2.5), and a less acidic site (pH 5.0), fewer 16S genes were detected in the less acidic sample and higher functional diversity was observed in lower pH samples.

GeoChip functional gene arrays

The GeoChip arrays (He *et al.*, 2007, 2010a) are the most comprehensive FGA to date and were developed in the Zhou lab. The prototype GeoChip array was also the first reported FGA (Wu *et al.*, 2001). This array used PCR amplicons as probes and was developed as a proof of concept and demonstrated that FGAs could be used to study microbial communities and to link microbial structure and functions to environmental geochemistry. However, the use of PCR amplicon probes severely limited the number and diversity of genes that could be covered by the array since PCR primers are required for each gene and a large number of clones and/or isolates would be required. In order to overcome these technical challenges, arrays are now designed using oligonucleotide probes instead of the PCR amplicons (Rhee *et al.*, 2004; Tiquia *et al.*, 2004). While PCR-based probes provide a higher degree of sensitivity, oligonucleotide probes provide higher specificity (Zhou, 2003), are easily customized to afford a more targeted probe design (Denef *et al.*, 2003; Zhou, 2003; Gentry *et al.*, 2006), and are relatively inexpensive.

The next generation of GeoChip array covered a larger number of genes and gene variants and was reported in two studies in 2004. This version contained 50-mer oligonucleotide probes for nitrification, denitrification, nitrogen fixation, methane oxidation, and sulfate reduction (Tiquia *et al.*, 2004), organic contaminant degradation and metal resistance (Rhee *et al.*, 2004). These probes were combined to form what is now referred to as GeoChip 1.0 and was comprised of 2006 oligonucleotide probes (50-mers), 10 negative control probes (six human and four plant derived), and two 16S rRNA gene probes as positive controls (Rhee *et al.*, 2004; Tiquia *et al.*, 2004). Optimal probe concentrations was determined to be 30–40 pmol/μl based on retention of probe to the array after printing, hybridization and washing (Tiquia *et al.*, 2004). To determine appropriate hybridization conditions, hybridizations were carried out at 45–75°C and 50% formamide (Tiquia *et al.*, 2004). The optimum hybridization temperature was found to be 50°C and 50% formamide since this produced the highest signal intensities and fewest false positives; signal intensities decreased sharply at higher temperatures. High signal intensities were noted for target sequences with similarities to probe sequences >86% similarity. Hybridization sensitivity was determined to be approximately 60 ng of DNA.

The array had a sensitivity of 50–100 ng of DNA in the presence of pure culture DNA, the equivalent of ~5% of the community (Rhee *et al.*, 2004). Microbial communities from several contaminated sites and from contaminant degradation-based enrichment cultures were examined using the organic contaminant degradation and metal resistance array and genes involved in naphthalene degradation were detected in naphthalene-fed enrichment cultures and microcosms (Rhee *et al.*, 2004). The GeoChip 1.0 has been used in several studies which have demonstrated the usefulness of a comprehensive FGA (Wu *et al.*, 2006, 2008b; Waldron *et al.*, 2009).

While GeoChip 1.0 provided a broader coverage of functional genes than previous FGAs, the probe set was still relatively small. To provide a truly comprehensive probe set and to increase probe specificity for genes with high homologies, a new version, GeoChip 2.0, was developed (He *et al.*, 2007). GeoChip 2.0 contains 24,243 50-mer oligonucleotide probes, targeting ~10,000 functional genes from 150 gene families involved in the geochemical cycling of C, N, and P cycling, sulfate reduction, metal reduction and resistance, and organic contaminant degradation (He *et al.*, 2007). Approximately 80% of the probes are gene-specific with the remainder being group-specific. This version of GeoChip also contained control spots, such as 16S rRNA probes as positive controls, human gene probes as negative controls, and blanks. Computationally and experimentally determined specificity tests indicated that the array probes are expected to be highly specific to their matching target(s).

GeoChip 3.0, which covers 56,990 sequences from 292 gene families, has been reported and has several new and improved features (He *et al.*, 2010a). These features include expanded functional gene probe coverage, including addition of antibiotic resistance gene probes, the addition of probes for the phylogenetic marker gyrB, an expanded control probe set with positive controls specific to different regions of the 16S rRNA gene to assist with grid alignment, and negative control probes specific for human, plant, or hyperthermophiles genes. This array also includes a common oligo reference standard (CORS), which is co-spotted with all probes on the GeoChip (Liang *et al.*, 2010a). CORS is an artificial sequence specific only to a complementary oligonucleotide that is labelled with a fluorescent dye (different from that used for samples) and is added to all samples. The target CORS then hybridizes to the CORS probe which is present on each spot of the array and provides a way to normalize signal intensity and to objectively compare different microbial communities across various conditions and habitats. A supporting computational pipeline has also been developed for GeoChip probe design and data analysis.

A newer version, GeoChip 4.0, has been developed and is being tested. This version is being produced using the Nimblegen array format. This GeoChip version further expands the functional gene probe set and includes probes for virulence genes, stress response, and bacterial phage genes and now covers 410 functional gene families.

The GeoChip technology has been recognized as a groundbreaking and cutting edge technology. GeoChip was cited as an important tool for linking microbial communities to ecosystem functioning by the National Ecology Observatory Networks (NEON) Program in 2006. The Nature Publishing Group issued a Press Release to highlight this novel technology when the paper describing GeoChip 2.0 was published in the ISME Journal in May 2007. GeoChip was also highlighted in the Metagenomics Report by the National Academy of Sciences in 2007. Based on GeoChip 3.0, the OU GeoChip was also awarded a 2009 R&D100 Award by *R&D Magazine*, which recognizes the 100 most innovative scientific and technical breakthroughs of the year that show great promise for commercialization.

GeoChip development

Probe design

The first step in probe design is selection of genes to be included. The GeoChip arrays, for example, cover a wide variety of gene groups including those involved in biogeochemical cycling of C, N, S, P and metals, bioremediation of environmental contaminants, and antibiotic resistance. Global climate change is currently an important area of study in microbial ecology. GeoChip 3.0 has several gene sets that appear to play important roles in the response of microorganisms to climate change including several involved in N cycling: (i) *nifH* (dinitrogenase reductase) for N_2 fixation, (ii) *amoA* (ammonia monooxygenase) and *hao* (hydroxylamine oxidoreductase) for nitrification, (iii) *gdh* (glutamate dehydrogenase) and *ureC* (urease) for ammonification, (iv) *napA* (nitrate reductase) and *nrfA* (c-type cytochrome nitrite reductase) for dissimilatory N reduction to ammonium (DNRA), (v) *nasA* (nitrate reductase) and *nirA*/*nirB* (nitrite reductase) for assimilatory N reduction to ammonium, (vi) *narG* (nitrate reductase), *nirS*/*nirK* (nitrite reductase), norB (nitric oxide reductase), and *nosZ* (nitrous oxide reductase) for denitrification, and (vii) *hzo* (hydrazine oxidoreductase) for anaerobic ammonium oxidation (anammox) (He *et al.*, 2010a).

In addition, 41 genes for enzymes involved in C cycling are also present (He *et al.*, 2010a). These include those for:

1. CO_2 fixation: ribulose-l,5-bisphosphate carboxylase/oxygenase (Rubisco), carbon monoxide dehydrogenase (CODH), propionyl-CoA/acetyl-CoA carboxylase (PCC/ACC), ATP citrate lyase (AclB) for the reductive acetyl-CoA pathway.
2. Carbon degradation: eight enzymes for starch degradation, four enzymes for cellulose degradation, five enzymes for hemicellulose degradation, three enzymes for chitin degradation, one enzyme for pectin degradation and four lignin degradation.
3. Methanogenesis: *mcrA*, which encodes methyl coenzyme M reductase, a key enzyme for methanogenesis.
4. Methane oxidation: two enzymes for methanogenesis, methane monooxygenase (mmoX) and particulate methane monooxygenase (pmoA), are included.
5. Acetogenesis: formyltetrahydrofolate synthetase was included for autotrophic CO_2 fixation.
6. Other: eight miscellaneous carbon cycling enzymes are also covered, including isocitrate lyase (aceA), malate synthase (aceB) and alkylsuccinate synthase (assA).

While any functional gene can be included on an FGA, in general, genes for enzymes or proteins key to the function of interest are best and provide the greatest amount of information. If a multi-step process or protein complex is involved, genes for catalytic subunits or with the active site(s) should be chosen. As an example, the CnrCBA transporter complex, which confers Ni and Co resistance via efflux, is made up of three proteins (Liesegang *et al* 1993; Grass *et al* 2000; Tibazarwa *et al* 2000). The transporter itself has an outer membrane factor (OMF) (CnrC), which passes through the outer membrane, a trimer of resistance–nodulation–cell division (RND) protein family members (CnrA) which form a tube and a membrane fusion protein (MFP) (CnrB), which maintains the alignment of the OMF and the RND trimer (Saier *et al* 1994; Paulsen *et al* 1997; Nies *et al.*, 2003). Since the MFP (CnrB) is not actively involved in Ni or Co efflux or recognition, this gene would not be included, but the OMF (CnrC) and the RND proteins (CnrA) responsible for Ni and Co transportation and specificity would be included.

With the data analysis pipeline developed for the GeoChip 3.0, retrieval of sequences from a public database, such as NCBI GenBank and subsequent verification steps are automated. The first step is to select appropriate keywords to search. It is important to remember that genes may be annotated differently in different microorganisms or may have a more general or specific description so using very broad key words is often best. Examples of potential keywords include the name or abbreviation of the target gene/enzyme or the enzyme commission number. Once potential sequences are downloaded, seed sequences (SS) are selected. SS are those sequences which have had protein identity and function experimentally confirmed. The selection of SS is a critical step in the probe design process and should be chosen with extreme care. The SS are then used to evaluate the remaining downloaded sequences using HMMER alignment (http://hmmer.janelia.org; Eddy *et al.*, 1998). Based on seed sequence alignment, ClustalW (Thompson *et al* 1994) and HMM building software (Eddy *et al.*, 1998) are used to construct a profile hidden Markov model (HMM), which is then used to search against the remaining candidate sequences with both HMMER local and global algorithms. The resulting e-values are used to determine whether a sequence will be including using a cut-off *e*-value (e.g. 0.1) as these sequences are considered high-confidence targets for probe design. Sequences that do not meet the cut-off can be manually checked. Third, gene-specific or group-specific 50-mer oligonucleotide probes are designed with CommOligo (Li *et al.*, 2005) using experimentally determined criteria based on sequence homology (≤90% identity for gene-specific; ≥96% for group-specific), continuous stretch length (≤20 bases for gene-specific; ≥35 for group-specific), and free energy (≥−35 kJ/mol for gene-specific; ≤−60 kJ/mol for group-specific) (He *et al.*, 2005b; Liebich *et al.*, 2006). The designed probes are then screened against the GenBank database to ensure specificity. Lastly, the resultant probes are commercially synthesized and used for array construction.

The GeoChip 3.0 pipeline was executed using Perl scripts integrated with a web standard protocol, the Common Gateway Interface (CGI), and other open-source modules, including DBI and Bioperl (He *et al.*, 2010a). All sequence and probe data (e.g. retrieved sequences, verified sequences, seed sequences, key words, probes) are stored and managed in databases integrated with the pipeline.

Array construction

There are several methods of array printing that can be used. Most of these transfer probes to glass slides (Taroncher-Oldenburg *et al.*, 2003; Tiquia *et al.*, 2004; Rhee *et al.*, 2004) since they produce less background fluorescence (Schena *et al.*, 1995, 1996) and allow higher probe density (Ehrenreich, 2006), although nylon membranes can also be used (Steward *et al.*, 2004). Bubble Jet printing has been used to produce an array designed to detect single-nucleotide polymorphisms (Okamoto *et al.*, 2000). Probe (10- to 300-mer) solutions at concentrations of 0.02 to 1.6 mg/ml could be printed using this method with no loss of DNA or shearing of probes. Small spot sizes were obtained by using very small droplet sizes (24 pl). Laser-induced forward transfer (LIFT) can also be used for array fabrication (Serra *et al.*, 2004). With LIFT, lasers transfer material (i.e. array probes) from a film support, containing a thin film of the material, onto the array surface. Spot size could be varied by modulating the laser power. Spots as small as 40 μm could be produced using LIFT. Some commercially available array production incorporates probe synthesis with array printing. NimbleGen uses micromirror array synthesis, which synthesizes the probes directly onto the glass array surface; 786,000 micromirrors are used to deprotect the nascent oligonucleotide and allow addition of the next base (http://www.nimblegen.com/technology/manufacture.html). In addition, contact printing, which uses printing pins is often used in laboratory array printing.

Target preparation

High-quality DNA or RNA is extremely important for obtaining reliable microarray data. Microbial community DNA from environmental samples is generally extracted and purified using a well-established freeze-grind method since it results in large fragments of genomic DNA (Zhou *et al.*, 1996; Hurt *et al.*, 2001) which are important in subsequent DNA amplification. Ideally, the purified DNA should have absorbance ratios of A_{260}:A_{280} > 1.8 and A_{260}:A_{230} > 1.7. The A_{260}:A_{230} ratio has the most influence on hybridization success (Ning *et al.*, 2009). If sufficient quantities of high molecular weight DNA are available, agarose gel purification followed by phenol-chloroform–butanol extraction can provide high quality DNA (Liang *et al.*, 2009 a). Impurities in the DNA, such as humic acids or polysaccharides, can inhibit subsequent amplification, labelling and hybridization steps, so DNA purity is very important.

For GeoChip 3.0 and earlier versions, 2.0 μg of DNA is necessary for hybridization. If less DNA is obtained from a given sample, WCGA can be used to increase the amount of DNA available for hybridization using a small quantity of initial DNA (1–100 ng) (Wu *et al.*, 2006). WCGA provides a sensitive (10 fg detection limit) and representative amplification (<0.5% of amplified genes showed >2-fold different from unamplified) (Wu *et al.*, 2006). Less DNA is required for GeoChip 4.0 (~0.5 μg).

While DNA provides information on the community structure and functional potential of the microbial community, it does not indicate whether those functions are actively occurring. In order to monitor microbial community activity, mRNA can also be used. However, there are challenges in using RNA, including the relatively low abundance of mRNA in environmental samples and the short turnover rates of mRNA. Different methods can be used for RNA extraction from environmental samples, such as simultaneous extraction of DNA and RNA (Hurt *et al.*, 2001), RNA extraction via a bead-beating method (Burgmann *et al.*, 2003), or use of gel electrophoresis to isolate mRNA from total RNA (McGrath *et al.*, 2008). As with DNA, the purity of RNA is critical for microarray success and the purified RNA should have A_{260}:A_{280} > 1.90 and A_{260}:A_{230} > 1.70. A large quantity of RNA (10–20 μg) is required for hybridization since only a small portion of the total RNA is mRNA. However, environmental samples often do not yield a sufficient quantity of RNA, so whole community RNA amplification (WCRA) (Gao *et al.*, 2007) may be required. WCRA uses a fusion primer comprised of a short (6–9) set of random nucleotides and a T7 promoter. Amplification of 50–100 ng of total RNA provides a representative amplification that maintains the original relationship of mRNA abundance (Gao *et al.*, 2007). Another option to examine microbial community activity is to use stable isotope probing of active community members (Leigh *et al.*, 2007; see also Chapter 6).

After extraction, purification, and amplification, if necessary, the nucleic acids are labelled with fluorescent dye (e.g. Cy3, Cy5). For DNA, random priming with the Klenow fragment of DNA polymerase is used (Wu *et al.*, 2006) while RNA is labelled using Superscript™ II/III RNase H-reverse transcriptase (He *et al.*, 2005b). The labelled nucleic acids are then purified and dried for hybridization.

Hybridization

Labelled DNA or RNA is suspended in a hybridization buffer composed of salts, detergent, and formamide for hybridization. GeoChips are hybridized at 42–50°C and 40–50% formamide (He *et al.*, 2007; Mason *et al.*, 2008; Liang *et al.*, 2009a,b; Van Nostrand *et al.*, 2009; Waldron *et al.*, 2009). The hybridization temperature and formamide concentration can be adjusted to alter stringency in order to detect more or less diverse sequences. The effective hybridization temperature increases 0.6°C for every 1% of formamide.

Hybridizations can be performed manually or with automated or semiautomated hybridization stations. Manual hybridizations are performed using a water bath or hybridization oven and specially designed hybridization chambers that help maintain humidity levels around the microarray. Several hybridization stations provide incubation at controlled temperatures with mixing (e.g. Mai Tai® from SciGene, SlideBooster from Advalytix, Maui from BioMicro Systems). Post-hybridization washes can be performed manually or with an automated wash station (e.g. Maui Wash Station, BioMicro Systems). Other systems provide a completely automated system from pre-hybridization array preparation through post-hybridization washes (e.g. Tecan HS4800Pro, TECAN US).

Image analysis

Once hybridized, arrays that are constructed using contact printing are imaged using a microarray scanner with a resolution of 10 µm or better. Nimblegen arrays or other ultra high density arrays require a higher resolution (2–3 µm). The image is then digitally analysed by quantifying the pixel density (intensity) of each spot using microarray-analysis software. The analysis software can also be used to evaluate spot quality using predetermined criteria and flag poor or low quality spots for later removal. Distinguishing a positive spot from background noise is generally based on signal-to-noise ratio [SNR; SNR = (signal mean – background mean)/background standard deviation]. Other calculations can also be used such as the signal-to-background ratio (SBR; SBR = signal mean/background). Another ratio, the signal-to-both-standard-deviations ratio [SSDR; SSDR = (signal mean – background mean)/(signal standard deviation – background standard deviation)] has been developed which resulted in fewer false positives and negatives than the SNR calculation (He and Zhou, 2008).

This raw microarray data is then uploaded to the GeoChip data analysis pipeline (http://ieg.ou.edu/) for further evaluation. Spot quality, evenness of control spot hybridization signals across the slide surface, and background levels are determined. Poor and low quality spots are removed along with outliers [defined as those positive spots with (signal – mean signal intensity of all replicate spots)] greater than three times the replicate spots' signal standard deviation (He and Zhou, 2008)). The signal intensities are then normalized and the data is stored in an experiment database for further statistical analysis using the data analysis pipeline.

Data analysis

While achieving successful hybridizations can be difficult, the most difficult and time consuming task with FGAs, is data analysis due to the large amount of data obtained. Several data analysis methods have been used frequently with GeoChip data sets and include relative abundance of gene groups based on gene number or total signal intensity, richness and diversity indices based on gene number, per cent of gene overlap between samples, and response ratios. Methods commonly used for statistical analysis of microarray data include principal component analysis (PCA) or detrended correspondence analysis (DCA), cluster analysis (CA), and neural network analysis (NNA) (He *et al.*, 2008). PCA is a multivariate statistical method which reduces the dimensionality of variables to maximize the visible variability of the data. The major advantage of PCA is that it identifies outliers (e.g. genes) in the data set that behave differently from most of the genes across a set of experiments. DCA is an ordination technique that uses detrending to remove the arch effect typical in correspondence analysis (Hill and Gauch, 1980). CA is used to identify groups with similar gene profiles, and it can help establish functionally related groups of genes to gain insights into structure and function of a given microbial community. NNA is a relatively new analysis technique for GeoChip data but can be used to examine gene relationships based on the random matrix theory (RMT), a novel mathematical theory used to distinguish random and system-specific properties of complex systems (Mehta, 1990). In nature, biological community members form complex ecological networks through their interaction and exchange of energy, matter, and information (Montoya *et al.*, 2006). While these networks have been extensively studied in macroecology, little is known regarding ecological networks of microbial communities (Fuhrman and Steele, 2008; Raes and Bork, 2008). With the availability of relatively inexpensive metagenomic technologies like high-throughput sequencing and microarrays

(Handelsman *et al.*, 2007; He *et al.*, 2007, 2010a), scientists now have the extraordinary opportunity to examine network interactions within microbial communities (Raes and Bork, 2008). Response ratios, which compare community response (e.g. gene levels or signal intensity) between conditions (e.g. control vs. treatment, contaminated vs. uncontaminated) (Luo *et al.*, 2006), has been used to compare the community response to varying levels of oil contamination (Liang *et al.*, 2010b). In addition, if environmental variables are available, canonical correspondence analysis (CCA) (ter Braak, 1986), variation partitioning analysis (VPA) (Økland and Eilertsen, 1994; Ramette and Tiedje, 2007), and other correlation analyses (e.g. Mantel test) can be used to correlate environmental conditions with the community structure for further understanding of the relationship between the microbial community and ecosystem functioning. CCA has been used in several GeoChip-based studies to better understand how environmental factors are affecting community structure (Yergeau *et al.*, 2007; Wu *et al.*, 2008; Zhou *et al.*, 2008; Waldron *et al.*, 2009; Van Nostrand *et al.*, 2009). VPA is used to determine the relative influence of environmental parameters on the microbial community structure and is based on results of the CCA. The Mantel test has been used to correlate environmental factors with functional genes detected with GeoChip (He *et al.*, 2007; Wu *et al.*, 2008; Van Nostrand *et al.*, 2009; Waldron *et al.*, 2009).

Issues in microarray application

While GeoChips have allowed for the linking of microbial function with ecosystem processes, there are still limitations and challenges that must be overcome.

Nucleic acid quality

One of the most important steps for successful microarray analysis is obtaining high-quality DNA or/and RNA from environmental samples. Our lab has successfully used an established freeze-grind extraction method (Zhou *et al.*, 1996; Hurt *et al.*, 2004) followed by agarose gel purification (Liang *et al.*, 2009a). However, some samples are still difficult to purify to the necessary level. In addition, gel purification only works for fresh DNA and for samples which yield a relatively large amount of high molecular weight DNA, since portions of the DNA are lost with this method. Use of mRNA presents an even greater challenge as isolation of mRNA from environmental samples is difficult due to the low abundance and instability of the mRNA. A few studies have used FGAs for environmental mRNA analysis (Hurt *et al.*, 2001; Dennis *et al.*, 2003; Gao *et al.*, 2007).

Probe coverage

The GeoChip arrays have greatly expanded probe coverage in terms of both the number of genes covered and the number of gene variants (He *et al.*, 2007, 2010a). However, regardless of how comprehensive any array is now, sequences are constantly being added to the public databases, leading to an exponential increase in the number of functional genes and gene variants. As such, periodic updates of the GeoChip are necessary to maintain its comprehensive status. The GeoChip design pipeline has incorporated an automated update feature using predetermined keywords and seed sequences. Additional genes and gene categories are also continually being added in order to remain relevant to current research needs and interests. However, even with the advances in probe design software, this process is still time consuming due to the large number of sequences and probes that must be designed and tested.

Sensitivity

Sensitivity is a major concern with GeoChip or any FGA microarray, especially since these arrays are frequently used for analysis of complex environments where many gene variants are expected to be in low abundance. The current level of sensitivity for oligonucleotide arrays using environmental samples is approximately 50–100 ng or 10^7 cells (Wu *et al.*, 2001; Bodrossy *et al.*, 2003; Rhee *et al.*, 2004; Tiquia *et al.*, 2004; Chen *et al.*, 2009) or approximately 5% of the microbial community (Bodrossy *et al.*, 2003), providing coverage of only the most dominant community members. This is an important issue especially for most environmental samples from soils, groundwater, and marine water columns since the abundance in these samples is generally lower than the detection limit. As such, the use of microarrays for many environmental studies is very limited.

Several strategies have been utilized to increase sensitivity, the most important being the use of target DNA or RNA amplification (Wu *et al.*, 2006; Palka-Santini *et al.*, 2009, Goa *et al.*, 2007). As the name implies, whole community genome amplification (WCGA) amplifies all of the community DNA, including sequences in low abundance (Wu *et al.*, 2006a). WCGA has been shown to provide a representative, sensitive, and quantitative amplification of microbial community DNA using starting amounts of 1–250 ng. Use of amplified DNA for hybridization increased the sensitivity of GeoChip hybridization to 10 fg (equivalent to one or two cells) (Wu *et al.*, 2006a). Multiplex PCR, which uses primers for all genes covered by an array, has been used for some FGAs (Palka-Santini *et al.*, 2009), although this would be impractical for FGAs covering larger numbers of genes.

Several array modifications can increase sensitivity, although there is generally a payoff in decreased specificity, another important concern with microarrays. Increasing the length of oligonucleotide probes increases sensitivity (Denef *et al.*, 2003; He at el. 2005a), although this does decrease specificity (Relogio *et al.*, 2002). Another option is to increase the amount of probe per spot (Cho and Tiedje, 2002; Relogio *et al.*, 2002; Zhou and Thompson, 2002). This strategy is based on the fact that membrane-based arrays are more sensitive than glass-based arrays, probably because of the higher probe concentration (> 1 μg/spot for membranes; < 20 pg/spot for glass slides) (Cho and Tiedje, 2002). However, again, this decreases the array density and may actually decrease sensitivity by reducing overall signal intensity (Denef *et al.*, 2003).

Decreasing the volume of hybridization solution can also increase sensitivity. Wu *et al.* (2004) demonstrated that decreasing the hybridization buffer to 3 μl from the typical 15 μl increased the sensitivity from 5 ng in the presence of multiple DNA to 0.2 ng for pure culture DNA. Mixing during hybridization can also increase sensitivity with SNR values three times that of static hybridizations (Schaupp *et al.*, 2005). Hybridization results were consistent over a range of RNA concentrations (5–20 μg), indicating mixing during hybridization allows for detection of lower amounts of nucleic acid (Schaupp *et al.*, 2005).

Different labelling techniques can also increase the detection limit of GeoChip and other microarrays. Labelling with cyanine dye-doped nanoparticles (Zhou and Zhou, 2004) or tyramide signal amplification (Denef *et al.*, 2003) has been shown to provide up to a 10-fold increase in sensitivity. Improving air quality where sample preparation, hybridization and array scanning occur may improve sensitivity since ozone has been implicated in loss of cy-dye signal, especially Cy5 (Branham *et al.*, 2007; Byerly *et al.*, 2009). Ozone control systems are available for equipment, work spaces or entire rooms. Development of signal detection systems with higher sensitivities would also increase the detection sensitivity of GeoChips (Cho and Tiedje, 2002; Zhou and Thompson, 2002). In addition, because positive signals must be distinguished from background noise, increased sensitivity can be accomplished by reducing background signal on arrays. The use of unmodified glass slides for array construction (Kumar *et al.*, 2000; Gudnason *et al.*, 2008) may decrease background levels since unmodified glass has 4- to 12-fold less background intensity (Kumar *et al.*, 2000). Modifications can also be made to the calculations used to distinguish true signals from background (He and Zhou, 2008).

Specificity

Specificity is an important attribute of gene probes, especially for those designed to analyse environmental samples. However, a difficulty in designing specific probes is that so many environmental sequences are unknown, and that many homologous genes are highly similar. Careful design of oligonucleotide probes can provide highly specific hybridizations. Criteria based on similarity, stretch and free energy have been used to design specific probes (He *et al.*, 2005a, 2007; Liebich *et al.*, 2006). Our evaluation of GeoChip 2.0 probes designed using these criteria revealed that a minimal number of false positives (0.002–0.004%) were observed (He *et al.*, 2007).

In addition to probe design criteria, specificity

can be adjusted by changing the hybridization conditions to increase or decrease stringency. Hybridization stringency is generally controlled by temperature or formamide concentration. At 65°C hybridization occurred for sequences with similarities >87% (Wu *et al.*, 2001). At 45°C, hybridization occurred for sequences with similarities as low as 70–75% (Wu *et al.*, 2001). Hybridizations using 50-mer oligonucleotide FGAs at 50°C and 50% formamide (effective temperature, 80°C) were able to discriminate sequences with <88 to 94% similarities (Rhee *et al.*, 2004; Leibeich *et al.*, 2006; Deng *et al.*, 2008).

Signal intensity can be affected by sequence divergence as well as by sequence abundance; therefore, strategies need to be developed to determine which condition is occurring. One option to determine a true signal is the use of mismatched probes. Deng *et al.* (2008) found that probes with mismatches (3 to 5) distributed evenly across the probe were better able to distinguish perfect-matched targets versus mismatched targets than randomly distributed mismatched nucleotides. In addition, using relative comparisons across samples (i.e. comparing signal intensities from test samples to a control or background sample) rather than absolute comparisons will minimize or eliminate the effects of potential cross-hybridization (He *et al.*, 2007). Assuming that test and control samples have similar community composition, using the ratios of test to control samples will cancel out any cross-hybridization (He *et al.*, 2007).

Quantitative

A major goal for microarray analysis is to be able to provide quantitative information. Some studies have shown a correlation between signal intensity and DNA concentration. PCR-based probes showed a correlation ($r=0.94$) between signal intensity and DNA quantity over the concentration range of 0.5 to 100 ng (Wu *et al.*, 2001). Oligonucleotide probes (50-mer) provided a linear relationship ($r=0.98$–0.99) over a concentration range of 8–1000 ng (Tiquia *et al.*, 2004). Hybridizations with RNA have also been shown to be linear over the range of 50–100 ng (Gao *et al.*, 2007).

Activity

Using DNA with FGAs provides information on population changes which can be used to infer microbial activity, but it does not provide absolute evidence of that activity. The use of community mRNA would provide information on which community members or functional processes are active, similar to transcriptional arrays for pure cultures (Dennis *et al.*, 2003; Bodrossy *et al.*, 2006; Gao *et al.*, 2007). However, our current difficulties with community mRNA are the low amount and the stability of environmental mRNA. WCRA overcomes the problem of low abundance (Gao *et al.*, 2007) although the current amplification protocol is complex and time consuming. Therefore, improved methods for RNA extraction from environmental samples and mRNA amplification are needed. Another option for determining microbial activity is SIP analysis (Leigh *et al.*, 2007; see also Chapter 6).

Application of GeoChip

GeoChips have been used to examine a variety of microbial habits including groundwater, lakes, oceans, river sediments, forest and agricultural soils, grasslands, and metal, oil, or pesticide contaminated sites. While the majority of these studies have focused on community DNA to examine changes in community structure and function, RNA or stable isotope probing (SIP) have been examine community activity.

Metal contamination

GeoChip has been used in several studies to examine uranium contaminated groundwater at the US DOE's Oak Ridge, TN Integrated Field Research Challenge (OR-IFRC) site. GeoChip 1.0 (Rhee *et al.*, 2004; Tiquia *et al.*, 2004), was used to examine communities within the OR-IFRC, which is contaminated with high amounts of nitrate, uranium, and organic compounds. This study observed higher gene numbers in uncontaminated sites compared to contaminated sites, indicating the deleterious effect of the contaminants on the microbial communities (Wu *et al.*, 2006). Using the same array, a later study examined samples with a range of contamination levels and similarly found that more genes were detected from the uncontaminated control site compared to contaminated wells (Waldron *et al.*, 2009). This site has also been used to study various remediation techniques and GeoChip 2.0 was used to monitor changes in the microbial communities within a pilot-scale test system established for the biostimulation of U(VI) (soluble uranium) reduction in the subsurface by injection of ethanol. A significant correlation ($r = 0.73$, $P < 0.05$) was observed between the U(VI) concentration and the amount of cytochrome genes detected, indicating the importance of cytochrome-containing microorganisms in U(VI) reduction at this site (He *et al.*, 2007). In the same bioremediation system, the effects of dissolved oxygen (DO) and ethanol amendment on the microbial community were examined, and the results showed that ethanol was a much stronger driver in controlling community structure than U(VI) or DO (Van Nostrand *et al.*, 2009).

Amplified community mRNA was used to monitor the activity of microbial communities within a denitrifying fluidized bed reactor (FBR) designed to reduce the nitrate concentration in order to provide conditions conducive to U(VI) reduction (Gao *et al.*, 2007). Genes for nitrate and nitrite reduction, several genes for organic contaminant degradation, dsr, and polyphosphate kinase were detected as would be expected from this site.

GeoChip 3.0 was used to understand how the arsenic accumulator *Pteris vittata* rhizosphere microbial communities respond to arsenic contamination (Xiong *et al.*, 2010). Populations of arsenic resistance, sulfur reduction, phosphorus utilization and denitrification genes were markedly different between *P. vittata* rhizosphere and non-rhizosphere soils, suggesting a strong linkage between the level of arsenic contamination in the soil and the rhizosphere community and functional gene distribution. CCA revealed that arsenic is the primary driver in reducing the functional gene diversity in the soil community, suggesting that rhizobacteria may play an important role in soil arsenic uptake and hyperaccumulation in *P. vittata* (Xiong *et al.*, 2010).

GeoChip 2.0 and Phylochip were used to examine horizontal gene transfer (HGT) among microorganisms from a hypersaline lake [Great Salt Lake (GSL), UT, USA] (Parnell *et al.*, 2010). GSL contains high concentrations of sulfate and heavy metals and has a range of NaCl concentrations from near seawater to saturation. A high diversity of dsr and chromium resistance gene variants were observed in all samples, suggesting that high selection pressure from the high levels of chromium and sulfate across the lake resulted in an increase in gene transfer (Parnell *et al.*, 2010).

Hydrocarbons and other organic contaminants

The microbial community of a diesel fuel bioremediation system in Vega Baja, Puerto Rico was examined with GeoChip 2.0 (Rodríguez-Martínez *et al.*, 2006). Genes involved in the degradation of diesel and organic contaminants (phthalate, biphenyl, cyclohexanol, benzoate, and naphthalene degradation genes) were detected. The amount of anaerobic degradation genes increased over time, suggesting that, and consistent with other evidence, the system shifted to an anaerobic process. Liang *et al.* (2009b) used GeoChip 2.0 to examine the microbial community of contaminated oil fields before and after bioremediation treatment. Ozonation treatment resulted in a decrease in most functional gene categories, including carbon, nitrogen, and sulfur cycling and organic contaminant degradation genes, but all gene categories recovered after treatment. Another study examined microbial communities from a contaminated oil field in China and found that higher levels of oil contamination resulted in lower diversity and a decreased number of functional genes detected (Liang *et al.*, 2009a). In addition, genes associated with oil degradation, including catechol, protocatechuate, biphenyl degradation, increased under a moderate oil contamination level.

Another study related to contaminated oil fields examined microbial communities along a contaminant gradient. Functional gene diversity and number decreased as the contaminant level increased (Liang *et al.*, 2009a). However, under a moderate level of contamination, the number of genes related to organic contaminant degradation did increase. Genes within this group included those involved in degradation of catechol, protocatechuate, and biphenyl. When environmental parameters were evaluated, oil concentration and soil available nitrogen were strong drivers in determining community structure.

Leigh *et al.* (2007) used GeoChip 2.0 in combination with stable isotope probing to monitor active microbial populations from a polychlorinated biphenyl (PCB) contaminated site. Microcosms were set up from soil samples and fed ^{13}C-labelled or unlabelled biphenyl and sampled at 1, 4 and 14 days. The extracted DNA was then used for hybridization onto the GeoChip. Genes involved in biphenyl degradation were detected as were other organic contaminant degradation genes including those for degradation of benzoate, catechol, naphthalene, and phenol.

Liebich *et al* (2009) used GeoChip 2.0 to examine a background and three atrazine contaminated aquifers. More genes were detected in the background site than in the contaminated sites. Among the contaminated sites, the aquifer with the highest level of contamination had the highest number of genes detected, with a majority involved in contaminant degradation. Genes involved in atrazine degradation were detected in all contaminated wells and the presence of these genes was validated by PCR. These results indicated that even small amounts of contaminants were enough to select for specific degrading populations.

Industrial pollutant and pesticide contaminated river sediments were characterized with the GeoChip 2.0 and the results indicated that contaminant level was not a major driver in these systems (Taş *et al.*, 2009). Rather, C/N ratio, depth, total Kjeldahl N, and location were the strongest drivers in determining the community structure. Results also indicated that most reductive dehalogenation genes detected from the river sediments were from *Dehalococcoides* spp., suggesting that this microorganism may play an important role in contaminant degradation in this system.

GeoChip 4.0 was used, along with the PhyloChip, to examine microbial communities in oil-contaminated water associated with the 2010 oil spill in the Gulf of Mexico (Hazen *et al.*, 2010). Results from both arrays showed that the dispersed hydrocarbon plume stimulated deep-sea (1100 m) indigenous microorganisms closely related to known petroleum-degraders. The presence of the oil resulted in a significant shift in the microbial community functional composition and structure after only ~ 40 days. Additionally, metabolic genes involved in aerobic and anaerobic hydrocarbon degradation were enriched in plume samples, indicating a potential for intrinsic bioremediation of the oil in the deep-sea (Lu *et al.*, unpublished). These findings suggest that the metabolically diverse marine microbial communities appeared to be undergoing a rapid and dynamic adaptation in response to the oil spill. These microorganisms may play significant roles in controlling the fate of deep-sea oil spills.

Marine environments

Sediments from the Gulf of Mexico were examined using GeoChip 1.0 (Wu *et al.*, 2008). Genes for carbon degradation, nitrification, denitrification, nitrogen fixation, sulfur reduction, phosphorus utilization, contaminant degradation, and metal resistance were detected and communities become more distinctive as depth increased. The environmental parameters, depth, porosity, and concentrations of ammonium, phosphate, Mn(II), and silicic acid appeared to be important drivers in determining the structure of the microbial communities in this environment. Another study characterized microbial communities from deep sea hydrothermal vents, including a mature chimney and the inner and outer portions of a 5-day-old chimney with GeoChip 2.0, and the results showed communities from the inner chimney were less diverse than those from the outer portion of the 5-day-old chimney or the mature chimney (Wang *et al.*, 2009). GeoChip 2.0 has also been used to examine microbial communities from deep sea basalt and genes involved in carbon fixation, methane oxidation, methanogenesis, and nitrogen fixation, processes not previously associated with this environment, were detected (Mason *et al.*, 2009).

GeoChip 2.0 was used to examine the functional structure of microbial communities associated with the surface mucopolysaccharide layer and tissue of healthy and yellow band diseased coral, *Montastraea faveolata* (Kimes *et al.*, 2010). The number and signal intensity of genes involved in cellulose degradation and nitrification were increased in diseased corals, suggesting that these processes may provide a competitive advantage to coral pathogens.

In addition to these community DNA-based studies, several other studies have used GeoChip 2.0 to examine pure cultures and microbial activity within communities. Van Nostrand *et al.* (2007) used GeoChip 2.0 to probe four metal-resistant (Ni, Co, Cd, Zn) actinomycetes for metal resistance genes. Multiple metal resistance genes were detected including some implicated in Ni, Co, Cd, and Zn resistance. Leigh *et al.* (2007) used stable isotope probing (biphenyl) in conjunction with GeoChip to detect active PCB-degrading microbial populations within a hydrocarbon-contaminated aquifer. Gao *et al.* (2007) used amplified community mRNA to examine the activity of microbial communities from a denitrifying fluidized bed reactor at a uranium contaminated site. Genes expected to be active at this site were detected including nitrate reduction genes and several organic contaminant degradation genes.

Soil microbial communities

GeoChip 1.0 was used to examine functional changes of microbial communities under different land use strategies (a primitive fir forest, two spruce plantations, established for ~20 or ~ 40 years, and cropland) and found that functional diversity and functional gene numbers of the disturbed sites (spruce plantations and cropland) increased as soil organic carbon increased (Zhang *et al.*, 2007). Microbial communities associated with Antarctic sediments were characterized using GeoChip 2.0 to better understand N and C cycling in this environment (Yergeau *et al.*, 2007). Canonical correspondence analysis (CCA) was used on individual gene groups (C-cycling, CH_4-cycling, and N-cycling genes) to determine what environmental factors had the most significant effect on the community structure. For all gene groups pH and the number of freeze–thaw cycles were significant factors. In addition, the C:N ratio had a significant effect on the C-cycling and N-cycling functional genes. In addition soil temperature was found to be positively correlated with cellulose degradation and denitrification genes.

GeoChip has also been used to address high-level ecological issues, such as understanding the spatial pattern of organisms and the underlying mechanisms shaping biotic communities. GeoChip 2.0 was used to determine spatial scaling differences across various functional and phylogenetic groups at the whole-community level (Zhou *et al.*, 2008). The results indicated that the forest soil microbial community exhibited a relatively flat gene–area relationship, but varied considerably across different functional and phylogenetic groups. The results revealed that the turnover in space of microorganisms might be, in general, lower than that of plants and animals.

Other systems

GeoChip was used to examine the microbial community functional structure of several microbial electrolysis cells (MEC), a bioelectrochemical system that can produce hydrogen from acetate using wastewater as inoculum, operated under different conditions (Liu *et al.*, 2010). Highest hydrogen yields were positively correlated with microbial functional diversity. Additionally, several well-known exoelectrogenic bacteria (e.g. *Geobacter, Shewanella, Desulfovibrio, Anaeromyxobacter*) were detected in all reactors. Results indicated that reactor operation altered the microbial community composition and resulted in community shifts that were related to Coloumbic efficiency and COD removal, suggesting that conditions affecting MEC performance resulted in significant changes to the microbial community composition (Liu *et al.*, 2010).

Climate and global change science

A central issue in ecology is the understanding of the response of biological systems to elevated atmospheric CO_2. This understanding also has important consequences for society as we face the current and future effects of global climate change. One poorly understood area in the response of below-ground microbial communities to elevated CO_2 and how treatment with fertilizers will ultimately impact carbon loss or sequestration in terrestrial ecosystems. To better understand this system, the response of soil microbial communities to elevated CO_2 in a grassland system was examined using GeoChip

3.0 (He *et al.*, 2010b). The study was conducted at a multifactor grassland experiment site, BioCON (Biodiversity, CO_2, and Nitrogen deposition) in the Cedar Creek Ecosystem Science Reserve, MN (Reich *et al.*, 2001). After a 10-year field exposure of grass species to elevated CO_2, results of GeoChip analysis indicated that elevated CO_2 significantly altered the below-ground microbial community functional structure based on DCA ordination plots of GeoChip data as well as 16S rRNA gene-based pyrosequencing data. Genes involved in nitrogen and carbon fixation significantly increased under elevated CO_2, while genes involved in degradation of recalcitrant carbon remained unchanged. These community changes were significantly correlated to soil carbon and nitrogen content and to plant productivity. These results provide important information regarding the feedback response of ecosystems to atmospheric CO_2 and to global climate change modelling. Analyses of the network structure of these microbial communities indicated that elevated CO_2 dramatically altered the network interactions of the below-ground microbial communities and that the network connectivity significantly correlated with soil carbon and nitrogen dynamics and plant productivity (Zhou *et al.*, unpublished).

Another concurrent study at the BioCON site examined how plant mono-cultures, an important global agricultural trend, affects the functional diversity, composition and structure of soil microbial communities (He *et al.*, 2010b). This study indicated that microbial community structure was significantly different between mono-species (one species) and multispecies (four, 9 and 16 species) plots as indicated by DCA ordination plots of all genes detected by GeoChip), suggesting that plant mono-cultures may lead to a decrease in soil microbial diversity.

Another study was conducted at a long-term experimental warming site in a tall grass prairie ecosystem in the US Great Plains of Central Oklahoma to explore microbial-mediated carbon cycle feedback in response to an 8-year period of warming by approximately 2°C, and the results indicated that microorganisms are critical to the regulation of ecosystem carbon cycling in response to climate warming (Zhou *et al.*, unpublished).

Conclusions

The past decade has brought great advances in microarray technology and in FGA development. The GeoChip arrays have garnered a great deal of attention, and numerous studies over the last several years have demonstrated its applications in the study of microbial ecology and linking microbial communities with geochemistry. Several recent studies have also shown that the GeoChip can be used to address high level ecological questions pertaining to global climate change and to ecological theory. The GeoChip is able to provide sensitive, specific, and potentially quantitative information concerning microbial communities from a wide range of environments and is expected to revolutionize the field of microbial ecology and the study of microbial community functional structure and dynamics.

However, there are still many challenges in regards to technical, experimental and analysis that need to be overcome. These include increasing sensitivity either through new technologies and methods in array printing or novel and/or improved target labelling methods. Strategies must be developed to improve the quantitative accuracy of FGA hybridizations. Bioinformatic tools and techniques are needed to assist in the analysis, evaluation, and interpretation of the vast amounts of data obtained from GeoChip analysis. New tools are also needed for sequence retrieval, evaluation, and probe design to speed these processes up. Novel analytical techniques are also needed to fully utilize the FGA data. Finally, strategies and techniques must be developed in order to be able to compare data sets across samples, experiments, and labs.

Future trends

An important application of FGAs is in the field of pathogen detection either in clinical settings as a diagnostic tool or in food and water safety testing. As described earlier several preliminary pathogen detection arrays have been designed to detect virulence genes (Kostić *et al.*, 2005; Cleven *et al.*, 2006; Miller *et al.*, 2008; Palka-Santini *et al.*, 2009), antibiotic resistance (Call *et al.*, 2003), or a DNA fingerprinting array using short (nonamer), randomly generated probes (Kingsley *et al.*, 2002). These arrays show promise as future diagnostic and testing tools but much work is needed to expand the array coverage and to increase sensitivity. There have also been advances in hybridization technology. A microfluidic device has been developed which allows hybridization in 15 min which would easily allow for clinical applications (Peytavi *et al.*, 2005). Hybridization with this device allowed for discrimination between 4 Staphylococcus strains (Peytavi *et al.*, 2005).

As a general rule, microarrays cannot be used to find new sequences. Samples are interrogated against a set of known sequences so sequences that were not covered by the array could not be detected. This is the microarray's major weakness when compared to other high-throughput metagenomics technologies like pyrosequencing. However, several recent publications have described methods whereby microarrays can be used to identify novel sequences. The first is special sequence capture arrays, which are designed to hybridize sequences. The first sequence capture array was developed for human genomes (Albert *et al* 2007; Okou *et al.*, 2007). The array contained specially designed oligonucleotide probes (>60-mer) covering exon regions of the human genome. With these arrays, fragmented DNA is hybridized to the array for >60 hours and the hybridized or 'captured' DNA is eluted using water, amplified, and sequenced using 454 pyrosequencing. This method did result in the detection of gene variants. Since this first report, several studies have been published using this method, although most have focused on human of mouse DNA. The capture array technique has also been tested for the bacterium *Escherichia coli* K12 MG1655 (Schracke *et al.*, 2009). The researchers used a microfluidic DNA array (Bau *et al.*, 2009), designed to capture sequence variants from 68 *E. coli* genes and used a 6 bp tiling approach (Schracke *et al.*, 2009). Results indicated that this method was reproducible and provided high data fidelity. This is an intriguing technology which has the potential to overcome one of the disadvantages of microarray technology – the inability to detect new sequences.

A virus array has also been developed that was able to identify a novel viral pathogen (Ksiazek *et al.*, 2003; Wang *et al.*, 2002). The array is comprised of 70-mer oligonucleotide probes specific for various sequenced viral families and covers hundred of viruses (Wang *et al.*, 2002). Novel viruses can be detected based on hybridization patterns.

Acknowledgements

This work has been partially supported through ENIGMA, which is supported by the Office of Science, Office of Biological and Environmental Research, of the US Department of Energy under Contract No. DE-AC02-05CH11231, Environmental Remediation Science Program (ERSP) Programs, Oklahoma Applied Research Support (OARS), Oklahoma Centre for the Advancement of Science and Technology (OCAST), the Oklahoma Bioenergy Centre (OBC), the State of Oklahoma through the Project AR062-034.

Web resources

http://ieg.ou.edu/

References

Albert, T.J., Molla, M.N., Muzny, D.M., Nazareth, L., Wheeler, D., Song, X., Richmond, T.A., Middle, C.M., Rodesch, M.J., Packard, C.J., Weinstock, G.M., and Gibbs, R.A. (2007). Direct selection of human genomic loci by microarray hybridization. Nature Meth. 4, 903–905.

Amann, R., Ludwig, W., and Schleifer, K. (1995). Phylogenetic identification and *in situ* detection of individual microbial cells without cultivation. Microbiol. Rev. 59, 143–169.

Bau, S., Schracke, N., Kränzle, M., Wu, H., Stähler, P.F., Hoheisel, J.D., Beier, M., and Summerer, D. (2002). Targeted next-generation sequencing by specific capture of multiple genomic loci using low-volume microfluidic DNA arrays. Anal. Bioanal. Chem. 393, 171–175.

Bodrossy, L., Stralis-Pavese, N., Konrad-Köszler, M., Weilharter, A., Reichenauer, T.G., Schöfer, and Sessitsch, A. (2006). mRNA-based parallel detection of active methanotroph populations by use of a diagnostic microarray. Appl. Environ. Microbiol. 72, 1672–1676.

Bodrossy, L., Stralis-Pavese, N., Murrell, J.C., Radajewski, S., Weilharter, A., and Sessitsch, A. (2003). Development and validation of a diagnostic microbial microarray for methanotrophs. Environ. Microbiol. 5, 566–582.

Bontemps, C., Goldier, G., Gris-Liebe, C., Carere, S., Talini, L., and Boivin-Masson, C. (2005). Microarray-based detection and typing of the rhizobium nodulation gene nodC: Potential of DNA arrays to diagnose biological functions of interest. Appl. Environ. Microbiol. *71*, 8042–8048.

Branham, W., Melvin, C., Han, T., Desai, V., Moland, C., Scully, A., and Fuscoe, J. (2007). Elimination of laboratory ozone leads to a dramatic improvement in the reproducibility of microarray gene expression measurements. BMC Biotechnol. *7*, 8.

Brodie, E.L., DeSantis, T.Z., Joyner, D.C., Baek, S.M., Larsen, J.T., Andersen, G.L., Hazen, T.C., Richardson, P.M., Herman, D.J., Tokunaga, T.K., *et al.* (2006). Application of a high-density oligonucleotide microarray approach to study bacterial population dynamics during uranium reduction and reoxidation. Appl. Environ. Microbiol. *72*, 6288–6298.

Brodie, E.L., DeSantis, T.Z., Moberg Parker, J.P., Zubietta, I.X., Piceno, Y.M., and Andersen, G.L. (2007). Urban aerosols harbor diverse and dynamic bacterial populations. Proc. Nat. Acad. of Sci. USA. *104*, 299–304.

Burgmann, H., Widmer, F., Sigler, W.V., and Zeyer, J. (2003). mRNA extraction and reverse transcription-PCR protocol for detection of nifH gene expression by Azotobacter vinelandii in soil. Appl. Environ. Microbiol. *69*, 1928–1935.

Byerly, S., Sundin, K., Raja, R., Stanchfield, J., Bejjani, B.A., and Shaffer, L.G. (2009). Effects of ozone exposure during microarray posthybridization washes and scanning. J. Mol. Diag. *11*, 590–597.

Call, D.R., Bakko, M.K., Krug, M.J., and Roberts, M.C. (2003). Identifying antimicrobial resistance genes with DNA microarrays. Antimicrob. Agents Chemother. *47*, 3290–3295.

Chen, Q., Yin, H., Luo, H., Xie, M., Qiu, G., and Liu, X. (2009). Micro-array based whole-genome hybridization for detection of microorganisms in acid mine drainage and bioleaching systems. Hydrometallurgy. *95*, 96–103.

Cho, J.C., and Tiedje, J.M. (2002). Quantitative detection of microbial genes by using DNA microarrays. Appl. Environ. Microbiol. *58*, 1425–1430.

Cleven, B.E.E., Palka-Santini, M., Gielen, J., Meembor, S., Kronke, M., and Krut, O. (2006). Identification and characterization of bacterial pathogens causing bloodstream infections by DNA microarray. J. Clin. Microbiol. *44*, 2389–2397.

Denef, V.J., Park, J., Rodrigues, J.L.M., Tsoi, T.V., Hashsham, S.A., and Tiedje, J.M. (2003). Validation of a more sensitive method for using spotted oligonucleotide DNA microarrays for functional genomics studies on bacterial communities. Environ. Micrbiol. *5*, 933–943.

Deng Y., He, Z., Van Nostrand J.D., and Zhou, J. (2008). Design and analysis of mismatch probes for long oligonucleotide microarrays. BMC Genomics. *9*, 491–503.

Dennis, P., Edwards, E.A., Liss, S.N., and Fulthorpe. (2003). Monitoring gene expression in mixed microbial communities by using DNA microarrays. Appl. Environ. Microbiol. *69*, 769–778.

DeSantis, T.Z., Brodie, E.L., Moberg, J.P., Zubieta, I.X., Piceno, Y.M., and Andersen, G.L. (2007). High-density universal 16S rRNA microarray analysis reveals broader diversity than typical clone library when sampling the environment. Microbial Ecol. *53*, 371–383.

Dong, Y., Glasner, J.D., Blattner, F.R., and Triplett, E.W. (2001). Genomic interspecies microarray hybridization: rapid discovery of three thousand genes in the maize Endophyte, Klebsiella pneumoniae *342*, by microarray hybridization with *Escherichia coli* K-12 open reading frames. Appl. Environ. Microbiol. *67*, 1911–1921.

Dunbar, J., Barns, S.M., Ticknor, L.O., and Kuske, C.R. (2002). Empirical and theoretical bacterial diversity in four arizona soils. Appl. Environ. Microbiol. *68*, 3035–3045.

Eddy, S.R. (1998). Profile hidden markov models. Bioinformatics *14*, 755–63.

Ehrenreich, A. (2006). DNA microarray technology for the microbiologist: an overview. Appl. Microbiol. Biotechnol. *73*, 255–273.

Fitter, A.H., Gilligan, C.A., Hollingworth, K., Kleczkowski, A., Twyman, R.M., Pitchford, J.W., and The Members of the NERC Soil Biodiversity Programme. (2005). Biodiversity and ecosystem function in soil. Funct. Ecol. *19*, 369–377.

Flanagan, J.L., Brodie, E.L., Weng, L., Lynch, S.V., Garcia, O., Brown, R., Hugenholtz, P., DeSantis, T.Z., Andersen, G.L., Wiener-Kronish, J.P., *et al.* (2007). Loss of bacterial diversity during antibiotic treatment of intubated patients colonized with *Pseudomonas aeruginosa*. J. Clin. Microbiol. *45*, 1954–1962.

Fuhrman, J.A., and Steele, J.A. (2008). Community structure of marine bacterioplankton: patterns, networks, and relationships to function. Aquat. Microb. Ecol. *53*, 69–81.

Gao, H., Yang, Z.K., Gentry, T.J., Wu, L., Schadt, C.W., and Zhou, J. (2007). Microarray-based analysis of microbial community RNAs by whole-community RNA amplification. Appl. Environ. Microbiol. *73*, 563–571.

Gentry, T.J., Wickham, G.S., Schadt, C.W., He, Z., and Zhou, J. (2006). Microarray application in microbial ecology research. Microbial. Ecol. *52*, 159–175.

Grass, G., Groe, C., and Nies, D.H. (2000). Regulation of the cnr cobalt and nickel resistance determinant from Aalstonia sp. Strain CH34. J. Bacteriol. *182*, 1390–1398.

Gudnason, H., Dufva, M., Bang, D.D., Wolff, A. (2008). An inexpensive and simple method for thermally stable immobilization of DNA on an unmodified glass surface: UV linking of poly(t)10-poly©10-tagged DNA probes. BioTechniques. *45*, 261–271.

Guschin, D.Y., Mobarry, B.K., Proudnikov, D., Stahl, D.A., Rittmann, B.E., and Mirzabekov, A.D. (1997). Oligonucleotide microchips as genosensors for determinative and environmental studies in microbiology. Appl. Environ. Microbiol. *63*, 2397–2402.

Handelsman, J., Tiedje, J.M., Alvarez-Cohen, L., Ashburner, M., Cann, I.K.O., DeLong, E.F., Doolittle, W.F., Fraser-Liggett, C.M., Godzik, A., Gordon, J.I., *et al.* (2007). Committee on Metagenomics: Challenges and functional applications. (Washington, D.C.: National Academy of Sciences).

Hazen, T.C., Dubinsky, E.A., DeSantis, T.Z., Andersen, G.L., Piceno, Y.M., Singh, N., Jansson, J.K., Probst, A., Borglin, S.E., Fortney, J.L. *et al.* (2010). Deep-sea oil plume enriches indigenous oil-degrading bacteria. Science. *330*, 204–208.

He, Z., Wu, L., Fields, M.W., and Zhou, J. (2005a). Use of microarrays with different probe sizes for monitoring gene expression. Appl. Environ. Microbiol. *71*, 5154–5162.

He, Z., Wu, L.Y., Li, X.Y., Fields, M.W., and Zhou, J.Z. (2005b). Empirical establishment of oligonucleotide probe design criteria. Appl. Environ. Microbiol. *71*, 3753–3760.

He, Z., Gentry, T.J., Schadt, C.W., Liyou Wu, L., Jost Liebich, J., Song C Chong, S.C., Zhijian Huang, Z., Weimin Wu,W., Baohua Gu, B., Phil Jardine P. *et al.* (2007). GeoChip: a comprehensive microarray for investigating biogeochemical, ecological and environmental processes. ISME J. *1*, 67–77.

He, Z., and Zhou, J. (2008). Empirical evaluation of a new method for calculating signal to noise ratio (SNR) for microarray data analysis. *Appl. Environ. Microbiol. 74, 2957–2966.*

He, Z., Van Nostrand, J.D., Wu, L., Zhou, J. (2008). Development and application of functional gene arrays for microbial community analysis. T. Nonferr. Metal Soc. *18*, 1319–1327.

He, Z., Deng, Y., Van Nostrand, J.D., Tu, Q., Xu, M., Hemme, C.L., Li, X., Wu, L., Gentry, T.J., Yin, Y., Liebich, J., Hazen, T.C., and Zhou, J. (2010a). Geochip 3.0 as a high-throughput tool for analyzing microbial community composition, structure and functional activity. ISME J. *4*, 1167–1179.

He, Z., Xu, M., Deng, Y., Kang, S., Kellogg, L., Wu, L., Van Nostrand, J.D., Hobbie, S.E., Reich, P.B., and Zhou, J. (2010b). Metagenomic analysis reveals a marked divergence in the structure of below-ground microbial communities at elevated CO_2. Ecol. Lett. *13*, 564–575.

Hill, M.O., and Gauch, H.G. (1980). Detrended correspondence analysis: an improved ordination technique. Vegetatio. *42*, 47–58.

Hong, S.-H., Bunge, J., Jeon, S.-O., and Epstein, S.S. (2006). Predicting microbial species richness. Proc. Nat. Acad. Sci. USA. *103*, 117–122.

Hugenholtz, P., Goebel, B.M., and Pace, N.R. (1998). Impact of culture-independent studies on the emerging phylogenetic view of bacterial diversity. J. Bacteriol. *180*, 4765–4774.

Hurt, R.A., Qiu, X., Wu, L., Roh, Y., Palumbo, A.V., Tiedje, J.M., and Zhou, Z. (2001). Simultaneous recovery of RNA and DNA from soils and sediments. Appl. Environ. Microbiol. *67*, 4495–4503.

Jenkins, B.D., Steward, G.F., Short, S.M., Ward, B.B., and Zehr, J.P. (2004). Fingerprinting diazotroph communities in the Chesapeake Bay by using a DNA macroarray. Appl. Environ. Microbiol. *70*, 1767–1776.

Kingsley, M.T., Straub, T.M., Call, Daly, D.R., D.S., Wunschel, S.C., and Chandler, D.P. (2002). Fingerprinting closely related Xanthomonas pathovars with random nonamer oligonucleotide microarrays. Appl. Environ. Microbiol. *68*, 6361–6370.

Kimes, N.E., Van Nostrand, J.D., Weil, E., Zhou, J., and Morris, P.J. (2010). Microbial functional structure of montastraea faveolata, an important caribbean reef-building coral, differs between healthy and yellow-band diseased colonies. Environ. Microbiol. *12*, 541–556.

Kostic, T., Weilharter, A., Sessitsch, A., and Bodrossy, L. (2005). High-sensitivity, polymerase chain reaction-free detection of microorganisms and their functional genes using 70-mer oligonucleotide diagnostic microarray. Analyt. Biochem. *346*, 333–335.

Ksiazek, T.G., Erdman, D., Goldsmith, C.S., Zaki, S.R., Peret, T., Emery, S., Tong, S., Urbani, C., Comer, J.A., Lim, W. *et al.* (2003). A novel coronavirus associated with severe acute respiratory syndrome. N. Engl. J. Med. *348*, 1953–1966.

Kumar, A., Larsson, O., Parodi, D., and Liang, Z. (2000). Silanized nucleic acids: A general platform for DNA immobilization. Nucleic Acids Res. *28*, e71.

Lee, D.Y., Shannon, K., and Beaudette, L.A. (2006). Detection of bacterial pathogens in municipal wastewater using an oligonucleotide microarray and real-time quantitative PCR. J. Microbiol. Meth. *65*, 453–467.

Leigh, M.B., Pellizari, V.H., Uhlík, O., Sutka, R., Rodrigues, J., Ostrom, N.E., Zhou, J., and Tiedje, J.M. (2007). Biphenyl-utilizing bacteria and their functional genes in a pine root zone contaminated with polychlorinated biphenyls (PCBs). ISME J. *1*, 134–148.

Li, X., He, Z., and Zhou, J. (2005). Selection of optimal oligonucleotide probes for microarrays using multiple criteria, global alignment and parameter estimation. Nucleic Acids Res. *33*, 6114–6123.

Liang, Y., He, Z., Wu, L., Deng, Y., Li, G., and Zhou, J. (2010). Development of a common oligo reference standard (CORS) for microarray data normalization and comparison across different microbial communities. Appl. Environ. Microbiol. *76*, 1088–1094.

Liang, Y., Li, G., Van Nostrand, J.D., He, Z., Wu, L., Deng, Y., Zhang, X., and Zhou, J. (2009a). Microarray-based analysis of microbial functional diversity along an oil contamination gradient in oil field. FEMS Microbiol. Ecol. *70*, 324–333.

Liang, Y., Wang, J., Van Nostrand, J.D., Zhou, J., Zhang, X., and Li, G. (2009b). Microarray-based functional gene analysis of soil microbial communities in ozonation and biodegradation of crude oil. Chemosphere. *75*, 193–199.

Liang, Y., Van Nostrand, J.D., Deng, Y., He, Z., Wu, L., Zhang, X., Li, G., and Zhou, J. (2010). Functional gene diversity of soil microbial communities from five oil-contaminated fields in China. ISME J. *5*, 403–413.

Liebich, J., Schadt, C.W., Chong, S.C., He, Z., Rhee, S.K., and Zhou, J. (2006). Improvement of oligonucleotide probe design criteria for functional gene microarrays in environmental applications. Appl. Environ. Microbiol. *72*, 1688–1691.

Liesegang, H., Lemke, K., Siddiqui, R.A., and Schlegel, H.G. (1993). Characterization of the inducible nickel and cobalt resistance determinant cnr from pMOL28 of Alcaligenes eutrophus CH34. J. Bacteriol. *175*, 767–778.

Liu, W., Wang, A., Cheng, S., Logan, B.E., Yu, H., Deng, Y., Van Nostrand, J.D., Wu, L., He, Z., and Zhou, J. (2010). Geochip-based functional gene analysis of anodophilic communities in microbial electrolysis cells under different operational modes. Environ. Sci. Technol. *44*, 7729–7735.

Loy, A., Lehner, A., Lee, N., Adamczyk, J., Meier, H., Ernst, J., Schleifer, K.H., and Wagner, M. (2001). Oligonucleotide microarray for 16S rRNA gene-based detection of all recognized lineages of sulfate-reducing prokaryotes in the environment. Appl. Environ. Microbiol. *68*, 5064–5081.

Lueders, T., and Friedrich, M.W. (2003). Evaluation of PCR amplification bias by terminal restriction fragment length polymorphism Analysis of small-subunit rRNA and mcrA genes by using defined template mixtures of methanogenic pure cultures and soil DNA extracts. Appl. Envrion. Microbiol. *69*, 320–326.

Luo, Y., Hui, D., and Zhang, D. (2006). Elevated CO_2 stimulates net accumulations of carbon and nitrogen in land ecosystems: A meta-analysis. Ecology. *87*, 53–63.

Mason, O.U., DiMeo-Savoie, C.A., Van Nostrand, J.D., Zhou, J., Fisk, M.R., and Giovannoni, S.J. (2009). Prokaryotic diversity, distribution, and preliminary insights into their role in biogeochemical cycling in marine basalts. ISME J. *3*, 231–242.

McGrath, K.C., Thomas-Hall, S.R., Cheng, C.T., Leo, L., Alexa, A., Schmidt, S., and Schenk, P.M. (2008). Isolation and analysis of mRNA from environmental microbial communities. J. Microbiol. Meth. *75*, 172–176.

McKenna, P., Hoffmann, C., Minkah, N., Aye, P.P., Lackner, A., Liu, Z., Lozupone, C.A., Hamady, M., Knight, R., and Bushman, F.D. (2008). The Macaque Gut Microbiome in Health, Lentiviral Infection, and Chronic Enterocolitis. PLOS Pathog. *4*, e20.

Mehta, M.L. (1990). Random Matrix Theory. (NewYork, USA: Springer).

Miller, S.M., Tourlousse, D.M., Stedtfeld, R.D., Baushke, S.W., Herzog, A.B., Wick, L.M., Rouillard, J.M., Gulari, E., Tiedje, J.M., and Hashsham, S.A. (2008). In situ-synthesized virulence and marker gene biochip for detection of bacterial pathogens in water. Appl. Environ. Microbiol. *74*, 2200–2209.

Montoya, J.M., Pimm, S.L., and Sole, R.V. (2006). Ecological networks and their fragility. Nature 442:259–264.

Murray, A.E., Lies, D., Li, G., Nealson, K., Zhou, J., and Tiedje, J.M. (2001). DNA–DNA hybridization to microarrays reveals gene-specific differences between closely related microbial genomes. Proc. Natl. Acad. Sci. U.S.A. *98*, 9853–9858.

Nies, D.H. (2003). Efflux-mediated heavy metal resistance in prokaryotes. FEMS Microbiol. Rev. *27*, 313–39.

Ning, J., Liebich, J., Kästner, M., Zhou, J., Schäffer, A., and Burauel, P. (2009). Different influences of DNA purity indices and quantity on PCR-based dgge and functional gene microarray in soil microbial community study. Appl. Microbiol. Biotechnol. *82*, 983–993.

Okamoto, T., Suzuki, T., and Yamamoto, N. (2000). Microarray fabrication with covalent attachment of DNA using Bubble Jet technology. Nature Biotechnol. *18*, 438–441.

Økland, R.H., and Eilertsen, O. (1994). Canonical correspondence analysis with variation partitioning: Some comments and an application. J. Veg. Sci. *5*, 117–126.

Okou, D.T., Steinberg, K.M., Middle, C., Cutler, D.J., Albert, T.J., and Zwick, M.E. (2007). Microarray-based genomic selection for high-throughput resequencing. Nature Meth. *4*, 907–909.

Palka-Santini, M., Cleven, B.E., Eichinger, L., Krönke, M., and Krut, O. (2009). Large scale multiplex PCR improves pathogen detection by DNA microarrays. BMC Microbiol. *9*, 1.

Palmer, C., Bik, E.M., Digiulio, D.B., Relman, D.A., and Brown, P.O. (2007). Development of the human infant intestinal microbiota. PLoS Biol. *5*, e177.

Parnell, J.J., Rompato, G., Latta, L.C.I.V., Pfrender, M.E., Van Nostrand, J.D., He, Z., Zhou, J., Andersen, G., Champine, P., Ganesan, B., and Weimer, B.C. (2010). Functional biogeography as evidence of gene transfer in hypersaline microbial communities. PLoS ONE. *5*, e12919.

Paulsen, I.T., Park, J.H., Choi, P.S., and Saier, M.H. (1997). A family of gram-negative bacterial outer membrane factors that function in the export of proteins, carbohydrates, drugs and heavy metals from gram-negative bacteria. FEMS Microbiol. Lett. *156*, 1–8.

Peytavi, R., Raymond, F.R., Gagné, Picard, D., F.J., Jia, G., Zoval, J., Madou, M., Boissinot, K., Boissinot, M., Bissonnette, L. *et al.* (2005). Microfluidic device for rapid (<15 min) automated microarray hybridization. Clin. Chem. *51*, 1836–1844.

Raes, J., and Bork, P. (2008). Molecular eco-systems biology: towards an understanding of community function. Nat. Rev. Microbiol. *6*, 693–9.

Ramette, A., and Tiedje, J.M. (2007). Multiscale responses of microbial life in spatial distance and environmental heterogeneity in a patchy ecosystem. Proc. Natl. Acad. Sci. USA. *104*, 2761–2766.

Reich, P.B., Knops, J., Tilman, D., Craine, J., Ellsworth, D., Tjoelker, M., Lee, T., Wedin, D., Naeem, S., Bahauddin, D. *et al.* (2001). Plant diversity enhances ecosystem responses to elevated CO_2 and nitrogen deposition. Nature. *410*, 809–12.

Relógio, A., Schwager, C., Richter, A., Ansorge, W., and Valcárcel, J. (2002). Optimization of oligonucleotide-based DNA microarrays. Nucleic Acids Res. *30*, e51.

Rhee, S.K., Liu, X., Wu, L., Chong, S.C., Wan, X., and Zhou, J. (2004). Detection of genes involved in biodegradation and biotransformation in microbial communities by using 50-mer oligonucleotide microarrays. Appl. Environ. Microbiol. *70*, 4303–4317.

Rich, V.I., Konstantinidis, K., and DeLong, E.F. (2008). Design and testing of 'genome-proxy' microarrays to profile marine microbial communities. Environ. Microbiol. *10*, 506–521.

Rodríguez-Martínez, E.M., Pérez, E.X., Schadt, C.W., Zhou, J., and Massol-Deyá, A.A. (2006). Microbial diversity and bioremediation of a hydrocarbon-contaminated aquifer (Vega Baja, Puerto Rico). Int. J. Environ. Res. Public Health. *3*, 292–300.

Roesch, L.F., Fulthorpe, R.R., Riva, A., Casella, G., Hadwin, A.K., Kent, A.D., Daroub, S.H., Camargo, F.A., Farmerie, W.G., and Triplett, E.W. (2007). Pyrosequencing enumerates and contrasts soil microbial diversity. ISME J. *1*, 283–90.

Saier, M.H., Tam, R., Reizer, A., and Reizer, J. (1994). Two novel families of bacterial membrane proteins concerned with nodulation, cell division and transport. Mol. Microbiol. *11*, 841–847.

Sarkar, S.F., Gordon, J.S., Martin, G.B., and Guttman, D.S. (2006). Comparative genomics of host-specific virulence in *Pseudomonas syringae*. Genetics. *147*, 1041–1056.

Schaupp, C.J., Jiang, G., Myers, T.G., and Wilson, M.A. (2005). Active mixing during hybridization improves the accuracy and reproducibility of microarray results. BioTechniques. *38*, 117–119.

Schena, M., Shalon, D., Davis, R.W., and Brown, P.O. (1995). Quantitative monitoring of gene ecpression patterns with a complementary DNA microarray. Science. *270*, 467–470.

Schena, M., Shalon, D., Heller, R., Chai, A., and Brown, P.O. (1996). Parallel human genome analysis: Microarray-based expression monitoring of 1000 genes. Proc. Natl. Acad. Sci. USA. 93, 10614–10619.

Schloss, P.D., and Handelsman, J. (2006). Toward a census of bacteria in soil. PLoS Comput. Biol. *2*, e92.

Schracke, N., Kornmeyer, T., Kränzle, M., Stähler, P.F., Summerer, D., and Beier, M. 2009. New Biotechnol. *26*, 229–233.

Sebat, J.L., Colwell, F.S., and Crawford, R.L. (2003). Metagenomic profiling: Microarray analysis of an environmental genomic library. Appl. Environ. Microbiol. *69*, 4927–4934.

Serra, P., Colina, M., Fernández, J.M., Sevilla, L., and Morenza, J.L. (2004). Preparation of functional DNA microarrays through laser-induced forward transfer. Appl. Physics Lett. *85*, 1639–1641.

Small, J., Call, D.R., Brockman, F.J., Straub, T.M., and Chandler, D.P. (2001). Direct detection of 16S rRNA in soil extracts by using oligonucleotide microarrays. Appl. Environ. Microbiol. *67*, 4708–4716.

Steward, G.F., Jenkins, B.D., Ward, B.B., and Zehr, J.P. (2004). Development and Testing of a DNA Macroarray To Assess Nitrogenase (nifH) Gene Diversity. Appl. Environ. Microbiol. *70*, 1455–1465.

Stralis-Pavese, N., Sessitsch, A., Weilharter, A., Reichenauer, T., Riesing, J., Csontos, J., Murrell, J.C., and Bodrossy, L. (2004). Optimization of diagnostic microarray for application in analysing landfill methanotroph communities under different plant covers. Environ. Microbiol. *6*, 347–363.

Suzuki, M.T., and Giovannoni, S.J. (1996). Bias caused by template annealing in the amplification of mixtures of 16S rRNA genes by PCR. Appl. Envrion. Microbiol. *62*, 625–630.

Taroncher-Oldenburg, G., Griner, E.M., Francis, C.A., and Ward, B.B. (2003). Oligonucleotide microarray for the study of functional gene diversity in the nitrogen cycle in the environment. Appl. Environ. Microbiol. *69*, 1159–1171.

Tas, N., van Eekert, M.H., Schraa, G., Zhou, J., de Vos, W.M., and Smidt, H. (2009). Tracking functional guilds: 'Dehalococcoides' Spp. In european river basins contaminated with hexachlorobenzene. Appl. Environ. Microbiol. *75*, 4696–704.

ter Braak, C.J.F. (1986). Canonical correspondence analysis: A new eigenvector technique for multivariate direct gradient analysis. Ecology. *67*, 1167–1179.

Tibazarwa, C., Wuertz, S., Mergeay, M., Wyns, L., and van der Lelie, D. (2000). Regulation of the cnr cobalt and nickel resistance determinant of Ralstonia eutropha (Alcaligenes eutrophus) CH34. J. Bacteriol. *182*, 1399–1409.

Tiquia, S.M., Wu, L., Chong, S.C., Passovets, S., Xu, D., Xu, Y., and Zhou, J. (2004). Evaluation of 50-mer oligonucleotide arrays for detecting microbial populations in environmental samples. Biotechniques. *36*, 1–8.

Thompson, J.D., Higgins, D.G., and Gibson, T.J. (1994). Clustal w: Improving the sensitivity of progressive multiple sequence alignment through sequence weighting, position-specific gap penalties and weight matrix choice. Nucleic Acids Res. *22*, 4673–80.

Torsvik, V., Ovreas, L., and Thingstad, T.F. (2002). Prokaryotic diversity – magnitude, dynamics, and controlling factors. Science. *296*, 1064–6.

Van Nostrand, J.D., Khijniak, T.V., Gentry, T.J., Novak, M.T., Sowder, A.G., Zhou, J.Z., Bertsch, P.M., and Morris, P.J. (2007). Isolation and characterization of four Gram-positive nickel-tolerant microorganisms from contaminated sediments. Microb. Ecol. *53*, 670–682.

Van Nostrand, J.D., Wu, W.M., Wu, L., Deng, Y., Carley, J., Carroll, S., He, Z., Gu, B., Luo, J., Criddle, C.S. *et al.* (2009). GeoChip-based analysis of functional microbial communities during the reoxidation of a bioreduced uranium-contaminated aquifer. Environ. Microbiol. *11*, 2611–2626.

Wagner, M., Smidt, H., Loy, A., and Zhou, Z. (2007). Unravelling microbial communities with DNA-microarrays: challenges and future directions. Microb. Ecol. 53, 498–506.

Waldron, P.J., Wu, L., Van Nostrand, J.D., Schadt, C., Watson, D., Jardine, P., Palumbo, T., Hazen, T.C., and Zhou, J. (2009). Functional gene array-based analysis of microbial community structure in groundwaters with a gradient of contaminant levels. Environ. Sci. Technol. *43*, 3529–3534.

Wang, D., Coscoy, L., Zylberberg, M., Avila, P.C., Boushey, H.A., Ganem, D., and DeRisi, J.L. (2002). Microarray-based detection and genotyping of viral pathogens. Proc. Natl. Acad. Sci. USA. *99*, 15687–15692.

Wang, F., Zhou, H., Meng, J., Peng, X., Jiang, L., Sun, P., Zhang, C., Van Nostrand, J.D., Deng, Y., He, Z., Wu, L., Zhou, J., and Xiao, X. (2009). GeoChip-based analysis of metabolic diversity of microbial communities at the Juan de Fuca Ridge hydrothermal vent. Proc. Natl. Acad. Sci. USA. *106*, 4840–4845.

Warnecke, P.M., Stirzaker, C., Melki, J.R., Millar, D.S., Paul, C.L., and Clark, S.J. (1997). Detection and measurement of PCR bias in quantitative methylation analysis of bisulphite-treated DNA. Nucleic Acids Res. *25*, 4422–4426.

White, J.R., Escobar-Paramo, P., Mongodin, E.F., Nelson, K.E., and DiRuggiero, J. (2008). Extensive genome rearrangements and multiple horizontal gene transfers in a population of Pyrococcus isolates from Vulcano Island, Italy. Appl. Environ. Microbiol. *74*, 6447–6451.

Whitman, W.B., Coleman, D.C., and Wiebe, W.J. (1998). Prokaryotes: the unseen majority. Proc. Natl. Acad. Sci. U.S.A. *95*, 6578–6583.

Wilson, K.H., Blitchington, R.B., and Greene, R.C. 1990. Amplification of bacterial 16S ribosomal DNA with polymerase chain reaction. J. Clin. Microbiol. *28*, 1942–1946.

Wilson, M., DeRisi, J., Kristensen, H.H., Imboden, P., Rane, S., Brown, P.O., and Schoolnik, G.K. (1999). Exploring drug-induced alterations in gene expression in Mycobacterium tuberculosis by microarray hybridization. Proc. Natl. Acad. Sci. U.S.A. *96*, 12833–12838.

Wilson, W.J., Strout, C.L., DeSantis, T.Z., Stilwell, J.L., Carrano, A.V., and Andersen, G.L. (2002). Sequence-specific identification of 18 pathogenic microorganisms using microarray technology. Mol. Cell. Probes. *16*, 119–127.

Wu, L., Kellogg, L., Devol, A.H., Tiedje, J.M., and Zhou, J. (2008). Microarray-based characterization of microbial community functional structure and heterogeneity in marine sediments from the Gulf of Mexico. Appl. Environ. Microbiol. *74*, 4516–4529.

Wu, L., Liu, X., Fields, M.W., Thompson, D.K., Bagwell, C.E., Tiejde, J.M., Hazen, T.C., and Zhou, J.Z. (2008b). Microarray-based whole–genome hybridization as a tool for determining prokaryotic species relatedness. ISME J. *6*, 642–655.

Wu, L., Liu, X., Schadt, C.W., and Zhou, J. (2006). Microarray-based analysis of submicrogram quantities of microbial community DNAs by using whole-community genome amplification. Appl. Environ. Microbiol. *72*, 4931–4941.

Wu, L., Thompson, D.K., Li, G., Hurt, R.A., Tiedje, J.M., and Zhou, J. (2001). Development and evaluation of functional gene arrays for detection of selected genes in the environment. Appl. Environ. Microbiol. *67*, 5780–5790.

Wu, L., Thompson, D.K., Liu, X., Fields, M.W., Bagwell, C.E., Tiedje, J.M., and Zhou, J. (2004). Development and evaluation of microarray-based whole genome hybridization for detection of microorganisms within the context of environmental applications. Environ. Sci. Technol. *38*, 6775–6782.

Xie, J., He, Z., Liu, X., Liu, X., Van Nostrand, J.D., Deng, Y., Wu, L., Qiu, G., and Zhou, J. (2010). Geochip-based analysis of the functional gene diversity and metabolic potential of microbial communities in acid mine drainage. Appl. Environ. Microbiol. 77, 991–999.

Xiong, J., Wu, L., Tu, S., Van Nostrand, J.D., He, Z., Zhou, J., Wang, G. (2010). Microbial communities and functional genes associated with soil arsenic contamination and rhizosphere of the arsenic hyper-accumulating plant *Pteris vittata* L. Appl. Environ. Microbiol. *76*, 7277–7284.

Xu, M., Wu, W.-M., Wu, L., He, Z., Van Nostrand, J.D., Deng, Y., Luo, J., Carley, J., Ginder-Vogel, M., Gentry, T.J., Gu, B., Watson, D., Jardine, P.M., Marsh, T.L., Tiedje, J.M., Hazen, T., Criddle, C.S., Zhou, J. (2010). Responses of microbial community functional structures to pilot-scale uranium *in situ* bioremediation. ISME J. *4*, 1060–1070.

Yamamoto, S., and Harayama, S. (1995). PCR amplification and direct sequencing of gyrB genes with universal primers and their application to the detection and taxonomic analysis of *Pseudomonas* putida strains. Appl. Environ. Microbiol. *61*, 1104–1109.

Yin, H., Cao, L., Qiu, G., Wang, D., Kellogg, L., Zhou, J., Dai, Z., and Liu, X. (2007). Development and evaluation of 50-mer oligonucleotide arrays for detecting microbial populations in Acid Mine Drainages and bioleaching systems. J. Microbiol. Meth. *70*, 165–178.

Yergeau, E., Kang, S., He, Z., Zhou, J., and Kowalchuk, G.A. (2007). Functional microarray analysis of nitrogen and carbon cycling genes across an Antarctic latitude transect. ISME J. *1*, 1–17.

Zhang, L., Srinivasan, U., Marrs, C.F., Ghosh, D., Gilsdorf, J.R., and Foxman, B. (2004). Library on a slide for bacterial comparative genomics. BMC Microbiol. *4*, 12–18.

Zhang, K., Martiny, A.C., Reppas, N.B., Barry, K.W., Malek, J., Chisholm, S.W., and Church, G.M. (2006). Sequencing genomes from single cells by polymerase cloning. Nature Biotechnol. *24*, 680–686.

Zhang, Y., Zhang, X., Liu, X., Xiao, Y., Qu, L., Wu, L., and Zhou, J. (2007). Microarray-based analysis of changes in diversity of microbial genes involved in organic carbon decomposition following land use/cover changes. FEMS Lett. *266*, 144–151.

Zhou, J., Bruns, M.A., and Tiedje, J.M. (1996). DNA recovery from soils of diverse composition. Appl. Environ. Microbiol. *62*, 316–322.

Zhou, J., and Thompson, D.K. (2002). Challenges in applying microarrays to environmental studies. Curr. Opin. Biotech. *13*, 204–207.

Zhou, J. (2003). Microarrays for bacterial detection and microbial community analysis. Curr. Opin. Microbiol. *6*, 288–294.

Zhou X., and Zhou J. (2004). Improving the signal sensitivity and photostability of DNA hybridizations on microarrays by using dye-doped core-shell silica nanoparticles. Anal. Chem. *76*, 5302–5312.

Zhou, J., Kang, S., Schadt, C.W., and Garten, C.T. Jr. (2008). Spatial scaling of functional gene diversity across various microbial taxa. Proc. Natl. Acad. Sci. U.S.A. *105*, 7768–7773.

Topic Index

Taxonomic Index

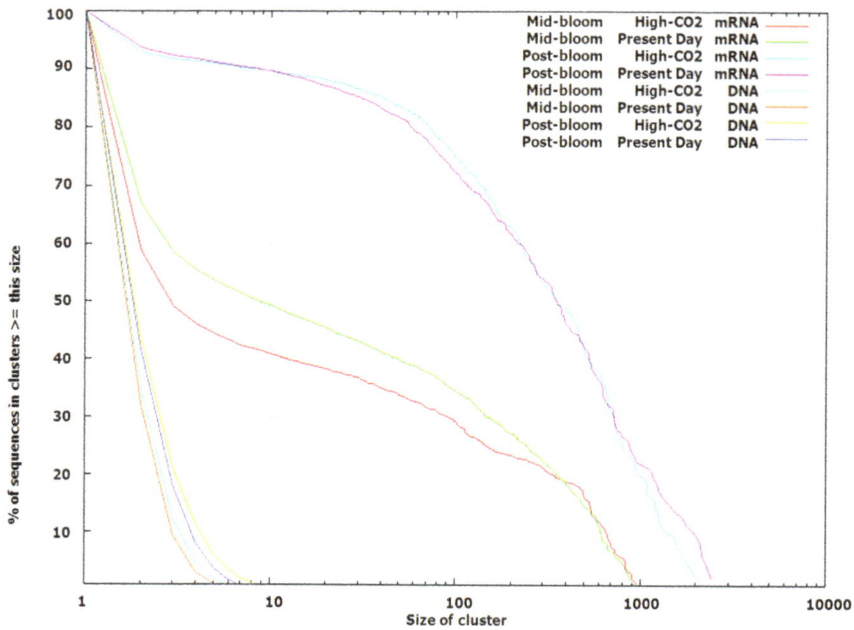

Figure 8.6 Distribution of clusters from eight samples. The *x*-axis is cluster size. The *y*-axis is the percentage of total sequences within clusters ≥ this size. The clustering analysis was made for four metagenomic and four metatranscriptomic samples from two time points and two conditions (high CO_2 and normal CO_2) (Gilbert *et al.*, 2008). Clear patterns are observed for different type of samples.

www.ingramcontent.com/pod-product-compliance
Lightning Source LLC
Chambersburg PA
CBHW081212180926
51613CB00003B/218
9781904455875